Coastal and Deep Ocean Pollution

Editors

Andrés Hugo Arias
Instituto Argentino de Oceanografía (CONICET/UNS)
Bahía Blanca, Argentina
AND
Departamento de Química, Area III, Química Analítica.
Universidad Nacional del Sur
Bahía Blanca, Argentina

Sandra Elizabeth Botté
Instituto Argentino de Oceanografía (CONICET-UNS)
Bahía Blanca, Argentina
AND
Universidad Nacional del Sur
Bahía Blanca, Argentina

CRC Press
Taylor & Francis Group
Boca Raton London New York

CRC Press is an imprint of the
Taylor & Francis Group, an **informa** business
A SCIENCE PUBLISHERS BOOK

CRC Press
Taylor & Francis Group
6000 Broken Sound Parkway NW, Suite 300
Boca Raton, FL 33487-2742

© 2020 by Taylor & Francis Group, LLC
CRC Press is an imprint of Taylor & Francis Group, an Informa business

No claim to original U.S. Government works

Version Date: 20200118

International Standard Book Number-13: 978-1-138-56939-3 (Hardback)

International Standard Book Number-13: 978-0-367-50467-0 (Paperback)

Visit the Taylor & Francis Web site at
http://www.taylorandfrancis.com

and the CRC Press Web site at
http://www.crcpress.com

Preface

The marine environment is subjected to many forms of pollution; as defined by the United Nations Convention (1982) it involves the direct/indirect introduction of substances or energy into the marine environment which ultimately results in a deleterious effect to living resources and marine life. The main sources of marine pollution include the land-based activities or human activities at the sea, runoff (rivers, streams, land run-off) and atmospheric deposition, which ultimately impact the coastal and deep ocean health. Due to the extension and the difficulty in making quantitative estimates of these inputs, there is a general uncertainty and patchily distributed information with regards to the status of the ocean health. To address the state of the art within the coastal and deep ocean pollution, the authors of this book have made an extensive assessment of the scientific basis of each pollution topic, covering general pollution processes, from inorganic pollutants to several organic pollutants.

This book includes contributions from an impressive group of international scientists, and therefore presents a comprehensive analysis of marine pollution. The book begins with a very interesting glance at the general pollution processes in coastal and deep ocean zones. Chapter 1 addresses the black tides issue in coastal and offshore locations. Chapter 2 covers the oxygenation of coastal and open waters, reviewing hypoxia and anoxia cases and their consequences for the biogeochemical cycling and marine life. The last chapter in this section, Chapter 3 explores the causes and effects of harmful algal blooms occurrence in both coastal and offshore waters.

The section on inorganic pollutants impacting the marine environment starts with Chapter 4 which tackles the issue of mercury in the ocean, addressing the health associated risks from a global perspective. In Chapter 5, both scientists and the general public will find a review on the essentials of Copper in the ocean; also, Chapter 6 will serve as a link which explores the interaction between marine animals and several pollution processes: the flux of metals, acidification, hypoxia and noise pollution. Chapter 7 covers the issue of remediation in the marine environments and reviews the essentials of bio and phytoremediation approaches. To end the inorganic pollutant section, a very interesting Chapter 8 will take the readers through the less-known metallic pollutants impacting the marine environment.

The third section of the book covers the general aspects and features of organic pollutants, highlighting the strong need of sustainable developments as a primary goal for reaching healthy ecosystems and environments. On the one hand, Chapter 9 addresses the issue of organophosphorus compounds reviewing their general chemistry, monitoring and cytotoxicity assessment in marine species. On the other hand, Chapter 10 assesses the chemistry and occurrence of halogenated compounds: PCBs, organochlorinated pesticides and PCDD/PCDFs in the marine environment. Chapter 11 discusses the polycyclic aromatic hydrocarbons world exploring their occurrence, sources, levels and global marine distribution. Finally, Chapter 12 covers the brominated flame retardants issue, offering to the general reader a precise historical overview, uses, classification, matrix distribution in the environment and marine global trends.

The preparation of this book was significantly facilitated by the collaborative efforts of each of the world-class authors and their respective Scientific Institutions. We are indebted to them, main players in the general building of this project, and to many other colleagues who provided suggestions and help during the entire process of development of the book. The editors especially thank the National Council of Scientific and Technological Research of Argentina (CONICET) and the National South University (UNS) which made available to us vibrant research experience. An acknowledgement is also given to the main editorial board and all the editorial staff who provided us with the confidence and help to accomplish this project which started in 2017.

January 2020

Andrés Hugo Arias
Sandra Elizabeth Botté

Contents

Section III: Organic Pollutants

Section I

General Pollution in Coastal and Deep Ocean Environments

1

Black Tides
Petroleum in the Ocean

Alfonso Vazquez-Botello,* Guadalupe Ponce-Velez, Luis A. Soto
and *Susana Villanueva*

1. Introduction

Crude oil is a result of the transformation of organic (animal and plant) debris from marine populations, under great pressure and in the absence of oxygen (Fig. 1) (http://www.black-tides.com/).

Crude oil enters the marine environment by two principal processes. One process involves human activities related to the extraction, transportation, refining, storage, and use of petroleum (crude oil and natural gas). An example is marine oil spills, caused by failures in human-designed transportation systems such as tankers and pipelines, which are built to move crude oil from one place to another. The second process involves natural oil seepage.

The term "oil seep" is used here to mean naturally occurring seepage of crude oil and tar. Crude-oil seeps are geographically common and have likely been active through much of geologic time (Hunt 1996).

An examination of reports from a variety of sources, including industry, government, and academic sources, indicates that although the origins of petroleum input to the sea are diverse, they can be categorized effectively into four major groups: natural seeps, petroleum extraction, petroleum transportation, and petroleum consumption. Natural seeps are purely natural phenomena that occur when crude oil seeps from the geologic strata beneath the seafloor to the overlying water column. Recognized by geologists for decades as indicating the existence of potentially economic reserves of petroleum, these seeps release vast amounts of crude oil annually. Yet these large volumes are released at a rate low enough that the surrounding ecosystem can adapt and even thrive in their presence (NRC 2003).

Instituto de Ciencias del Mar y Limnología. Universidad Nacional Autónoma de México, Av. Universidad 3000. Ciudad Mexico 04510, México.
  Email: gatoponcho2015@gmail.com
* Corresponding author: pomito69@gmail.com

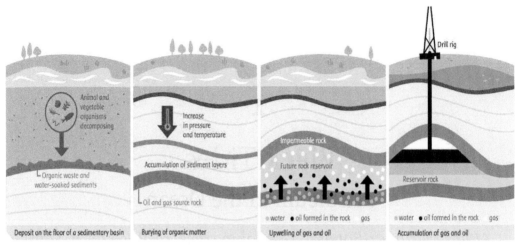

Fig. 1. Origins of oil. Source: Understanding Black Tides. July 2017. Michel Girin (http://www.black-tides.com/).

2. Petroleum Transportation

Practically all crude oil and vast quantities of refined products are transported over long distances. All transportation, whether by sea or by land, involves the risk of accidents (http://www.black-tides.com/).

Transportation of oil via pipelines, from the exploitation area to the consumption destination, is certainly safer than transportation by vessel, train or truck. However, it is not exempt from risks. There have been many cases of leaks due to negligence, carelessness or even malicious attacks. Moreover, pipelines cannot fulfil all demands since it is not always possible to build them because of physical or political constraints.

The transportation (including refining and distribution activities) of crude oil or refined products results in the release, on average, of an estimated 9,100 tons (2,700,000 gallons) of petroleum to North American waters, and 150,000 tons (44,000,000 gallons) worldwide, each year. Releases due to the transportation of petroleum, therefore, make up roughly 9% of the total petroleum input through anthropogenic activities to North American waters and less than 22% worldwide. Similar to releases from petroleum extraction, these volumes are dwarfed by those from other sources of petroleum to the marine environment. And like releases from extraction activities, these inputs are not trivial, as they can occur as large spills. Unlike releases associated with extraction, which tend to be concentrated in production fields in the Gulf of Mexico or coastal areas off California and Alaska, these spills can occur anywhere tanker vessels may travel or where pipelines are located statistically. Areas near major petroleum handling facilities face the greatest threat.

Studies completed in the last 20 years again bear out the significant environmental damage that can be caused by spills of petroleum into the marine environment. No spill is entirely benign. Even a small spill at the wrong place, at the wrong time, can result in significant damage to individual organisms or entire populations. With a few notable exceptions (e.g., the Exxon Valdez, North Cape, and Panama spills), there has been a lack of resources to support studies of the fates and effects of spilled oil. Much of what is known about the fate and effect of spilled oil has been derived from a very few, well-studied spills. Federal agencies, especially the U.S. Coast Guard, the National Oceanic and Atmospheric Administration (NOAA), and the Environmental Protection Agency (EPA), should work with industry to develop and implement a rapid response system to collect *in situ* information about spill behavior and impacts (Fig. 2).

To meet the increased demand, an increase in world oil supply of roughly 45 mb/d (6.4 mt/d) is projected for the next two decades. It is expected that Organization of the Petroleum Exporting

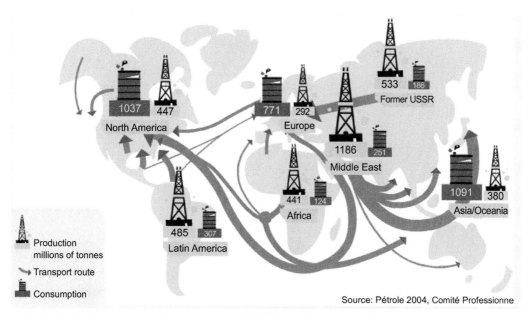

Fig. 2. Worldwide production and consumption in 2004 (millions of tonnes) (https://www.eia.gov/beta/international/).

Color version at the end of the book

Countries (OPEC) producers will be responsible for more than two-thirds of this increase. Imports into industrialized countries are therefore expected to increase from 34.0 to 43.7 mb/d (4.9 to 6.2 mt/d), and imports into developing countries from 19.3 mb/d to 42.8 mb/d (2.8 to 6 mt/d). Much of these imports will move by sea. Major importers and exporters are shown in Fig. 3.

Increased imports into China and Pacific Rim countries will come largely from the Persian Gulf. North American imports are projected to increase from 11.0 mb/d (1.57 mt/d) in 1998 to about 18.0 mb/d (2.6 mt/d) in 2020. More than half of the North American imports are expected to come from the Atlantic Basin, principally Latin American and West African producers. Imports into North America from the Persian Gulf are expected to double, from 2.2 to 4.2 mb/d (0.3 to 0.6 mt/d) (NRC 2003) (http://www.nap.edu/catalog/10388.html).

3. Quantities and Modes of Transport

It is apparent that greater and greater amounts of oil will be transported by vessel, refineries will have to increase capacity, and more coastal petroleum handling facilities will be needed. These have the potential to increase the input of hydrocarbons into the oceans. However, the operational and accidental discharge of oil from vessels and platforms has declined substantially over the past three decades, and it is reasonable to expect continued improvements in these areas in future years as the benefits from recently enacted regulations and improved operational practices are fully realized. The expected growth in worldwide consumption, with much of the increase concentrated in the transportation sector, is of concern. Land-based runoff of petroleum hydrocarbons can be expected to increase with consumption unless steps are taken to reduce the release of petroleum from consumption-related activities (Fig. 4).

Of the 3.5 billion tons produced annually worldwide, about half is exported from the Middle East, Africa and Latin America to North America, Europe and South East Asia (https://www.focus-economics.com/blog/economic-outlook-for-the-top-oil-producing-countries).

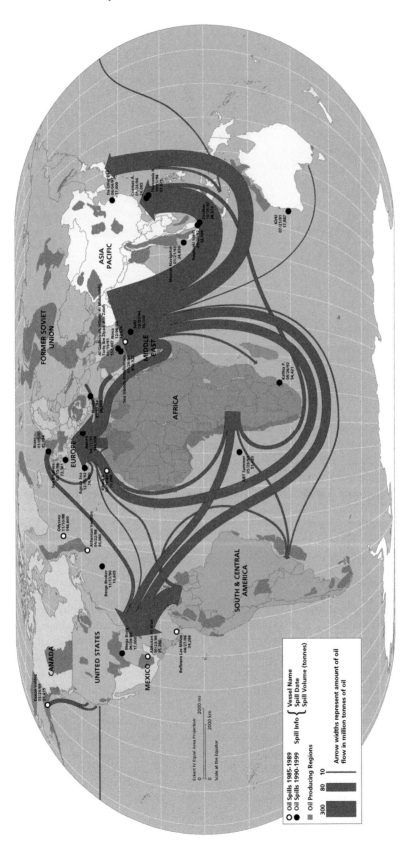

Fig. 3. Worldwide sea-borne flow of oil in 2000 (modified from Newton 2002; other information sources include U.S. Geological Survey, U.S. Coast Guard, Minerals Management Service). Solid black dots indicate spills included in the average, annual (1990–1999) (NRC 2003).

Fig. 4. How will top oil producers perform in 2018? (https://www.focus-economics.com/blog/economic-outlook-for-the-top-oil-producing-countries).

4. Oil Spills and Black Tides

An oil spill is the accidental large release of oil or other petroleum products, usually into freshwater or marine ecosystems, and usually in large quantities. It can be controlled by chemical dispersion, combustion, mechanical containment, and absorption (EEA multilingual environmental glossary, http://www.glossary.eea.europa.eu/EEAGlossary). A black tide is a major oil slick that threatens to reach and pollute the shore (Zilberberg 1998) or large-scale pollution at sea (American Dialect Society) (http://www.glossary.eea.europa.eu/EEAGlossary).

To develop suitable prevention and response strategies, it is essential to identify the sources of oil discharge into the marine environment, as well as their contribution to the phenomenon of black tides. This is no easy task. Although accidents involving oil spills may be documented, the same cannot be said for other sources of discharge.

In a diagram published by the International Maritime Organization and reproduced in Fig. 5, only direct discharges into the water are included, and not the share of atmospheric emissions that rain and streams transfer into the water. The method used to estimate such sources varies from one documentary source to another, generating estimates that are not always consistent.

In 2003, 1,700 million tons of crude oil and nearly 500 million tons of refined products (e.g., petrol, kerosene, fuel oil, bitumen) were transported by sea. With an average capacity of 100,000 tons of oil per tanker, this equates to around 22,000 journeys made by oil tankers between oil-producing countries and oil-consuming countries, over considerable distances. The average voyage for an oil tanker lasts two weeks and includes at least one passage in a high-risk zone.

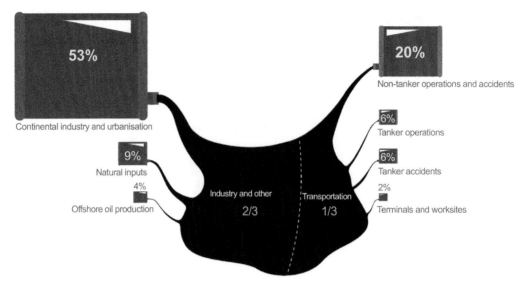

Fig. 5. Sources of black tides (International Maritime Organization, http://www.black-tides.com/).

In addition, coastal tankers, barges and canal boats cover a multitude of coastal and river routes, with several hundreds or thousands of cubic meters of refined products on board.

5. High-risk Zones

High-risk zones tend to be areas such as straits and capes where several vessel routes meet. Examples are the Pas-de-Calais (between France and Great Britain), the Strait of Gibraltar (Spain), the Strait of Malacca (between Malaysia and Indonesia) and the Bosphorus (Turkey). The increase in exportation of Russian oil across the Baltic Sea has created heavy traffic in that area, increasing the risks.

In France, every day more than 300 vessels sail by the furthest point of Brittany in one direction or another, transporting more than 600,000 tons of dangerous goods (petroleum products, chemicals, radioactive or explosive substances). This area is one of the most dangerous in the world. It holds the tragic world record of the greatest tonnage of hydrocarbons spilled in accidents involving vessels.

High concentrations of vessels produce an increased risk of oil spills, which could be due to defects in a vessel's structure, collision, or grounding (https://www.eia.gov/todayinenergy/detail.php?id=35332#).

Conclusion of Quigley et al. (1999) indicates that about 10% of crude oil entering the oceans during the early 1980s came from natural oil seeps, whereas about 27% came from oil production, transportation, and refining. The remaining 63% came from atmospheric emissions, municipal and industrial sources, and urban and river runoff. Crude-oil seeps are natural phenomena over which humankind has little direct control, although oil production probably has reduced seepage rates (Quigley et al. 1999).

Worldwide consumption of crude oil has been continuously increasing since the end of the 1970s. Consumption has now reached some 3.5 billion tons per year and represents 40% of world energy consumption. This level of consumption is accounted for by the energy needs of industrialized countries and vastly surpasses their own resources. The oil industry can be broken down into two distinct parts, which have arisen from geographic, economic and technical factors:

• Exploration and production, located in regions of the world where oil reservoirs can be found.

• Refinery and distribution closely linked to the geographical location of the consumption zones.

Oceanic oil spills became a major environmental problem in the 1960s, chiefly as a result of intensified petroleum exploration and production on continental shelves and the use of supertankers capable of transporting more than 500,000 metric tons of oil. Spectacular oil spills from wrecked or damaged supertankers are now rare because of stringent shipping and environmental regulations. Nevertheless, thousands of minor and several major oil spills related to well discharges and tanker operations are reported each year, with the total quantity of oil released annually into the world's oceans exceeding one million metric tons. The unintentional or negligent release of used gasoline solvents and crankcase lubricants by industries and individuals greatly aggravates the overall environmental problem. Combined with natural seepage from the ocean floor, these sources add oil to the world's waterways at the rate of 3.5 million to 6 million metric tons a year (Fig. 6a).

6. Largest Oil Tanker Spills in History

Two enormously important oil tanker spills that took place in European waters were the Torrey Canyon disaster off Cornwall, England, in 1967 (121,000 metric tons of crude oil were spilled) and the Amoco Cadiz disaster off Brittany, France, in 1978 (227,000 metric tons of crude oil and ship fuel were spilled). Both events led to lasting changes in the regulation of shipping and in the organization of responses to ecological emergencies such as oil spills. In the Gulf of Mexico in 1979, from Ixtoc-I about 500,000 metric tons of crude oil were spilled. In North America the Exxon Valdez oil spill of 1989 in Prince William Sound, Alaska, caused great ecological and economic damage, though it ranks well below the largest oil tanker spills in history if measured by the amount of oil spilled (37,000 metric tons) (Figs. 6a, 6b).

6.1 Torrey Canyon, March 18, 1967 Wales, 121,000 metric tons

The Torrey Canyon was one of the first big supertankers, and it was also the source of one of the first major oil spills (http://www.robindesbois.org/en/torrey-canyon-18-mars-1967-la-mere-des-marees-noires/).

On February 19, 1967, it left the oil refinery of Al Ahmadi in Kuwait. The supertanker was heading for Milford Haven, Wales, after circumventing South Africa. It was transporting 121,000 tons of crude oil. Built in 1959 in the United States by Newport News Shipbuilding, it was jumboized in 1965, and lengthened from 247 to 297 m; its initial capacity of 60,000 tons was doubled. It was the pride of oil shipping companies (Fig. 7).

On March 18, 1967, because of faulty navigation, the Torrey Canyon was impaled on the reefs between the Isles of Scilly and Cornwall.

Two months later, 150 km of coastline was polluted in the southeastern United Kingdom. Thousands of birds covered in oil came across the English Channel and were washed up dead or dying from Calais to the Ile d'Yeu, France. The coastline was in mourning from the peninsula of La Hague to the tip of Brittany. The Channel Islands wore black. The expression "black tide" appeared on the front pages of newspapers.

At the end of March 1967, confusing the Torrey Canyon with the Scharnhorst or the Bismarck, the Royal Air Force dropped firebombs and napalm rockets on the wreck and spread tons of kerosene over the ocean so that the crude would burn and vanish.

A wide variety of methods to mitigate the spill were tried. Burning the slick proved unsuccessful, and eventually the British Government gave orders for TORREY CANYON to be destroyed by aerial bombardment in the hope that all the oil still remaining on board would be burnt off. This operation was partially successful, but did not prevent escaping oil from polluting many parts of the south west of England, causing the deaths of thousands of seabirds and threatening the livelihoods of many local people in the forthcoming summer tourist season. Later the drifting oil polluted beaches and harbours in the Channel Islands and Brittany.

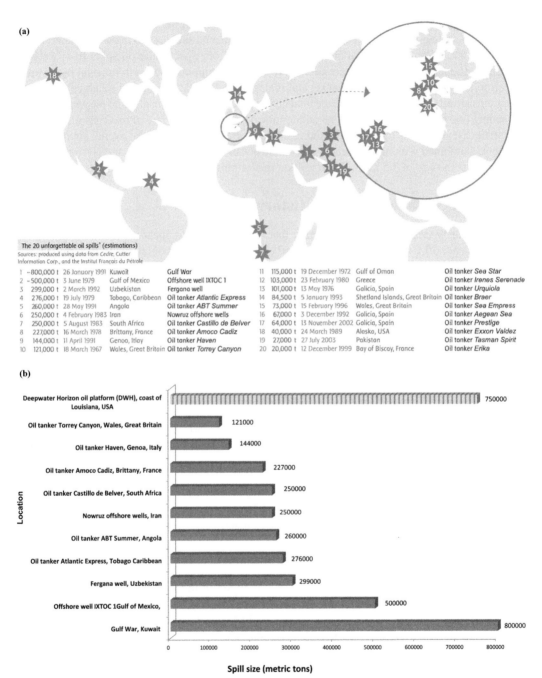

Fig. 6. (a) The 20 unforgettable oil spills (estimations). Source of ranking and spill sizes: International Tanker Owners Pollution Federation. Sources of notes on damage: International Tanker Owners Pollution Federation; Centre of documentation, Research and Experimentation on Accidental Water Pollution; U.S. National Oceanic and Atmospheric Administration (http://www.black-tides.com/). (b) The largest oil tanker spills in history.

A distinguishing feature of the TORREY CANYON response operation was the excessive and indiscriminate use of early dispersants and solvent based cleaning agents, which caused considerable environmental damage. The dispersants were generally successful at their task of reducing the amount of oil arriving ashore and subsequently expediting onshore cleanup operations, but they were

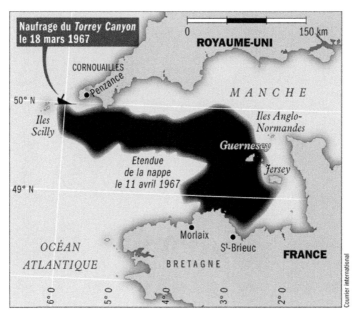

Fig. 7. Shipwreck of the Torrey Canyon (http://4.bp.blogspot.com/LKoATQJXsMo/UI8oXo_9mEI/AAAAAAAAAB4/Ol9EVqGYy6s/s1600/1029-TorreyCanyon-655.gif).

considerably more toxic than those used today and were applied in far greater concentrations, often being poured undiluted on slicks and beaches. Many of the detrimental impacts of the spill were later related to the high volume, high concentration and high toxicity of the dispersant and detergents used. https://www.itopf.org/in-action/case-studies/case-study/torrey-canyon-united-kingdom-1967/.

The United Kingdom used 10,000 tons of chemical dispersants to divide, disperse, and "erase" the crude oil slicks. This practice, which only adds pollution to pollution, has widely been abandoned, although in 2010 chemical dispersants toxic to fauna and flora, fish farming and oyster farming, were spread following the oil spill of the Deepwater Horizon platform, owned by British Petroleum, in the Gulf of Mexico.

A very different approach to black tide management was taken following the Erika shipwreck, 33 years after the Torrey Canyon wreck. The stockpiles were inventoried from the beginning of the crisis. Wastes were then regrouped and treated in Donges in the estuary of the Loire River under the control of State services and a Local Commission of Information and Surveillance and under the technical and financial responsibility of the cargo owner, the company Total.

6.2 Amoco Cadiz, March 16, 1978, Brittany, France, 227,000 metric tons

The tanker Amoco Cadiz ran aground off the coast of Brittany on March 16, 1978, following a steering gear failure. Over a period of two weeks the entire cargo of 223,000 tons of light Iranian and Arabian crude oil and 4,000 tons of bunker fuel was released into heavy seas. Much of the oil quickly formed a viscous water-in-oil emulsion, increasing the volume of pollutant by up to five times. By the end of April, oil and emulsion had contaminated 320 km of the Brittany coastline, and had extended as far east as the Channel Islands (http://www.itopf.com/in-action/case-studies/case-study/amoco-cadiz-france-1978/) (Fig. 8).

Strong winds and heavy seas prevented an effective offshore recovery operation. All told, less than 3,000 tons of dispersants were used. Some chalk was also used as a sinking agent, but with the consequence of transferring part of the problem to the sea bed. The at-sea response did little to reduce shoreline oiling. A wide variety of shores were affected, including sandy beaches, cobble and shingle

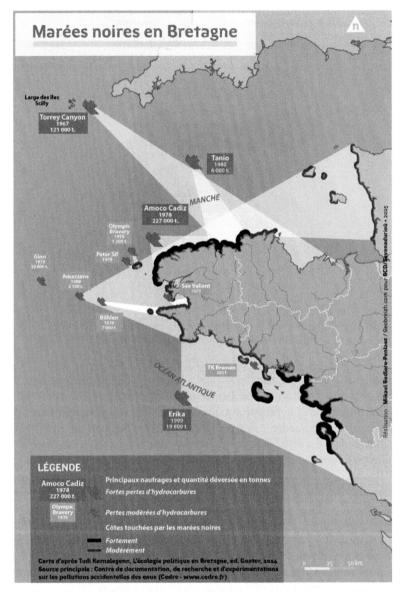

Fig. 8. Amoco Cadiz (http://bcd.bzh/becedia/fr/la-maree-noire-de-l-amoco-cadiz).

Color version at the end of the book

shores, rocks, seawalls and jetties, mudflats and salt marshes. Removal of bulk free oil trapped against the shore using skimmers proved difficult, largely because seaweed and debris were mixed with the oil. Greater success was achieved with vacuum trucks and agricultural vacuum units, although much of the free oil was simply removed by hand by more than 7,000 personnel (mainly military). A considerable portion of the oil that did come ashore eventually became buried in sediments and entrapped in the low-lying salt marshes and estuaries.

At the time, the Amoco Cadiz incident resulted in the largest loss of marine life ever recorded after an oil spill. Two weeks after the accident, millions of dead mollusks, sea urchins and other benthic species washed ashore. Diving birds constituted the majority of the nearly 20,000 dead birds that were recovered. Oyster cultivation in the estuaries was seriously affected and an estimated 9,000 tons were destroyed because of contamination and to safeguard market confidence. Other shell and fin fisheries as well as seaweed gathering were seriously affected in the short term, as was tourism.

Cleanup activities on rocky shores, such as pressure-washing, as well as trampling and sediment removal on salt marshes caused biological impacts. While rocky shores recovered relatively quickly, the salt marshes took many years. Failure to remove oil from temporary oil collection pits on some soft sediment shorelines before inundation by the incoming tide also resulted in longer-term contamination. Numerous cleanup and impact lessons were learned from the Amoco Cadiz incident, and it still remains one of the most comprehensively studied oil spills in history.

6.3 Exxon Valdez, March 24, 1989, Alaska, USA, 40,000 metric tons

It is over 25 years since Alaska experienced the worst ecological disaster in its history. On March 24, 1989, the Exxon Valdez oil tanker, owned by ExxonMobil, set out from the Valdez oil terminal to Long Beach, California. It had on board 41 million liters of crude oil. The captain is said to have left the command to two crew members and the autopilot system. The ship ran aground in Prince William Sound, and almost all the oil in the ship was spilled (Fig. 9).

Approximately 2,000 km of coastline was affected and with it thousands of aquatic species that to date are still recovering from that accident. The spill endangered 10 million migratory birds and waterbirds, otters, sea lions, porpoises, whales and fish.

This was the second largest oil spill in the history of the United States and ranked number 54 worldwide. The cause of the accident was said to be the captain's error. Exxon spent over $3.8 billion to clean up the site, compensate the 11,000 residents, and pay fines. But it could have been $4.5 billion more. The Alaskan court ordered Exxon to pay $5 billion in punitive damages in 1994. After 14 years of lawsuits and appeals, the U.S. Supreme Court ruled that Exxon only owed $507.5 million. That was only about 12 hours of revenue for the giant oil company. https://www.thebalance.com/exxon-valdez-oil-spill-facts-effects-on-economy-3306206.

The freezing conditions on the coasts of Alaska make it difficult for oil to dissolve and for microorganisms to absorb it. It is probable that life in this place will never be the same and there are no guilty parties interested in repairing their error or authorities that demand it.

Fig. 9. Exxon Valdez. Field studies continue to examine the effects of the Exxon supertanker's disastrous grounding on Bligh Reef in Alaska's Prince William Sound in 1989. Photo courtesy Erik Hill, Anchorage Daily News (https://response.restoration.noaa.gov/about/media/10-photos-tell-story-exxon-valdez-oil-spill-and-its-impacts.html; https://aoghs.org/transportation/exxon-valdez-oil-spill/).

Color version at the end of the book

Although the final number of dead species will never be known, it is estimated that in the months after the accident about 250,000 seabirds, 2,800 otters, 300 seals, 250 bald eagles, some 22 orcas and billions of specimens and eggs died, causing, in addition to ecological damage, a disaster for the fishing industry.

The Oil Spill Commission's report found the capabilities to respond to the spill "unexpectedly slow and woefully inadequate." The report declared, "The worldwide capabilities of Exxon Corporation would mobilize huge quantities of equipment and personnel to respond to the spill—but not in the crucial first few hours and days when containment and cleanup efforts are at a premium."

Alaskan weather conditions –33 degrees with a light rain—and the remote location added to the 1989 disaster, the report continues.

With the captain not present, the third mate made a navigation error, adds another 1990 report, Practices that relate to the Exxon Valdez by the National Transportation and Safety Board. "The third mate failed to properly maneuver the vessel, possibly due to fatigue or excessive workload," it concludes. https://aoghs.org/transportation/exxon-valdez-oil-spill/.

Experts continue to review the effects of the Exxon Valdez grounding on Bligh Reef. Most scientists today say the ecosystem in Prince William Sound, although still recovering, is healthy. Since the supertanker's accident, ExxonMobil has spent more than $5.7 billion in compensatory and cleanup payments, settlements and fines. Field and laboratory studies still examine the spill, which resulted in the Oil Pollution Act of 1990.

6.4 *Prestige, November 13, 2002, Galicia, Spain, 64,000 metric tons*

On November 13, 2002, the tanker Prestige, carrying a cargo of 77,000 tons of heavy fuel oil, suffered hull damage in heavy seas off northern Spain (http://www.itopf.com/in-action/case-studies/case-study/prestige-spainfrance-2002/). It developed a severe list and drifted towards the coast, and was eventually taken in tow by salvage tugs. The casualty was reportedly denied access to a sheltered, safe haven in either Spain or Portugal and so had to be towed out into the Atlantic. Although attempts were made by salvors to minimize the stresses on the vessel, it broke in two early on November 19 some 170 miles west of Vigo, and the two sections sank some hours later in water two miles deep. In all, it is estimated that 63,000 tons were lost from the Prestige (Fig. 10).

Owing to the highly persistent nature of Prestige's cargo, the released oil drifted for extended periods with winds and currents, travelling great distances. Oil first came ashore in Galicia, where the

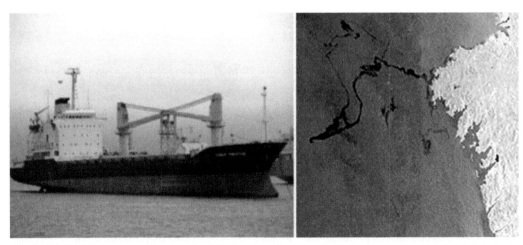

Fig. 10. The Prestige was an oil tanker ship from Liberia that on November 19, 2002, sank near the coast of Galicia in Spain. Satellite picture of the Prestige oil spill (ESA) (http://www.latimes.com/news/nationworld/nation/la-na-oil-spill-history-html,0,3901663.htmlstory#axzz2zFv2Yvcm; https://wildlife-oilspills.wikispaces.com/Oil+Spills).

predominantly rocky coastline was heavily contaminated. Remobilization of stranded oil and fresh strandings of increasingly fragmented weathered oil continued over the ensuing weeks, gradually moving the oil into the Bay of Biscay and affecting the north coast of Spain and the Atlantic coast of France, as far north as Brittany. Some light and intermittent contamination was also experienced on the French and English coasts of the English Channel. Although oil entered Portuguese waters, there was no contamination of the coastline.

A major offshore cleanup operation was carried out using vessels from Spain and nine other European countries. The response, which was probably the largest international effort of its kind ever mounted, was hampered by severe weather and by the inability of those vessels that lacked cargo heating to discharge recovered oil (Albaiges et al. 2006).

Despite the enormous effort at sea and the extensive booming of estuaries and sensitive areas with over 20 km of boom, extensive coastal contamination occurred. Altogether, approximately 1,900 km of shoreline was affected. The shorelines of Spain were largely cleaned manually by a workforce of over 5,000 military and local government personnel, contractors and volunteers.

Fisheries exclusion zones were put in place in Galicia shortly after the incident, banning virtually all fishing along about 90% of the coastline. All bans had been lifted by October 2003. The impact on fisheries in France was less extensive. In both countries, an impact on tourism was reported for 2003.

The Spanish authorities decided to remove the oil remaining in the wreck. The work commenced in May 2004 using bespoke shuttle tanks and was completed in September 2004, at an estimated cost of €100 million.

The incident qualified under the Civil Liability and Fund Conventions. The vessel's P&I Club and the 1992 International Oil Pollution Compensation Fund established a joint claims office in La Coruña to facilitate the receipt and handling of claims in Spain. Payments for compensation to France, Spain and Portugal are expected to reach the Fund limit of £171 million. Legal proceedings in Spain and France are on-going.

7. Massive Oil Spills in the Gulf of Mexico

The Gulf of Mexico (GoM) is broadly recognized for its macro biodiversity and fossil fuel reserves. This large marine ecosystem provides invaluable ecological services that have traditionally supported the socio-economic development of the bordering countries. For many years, maritime operations, recreation activities, and the exploitation of living resources were the primary activities conducted in the GoM waters. However, since the early 1950s, the oil industry irrupted in the GoM domain, first in the offshore waters and later in deep waters of the northern and southwestern regions. Since then, the ongoing exploitation of oil and gas reserves has sustained a flourishing oil industry in the USA and Mexico (Fig. 11).

Interestingly, in coastal and oceanic surface waters of the GoM the natural emission of methane gas bubbles and oil slicks seeping from the seabed is a common event (MacDonald et al. 2015); presumably, the Gulf ecosystem has adjusted itself to cope with this chemical disturbance produced by such a long-standing phenomenon without endangering its ecological equilibrium. However, massive oil spills like the Ixtoc-I in the Campeche Sound in 1979 and the Deepwater Horizon off the Mississippi Delta in 2010 have seriously jeopardized the stability and resilience capacity of the GoM. These accidents have highlighted the void of our knowledge concerning the fate of hydrocarbon compounds in the marine ecosystem, and the environmental response of a large marine ecosystem. Marine oil pollution continues to be a controversial issue among Gulf's stockholders, whose short-time profit expectations seem to prevail over the conservation of the health of the oceans.

Fortunately, prevention measurements and bioremediation mechanisms have recently been fostered to mitigate harmful effects caused to marine ecosystem by oil and gas exploitation. Alabresm

Fig. 11. Gulf of Mexico.

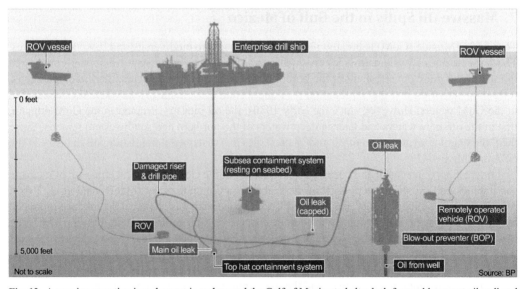

Fig. 12. A massive operation is under way in and around the Gulf of Mexico to halt a leak from a blown-out oil well and prevent the spread of the slick (http://news.bbc.co.uk/2/hi/americas/8651333.stm).

et al. (2018) have recently summarized the pros and cons of the oil removal techniques currently in use, advocating a novel procedure that combines nanotechnology and microbial remediation as a viable *in situ* oil remediation method (Fig. 12).

7.1 IXTOC-I Blowout, June 3, 1979, to March 23, 1980, Bay of Campeche, Gulf of Mexico, ~ 500,000 metric tons

Ever since the oil spill caused by the Ixtoc-I blowout in Campeche Sound, there has been a growing concern about deleterious environmental effects upon a marine ecosystem known for its reasonably pristine conditions prior to the accelerated exploration and extraction of fossil fuels in the area. This accident was the world's first massive oil spill occurring in offshore waters of a tropical environment. More than 3.4 million barrels of crude oil were liberated by the Ixtoc-I blowout into the southwestern GoM, for nearly nine months (Soto et al. 2014). Oil mousse, sheen, and tar balls were advected towards the coastal zone by the dominant surface circulation pattern in the impacted area. The initial appraisal of the ecological damages caused by the spill was precluded by the lack of pre-spill baseline information. The Mexican authorities convened the leading national researchers to execute a short- and mid-term observational plan to assess the environmental disturbances caused by the crude oil in the atmosphere, water, sediments, and biota. Prevention plans were also implemented to protect potential coastal areas susceptible of being contaminated by fresh or weathered fossil hydrocarbons from the Ixtoc-I. Given the dearth of knowledge on pre-spill conditions in the Campeche Sound it was difficult for the experts to differentiate the natural variability of biotic community parameters (e.g., biomass, diversity, abundance, and density) from that caused by anthropogenic disturbances. The technical reports emanating from the field surveys conducted in Campeche Sound from 1979 to 1980 concurred that the evaporation, dispersion, photo-oxidation and biodegradation processes in this zone played a significant role in alleviating the harmful environmental effects of the oil spill. For instance, Boehm and Fiest (1982) reported, after the Ixtoc-I blowout, total hydrocarbon levels above 100 μg g^{-1}, which returned to the background level (~ 70 μg g^{-1}) 12 months later (Botello and Villanueva 1987). Additionally, the tropical shelf ecosystem displayed a remarkable resilience capacity by absorbing, in time and space, the severe anthropogenic disturbance. However, numerous questions arose afterward concerning the oil degradation rates and the toxicity levels of the residual Ixtoc-I hydrocarbons in sediments and biota. Similarly, there were no clear ideas as to the mid- and long-term post-spill environmental consequences.

In the light of this brief recounting of facts at the outset of the oil spill, an irrefutable truth emerged: in the southwestern Gulf, particularly in the Campeche Sound, there was a biological economy prior to the Ixtoc-I blowout, and a different one afterwards. Many changes have occurred in this region of the Gulf since this accident took place. The Mexican oil consortium (PEMEX) has continued expanding its inshore and offshore operations causing chronic accidental oil spills and the government legislation concerning the exploitation of Mexico's energy resources has also been modified. In 2013, the Mexican federal government introduced a major amendment in its constitution—the Energy Reform—that allowed continental and outer-shelf oil and gas leasing in Mexico's territorial waters. This means that transnational oil corporations may participate in the exploration, extraction, and trading of national fossil fuel reserves. The expansion of such operations both in shallow and deep waters of the GoM would presumably be conducted under a better legal framework focused not only on economic profit but also on preserving the health of the marine ecosystem. In spite of these actions, the fate of the coastal and marine resources in the GoM remains uncertain (Fig. 13).

One of the earliest and most objective environmental assessments of the devastating Ixtoc-I oil spill was the United Nations Environment Programme (UNEP) report elaborated by Arne Jernelöv (see Jernelöv and Linden 1981). This report drew important conclusions related to the total volume of crude oil released by the blowout, its fate, and the weathering processes in the warm waters of the Gulf (> 25.0°C); but more importantly, it offered the first evaluation of the acute effects of fossil hydrocarbons upon local fisheries and coastal environments adjacent to Ixtoc-I. Later on, other authors (e.g., Botello and Castro 1980, Soto et al. 1981, Yáñez-Arancibia 1986, Licea-Duran et al. 1982, Guzmán del Próo et al. 1986) added further evidence concerning the effects of crude oil and its possible bioaccumulation in different ecosystem compartments of Campeche Sound.

Fig. 13. Ixtoc-I oil well releasing thousands of metric tons of oil into the Gulf of Mexico. Ixtoc-I oil well (photo courtesy NOAA) (http://www.whoi.edu/oil/ixtoc-I, http://counterspill. org/disaster/ixtoc-oil-spill).

According to Jernelöv and Linden (1981), a considerable percentage of the Ixtoc-I crude oil was either evaporated (48%) or sunk to the sea floor (24%); other oil remains were washed ashore in adjacent coastal areas (6%), and on USA beaches (3%), 12% was biodegraded, and 7% was burned or directly recovered at the site. Naturally, the new findings of high concentrations of total hydrocarbons in surficial sediments in both coastal and estuarine systems (Botello et al. 1991, Botello 1996) ignited once more the controversy on the fate and toxicity of the Ixtoc-I oil spill versus the chronic pollution caused by PEMEX's offshore operations in Campeche Sound.

In 1982, PEMEX authorities publicly announced, on the basis of a two-year multidisciplinary study of Campeche Sound, that "...the marine ecosystem did not suffer any damage from the Ixtoc-I oil spill." With this uncalled-for statement, it seemed as if the final chapter on the Ixtoc-I oil spill had come to an end. No intentions from the Mexican government to support a long-term research program to assess the ecological damages derived from a massive oil spill were ever expressed. Since then, only the individual efforts of a few scientists (e.g., Yáñez-Arancibia 1986, Lizárraga-Partida et al. 1991, Soto and Escobar-Briones 1995, García-Cuellar et al. 2004, Botello et al 2004, Soto and González, 2009, Soto et al. 2009, Yáñez-Arancibia et al. 2013a, b, Baltz and Yáñez-Arancibia, 2013) concerned with the environmental health of one of the most productive regions in the Gulf kept alive the commitment to maintain permanent observations on stressor factors that may upset the fragile ecosystem integrity.

In the aftermath of Ixtoc-I, several revealing facts came to light: (1) tropical environments respond differently from temperate regions to anthropogenic disturbances; (2) both physical and chemical reactions are accelerated at twice their normal rate by warm ambient temperatures, biological processes having a similar metabolic response (Q_{10}); (3) given their rich biodiversity, tropical systems are more resilient and perhaps can recover faster from a disturbance than a temperate system; (4) biodegradation processes can also be promoted by a rich microbiota; (5) in regions such as Campeche Sound, with a high input of biogenic hydrocarbons from estuarine systems, the detection of pyrogenic hydrocarbons in sediments or animal tissues can have a masking effect; (6) natural biological variability in a tropical ecosystem may impede the production of unequivocal evidence of pollution from offshore oil operations.

From a strictly environmental perspective, the Ixtoc-I blowout event has served as an example of the imperative need to strengthen prevention and emergency procedures of the international oil and gas operations in offshore waters.

Many controversial issues arise among authors confronted with the predicament of assessing the magnitude of damage caused by the oil industry in marine ecosystems. Holdway (2002) proposed as a possible solution long-term environmental observation (10 to 20 years) to determine chronic and sublethal effects attributable to offshore oil exploitation. Nonetheless, there is a consensus that acute effects are space-restricted and of short duration, whereas the mid- and long-term or sublethal effects are more difficult to assess. Another significant finding from the Ixtoc-I blowout is that the pelagic realm recovers faster from a massive oil spill than the benthic compartment, particularly the deep inhabitants. Oil degradation rates and weathering processes in sediments may take a much longer time (see Boehm and Fiest 1982, Oudot and Chaillan 2009), in the scale of several decades.

7.2 Deepwater Horizon, April 20 to July 2010, offshore Louisiana, ~ 750,000 metric tons

Because of its geological origin, the Gulf of Mexico represents a favorable basin for the accumulation of fossil hydrocarbon and gas deposits. Historically, the gulf as a large marine ecosystem has been exposed to natural hydrocarbon and gas leaks through fault zones of the marine subsoil. These natural emanations have been recorded in several sectors of the gulf and represent an important source of pollution (Wilson et al. 1974). The spills of crude oil produced by the Ixtoc-I and the sinking of the Deepwater Horizon oil platform (DWH) off the coast of Louisiana in 2010 are examples that have caused severe damage to the environmental health of the gulf (Sammarco et al. 2013, Soto et al. 2014).

On April 20, 2010, a serious accident occurred on the DWH oil platform 50 nautical miles southeast of the Mississippi River Delta, in the northern Gulf of Mexico. Eleven lives were lost and 4.9 million barrels of crude oil were spilled at 1650 m depth. This accident, recognized by experts as the biggest disaster of the oil industry in the USA (Joye et al. 2011, McNutt et al. 2011), caused severe environmental damage to both coastal and deep-water habitats in the vicinity of the wellhead (Camilli et al. 2010, Griffiths 2012, Joye et al. 2014, White et al. 2012). Persistent surface and subsurface plumes of oil and gas were detected in that region even four months after the DWH blowout.

The British firm British Petroleum (BP), responsible for the operation of the DWH, jointly with the NOAA, implemented a series of immediate remediation actions to mitigate the damage to the marine ecosystem caused by the escape of 12,000 to 19,000 barrels of oil per day. Among these actions was the direct recovery of oil, the selective burning of hydrocarbon slicks in surface waters and the use of 1.85 million gallons of chemical dispersants (Corexit®) both at the surface and on the seabed. The hydrographic conditions prevailing in the Gulf of Mexico during the summer season, combined with the arrival of Hurricane Alex in July, contributed to contain the stain of crude oil in the vicinity of the Mississippi Canyon, with trajectories to the northeast. Satellite images obtained by NOAA, complemented by the circulation simulation models generated by the Consortium of North American Universities of the Gulf of Mexico, confirmed those trajectories, with small filaments detached toward the Texas coasts. Similarly, remote sensors made possible the detection of crude oil slicks trapped in the Eddy Franklin, an important component of the Loop Current; some of these images revealed oil slicks flowing into the Exclusive Economic Zone waters of Mexico, just north of the Yucatan Peninsula.

The volume of crude oil spilled and the amount of chemical dispersants used constituted severe alterations to the ecological balance and environmental health of the GoM. The evaluation of the environmental damage caused by the massive leakage of gas and hydrocarbons from the Macondo well off the coast of Louisiana, in the northern Gulf of Mexico, represented a huge challenge for the scientific communities of the countries that share their waters and resources. The precise calculation of the volume of oil spilled, the trajectory of the hydrocarbon slicks at sub- and superficial levels, as well as the degradation rates of crude oil and its derivative compounds, have represented controversial issues among specialists. This problem is further magnified by the chemical complexity of crude oil. It includes approximately 17,000 organic compounds (Bjorlykke 2011), each with its properties of volatility, density, solubility, and degrees of toxicity to marine biota and humans.

At the outset of the oil spill, the North American agencies, NOAA and the EPA, in an attempt to minimize the seriousness of the accident, claimed that a considerable percentage of crude oil was directly recovered, and the rest was burned or lost by evaporation. Nowadays, there are still many uncertainties concerning the final fate of the spilled oil and its persistence in both shallow and deep habitats. The stability and resilience capacities of the GoM were exposed to severe environmental changes by the DWH incident and its long-term effects remain unknown.

8. Trans-boundary Pollutants

Presumably, pollutants originating in the Gulf can be easily transported to a remote region within or outside of the Gulf. The Ixtoc-I blowout is perhaps the best example of an oil spill affecting outlying areas as far north as the Texas shoreline. In the ocean, there are no physical barriers that impede the dispersion of pollutants into the waters of a neighboring country. Presently, we know that due to the high connectivity within the GoM, trans-boundary mechanisms facilitate the dispersal of larvae and pollutants from the north to the south and vice versa.

Oil trapped in sediments is known for causing a series of ecological disturbances in benthic communities ranging from the decrease of density and biomass to the loss of diversity (see Lee and Li 2013). A few months after the DWH event in the northern Gulf of Mexico, Montagna et al. (2013) documented both its near and farther-field deep-sea benthic footprint. Total hydrocarbons, polycyclic aromatic hydrocarbons (PAH), and barium concentrations were identified by these authors as the leading environmental drivers affecting macrofauna and meiofauna abundance and diversity. Other authors (Schwing et al. 2015) added further evidence concerning the environmental disturbance caused by the oil spill upon foraminiferal benthic density in the northeastern Gulf. The significant decline (80–93%) was attributed to abrupt increases in sedimentary rates, PAH concentrations and changes in redox conditions.

Indeed, the study of fossil fuels trapped in marine surface sediments constitutes a complex problem. The source identification and the spilled oil potential trajectory in the marine environment are further complicated by the presence of biogenic hydrocarbons, natural gas and oil seeping processes, and weathering factors (Elmgren et al. 1983, Liu et al. 2012).

Presently, many uncertainties remain concerning the final fate of roughly 2 million barrels of submerged oil presumably sequestered in deep ocean sediments of the northern Gulf (Valentine et al. 2014). Valentine et al. (2014) hypothesized that, as a consequence of the patchy spatial distribution of oil particles deposited on the ocean floor and the distance gradient from the Macondo's wellhead, their detection becomes problematical outside an area of 3,200 km^2 that exposes clear signs of ecological damage. In remote areas such as the NW Gulf, the uncovering of Macondo's footprint also seems an elusive process (Fig. 14).

9. Biological Effects of Oil Spills

Since the 1980s, research results have been reported on the possible effects of oil spills in the sea, considering scenarios of acute effects represented by the mortality of exposed organisms as well as those longer-term chronic exposures causing sublethal consequences in the surviving species (Sanders et al. 1980, Southward 1982). In recent years there has been an increase in the number of studies aimed at evaluating the chronic biological effects of large oil spills that are extraordinary events in time and magnitude, as well as minor discharges mainly derived from offshore drilling and extraction platform hydrocarbons, from world transport through large tankers as well as from port activities (Boesch et al. 1987, NRC 2003).

The effects of an oil spill begin when the organisms come into contact with it, from the species that inhabit the surface to those found on the seabed, affecting the trophic chains and the integral functioning of the ecosystem; for example, oil clogs fish gills, causing the fish to die by asphyxiation (NRC 2003, Centeno-Chalé et al. 2015); seabirds, aquatic mammals, and reptiles die from loss of shelter, starvation, ingestion of oil or drowning after being saturated with oil (Baker et al. 1991,

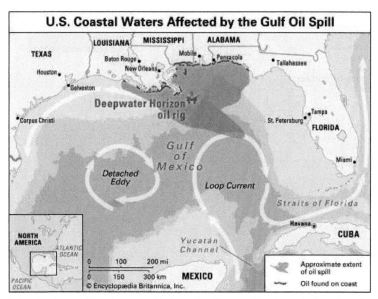

Fig. 14. Map depicting the effects of the Deepwater Horizon oil spill, caused by the explosion of an oil rig off the coast of Louisiana on April 20, 2010. Scientists noted that the prevailing paths of the Gulf of Mexico's Loop Current and a detached eddy located to the west kept much of the oil, which covered a sizable part of the gulf some three months after the accident, from reaching the shore (https://www.britannica. com/ event/Deepwater-Horizon-oil-spill-of-2010).

Maki 1991); organisms on the sea floor are affected or die as a result of oil residues, which sink to the bottom over time (Khan and Nag 1993, Khan 2010).

Oil can reduce animal fitness through sublethal effects and disrupt the structure and function of marine communities and ecosystems. Multiple temporal and spatial variables make deciphering the effects extremely difficult, especially when considering the time and space scales at which marine populations and ecosystems change. In Fig. 15, the processes that must be quantified in the biological compartment are illustrated. Basically, the fate of petroleum hydrocarbons in the biota results from (1) ingestion (uptake) of a complex range or variety of petroleum hydrocarbons by an organism and subsequent excretion to the sediment or the water column of a different suite after digestion and metabolism; (2) transfer of petroleum hydrocarbons up the food chain through the predator-prey relationship; and (3) sorption or ingestion of petroleum hydrocarbons by a marine organism followed by death of the organism and return of the petroleum hydrocarbon to the water column or sediment through biochemical decay. In addition, certain petroleum hydrocarbons will be carried away from the sea by birds and land-based animals, when oil adsorbs on their bodies or they ingest seafood and then excrete products or die on shore (NRC 2003).

The biological effects of oil pollution are often referred to as acute or chronic. Spills are commonly thought of as having short-term effects from high concentrations of petroleum. Chronic pollution, such as might occur from urban runoff into coastal embayments, may have continuous effects at low exposures; effects are known to occur for long periods after some spills (Vandermeulen et al. 1982, Spies 1987, Teal et al. 1992, Burns et al. 1993), and chronic exposures can be quite high, as is the case near petroleum seeps (Spies et al. 1980, Steurmer et al. 1982).

The responses of organisms to petroleum hydrocarbons can be manifested at four levels of biological organization: (1) biochemical and cellular; (2) organism, including the integration of physiological, biochemical and behavioral responses; (3) population, including alterations in population dynamics; and (4) community, resulting in alterations in community structure and dynamics. Impairment of behavioral, developmental, and physiological processes may occur at concentrations significantly lower than acutely toxic levels; such responses may alter the long-term survival of affected populations. Thus, the integration of physiological and behavioral disturbances may result

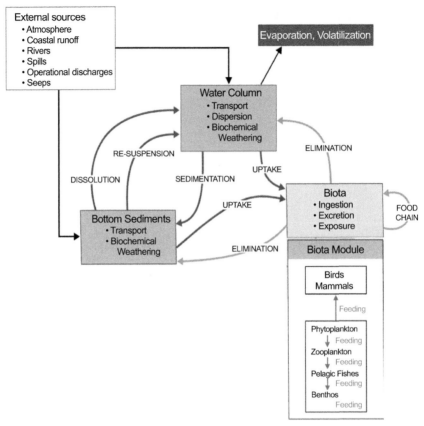

Fig. 15. Detailed interactions of a conceptual model for the fate of petroleum in the marine environment with emphasis on the biological compartment (NRC 2003).

in alterations at the population and community levels. The effects of petroleum hydrocarbons in the marine environment can be either acute or chronic. Acute and chronic toxicity of petroleum hydrocarbons to marine organisms is dependent upon the following (NRC 2003):

- Concentration of petroleum hydrocarbons and length of exposure,
- Persistence and bioavailability of specific hydrocarbons,
- The ability of organisms to accumulate and metabolize various hydrocarbons,
- The fate of metabolized products,
- The interference of specific hydrocarbons (or metabolites) with normal metabolic processes that may alter an organism's chances for survival and reproduction in the environment (Capuzzo 1987), and
- The specific narcotic effects of hydrocarbons on nerve transmission.

The bioavailability of petroleum hydrocarbons depends mainly on two processes: environmental availability and biological availability. Environmental availability is the physical and chemical form of the chemical in the environment and its accessibility to biological receptors. Generally, chemicals in true solution in the ambient water are considered more bioavailable than chemicals in solid or adsorbed forms. Petroleum hydrocarbons of the types found in the marine environment may be present in true solution, complexed with dissolved organic matter and colloids, as dispersed micelles, adsorbed on the surface of inorganic or organic particles, occluded within particles (e.g., in soot, coal, or tar), associated with oil droplets, and in the tissues of marine organisms (Readman et al. 1984, Gschwend and Schwarzenbach 1992). The other aspect of environmental availability is accessibility. Petroleum hydrocarbons that are buried deep in sediments or sequestered in solid, highly weathered oil deposits on the shore are not accessible to marine and terrestrial organisms and therefore have a low bioavailability.

Biological availability depends on the rate at which a chemical is assimilated into the tissues of the organism and accumulates at the sites of toxic action in the organism. This depends on the physical and chemical properties of the chemical in contact with the organism, the relative surface area of permeable epithelia in the organism, and the ability of the organism to excrete or detoxify the chemical. Nonpolar (hydrophobic) organic chemicals such as petroleum hydrocarbons have a low aqueous solubility and high lipid solubility. Hydrocarbons in solution in water diffuse down an activity or fugacity gradient from the water phase into lipid-rich tissues of marine organisms in contact with the water. According to equilibrium partitioning theory (Davies and Dobbs 1984, Bierman 1990), when an aquatic animal is exposed to a nonpolar organic chemical dissolved in the ambient water, the chemical partitions across permeable membranes into tissue lipids until an equilibrium, approximated by the octanol/water partition coefficient (K_{ow}) for the chemical, is reached. At equilibrium, the rates of absorption into and desorption from the lipid phase of the organism are equal. Toxic responses in the organism occur when the concentration of nonpolar organic chemicals in the tissues reaches a critical concentration (McCarty and Mackay 1993).

9.1 Fish

Large oil spills have been conceived as a major threat to fish stocks; however, very few studies have demonstrated increased mortality of fish as a result of oil spills (Fodrie et al. 2014). The fish stocks may be especially vulnerable to oil spills close to the spawning grounds or egg and larval drift areas (Hjermann et al. 2007, Rooker et al. 2013). Fish eggs and larvae are typically vulnerable to toxic oil compounds because of their small size, poorly developed membranes and detoxification systems as well as their position in the water column. A number of laboratory studies have shown that oil or oil compounds, mainly PAHs, at low concentrations can kill or cause sublethal damage to fish eggs and larvae (Hicken et al. 2011, Meier et al. 2010, Sørhus et al. 2015). Many of the toxic components in oil are multi-ring PAHs, which deposit in sediment and may bioconcentrate in marine species with low cytochrome P450 monooxygenase activity, such as certain invertebrates and even fish, vertebrates with a greater enzymatic activity (Smeltz et al. 2017). Altered enzymatic activity in fish serves as a biological monitor for certain xenobiotic exposures, including the components of petroleum oil (Nahrgang et al. 2010). Under appropriate conditions, biochemical markers such as ethoxyresorufin O-deethylase (EROD), benzo(a)pyrene hydroxylase (AHH), glutathione transferase (GST), and glutathione peroxidase (GPx) can indicate exposure to oil-related chemicals in the laboratory (Oliva et al. 2010, Kerambrun et al. 2012) and in the environment following oil spills, as documented for the Exxon Valdez spill (Jewett et al. 2002) and the Prestige spill (Morales-Caselles et al. 2006). A family of cytochrome P450 enzymes very important in oil spills is the CYP1 family; fish CYPs known to biotransform PAHs are mainly CYP1A and CYP1C isoforms (Stegeman et al. 2015). Catalytic activity of the CYP1 family is commonly measured through AHH activity or EROD activity, which have been shown to be very sensitive for assessment of PAH exposure that results in increased CYP1 expression in fish (Whyte et al. 2000). Other hepatic enzymes reported to be induced in fish, associated with exposure to PAH, include several forms of GSTs (Higgins and Hayes 2011). Additionally, GPx activity has been linked to exposure to oil-related chemicals, including PAHs, PAH metabolites and other hydrocarbons (Carvalho et al. 2012).

Sublethal effects include morphological deformities, reduced feeding and reduced growth rates, and are likely to increase vulnerability to predators and starvation. The few existing *in situ* studies of fish mortality at spill sites indicate sublethal effects or elevated mortality of eggs and larvae (deBruyn et al. 2007, Incardona et al. 2012). An oil spill mainly kills fish at the egg or larval stage. The impact of an oil spill on a fish stock depends on (1) the proportion of the eggs and larvae killed by the oil spill, and (2) the effect of early-stage mortality on cohort survival in subsequent stages.

Spatial variation in natural mortality may significantly alter the effect of an oil spill on the recruitment of marine fish; Langangen et al. (2017) demonstrated that the probability of extreme effects might be underestimated. At the same time, their results strongly indicated a general need for explicitly including spatial mortality in oil spill assessments to better capture the risk of accentuation of oil spill effects over time.

9.2 Benthos

Benthic organisms are used extensively as biotic indicators of environment because they generally have limited mobility and cannot avoid adverse environmental changes. The group of foraminifera is particularly useful for environmental monitoring (Foster et al. 2012); they are sensitive to changes in marine environments such as water temperature, salinity, pH, water mass, ocean current, and geographical variables (Murray 2006). Previous studies showed that foraminifera could serve as biotic indicators to evaluate the impacts of oil spills (Armynot du Châtelet et al. 2004). Durrieu et al. (2006) and Mojtahid et al. (2006) showed that benthic foraminifera could be used to estimate the pollution from oil drill mud disposal. The general responses of benthic foraminifera to pollutants include decreased diversity and increased dominance of tolerant or opportunistic species, or alteration of species morphology and reproduction, but different species may show differential response. In the investigations carried out on this subject, it has generally been observed that although the toxic hydrocarbon components appeared to be responsible for the observed changes in foraminiferal abundance and species composition (Armynot du Châtelet and Debenay 2010), species-specific responses to environmental stress induced by oil pollution were evident (Lei et al. 2015).

During oil spills, a process known as marine snow may occur; the discharge of the rivers towards the marine area where the spill has occurred contributes an enormous amount of suspended solids as well as a phytoplankton bloom (O'Connor 2013). Marine snow aggregates, made of dispersed oil, organic debris, phytoplankton, and suspended particles, glued together by the sticky extracellular polymeric substances, settle on the ocean floor in a process called MOSSFA: marine oil snow sedimentation and flocculent accumulation (Daly et al. 2016). MOSSFA can increase sedimentation rates and cause a downward flux of oil to the sediment. Estimates vary, but as much as 14% of the total oil released during the DWH oil spill may have ended up on the sediment due to MOSSFA (Daly et al. 2016). A review of large historical oil spills has indicated that the MOSSFA process may have occurred during other spills as well, such as the IXTOC-I blowout (Vonk et al. 2015). The MOSSFA-related oil contamination sparked interest in the potential long-term effects of sedimented oil for benthic habitats (Kinner et al. 2014). Benthic organisms, especially those that are sedentary, are particularly at risk for MOSSFA-related oil contamination (Fisher et al. 2016), since they cannot easily escape it. Moderate to severe reduction of macro- and meiofaunal abundance and diversity was found over an area of 172 km^2 around the wellhead in the DWH oil spill (Montagna et al. 2013). Persistent reducing conditions in the sediment and the two- to three-fold increase in PAHs reduced benthic foraminiferal diversity and density (Schwing et al. 2016). MOSSFA can affect benthic ecosystems via two mechanisms: (1) direct toxicity of the oil and (2) reduced oxygen availability caused by the microbial degradation of the marine snow. Direct oil toxicity to benthic organisms is widely reported, both in experimental studies (Bhattacharyya et al. 2003) and in oil spill observations (Lee and Li 2013). The availability of oxygen in deeper layers of the sediment can be increased by the bioturbation activity of many benthic organisms (Pelegrí and Blackburn 1994). Bioturbation enhances the sediment oxygenation, solute transport, and remineralization of organic matter, and mixes horizontal sediment layers (Levin 2003). Oxygen consumption by biodegradation of marine snow, the second mechanism, can impact benthic organisms after a MOSSFA event. The accumulation of organic material on the sea floor increased microbial respiration, resulting in decreased oxygen in sediment pore waters (Hastings et al. 2016). In laboratory studies, artificially produced marine snow was found to consume oxygen at a rapid rate (Rahsepar et al. 2017) and this lower oxygen concentration in the sediment is detrimental to benthic invertebrates living in the top layers of the sediment.

Benthic organisms in general are an important part of the marine food web, and many pelagic species are dependent on the benthic ecosystem for feeding or reproduction, while some have benthic life stages. Most oil effect studies are performed with pelagic species, and without marine snow or sediment (van Eenennaam et al. 2018).

9.3 Corals

Corals are a fundamental part of the marine reef system, one of the most important ecosystems in the world because of its great biodiversity and the environmental services it provides. They are at the same time among the most vulnerable to natural events of great magnitude such as hurricanes as well as anthropogenic wastes that alter their functioning and unbalance their biological composition. Modern societies have for several decades issued massive amounts of greenhouse gases that have contributed to accelerate global warming, causing serious consequences for this type of ecosystem, such as coral bleaching.

The black tides caused by oil spills have been one of the greatest environmental pressures on coral reefs; corals located in intertidal reef flats are exposed to oil slicks and are more susceptible to damage and death than corals in subtidal reefs. Coral located subtidally or in areas with high wave action are not directly exposed to the marine surface layer where oil slicks can coat them. Instead, only the water-soluble fraction of oil generally affects submerged coral. Acute and chronic exposures of oil on coral have been studied in the laboratory and field. Despite a wealth of information on the effects of petroleum pollutants on corals through laboratory studies, there is a scarcity of information on the levels of petroleum pollutant on corals in the field, and studies carried out in Taiwan (Ko et al. 2014), Gulf of Mexico (Sabourin et al. 2013), southwest Puerto Rico (Pait et al. 2009) and Kaneohe Bay in Oahu, Hawaii (Thomas and Li 2000), constitute the only background information we currently possess. Oil pollutants are extremely dangerous for coral health and survival because they alter the coral photo-physiological capacities (Meehan and Ostrander 1997), often characterized by zooxanthellae expulsion (Neff and Anderson 1981, Peters et al. 1981) or decrease of zooxanthellae primary production (Cook and Knap 1983, Neff and Anderson 1981, Rinkevich and Loya 1983), impair sediment clearance ability (Bak and Elgershuizen 1976) or coral fertilization (Negri and Heyward 2001), and may lead to tissue death (Neff and Anderson 1981, Wyers et al. 1986). Mixtures of dispersants and oil are more toxic to coral than the oil alone (Peters et al. 1997). It has also been observed that different types of coral suffer different effects. For example, branching coral such as *Acropora palmate* is considered more sensitive to oil exposure than massive coral such as *Montastraea* (Bak 1987). Ranjbar-Jafarabadi et al. (2018) conducted one of the last studies to evaluate the accumulation of various hydrocarbon groups and other persistent organic pollutants in corals of the Persian Gulf region, an area heavily impacted by oil spills for several decades; they conclude that the particular characteristics of the coral species studied are directly related to the bioaccumulation capacity of the toxic components of oil; for example, they recorded a greater amount of PAH in soft corals than in hard corals and also mention the influence of geochemical history of the place where these organisms live. Similarly, Turner and Renegar (2017) make one of the most complete reviews on the toxicity of petroleum hydrocarbons on corals, including oil sources and scenarios of exposure, composition and toxic effects of oil, incidents that have caused acute and chronic exposure of corals, evaluation of *in situ* effects as well as *in vitro* studies, and the limitations of carrying out this type of research.

9.4 Sand beach ecosystems

Another fundamental part of the coastal ecosystems are the sandy areas and the entire biological community that inhabits them. When an oil spill occurs, the oil slick often reaches this coastal region and significantly impacts the species that exist there, contributing to destabilize the ecological functioning of that coastal environmental component. Sand beaches provide multiple ecosystem functions including the regulation of biogeochemical cycles (e.g., nutrients, water, organic carbon), shoreline buffering, maintenance of genetic and biological diversity through the provision of habitat to many species, including vertebrate populations, many of which are of conservation concern, support of natural processes that promote energy flow among biological systems, and provision of several human-related services including recreation (McLachlan and Brown 2006, Schlacher et al. 2008). Over the last decades there have been reviews of the impact of oil on shoreline habitats, including mangroves (Getter 1982), marshes (Sell et al. 1995, Pezeshki et al. 2000, Michel and Rutherford 2014), and rocky shores (Sell et al. 1995), as well as reviews on general oil spill impacts (Gundlach and Hayes 1978, Vandermeulen et al. 1982, Teal and Howarth 1984, Kingston et al. 2003, NRC 2003). However, similar efforts have not been undertaken for sand beaches, despite the fact that these habitats are commonly oiled following an oil spill (Bejarano and Michel 2016). For example, following the 1991 Gulf War oil spills, where limited removal actions occurred on sand beaches, oil in exposed sand beaches showed some indication of weathering 2 years after the spills (Sauer et al. 1998), but substantial amounts of oil persisted for at least 12 years at < 60 cm depth (Bejarano and Michel 2010). Oil deposition along the coastline of Spain during the 2002 T/V Prestige oil spill coincided with a period of extensive beach erosion; although substantial cleanup was undertaken following the spill, oil remained buried to a depth of at least 2.38 m (Bernabeu et al. 2006, Fernández-Fernández et al. 2011), with biodegraded oil residues persisting in intertidal areas for 9 years (Bernabeu et al. 2013). Oil burial and exposure during the 2010 Deepwater Horizon oil spill were also influenced by coastal processes as the spill occurred at a time when beaches were in an erosional state; sand deposited from a combination of wind-driven and beach accretion processes resulted in oil burial at depth of 105 cm, with subsequent beach erosional periods re-exposing buried oil. Although evidence from several oil spills indicates that oil has the potential to persist in sand beaches, their links with direct effects to invertebrate communities are difficult to quantify. Figure 16 shows the data obtained from various studies on the time it takes to recover the benthos that inhabits sandy beaches impacted by oil spills that have occurred in different areas of the world; it is observed that the effect of light oils on the macro and meiobenthos requires less than 2 years for these communities to recover, while these same types of benthic biological groups under the pressure of heavy oils require up to 10 years to recover their ecological functionality, as in the case of the Gulf War (Bejarano and Michel 2016).

It is also important to take into account the factors that participate in the recovery of invertebrate communities on sandy beaches after an oil spill occurs; Fig. 17 shows some of these factors that intervene in different types of exposure due to spilled oil with different degree of extension and duration of the spill and the time it takes for these benthic communities to recover.

Most studies have shown some level of impacts in comparisons between oiled and unoiled beaches that often have encompassed two phases: (1) an impact phase, where the associated invertebrate community experiences a measurable reduction in abundance and species diversity caused mostly by mortality and oil fouling; and (2) a recovery phase, where there is an increase in dominance of opportunistic species, followed by the return of species characteristic of the assemblage signaling the start of the recovery.

As demonstrated by these limited studies, monitoring impacts of oil spills on benthic communities while simultaneously characterizing oil constituents in the exposure media (i.e., interstitial sediments) could provide quantitative data useful in environmental assessments. The large range of reported impacts from oil spills on sand beach invertebrates is likely the result of a combination of factors including oil type, amount of initial oil loading on the shoreline, oiling extent along shore, and re-oiling frequency, as well as the type and degree of cleanup and the rate of natural removal processes, and site-specific sand beach characteristics. Other factors related to study methods include knowledge of the initial composition of the invertebrate assemblage and their spatial and temporal variability, and sampling strategies (e.g., experimental design and sampling frequency). As a result, studies that experimentally control some of the natural variability that occurs in sand beaches may provide additional linkages between exposures and effects.

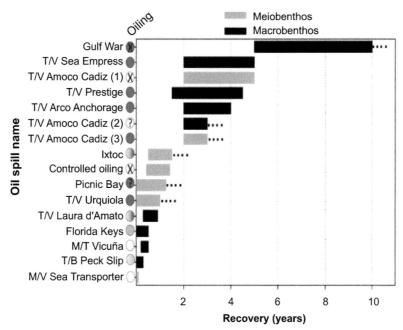

Fig. 16. Recovery (years) of the benthos on sand beaches impacted by oil spills, with dotted symbols indicating when additional time was required for a complete recovery. Symbols within circles indicate when cleaning was not undertaken, or when this information is unknown/not reported ("x" or "?", respectively) (Bejarano and Michel 2016).

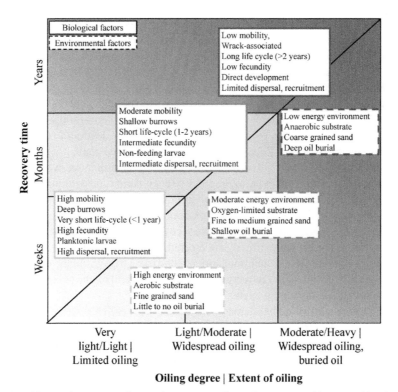

Fig. 17. Examples of factors that may contribute to the recovery of invertebrate communities on sand beaches following an oil spill. The anticipated recovery times are a continuum rather than exact times, and are a function of site-specific physical and biological factors and their continuum. Not all possible contributing factors are included (Bejarano and Michel 2016).

9.5 Mangroves

Mangroves are highly vulnerable to oil spills because oil is deposited on sensitive plant surfaces, affecting soils and dependent marine life and causing death and sublethal impacts (Swan et al. 1994, Duke et al. 1999, Duke and Burns 2003, NOAA 2014). This disruption affects ecosystem services of mangroves, such as fisheries production and shoreline protection worldwide. Oil spill impacts also persist for decades, and they can occur at any time, at any location. So, for as long as oil is extracted and transported around the world, the risks are ever present. Duke (2016) makes an excellent and comprehensive review of the impact of oil spills on mangroves. In this section we analyze the most important literature on aspects of geochemistry of the sediments of mangrove environments and the accumulation of various oil components, their toxicity, the physiological mechanisms of these plants to face the impact of these spills, and other factors.

When a spill occurs in the mangrove ecosystem, oil coats the breathing surfaces of mangrove roots, stems, and seedlings, and surrounding sediments, as well as fauna present in burrows and root hollows. When smothered with oil, shorter plants and animals die mostly within days. By contrast, taller mature trees and shrubbery, oiled only on their exposed roots and sediments, might persist for six or more months before dying. Plants are accordingly smothered, poisoned and starved by oil spills; and the lighter the oils are, the more damaging they are (Duke and Burns 1999, 2003, Michel and Rutherford 2014). These differences between lethal responses and their quantification are very informative, but there are also sublethal responses to consider. The sublethal responses are rarely ever quantified (Duke et al. 1997) but they are often very important since they help us better understand and define overall impacts and possible recovery trajectories. So, while oil type and concentration levels have a primary influence on the impacts observed, it is the responses of mangrove biota to all influential factors that better define their recovery potential, and the longer-term fate of oil-impacted habitat. For these reasons, and because deposited oil often becomes less visible after a few days and weeks, it is strongly recommended that deposited oil be surveyed and mapped accurately at the time the spill is taking place, or very soon afterwards. There are also further influential variables affecting habitat responses, such as differences in the sensitivity of individual mangrove species to oil and dispersed oil (Duke et al. 1998, Duke and Burns 2003, Lewis et al. 2011). In general, impact-level rankings corresponded to oil density, where light oils were more harmful than dense heavy oils. Oil toxicity trials were notably made during field trials and post-spill studies, and not during spill incidents (Swan et al. 1994). In field situations, oil-impacted mangrove ecosystems follow a reasonably ordered set of condition states with primary, secondary and residual effects, leading to recovery and/or loss. Figure 18 shows key trajectories followed by oil-affected mangrove plants.

The longer-term effects occur over decades with habitat either recovering via successful recruitment or lost permanently in a cycle of progressive degradation and deterioration. Because of the propensity of oil to clump, especially in cooler conditions, the areas oiled are often patchy. The result can be a combination of both oiled and unoiled patches occurring in close proximity. Accordingly, responses can be variable and complicated. Some recovery of unoiled undisturbed patches can be quite rapid (less than one year) with virtually undetected long-term effects. The fate of oiled patches, however, may follow one of two overall trajectories—one to recovery, and the other to deterioration and loss. These latter trajectories of recovery or loss further depend on the severity of oiling, and the severity of any subsequent damaging events.

Where oiling results in sublethal damage only, then recovery can be rapid (1–5 years), since affected trees only need to recover their canopy foliage, and regain their associated fauna. Where trees die, the recovery process will take much longer (5–25 years or more), involving seedling recruitment, establishment, and the growth of replacement plants to maturity with propagation. Such recovery processes are difficult to achieve in this physically dynamic and demanding environment. By most accounts, tree growth to maturity can take 25–30 years. Assessment protocols set out in the recommended operational guidelines below have been developed for assessments of both the extent and severity of impacts, and the status of recovery or degradation processes (Duke 2016).

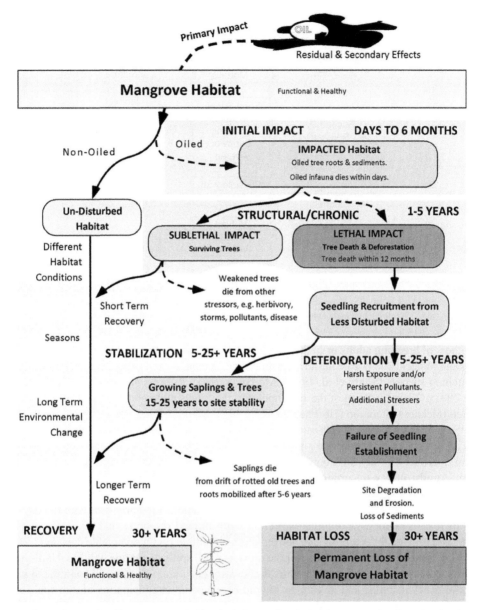

Fig. 18. Effects of large oil spills on mangrove habitat, including undisturbed and permanent loss from deterioration. Note the short-term effects where some trees die while others survive oiling. Adapted from Duke et al. (1999) and NOAA (2014).

9.6 Sea birds

Seabirds and coastal species are the most vulnerable fauna to exposure to crude oil released in marine environments (Fig. 19). Oil and its derivatives can have lethal effects on birds by poisoning them or causing hypothermia (Ferns 1992, Vidal and Domínguez 2015). Oil exposure can also have sublethal effects, which should not be ignored because of their potential consequences at the population level (Golet et al. 2002). For example, sublethal effects of oil and its derivatives on birds include the disruption of reproduction for one to several years after the spill (Walton et al. 1997), behavioral changes (Burger 1997, Burger and Tsipoura 1998) and a decrease in reproductive success (Andres 1997, Golet et al. 2002, Barros et al. 2014). During the reproductive period, small quantities of oil on

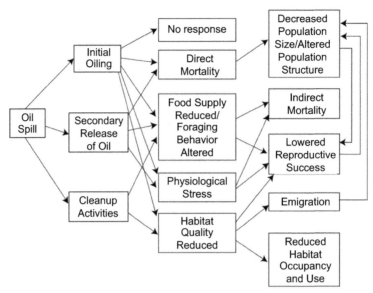

Fig. 19. The influences of an oil spill on seabirds. The three primary avenues of effects, on population size and structure, reproduction and habitat occupancy, are highlighted (Wiens 1995, NRC 2003).

the feathers of breeding adults or in the nest material or food can be quickly transferred to the eggs (Lewis and Malecki 1984), causing adverse effects on egg quality and structure, embryo development and hatching success (Shore et al. 1999). Moreover, oil spills are a source of PAHs (Latimer and Zheng 2003). When birds ingest these compounds, their eggs undergo teratogenesis, size alteration and shell thickness alteration (Hoffman and Gay 1981, Vanglider and Peterle 1981, Stubblefield et al. 1995, Shore et al. 1999). PAHs can also cause cancer in adult birds (Jarvis 1993).

Wild birds are also exposed to a range of other bioaccumulated pollutants from the food chain that can interfere with thyroid homoeostasis, raising concerns about synergistic effects on thyroid toxicity. Amidst policy uncertainties, international agreements, harmonization and cross-border issues, policy drivers have not been very effective at mitigating the impact of oil spills on seabird populations. Wildlife rehabilitators and non-profit organizations regularly deal with the day-to-day reality of mass strandings and animal suffering, with limited resources and minimal government support, and provide the scientific and conservation community with invaluable data and biological material to improve our understanding of the impacts of oil spills on marine wildlife. Until there is a transition to a sustainable energy economy, the severity of future oil spills will continue to increase and contingency planning and investment should be scaled up in anticipation of inevitable energy-driven disasters, ranging from accidents in ever busier shipping lanes to engineering safety failures in offshore rigs. Until there is a paradigm shift globally towards dependence on renewable clean energy, significant resources will continue to be depleted in an unsustainable manner (Troisi et al. 2016).

9.7 Marine mammals

Marine mammals, like humans, are located at the top of the ecological pyramid; they have efficient detoxification mechanisms to deal with potentially toxic agents of natural and anthropogenic origin such as hydrocarbons originating from oil spills; however, when these occur there is an immediate impact on the species of marine mammals that inhabit or transit the areas where the spill has occurred. Marine mammals may inhale toxic doses of petroleum vapor when at the surface in the vicinity of an oil spill (Geraci 1990, Geraci and Williams 1990, St. Aubin 1990a), although there appear to be few data indicating that this is an important source of mortality. In some cases, these predators of the upper trophic level may become exposed to oil by ingesting prey that have oil or its metabolites in their tissues. Marine mammals are also vulnerable to the toxic effects of ingested oil, and species of marine mammals such as sea otters that depend on a clean pelage for insulation are also vulnerable to surface oiling (Geraci and St. Aubin 1987, Geraci 1990, Geraci and Williams 1990, St. Aubin 1990a, b, St. Aubin and Lounsbury 1990). Effects may be exacerbated by stress resulting from handling during cleaning: among marine mammals, river otters (*Lutra lutra*) in the British Isles and Alaska, and sea otters (*Enhydra lutris*) and harbor seals (*Phoca vitulina*) in Prince William Sound, Alaska, all showed short-term population declines after oiling of their inshore marine habitats (Baker et al. 1981, Spraker et al. 1994, Monson et al. 2000, Peterson 2001). For some species, such as sea otters, these effects may have persisted over 10 years (Monson et al. 2000). However, in the case of the Exxon Valdez oil spill in Prince William Sound, Alaska, considerable controversy remains concerning the magnitude of the initial losses and the duration of population-level effects; these uncertainties stem from the lack of sufficient pre-spill data to characterize the population status of these species and difficulties in obtaining adequate post-spill data to distinguish between local movements of animals and area-wide population effects. Chronic or delayed responses of marine mammal populations to petroleum hydrocarbons in the sea can occur because of continued ingestion of oil via contaminated prey, or because of failure of prey populations to recover subsequent to injury. In the 30 years since the Exxon Valdez oil spill, several species of marine mammals have demonstrated indirect or delayed responses to the spill; these responses were found in several species that forage on small fish caught in inshore waters (Peterson 2001).

In the wake of the spill of the Macondo well in the northern Gulf of Mexico, the impact on various species of marine mammals among other affected biological groups in this marine area has been studied; Beyer et al. (2016) made a complete review of the various recorded data on the consequences of the DWH spill on a variety of marine mammals and other environmentally important and vulnerable vertebrates such as turtles and seabirds; for sea mammals, increased mortality rates were observed after an initial delay, but pre-spill data seem to indicate that the observed growth in the mortality rate was not entirely a consequence of the DWH spill (Antonio et al. 2011). For cetaceans, Williams et al. (2011) estimated that the carcass recovery rates may, on average, have been as low as 2% (range: 0–6.2%) of the actual cetacean deaths, suggesting that mortality after the DWH spill may have been severely underestimated. In 70% of dolphins exposed to the unusual mortality event (UME) that suffered from primary bacterial pneumonia, the condition either caused or contributed significantly to death, and the authors suggested that elevated exposure to petroleum compounds originating from the DWH spill had contributed to the observed decreased health conditions of dolphins within the UME area. Van Dolah et al. (2015) used transcriptomics of skin biopsies to assess dolphin health condition in the UME area, but without finding any links to possible contamination stress from the DWH spill. Although there are numerous knowledge gaps and research needs still remaining, as highlighted throughout this review, the broad collection of environmental research conducted after DWH contributes significantly to our understanding of ecological impacts of major marine oil spills, and this learning helps to clarify what measures are needed, both pre-spill and post-spill, to reduce the environmental effects of such events.

Throughout this section a series of studies has been reviewed on the various biological effects caused by oil spills in coastal and marine environments; the data of these investigations are conclusive evidence of the impacts caused by this type of environmental pressure at different biological and ecological levels, all of them with negative consequences, some reversible and others not. The pressures we have exerted on natural resources for many years both for obtaining energy and meeting social and economic needs and the ways in which we have altered those resources through chronic waste or the massive presence of harmful substances have resulted in human impoverishment and damage to health in different regions of the world as well as major environmental deterioration. It is time to make decisions on a large scale and weigh the cost against the benefit if we want to avoid reaching an ecological collapse that no energy resource can compensate and for which we will all pay a high price.

10. Conclusion

Anthropogenic disturbances can easily upset the delicate ecological balance among the different compartments (phytoplankton, zooplankton, benthos, nekton) of tropical marine ecosystems. Major oil spills in the Gulf of Mexico cause acute and long-term ecological damage. Nevertheless, there seems to be a consensus in accepting that persistent chronic oil pollution is eventually more harmful to the environment than accidental oil spills.

Over fifty years after the first oil spill, the remedial framework is reinforced by a set of measures: nautical, such as the required traffic separation scheme; legal, such as the affirmation and extension of the right of coastal states to intervene; financial, such as the International Oil Pollution Compensation Funds; international, such as the MARPOL convention, the International Convention for the Prevention of Pollution from Ships; and technical, such as the requirement for tankers to have a double hull.

In the specific case of the Ixtoc-I oil blowout, the reconstruction of ecological scenarios assessing the environmental damages detected in the southern Gulf of Mexico is a major challenge. The lack of a baseline data bank and a reliable source on the space-time variability on some environmental parameters precludes our reaching conclusions about the mid- and long-term ecosystem response to a catastrophic oil spill.

When the Ixtoc-I blowout took place, no one in our scientific community dared to anticipate the ecosystem response to a massive offshore oil spill. During the first months of the accident, it became clear that the weathering processes promoted by local atmospheric, hydrological, and biological conditions prevailing in the southwestern Gulf of Mexico would play a crucial role in the environmental impact. High water temperature, photocatalytic decomposition, and biodegradation promoted the degradation of the spilled crude oil. Contrary to the official position of Mexican agencies that adamantly insisted that no environmental damage was caused by Ixtoc-I, oil residues remained in the water and sediments. These were absorbed by filtering organisms and bio-accumulated by benthic fauna such as oysters (*Crassostrea virginica*) and clams (*Rangia cuneata* and *Polymesoda carolineana*), particularly in coastal lagoon systems of the states of Campeche and Tabasco. The primary fishery resource in this region of the Gulf, the penaeid shrimp, has shown a declining production trend since the late 1970s. In spite of the fishery management regulations implemented since 1989, the pink shrimp population, which sustains this fishery, is at high risk of suffering a collapse.

Presently, the environmental conditions in the southern Gulf of Mexico have been further aggravated by the rapid operation expansion of both PEMEX and the transnational oil companies in coastal and offshore waters. Accidental spills have again occurred in this area, raising new concerns about the fate of habitats known for their ecological value as spawning and nursery grounds for marine and estuarine-dependent species. One should understand the complications involved in offering unequivocal evidence of environmental damage caused by accidental oil spills knowing in advance the physicochemical and biological processes acting at different space and time scales. On the positive side, new research lines (genetics and molecular biology) are beginning to shed some light on this problem. Long-term oil exposure can cause loss of genetic variability and the disruption of the immune system in marine populations. Presently, we have at our disposal much better scientific tools (biomarkers and isotopic tracers) that can contribute to unraveling the complexity of oil pollution in the marine environment.

The Ixtoc-I legacy is a painful reminder of the fallibility of the offshore technology employed in the exploitation of fossil fuels contained in the seabed. It also accentuates the ecological fragility of coastal habitats, but more importantly, it emphasizes our inadequate knowledge on the mid- and long-term response of a tropical marine ecosystem exposed to accidental oil spills.

More than eight years after the DWH catastrophe, there is a consensus among the different stakeholders about the severe damage inflicted on the health of this vital marine ecosystem. The massive spill of gas, crude oil, and other toxic compounds had no precedent in the annals of the world oil industry. The primary concern in the early phase of the accident was containing the black tide at the surface; efficiency of direct recollection, selective burning, and chemical dispersion at the wellhead was not satisfactory. BP's decision to inject considerable quantities of chemical dispersant at the bore-hole only augmented the complexity of the spill. Extensive surface and subsurface oil plumes (at 500, 800, even 1200 m depth) originated in the northern Gulf and their trajectories threatened other sectors of the Gulf. Early predictions of the possible pathway of the oil slicks at the surface waters had a certain degree of accuracy; synoptic remote sensing techniques observations conducted by NOAA contributed to defining the emergency procedures and the field observations in USA waters.

The capping of the uncontrolled wellhead after nearly four months of emergency conditions proved a significant challenge for the underwater technology. The DWH incident left a painful experience for many sectors of our society. On the one hand, it revealed the gross negligence of a giant oil company operating in deep waters, and on the other, there continues to be great uncertainty concerning the immediate and long-term effects of the spilled oil.

Many factors determine the final destiny of complex oil molecules in the marine ecosystem. Coastal habitats (lagoons, coral reefs, marsh lands, mangroves), open waters and seabed are vulnerable to the lasting effects of toxic compounds incorporated in the trophic web or deposited in shallow and deep sediments.

More than 600 scientific studies have attempted to answer questions raised by the unusual injection of a high volume of gas and crude oil in the GoM. Some authors are optimistic about the rapid physical and ecological recovery shown by the GoM. Others have expressed concern about environmental injury over the next two or three decades.

Numerous authors have also voiced apprehension about extending future oil exploration and extractions of fossil fuel into deep-sea habitats. Even though there were significant deployments of human and technical resources to examining the environmental consequences of the DWH accident, particularly in the USA and to a lesser extent in Mexico, many questions remained unanswered. The total volume of oil released by the Macondo's blowout, its final destiny, and its long-term persistence in the ecosystem are not entirely known.

References

Alabresm, A., Y.P. Chen, A.W. Decho and J. Lead. 2018. A novel method for the synergistic remediation of oil-water mixtures using nanoparticles and oil-degrading bacteria. Science of the Total Environ. 630: 1292–1297.

Albaiges, J., F. Vilas and B. Morales-Nin. 2006. The Prestige: A scientific response. Mar. Pollut. Bull. 53: 5–7.

Andres, B.A. 1997. The Exxon Valdez oil spill disrupted the breeding of black oystercatchers. J. Wildl. Manag. 61: 1322–1328.

Antonio, F.J., R.S. Mendes and S.M. Thomaz. 2011. Identifying and modeling patterns of tetrapod vertebrate mortality rates in the Gulf of Mexico oil spill. Aquat. Toxicol. 105: 177–179.

Armynot du Châtelet, E., J.P. Debenay and R. Soulard. 2004. Foraminiferal proxies for pollution monitoring in moderately polluted harbors. Environ. Pollut. 127(1): 27–40.

Armynot du Châtelet, E. and J.P. Debenay. 2010. The anthropogenic impact on the western French coasts as revealed by foraminifera: A review. Revue de Micropaléontologie 53(3): 129–137.

Bak, R. and J. Elgershuizen. 1976. Patterns of oil-sediment rejection in corals. Mar. Biol. 37: 105–113.

Bak, R.P.M. 1987. Effects of chronic oil pollution on a Caribbean reef. Mar. Pollut. Bull. 18: 534–539.

Baker, J.R., A.M. Jones, T.P. Jones and H.C. Watson. 1981. Otter Lutralutra L. mortality and marine oil pollution. Biol. Conserv. 20: 311–321.

Baker, J.M., R.B. Clark and P.F. Kingston. 1991. Two Years After the Spill: Environmental Recovery in Price William Sound and the Gulf of Alaska. Institute of Offshore Engineering Heriot-Watt University, Edinburg, Scotland, p. 31.

Baltz, D.M. and A. Yáñez-Arancibia. 2013. Ecosystem-based management of coastal fisheries in the Gulf of Mexico: Environmental and anthropogenic impacts and essential habitat protection. pp. 337–370. *In*: Day, J.W. and A. Yáñez-Arancibia (eds.). The Gulf of Mexico Origin, Water, and Biota, Vol. 4 Ecosystem-Based Management. College Station, TX: Texas A&M University Press.

Barros, A., D. Álvarez and A. Velando. 2014. Long-term reproductive impairment in a seabird after the Prestige oil spill. Biol. Lett. 10: 20131041.

Bejarano, A.C. and J. Michel. 2010. Large-scale risk assessment of polycyclic aromatic hydrocarbons in shoreline sediments from Saudi Arabia: Environmental legacy after twelve years of the Gulf war oil spill. Environ. Pollut. 158: 1561–1569.

Bejarano, A.C. and J. Michel. 2016. Oil spills and their impacts on sand beach invertebrate communities: A literature review. Environ. Pollut. 218: 709–722.

Bernabeu, A., M. Nuez de la Fuente, D. Rey, B. Rubio, F. Vilas, R. Medina et al. 2006. Beach morphodynamics forcements in oiled shorelines: Coupled physical and chemical processes during and after fuel burial. Mar. Pollut. Bull. 52: 1156–1168.

Bernabeu, A., S. Fernández-Fernández, F. Bouchette, D. Rey, A. Arcos, J. Bayona et al. 2013. Recurrent arrival of oil to Galician coast: The final step of the Prestige deep oil spill. J. Hazard. Mater. 250: 82–90.

Beyer, J., H.C. Trannum, T. Bakke, P.V. Hodson and T.K. Collier. 2016. Environmental effects of the Deepwater Horizon oil spill: A review. Mar. Pollut. Bull. 110: 28–51.

Bhattacharyya, S., P.L. Klerks and J.A. Nyman. 2003. Toxicity to freshwater organisms from oils and oil spill chemical treatments in laboratory microcosms. Environ. Pollut. 122(2): 205–215.

Bierman, V.J.Jr. 1990. Equilibrium partitioning and biomagnification of organic chemicals in benthic animals. Environ. Sci. Technol. 24: 1407–1412.

Bjorlykke, K. 2011. Petroleum Geoscience: From Sedimentary Environments to Rock Physics. Springer, New York, 575 pp.

Boehm, P.D. and D.L. Fiest. 1982. Subsurface distribution of petroleum from an offshore well blowout. The Ixtoc-I blowout, Bay of Campeche. Env. Tech. 16: 67–74.

Boesch, D.F., J.N. Butler, D.A. Cacchione, J.R. Geraci, J.M. Neff, J.P. Ray et al. 1987. An assessment of the long-term environmental effects of U.S. offshore oil and gas development activities: Future research needs. pp. 1–53. *In*: Boesch, D.F. and N.N. Rabalais (eds.). Long-Term Environmental Effects of Offshore Oil and Gas Development. Elsevier Applied Science, London and New York.

Botello, A.V. and S. Castro. 1980. Chemistry and natural crude oil fractions from the Ixtoc-I oil spill. *In*: Proceedings Symposium Ixtoc-I Oil Spill. Key Biscayne, FL: U.S. Dept. Commerce, NOAA, 387–407.

Botello, A.V. and S.F. Villanueva. 1987. Vigilancia de los Hidrocarburos Fósiles en Sistemas Costeros del Golfo de México y Áreas Adyacentes: Sonda de Campeche. Vol. 14. México, D.F.: Anales del Instituto de Ciencias del Mar y Limnología. UNAM, 45–52.

Botello, A.V., C. González and G. Díaz. 1991. Pollution by petroleum hydrocarbons in sediments from Continent l shelf of Tabasco State, Mexico. Bull. Environ. Contam. Toxicol. 47: 565–571.

Botello, A.V. 1996. Características, composición y propiedades fisicoquímicas del petróleo. pp. 203–210. *In*: Botello, A.V., J.L. Rojas Galaviz, J.A. Benítez and D. Zárate Lomelí (eds.). Golfo de México, Contaminación e Impacto Ambiental: Diagnóstico y Tendencias. Serie Científica EPOMEX5. Campeche Univ. Autón.de Campeche.

Botello, A.V., S.F. Villanueva and L. Rosales Hoz. 2004. Distribución y contaminación de metales en el golfo de México. pp. 683–712. *In*: Caso, M., I. Pisanty and E. Ezcurra (eds.). Diagnóstico Ambiental del Golfo de México. Vol. 2. México, D.F. Instituto Nacional de Ecología INE, Instituto de Ecología A.C. INECOL, and Harte Research Institute for Gulf of Mexico Studies. Texas A&M University-Corpus Christi, SEMARNAT.

Burger, J. 1997. Effects of oiling on feeding behavior of Sanderlings and Semipalmated Plovers in New Jersey. Condor 99: 290–298.

Burger, J. and N. Tsipoura. 1998. Experimental oiling of Sanderlings (*Calidrisalba*): behavior and weight changes. Environ. Toxicol. Chem. 17: 1154–1158.

Burns, K.A., S.D. Garrity and S.C. Levings. 1993. How many years until mangrove ecosystems recover from catastrophic oil spills? Mar. Pollut. Bull. 26: 239–248.

Camilli, R., C.M. Reddy, D.R. Yoerger, B.A.S. Van Mooy, M.V. Jakuba, J.C. Kinsey et al. 2010. Tracking hydrocarbon plume transport and biodegradation at deepwater horizon. Science 330: 201–204.10.1126/science. 1195223.

Capuzzo, J.M. 1987. Biological effects of petroleum hydrocarbons: Assessments from experimental results. pp. 343–410. *In*: Boesch, D.F. and N.N. Rabalais (eds.). Long-term Environmental Effects of Offshore Oil and Gas Development. Elsevier Applied Science, London.

Carvalho, C., S. dos, V.A. Bernusso, H.S. de Araujo, E.L. Espindola and M.N. Fernandes. 2012. Biomarker responses as indication of contaminant effects in *Oreochromisniloticus*. Chemosphere 89: 60–69.

Centeno-Chalé, O.A., M.L. Aguirre-Macedo, G. Gold-Bouchot and V.M. Vidal-Martínez. 2015. Effects of oil spill related chemical pollution on helminth parasites in Mexican flounder *Cyclopsettachittendeni* from the Campeche Sound, Gulf of Mexico. Ecotox. Environ. Safe. 119: 162–169.

Cook, C. and A. Knap. 1983. Effects of crude oil and chemical dispersant on photosynthesis in the brain coral Diploriastrigosa. Mar. Biol. 78: 21–27.

Daly, K.L., U. Passow, J. Chanton and D. Hollander. 2016. Assessing the impacts of oil associated marine snow formation and sedimentation during and after the Deepwater horizon oil spill. Anthropocene 13: 18–33.

Davies, R.P. and A.J. Dobbs. 1984. The prediction of bioconcentration in fish. Water Resour. Bull. 18: 1253–1262.

deBruyn, A.M.H., B.G. Wernick, C. Stefura, B.G. McDonald, B.L. Rudolph, L. Patterson et al. 2007. *In situ* experimental assessment of lake whitefish development following a freshwater oil spill. Environ. Sci. Technol. 41: 6983–6989. http://dx.doi.org/10.1021/es0709425.

Duke, N.C., Z.S. Pinzón and M.C. Prada. 1997. Large-scale damage tomangrove forests following two large oil spills in Panamal. Biotropica 2: 2–14.

Duke, N.C., K.A. Burns and O. Dalhaus. 1998. Effects of oils and dispersed-oils on mangrove seedlings in planthouse experiments: A preliminary assessment of results two months after oil treatments. pp. 631–636. *In*: Beck, C. (ed.). APPEA Journal. APPEA, Canberra.

Duke, N.C. and K.A. Burns. 1999. Fate and effects of oil and dispersed oil on mangrove ecosystems in Australia. Final Report to the Australian Petroleum Production Exploration Association. Australian Institute of Marine Science and CRC Reef Research Centre.

Duke, N.C., K.A. Burns and R.P.J. Swannell. 1999. Research into the bioremediation of oil spills in tropical Australia: with particular emphasis on oiledmangrove and saltmarsh habitat. Report to the Australian Maritime Safety Authority and the Great Barrier Reef Marine Park Authority.

Duke, N.C. and K.A. Burns. 2003. Fate and effects of oil and dispersed oil on mangrove ecosystems in Australia. Environmental Implications of Offshore Oil and Gas Development in Australia: Further Research. A Compilation of Three Scientific Marine Studies.Australian Petroleum Production and Exploration Association (APPEA), Canberra, pp. 232–363 (521 pp.).

Duke, N.C. 2016. Oil spill impacts on mangroves: Recommendations for operational planning and action based on a global review. Mar. Pollut. Bull. 109: 700–715.

Durrieu, J., M. Mojtahid, L. Cazes, F. Galgani, F. Jorissen, D. Tran et al. 2006. Aged drilled cuttings offshore Gabon: New methodology for assessing their impact. Soc. Petrol. Eng. Int. (1): 170–177.

Elmgren, R., S. Hansson, U. Larsson, B. Sundellin and P.D. Boehm. 1983. The TSESIS oil spill: Acute and long-term impact on the benthos. Mar. Biol. 73: 51–65.

Fernández-Fernández, S., A. Bernabeu, F. Bouchette, D. Rey and F. Vilas. 2011. Beach morphodynamic influence on long-term oil pollution: The Prestige oil spill. J. Coast. Res. 64: 890–893.

Ferns, P. 1992. Bird life of Coasts and Estuaries, Ed. Cambridge University Press, Cambridge.

Fisher, C.R., P.A. Montagna and T.T. Sutton. 2016. How did the Deepwater Horizon oil spill impact deep-sea ecosystems? Oceanography 29(3): 182–195.

Fodrie, F.J., K.W. Able, F. Galvez, K.L. Heck, O.P. Jensen, P.C. Lopez-Duarte et al. 2014. Integrating organismal and population responses of estuarine fishes in Macondo spill research. Bioscience 64: 778–788. http://dx.doi.org/10.1093/biosci/biu123.

Foster, W.J., E. Armynot du Châtelet and M. Rogerson. 2012. Testing benthic foraminiferal distributions as a contemporary quantitative approach to biomonitoring estuarine heavy metal pollution. Mar. Pollut. Bull. 64(5): 1039–1048.

García-Cuéllar, J.A., F. Arreguín-Sánchez, S. Hernández-Vázquez and D.B. Lluch-Cota. 2004. Impacto ecológico de la industria petrolera en la Sonda de Campeche, México, tras tres décadas de actividad: una revisión. Interciencia 29: 311–319.

Geraci, J.R. and D.J. St. Aubin. 1987. Effects of offshore oil and gas development on marine mammals and turtles. pp. 587–617. *In*: Boesch, D.F. and N.N. Rabalais (eds.). Long Term Environmental Effects of Offshore Oil and Gas Development. Elsevier Applied Science Publishers, London, UK and New York.

Geraci, J.R. 1990. Physiologic and toxic effects on cetaceans. pp. 167–197. *In*: Geraci, J.R. and D.J. St. Aubin (eds.). Sea Mammals and Oil: Confronting the Risks. Academic Press. San Diego, CA.

Geraci, J.R. and T.D. Williams. 1990. Physiologic and toxic effects on sea otters. pp. 211–221. *In*: Geraci, J.R. and D.J. St. Aubin (eds.). Sea Mammals and Oil: Confronting the Risks. Academic Press. San Diego, CA.

Getter, C.D. 1982. Oil Spills and Mangroves: A Review of the Literature, Field and Lab Studies. Land and Water Issues Related to Energy Development. Ann Arbor Science, Ann Arbor, Michigan, USA, pp. 303–318.

Golet, G.H., P.E. Seiser, A.D. Mcguire, D.D. Roby, J.B. Fischer, K.J. Kuletz et al. 2002. Long-term direct and indirect effects of the 'Exxon Valdez' oil spill on pigeon guillemots in Prince William Sound, Alaska. Mar. Ecol. Prog. Ser. 241: 287–304.

Griffiths, S.K. 2012. Oil release from Macondo Well MC252 following the Deepwater Horizon Accident. Environ. Sci. Technol. 46: 5616–5622. Doi: 10.1021/es204569.

Gschwend, P.M. and R.P. Schwarzenbach. 1992. Physical chemistry of organic compounds in the marine environment. Mar. Chem. 39: 187–207.

Gundlach, E.R. and M.O. Hayes. 1978. Vulnerability of coastal environments to oil spill impacts. Mar. Technol. Soc. J. 12: 18–27.

Guzman del Próo, S.A., E.A. Chávez, F.M. Alatriste, S. de la Campa, G. de la Cruz, L. Gómez et al. 1986. The impact of the Ixtoc-1 oil spill on zooplankton. J. Plankton Res. 8: 557–581.

Hastings, D.W., P.T. Schwing, G.R. Brooks, R.A. Larson, J.L. Morford, T. Roeder et al. 2016. Changes in sediment redox conditions following the BP DWH blowout event. Deep-Sea Res. II Top. Stud. Oceanogr. 129: 167–178.

Hicken, C.E., T.L. Linbo, D.H. Baldwin, M.L. Willis, M.S. Myers, L. Holland et al. 2011. PNAS 108: 7086–7090. http://dx.doi.org/10.1073/pnas.1019031108.

Higgins, L.G. and J.D. Hayes. 2011. Mechanisms of induction of cytosolic and microsomal glutathione transferase (GST) genes by xenobiotics and proinflammatory agents. Drug Metab. Rev. 43: 92–137.

Hjermann, D.Ø., A. Melsom, G.E. Dingsør, J.M. Durant, A.M. Eikeset, L.P. Roed et al. 2007. Fish and oil in the Lofoten-Barents Sea system: Synoptic review of the effect of oil spills on fish populations. Mar. Ecol. Prog. Ser. 339: 283–299.

Hoffman, D.J. and M.L. Gay. 1981. Embryotoxic effects of benzo(a)pyrene, chrysene, and 7,12-dimethylbenz(a) anthracene in petroleum hydrocarbon mixtures in mallard ducks. J. Toxicol. Environ. Health 7: 775–787.

Holdway, D. 2002. The acute and chronic effects of wastes associated with offshore oil and gas production on temperate and tropical marine ecological processes. Mar. Pollut. Bull. 44: 185–203. Doi: 10.1016/S0025-326X (01) 00197-7.

Hunt, J.M. 1996. Petroleum Geochemistry and Geology. W.H. Freeman, New York.

Incardona, J.P., C.A. Vines, B.F. Anulacion, D.H. Baldwin, H.L. Day, B.L. French et al. 2012. Unexpectedly high mortality in Pacific herring embryos exposed to the 2007 Cosco Busan oil spill in San Francisco Bay. PNAS 109: E51–E58.

Jarvis, P.J. 1993. Environmental changes. pp. 42–85. *In*: Furness, R.W. and Greenwood, J.J.D. (eds.). Birds as Monitors of Environmental Change. Chapman & Hall, London.

Jernelöv, A. and O. Lindén. 1981. Ixtoc-I: A case study of the world's largest oil spill. Ambio 10: 299–306.

Jewett, S.C., T.A. Dean, B.R. Woodin, M.K. Hoberg and J.J. Stegeman. 2002. Exposure to hydrocarbons 10 years after the Exxon Valdez oil spill: evidence from cytochrome P4501A expression and biliary FACs in nearshore demersal fishes. Mar. Environ. Res. 54: 21–48.

Joye, S.B., I.R. MacDonald, I. Leifer and V. Asper. 2011. Magnitude and oxidation potential of hydrocarbon gases released from the BP oil well blowout. Nat. Geosci. 4: 160–164. Doi: 10.1038/NGEO1067.

Joye, S.B., A.P. Teske and J.E. Kostka. 2014. Microbial dynamics following the Macondo Oil Well Blowout across Gulf of Mexico environments. BioScience 64(9): 766–777. Doi: 10.1093/biosci/biu121.

Kerambrun, E., S. Le Floch, W. Sanchez, H.T. Guyon, T. Meziane, F. Henry et al. 2012. Responses of juvenile sea bass, *Dicentrarchuslabrax*, exposed to acute concentrations of crude oil, as assessed by molecular and physiological biomarkers. Chemosphere 87: 692–702.

Khan, R.A. and K. Nag. 1993. Estimation of hemosiderosis in seabird sand fish exposed to petroleum. Bull. Environ. Contam. Toxicol. 51: 125–131.

Khan, R.A. 2010. Two species of commercial flatfish, winter flounder, *Pleuronectes americanus*, and American plaice, *Hippoglossoides platessoides*, as sentinels of enviromental pollution. Bull. Environ. Contam. Toxicol. 85: 205–208.

Kingston, P., D. Runciman and J. McDougall. 2003. Oil contamination of sedimentary shores of the Galapagos Islands following the wreck of the Jessica. Mar. Pollut. Bull. 47: 303–312.

Kinner, N.E., L. Belden and P. Kinner. 2014. Unexpected sink for Deepwater Horizon oil may influence future spill response. Eos 95(21): 176.

Ko, F.-C., C.-W. Chang and J.-O. Cheng. 2014. Comparative study of polycyclic aromatic hydrocarbons in coral tissues and the ambient sediments from Kenting National Park, Taiwan. Environ. Pollut. 185: 35–43.

Langangen, Ø., E. Olsenb, L.C. Stige, J. Ohlberger, N.A. Yaragina, F.B. Vikebøb et al. 2017. The effects of oil spills on marine fish: Implications of spatial variation in natural mortality. Mar. Pollut. Bull. 119: 102–109.

Latimer, J.S. and J. Zheng. 2003. The Sources, Transport, and Fate of PAHs in the Marine Environment, ed. Wiley, Chichester.

Lee, L.H. and H.J. Li. 2013. Effects of an oil spill on benthic community production and respiration on subtropical intertidal sand flats. Mar. Pollut. Bull. 73: 291–299.

Lei, Y.L., T.G. Li, H. Bi, W.L. Cui, W.P. Song, J.Y. Li et al. 2015. Responses of benthic foraminifera to the 2011 oil spill in the Bohai Sea, PR China. Mar. Pollut. Bull. 96: 245–260.

Levin, L.A. 2003. Oxygen minimum zone benthos: Adaptation and community response to hypoxia. pp. 1–45. *In*: Gibson, R.N. and Atkinson, R.J.A. (eds.). Oceanography and Marine Biology: An Annual Review. Vol. 41. Taylor & Francis.

Lewis, M., R. Pryor and L. Wilking. 2011. Fate and effects of anthropogenic chemicals in mangrove ecosystems: A review. Environ. Pollut. 159: 2328–2346.

Lewis, S.J. and R.A. Malecki. 1984. Effects of egg oiling on larid productivity and population dynamics. Auk 101: 584–592.

Licea-Durán, S., P. Torres and C. Trejo. 1982. Informe Final del Proyecto de Investigación Evaluación de los Posibles Efectos del Derrame del Pozo Ixtoc-I, Sobre Las Comunidades del Fitoplancton y la Productividad Primaria. Ibíd. (Sexto informe): 1–44.

Liu, Z., J. Liu, Q. Zhu and W. Wu. 2012. The weathering of oil after the Deepwater Horizon oil spill: Insights from the chemical composition of the oil from the sea surface, salt marshes and sediments. Environ. Res. Letters 7: 1–14.

Lizárraga-Partida, M.L., F.B. Izquierdo-Vicuña and I. Wong-Chang. 1991. Marine bacteria on the Campeche bank oil field. Mar. Pollut. Bull. 22: 401–405.

MacCarty, L.S. and D. Mackay. 1993. Enhancing ecotoxicological modeling and assessment. Environ. Sci. Technol. 27(9): 1719–1728.

MacDonald, I.R., O. Garcia-Pineda, A. Beet, S. Daneshgar Asl, L. Feng, G. Graettinger et al. 2015. Natural and unnatural oil slicks in the Gulf of Mexico. J. Geophys. Res. Oceans 120: 8364–8380. Doi: 10.1002/2015JC011062.

McLachlan, A. and A.C. Brown. 2006. The Ecology of Sandy Shores, Second ed. Academic Press, New York.

McNutt, M.K., R. Camilli, T.J. Crone, G.D. Guthrie, P.A. Hsieh, T.B. Ryerson et al. 2011. Review of flow rate estimates of the Deepwater Horizon oil spill. Proc. Natl. Acad. Sci. USA. 109(50): 20260–7. Doi: 10.1073/pnas.1112139108.

Maki, A.W. 1991. The Exxon Valdez oil spill: Initial environmental impact assessment. Part 2 of a five-part series. Environ. Sci. Technol. 25: 24–29.

Meehan, W.J. and G.K. Ostrander. 1997. Coral bleaching: A potential biomarker of environmental stress. J. Toxicol. Environ. Health Part A 50: 529–552.

Meier, S., H. Craig Morton, G. Nyhammer, B.E. Grøsvik, V. Makhotin, A. Geffen et al. 2010. Development of Atlantic cod (*Gadusmorhua*) exposed to produced water during early life stages: Effects on embryos, larvae, and juvenile fish. Mar. Environ. Res. 70: 383–394. http://dx.doi.org/10.1016/j.marenvres.2010.08.002.

Michel, J. and N. Rutherford. 2014. Impacts, recovery rates, and treatment options for spilled oil in marshes. Mar. Pollut. Bull. 82: 19–25.

Mojtahid, M., F. Jorissen, J. Durrieu, F. Galgani, H. Howa, F. Redois et al. 2006. Benthic foraminifera as bio-indicators of drill cutting disposal in tropical east Atlantic outer shelf environments. Mar. Micropaleontol. 61(1-3): 58–75.

Monson, D.H., D.F. Doak, B.E. Ballachey, A. Johnson and J.L. Bodkin. 2000. Long-term impacts of the Exxon Valdez oil spill on sea otters, assessed through age-dependent mortality patterns. Proceedings of the National Academies of Science 97: 6562–6567.

Montagna, P.A., J.G. Baguley, C. Cooksey, I. Hartwell, L.J. Hyde, J.L. Hyland et al. 2013. Deep-sea benthic footprint of the Deepwater Horizon Blowout. PloS One 8(8): e70540. Doi: 10.1371/journal.pone.0070540.

Morales-Caselles, C., N. Jimenez-Tenorio, M.L. de Canales, C. Sarasquete and T.A. DelValls. 2006. Ecotoxicity of sediments contaminated by the oil spill associated with the tanker "Prestige" using juveniles of the fish *Sparusaurata*. Arch. Environ. Contam. Toxicol. 51: 652–660.

Murray, J.W. 2006. Ecology and Applications of Benthic Foraminifera. Cambridge University Press, pp. 1–426.

Nahrgang, J., L. Camus, P. Gonzalez, M. Jonsson, J.S. Christiansen and H. Hop. 2010. Biomarker responses in polar cod (*Boreogadus saida*) exposed to dietary crude oil. Aquat. Toxicol. 96: 77–83.

National Research Council (NRC). 2003. Oil in the sea III, Inputs, Fates and Effects. National Academy of Sciences. The National Academies Press. Washington, D.C., 119–157.

Neff, J.M. and J.W. Anderson. 1981. Response of Marine Animals to Petroleum and Specific Petroleum Hydrocarbons.

Negri, A. and A. Heyward. 2001. Inhibition of coral fertilisation and larval metamorphosis by tributyltin and copper. Mar. Environ. Res. 51: 17–27.

Newton, J. 2002. A Century of Tankers. INTERTANKO.

NOAA. 2014. Oil Spills in Mangroves. Planning & Response Considerations. US Department of Commerce, National Oceanic and Atmospheric Administration (NOAA), National Ocean Service, Office of Response and Restoration, Seattle, Washington (96 pp.).

O'Connor, B. 2013. Impacts of the Anomalous Mississippi River Discharge and Diversions on Phytoplankton Blooming in Northeastern Gulf of Mexico. MSc, University of South Florida.

Oliva, M., M.L. Gonzalez de Canales, C. Gravato, L. Guilhermino and J.A. Perales. 2010. Biochemical effects and polycyclic aromatic hydrocarbons (PAHs) in Senegal sole (*Soleasenegalensis*) from a Huelva estuary (SW Spain). Ecotoxicol. Environ. Saf. 73: 1842–1851.

Oudot, J. and F. Chaillan. 2009. Pyrolysis of asphaltenes and biomarkers for the finger printing of the *Amoco-Cadiz* oil spill after 23 years. Nat. Prec. 13: 548–552. Doi: 10.106/J.crci.2009.12.002.

Pait, A.S., C.F. Jeffrey, C. Caldow, D.R. Whitall, S.I. Hartwell, A.L. Mason et al. 2009. Chemical Contaminants in the Coral Porites astreoides From Southwest Puerto Rico.

Pelegrí, S.P. and T.H. Blackburn. 1994. Bioturbation effects of the amphipod *Corophiumvolutator* on microbial nitrogen transformations in marine sediments. Mar. Biol. 121(2): 253–258.

Peters, E.C., P.A. Meyers, P.P. Yevich and N.J. Blake. 1981. Bioaccumulation and histopathological effects of oil on a stony coral. Mar. Pollut. Bull. 12: 333–339.

Peters, E.C., N.J. Gassman, J.C. Firman, R.H. Richmond and E.A. Power. 1997. Ecotoxicology of tropical marine ecosystems. Environ. Toxicol. Chem. 16(1): 12–40.

Peterson, C.H. 2001. The Exxon Valdez oil spill in Alaska: Acute, indirect, and chronic effects on the ecosystem. Adv. Mar. Biol. 39: 1–103.

Pezeshki, S., M. Hester, Q. Lin and J. Nyman. 2000. The effects of oil spill and clean-up on dominant US Gulf coast marsh macrophytes: A review. Environ. Pollut. 108: 129–139.

Quigley, D.C., J.S. Hornafius, B.P. Luyendyk, R.D. Francis, J. Clark and L. Washburn. 1999. Decrease in natural marine hydrocarbon seepage near Coal Oil Point, California, associated with off- shore oil production. Geology 17: 1047–1050.

Rahsepar, S., A.A.M. Langenhoff, M.P.J. Smit, J.S. van Eenennaam, A.J. Murk and H.H.M. Rijnaarts. 2017. Oil biodegradation: Interactions of artificial marine snow, clay particles, oil and Corexit. Mar. Pollut. Bull. http://dx.doi.org/10.1016/j.marpolbul.2017.08.021.

Ranjbar-Jafarabadi, A., A. Riyahi-Bakhtiari, M. Aliabadian, H. Laetitia, A. Shadmehri-Toosi and Ch. Kong-Yap. 2018. First report of bioaccumulation and bioconcentration of aliphatic hydrocarbons (AHs) and persistent organic pollutants (PAHs, PCBs and PCNs) and their effects on alcyonacea and scleractinian corals and their endosymbiotic algae from the Persian Gulf, Iran: Inter and intra-species differences. Sci. Total Environ. 627: 141–157.

Readman, J.W., R.F.C. Mantoura and M.M. Read. 1984. The physicochemical speciation of polycyclic aromatic hydrocarbons (PAH) in aquatic systems. Z. Anal. Chem. 219: 126–131.

Rinkevich, B. and Y. Loya. 1983. Short-term fate of photosynthetic products in a hermatypic coral. J. Exp. Mar. Biol. Ecol. 73: 175–184.

Rooker, J.R., L.L. Kitchens, M.A. Dance, R.J.D. Wells, B. Falterman and M. Cornic. 2013. Spatial, temporal, and habitat-related variation in abundance of pelagic fishes in the Gulf of Mexico: Potential implications of the Deepwater Horizon oil spill. PLoS One 8: e76080. http://dx.doi.org/10.1371/journal.pone.0076080.s006.

Sabourin, D.T., J.E. Silliman and K.B. Strychar. 2013. Polycyclic aromatic hydrocarbon contents of coral and surface sediments off the South Texas coast of the Gulf of Mexico. International. J. Biol. 5: 1.

Sammarco, P.W., S.R. Kolian, R.A.F. Warby, J.L. Bouldin, W.A. Subra and S.A. Porter. 2013. Distribution and concentrations of petroleum hydrocarbons associated with the BP/Deepwater Horizon Oil Spill, Gulf of Mexico. Mar. Pollut. Bull. 73: 129–143.

Sanders, H.L., J.F. Grassle, G.R. Hampson, L.S. Morse, S. Garner-Price and C.C. Jones. 1980. Anatomy of an oil spill: Long-term effects from the grounding of the barge Florida off West Falmouth, Massachusetts. Journal of Marine Research 38: 265–380.

Sauer, T.C., J. Michel, M.O. Hayes and D.V. Aurand. 1998. Hydrocarbon characterization and weathering of oiled intertidal sediments along the Saudi Arabian coast two years after the Gulf War oil spill. Environ. Int. 24: 43–60.

Schlacher, T.A., D.S. Schoeman, J. Dugan, M. Lastra, A. Jones, F. Scapini et al. 2008. Sandy beach ecosystems: Key features, sampling issues, management challenges and climate change impacts. Mar. Ecol. 29: 70–90.

Schwing, P.T., I.C. Romero, G.R. Brooks, D.W. Hastings, R.A. Larson and D.J. Hollander. 2015. A decline of benthic foraminifer following the Deepwater Horizon Event in the Northeastern Gulf of Mexico. PLOS One. Doi: 10.1371/journal.pone. 0120565.

Schwing, P.T., B.J. O'Malley, I.C. Romero, M. Martínez-Colón, D.W. Hastings, M.A. Glabach et al. 2016. Characterizing the variability of benthic foraminifera in the northeastern Gulf of Mexico following the Deepwater Horizon event (2010–2012). Environ. Sci. Pollut. Res. 1–16.

Sell, D., L. Conway, T. Clark, G.B. Picken, J.M. Baker, G.M. Dunnet et al. 1995. Scientific criteria to optimize oil spill Cleanup. *In*: Proceedings of the 1995 International Oil Spill Conference. American Petroleum Institute, Washington, DC, pp. 595–610.

Shore, R.F., J. Wright, J.A. Horne and T.H. Sparks. 1999. Polycyclic aromatic hydrocarbon (PAH) residues in the eggs of coastal-nesting birds from Britain. Mar. Pollut. Bull. 38: 509–513.

Smeltz, M., L. Rowland-Faux, C. Ghiran, W.F. Patterson, S.B. Garner, A. Beers et al. 2017. A multi-year study of hepatic biomarkers in coastal fishes from the Gulf of Mexico after the Deepwater Horizon Oil Spill. Marine Environmental Research 129: 57–67.

Sørhus, E., R.B. Edvardsen, Ø. Karlsen, T. Nordtug, T. van der Meeren, A. Thorsen et al. 2015. Unexpected interaction with dispersed crude oil droplets drives severe toxicity in atlantic haddock embryos. PLoS One 10: e0124376. http://dx.doi.org/10.1371/journal.pone.0124376.s009.

Soto, L.A., A. Gracia and A.V. Botello. 1981. Study of the penaeid shrimp population in relation to petroleum hydrocarbons in Campeche Bank. Proc. Gulf Carib. Fish. Inst. 33: 100.

Soto, L.A. and E. Escobar-Briones. 1995. Coupling mechanisms related to benthic production in the SW Gulf of Mexico. pp. 233–242. *In*: Eleftheriou, A., A. Ansell and C.J. Smith (eds.). Biology and Ecology of Shallow Water. Olsen and Olsen, Denmark (Fredensborg).

Soto, L.A. and C. González. 2009. Pemex y La Salud Ambiental de la Sonda de Campeche. México, D.F.: IMP-UNAM-BATTELLE-UAM, 397.

Soto, L.A., A. Estradas, R. Herrera, A. Montoya, R. Ruiz, A. Corona et al. 2009. Biodiversidad Marina en la Sonda de Campeche. pp. 265–300. *In*: Soto y, L.A. and C. González (eds.). Pemex y La Salud Ambiental de la Sonda de Campeche. IMP-UNAM-BATTELLE-UAM. México, D.F.

Soto, L.A., A.V. Botello, S. Licea-Durán, M.L. Lizárraga-Partida and A. Yáñez-Arancibia. 2014. The environmental legacy of the Ixtoc-I oil spill in Campeche Sound, southwestern Gulf of Mexico. Front. Mar. Sci. 1: 57. Doi: 10.3389/fmars.2014.00057.

Spies, R.B. 1987. The biological effects of petroleum hydrocarbons in the sea: Assessments from the field and microcosms. pp. 411–467. *In*: Boesch, D.F. and N.N. Rabalais (eds.). Long-Term Environmental Effects of Offshore Oil and Gas Development. Elsevier Applied Science, London.

Spies, R.B., P.H. Davis and D. Stuermer. 1980. Ecology of a petroleum seep off the California coast. pp. 229–263. *In*: Geyer, R. (ed.). Marine Environmental Pollution. Elsevier, Amsterdam.

Stegeman, J.J., L. Behrendt, B.R. Woodin, A. Kubota, B. Lemaire, D. Pompon et al. 2015. Functional characterization of zebrafish cytochrome P450 1 family proteins expressed in yeast. Biochim. Biophys. Acta 1850: 2340–2352.

Spraker, T.R., L.F. Lowry and K.J. Frost. 1994. Gross necropsy and histopathological lesions found in harborseals. pp. 281–311. *In*: Loughlin, T.R. (ed.). Marine Mammals and the Exxon Valdez. AcademicPress, San Diego, CA.

St. Aubin, D.J. and V. Lounsbury. 1990. Oil effects on Manatees: Evaluating the risks. pp. 241–251. *In*: Geraci, J.R. and D.J. St. Aubin (eds.). Sea Mammals and Oil, Confronting the Risks. Academic Press, San Diego, CA.

St. Aubin, D.J. 1990a. Physiologic and toxic effects on polar bears. pp. 235–239. *In*: Geraci, J.R. and D.J. St. Aubin (eds.). Sea Mammals and Oil, Confronting the Risks. Academic Press, San Diego, California.

St. Aubin, D.J. 1990b. Physiologic and toxic effects on pinnipeds. pp. 103–127. *In*: Geraci, J.R. and D.J. St. Aubin (eds.). Sea Mammals and Oil, Confronting the Risks. Academic Press, San Diego, CA.

Steurmer, D.H., R.B. Spies, P.H. Davis, D.J. Ng, C.J. Morris and S. Neal. 1982. The hydrocarbons in the Isla Vista marine seep environment. Marine Chemistry 11: 413–426.

Stubblefield, W.A., G.A. Hancock, H.H. Prince and R.K. Ringer. 1995. Effects of naturally eathered Exxon Valdez crude oil on mallard reproduction. Environ. Toxicol. Chem. 14: 1951–1960.

Swan, J.M., J.M. Neff and P.C. Young. 1994. Energy Research and Development Corporation (Australia), Australian Petroleum Production Exploration Association. Environmental Implications of Offshore Oil and Gas Development in Australia: The Findings of an Independent Scientific Review. Sydney.

Teal, J.M. and R.W. Howarth. 1984. Oil spill studies: A review of ecological effects. Environ. Manag. 8: 27–43.

Teal, J.M., J.W. Farrington, K.A. Burns, J.J. Stegeman, B.W. Tripp, B. Woodin et al. 1992. The West Falmouth oil spill after 20 years: Fate of fuel oil compounds and effects on animals. Marine Pollution Bulletin 24: 607–614.

Thomas, S.D. and Q.X. Li. 2000. Immunoaffinity chromatography for analysis of polycyclic aromatic hydrocarbons in corals. Environ. Sci. Technol. 34: 2649–2654.

Troisi, G., S. Barton and S. Bexton. 2016. Impacts of oil spills on seabirds: Unsustainable impacts of non-renewable energy. International Journal of Hydrogen Energy 41: 16549–16555.

Turner, N.R. and D.A. Renegar. 2017. Petroleum hydrocarbon toxicity to corals: A review. Marine Pollution Bulletin 119: 1–16.

Valentine, D.L., G.B. Fisher, S.C. Bagby, R.K. Nelson, C.M. Reddy, S.P. Sylva et al. 2014. Fallout plume of submerged oil from Deepwater Horizon. Proc. Nat. Acad. Sci. 111(45): 15906–15911.

Vandermeulen, J.H., H.M. Platt, J.M. Baker and A.J. Southward. 1982. Some conclusions regarding long-term biological effects of some major oil spills [and discussion]. Philosophical Transactions of the Royal Society of London. B. Biol. Sci. 297: 335–351.

Van Dolah, F.M., M.G. Neely, L.E. McGeorge, B.C. Balmer, G.M. Ylitalo, E.S. Zolman et al. 2015. Seasonal variation in the skin transcriptome of common bottlenose dolphins (*Tursiopstruncatus*) from the Northern Gulf of Mexico. Plos One 10(21 pp.).

van Eenennaam, J.S., S. Rahsepara, J.R. Radovićb, T.B.P. Oldenburgb, J. Woninkc, A.A.M. Langenhoffa et al. 2018. Marine snow increases the adverse effects of oil on benthic invertebrates. Marine Pollution Bulletin 126: 339–348.

Vanglider, L.D. and T.J. Peterle. 1981. South Louisiana crude oil DDE in the diet of mallard hen-effects on egg quality. Bull. Environ. Contam. Toxicol. 26: 328–336.

Vidal, M. and J. Domínguez. 2015. Did the Prestige oil spill compromise bird reproductive performance? Evidences from long-term data on the Kentish Plover (*Charadriusalexandrinus*) in NW Iberian Peninsula. Biological Conservation 191: 178–184.

Vonk, S.M., D.J. Hollander and A.J. Murk. 2015. Was the extreme and wide-spread marine oil-snow sedimentation and flocculent accumulation (MOSSFA) event during the Deepwater Horizon blow-out unique? Mar. Pollut. Bull. 100(1): 5–12.

Walton, P., C.M.R. Turner, G. Austin, M.D. Burns and P. Monaghan. 1997. Sub-lethal effects of an oil pollution incident on breeding kittiwakes Rissa tridactyla. Mar. Ecol. Prog. Ser. 15: 261–268.

White, H.K., H. Pen-Yuan, C. Walter, T.M. Shank, E.E. Cordes, A.M. Quattrini et al. 2012. Impact of the Deepwater Horizon oil spill on a deep-water coral community in the Gulf of Mexico. Proc. Nat. Acad. Sci. 109(50): 20303–20308.

Whyte, J.J., R.E. Jung, C.J. Schmitt and D.E. Tillitt. 2000. Ethoxyresorufin-O-deethylase (EROD) activity in fish as a biomarker of chemical exposure. Crit. Rev. Toxicol. 30: 347–570.

Wiens, J.A. 1995. Recovery of seabirds following the Exxon Valdez oil spill: An overview. pp. 824–893. *In*: Wells, P.G., J.N. Butler and J.S. Hughes (eds.). Exxon Valdez Oil Spill: Fate and Effects in Alaskan Waters. STP 1219, American Society for Testing and Materials, Philadelphia, PA.

Wilson, R.D., P.H. Monogham, A. Osanik, L.C. Price and M.A. Rogers. 1974. Natural marine oil seepage. Science 184: 857–865.

Williams, R., S. Gero, L. Bejder, J. Calambokidis, S.D. Kraus, D. Lusseau et al. 2011. Underestimating the damage: Interpreting cetacean carcass recoveries in the context of the Deepwater Horizon/BP incident. Conserv. Lett. 4: 228–233.

Wyers, S.C., H. Frith, R.E. Dodge, S. Smith, A.H. Knap and T. Sleeter. 1986. Behavioural effects of chemically dispersed oil and subsequent recovery in *Diploriastrigosa* (Dana). Mar. Ecol. 7: 23–42.

Yáñez-Arancibia, A. 1986. Ecología, impacto ambiental y recursos pesqueros: El caso del Ixtoc-I y los peces. pp. 95–126. *In*: Yáñez-Arancibia, A. (ed.). Ecología de la Zona Costera Análisis de Siete Tópico. AGT Editorial. México, D.F.

Yáñez-Arancibia, A., J.W. Day, A.L. Lara-Domínguez, P. Sánchez-Gil, G.J. Villalobos-Zapata and J.A. Herrera-Silveira. 2013a. Ecosystem functioning, the basis for sustainable management of Terminos Lagoon, Campeche Mexico. pp. 167–199. *In*: Day, J.W. and A. Yáñez-Arancibia (eds.). The Gulf of Mexico Origin, Water, and Biota. Vol. 4 Ecosystem-Based Management. College Station, TX. Texas A&M University Press.

Yáñez-Arancibia, A., J.W. Day and E. Reyes. 2013b. Understanding the coastal ecosystem-based management approach in the Gulf of Mexico. pp. 244–262. *In*: Brock, J.C., J.A. Barras and S.J. Williams (eds.). Understanding and Predicting Change in the Coastal Ecosystems of the Northern Gulf of Mexico. Journal of Coastal Research, Special Issue *No*. 63. Coconut Creek, Fl: Coastal Education and Research Foundation.

Zilberberg, L.J. 1998. Elsevier's Dictionary of Marine Pollution. Copyright Elsevier's, 10 p.

https://aoghs.org/transportation/exxon-valdez-oil-spill/

http://bcd.bzh/becedia/fr/la-maree-noire-de-l-amoco-cadiz

http://www.black-tides.com/

http://4.bp.blogspot.com/LKoATQJXsMo/UI8oXo_9mEI/AAAAAAAAB4/Ol9EVqGYy6s/s1600/1029-TorreyCanyon-655.gif

https://www.britannica.com/event/Deepwater-Horizon-oil-spill-of-2010

https://www.eia.gov/beta/international/

https://www.eia.gov/todayinenergy/detail.php?id=35332#

http://www.glossary.eea.europa.eu/EEAGlossary

https://www.focus-economics.com/blog/economic-outlook-for-the-top-oil-producing-countries

http://www.itopf.com/in-action/case-studies/case-study/amoco-cadiz-france-1978/

http://www.itopf.com/in-action/case-studies/case-study/prestige-spainfrance-2002/

http://www.robindesbois.org/en/torrey-canyon-18-mars-1967-la-mere-des-marees-noires/

http://www.nap.edu/catalog/10388.html

http://news.bbc.co.uk/2/hi/americas/8651333.stm

https://response.restoration.noaa.gov/about/media/10-photos-tell-story-exxon-valdez-oil-spill-and-its-impacts.html

http://www.latimes.com/news/nationworld/nation/la-na-oil-spill-history-html,0,3901663.htmlstory#axzz2zFv2Yvcm

https://wildlife-oilspills.wikispaces.com/Oil+Spills

http://www.whoi.edu/oil/ixtoc-I

http://counterspill.org/disaster/ixtoc-oil-spill

https://www.thebalance.com/exxon-valdez-oil-spill-facts-effects-on-economy-3306206

2

Oxygen Depletion in Coastal Waters and the Open Ocean

Hypoxia and Anoxia Cases and Consequences for Biogeochemical Cycling and Marine Life

Shane O'Boyle

1. Introduction

Oxygen is critical to many marine organisms. They need it to breathe and live. Yet, oxygen depletion or deoxygenation in the world's oceans and coastal waters is one of the most serious environmental issues damaging the marine environment. The development of oxygen-deficient zones referred to as "dead zones" caused by nutrient enrichment has become a persistent feature of many coastal seas and estuarine systems. In the open ocean, increases in sea temperature caused by climate change are reducing the capacity of the ocean to hold oxygen, which is less soluble in warm water. Strengthening water column stratification and changes to large-scale ocean circulation are also reducing ocean ventilation, leading to a reduction in the oxygen content of the deep ocean.

Low oxygen levels, often referred to as either hypoxia (low oxygen) or anoxia (no oxygen), can have several detrimental effects on marine organisms. These can include slower growth rates, damage to the immune system, elevated stress levels, impaired reproductive capacity and, in severe cases, mortality. Acute and chronic oxygen depletion can therefore damage not only individual organisms but also populations and ecosystems. The continued decline in dissolved oxygen in coastal waters and the open ocean threatens not only the survival of individual marine organisms but also the functioning of entire ecosystems.

There have always been areas naturally low in oxygen. Bottom waters with long retention times or restricted flow tend to develop natural zones of hypoxia. Such areas include fjords, lagoons and enclosed seas with limited connection to coastal or ocean waters. Low-oxygen conditions may also

Environmental Protection Agency, Ecological Monitoring and Assessment Unit, Richview, Clonskeagh Road, Dublin 14, Ireland.

develop when the movement of oxygen from the well-oxygenated surface layer is prevented by the presence of a seasonal thermocline or an estuarine halocline. When the surface layer is naturally productive, the sinking of organic matter and its subsequent decomposition by aerobic bacteria can use up oxygen. In the open ocean, low-oxygen areas can naturally occur when phytoplankton production in the surface layer sinks and is consumed by bacteria. They typically occur where surface water productivity is high because of natural upwelling events. The extent of these oxygen minimum zones (OMZs), which can have a vertical dimension of many hundreds of metres (typically at depths between 200 and 1000 m) and a horizontal extent of thousands of kilometres, is largely controlled by physical conditions that lead to stagnant circulation and long residence times. Where these zones intercept the ocean floor at continental shelf margins and seamounts they can have damaging effects on benthic invertebrate communities and other organisms that live alongside these communities. The largest permanent OMZs are to be found over vast areas of the eastern North Pacific and eastern South Pacific as well as off the west coast of Africa, the Arabian Sea and the Bay of Bengal. Two seasonal OMZs have also been identified in the west Bering Sea and in the Gulf of Alaska.

It is not entirely surprising that these areas of low oxygen in the open ocean and in coastal waters are home to numerous species that have adapted to such conditions. For example, some species of fish (e.g., Dover sole off the coast of California) have adjusted their physiological requirements for oxygen. Other organisms can tolerate low-oxygen environments because of their size: smaller organisms, with higher ratios of surface area to volume, can transfer oxygen more effectively to where it is needed, helping respiration. Surprisingly, areas impacted by hypoxia are often rich in fisheries because of the enhanced levels of primary and secondary productivity that results from anthropogenic nutrient enrichment. Ocean OMZs also sustain vast populations of bacteria (e.g., mats of sulphur-oxidizing bacteria), protozoans (e.g., foraminifera) and metazoans (e.g., nematodes and polychaetes) adapted to survive in these environments and benefit from the absence of predators that require higher oxygen levels (although predatory fish are known to form dense aggregations at the edge of OMZs).

In general, though, hypoxic zones and OMZs are less productive and less biologically diverse than well-oxygenated zones. For a start, anaerobic respiration is far less efficient than aerobic respiration, which means anaerobic organisms must work harder to attain equivalent levels of biomass and productivity. Those aerobic organisms that have adapted to live in low-oxygen environments typically have a much lower growth rate than counterparts living in oxygen-rich environments. Biological diversity is also lower in these zones, as only species with a specific set of adaptations can survive, although densities and biomass of those that can are often high.

In this chapter, we look at five case studies of hypoxia in different types of marine areas around the world. These include a large estuarine system (Chesapeake Bay), a semi-enclosed inland sea (Baltic Sea), an enclosed inland sea (the Black Sea), an open shelf system (Gulf of Mexico) and an open ocean OMZ (the Arabian Sea OMZ) in the Indian Ocean. In each, we provide a description of the history and extent of hypoxia and the damage that hypoxia causes to marine life and the functioning of marine ecosystems. This includes examples of how hypoxia effects the composition and distribution of marine organisms as well as altering biogeochemical cycles and the structure of marine food webs. We also include examples of adaptations or behaviours that different organisms use to survive or even thrive in low-oxygen environments. There are also situations in which the cause of hypoxia, nutrient-enhanced biological production, allows biological communities on the border of hypoxic zones to flourish. Finally, we mention the actions being taken and the progress being made in addressing the problem of coastal and ocean hypoxia.

2. The Importance of Oxygen

Dissolved oxygen is the amount of gaseous oxygen dissolved in water. Oxygen plays a critical role in aerobic respiration. The oxygen molecule accepts electrons (i.e., it is the terminal electron acceptor) to form water and in doing so maintains an electron gradient that is used to drive metabolic processes including the formation of ATP, the "energy molecule" needed to drive cellular activity. Aerobic respiration is estimated to be up to 16 times more efficient in generating ATP than anaerobic fermentation (Catling and Claire 2005). The energy yield from aerobic metabolism is high relative to that from anaerobic metabolic pathways (Hochacka and Somero 2002).

When dissolved oxygen levels fall they can harm marine organisms and threaten their survival. When oxygen levels become depleted the rate of cellular activity decreases. This can cause sublethal effects such as slower growth rates, damage to the immune system, elevated stress levels, impaired reproductive capacity, reduced larval recruitment, lower fecundity and, in severe cases, mortality (Baden et al. 1990, Breitburg 1992, Diaz and Rosenberg 1995, Gray et al. 2002). Mobile species such as fish tend to avoid areas with depressed levels of dissolved oxygen, while less mobile species initiate survival behaviour including leaving their burrows to position themselves in waters with higher oxygen concentration (Fig. 1).

Low oxygen levels not only harm marine life, they can also alter the biogeochemical cycling of nutrients. When oxygen levels are low in bottom waters it is likely that sediments underlying them will be starved of oxygen. When sediments have little or no oxygen the rate at which some nutrients are released from them can increase many-fold. For example, phosphate binds to iron oxyhydroxides when oxygen is present; in the absence of oxygen, phosphate is released from the sediments. In Chesapeake Bay, for example, the rate of phosphate release from the sediments increases five-fold under anoxic conditions. This greater release of phosphate to the overlying water column reinforces the process of eutrophication by fuelling additional phytoplankton growth and subsequent oxygen depletion.

Many important biogeochemical processes occur in OMZs. For example, nitrification oxidizes ammonium into nitrate (this usually occurs at the OMZ boundary, where sufficient oxygen is present),

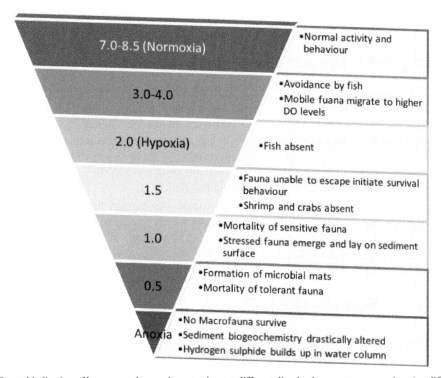

Fig. 1. General indication of harm caused to marine organisms at different dissolved oxygen concentrations (modified from Diaz et al. 2013).

and denitrification transforms nitrate into gaseous nitrogen (e.g., nitrogen gas and nitrous oxide), which is lost to the atmosphere and contributes to the oceanic nitrate deficit (Tyrrell 1999).

3. Thresholds of Oxygen Depletion

A number of numerical thresholds depicting different states of oxygen depletion have become established in the literature. Hypoxia is most often associated with a threshold value less than 2 mg L^{-1}, whereas anoxia is associated with a threshold value at or close to zero. For practical purposes, and to take into account instrumental precision, anoxia is often categorized as occurring when values are less than 0.2 mg L^{-1}. Normoxia, the prevalence of normal oxygen levels, is typically defined by threshold values greater than 6 but less than 9 mg L^{-1}. Oxygen concentrations above this may be considered to represent hyperoxia, the presence of too much oxygen, which can also cause harm to marine organisms.

While the threshold of 2 mg L^{-1} is widely used in the literature to define hypoxia, it is now recognized that levels of oxygen above this threshold can also be harmful. In an extensive study of different groups of marine organisms with different sensitivities to low oxygen concentrations, it was found that harm occurred at oxygen levels well above 2 mg L^{-1}. For example, a large meta-analysis of previous studies showed the median oxygen concentration causing sublethal effects in 62 invertebrate species (across 265 experiments) was 2.13 mg L^{-1} (close to the standard definition of hypoxia). However, the 90th percentile of the same dataset was as high as 4.6 mg L^{-1}, indicating significant sublethal effects at concentrations above 2.13 mg L^{-1} for a substantial number of species (Vaquer-Sunyer and Duarte 2008).

In studies of open ocean OMZs, a different unit of measurement is typically used (i.e., μmol L^{-1}), with 62.5 μmol L^{-1} being equivalent to 2 mg L^{-1}. In describing the spatial distribution of OMZs, a much lower concentration of less than 20 μmol L^{-1} is often used (Helly and Levin 2004, Paulmier and Ruiz-Pino 2008).

4. The Spread of Coastal Hypoxia

The number of low-oxygen coastal areas currently stands at over 500 (Diaz and Rosenberg 2008, Isensee et al. 2015), and one study suggests that the number of hypoxic zones globally has increased 10-fold since the 1950s (Breitburg et al. 2018) (Fig. 2). While this increase may be attributed to greater awareness and reporting, it is also likely that many areas remain unreported, especially in developing nations where monitoring of these areas does not exist or is not sufficient to fully describe their occurrence. Relatively little is known, for example, from some of the most populated parts on the planet, such as southeast Asia, and many of the Pacific Islands.

The development of seasonally persistent hypoxic or anoxic zones has been reported from the Baltic Sea, Black Sea, Kattegat, northern Adriatic Sea, Chesapeake Bay and the northern Gulf of Mexico (Breitburg 1992, Breitburg et al. 1997, Diaz and Rosenberg 1995, Gray et al. 2002, Justic et al. 1987, Pavela et al. 1983, Rabalais et al. 2001). The largest low-oxygen coastal areas are found in the Baltic Sea (60,000 km²), Gulf of Mexico (22,000 km²) and East China Sea.

5. Hypoxia in the Open Ocean

The largest OMZs are to be found over vast areas of the eastern North Pacific and eastern South Pacific as well as in the southern Atlantic Ocean off the coast of western Africa, in the Arabian Sea and Bay of Bengal (Fig. 3). These permanent open ocean OMZs cover an area of 30.4 million square kilometres (± 10%), which represents 8% of the present global ocean surface (Paulmier and Ruiz-Pino 2009). This is based on using a threshold of 20 μmol L^{-1} of oxygen to identify the boundary of an OMZ. With a mean vertical extent of 3360 m (± 800 m), the volume of these zones has been

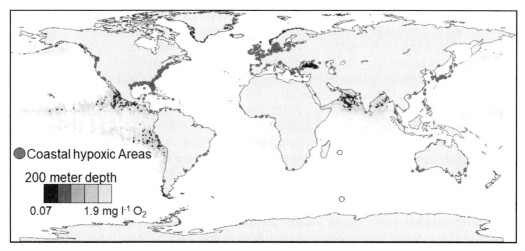

Fig. 2. Distribution of global coastal hypoxic areas (red circles) and oxygen minimum zones at a depth of 200 m (blue-shaded areas) (Breitburg et al. 2018).

Color version at the end of the book

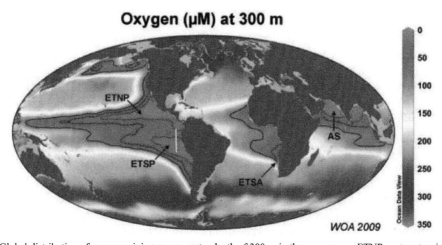

Fig. 3. Global distribution of oxygen minimum zones at a depth of 300 m in the open ocean. ETNP, eastern tropical North Pacific; ETSP, eastern tropical South Pacific; ETSA, eastern tropical South Atlantic; AS, Arabian Sea. Source: Max Planck Institute for Marine Microbiology, based on data from the World Ocean Atlas 2009.

Color version at the end of the book

estimated at 102 million cubic kilometres (± 15%), which represents 7% of the total ocean volume (Paulmier and Ruiz-Pino 2009). The core of OMZs, the layer with the lowest oxygen concentrations, is estimated to occupy a volume about one-tenth of that occupied by all OMZs as a whole. In volume, the biggest OMZs are in the eastern North Pacific and Arabian Sea.

Over the last half century, the open ocean has lost 2% of its dissolved oxygen or 77 billion metric tonnes (Schmidtko et al. 2017). This loss can be attributed to ocean warming, which results in declining oxygen solubility and reduced ventilation of the deep ocean (Keeling et al. 2010, Long et al. 2016). Ocean warming is estimated to account for 15% of current total global oxygen loss and more than 50% of the oxygen loss in the upper 1000 m of the ocean (Schmidtko et al. 2017, Helm et al. 2011). The phenomenon that explains most of the loss of oxygen from the ocean (i.e., 85%) is intensification of stratification. It reduces the vertical exchange of oxygen from the atmosphere to the ocean's surface, it reduces the diffusion of nutrients from deeper waters into sunlit surface waters, reducing primary productivity and the production of oxygen by photosynthesis, and lastly, it slows large-scale overturning circulation that brings oxygen to the deep ocean (Schmidtko et al. 2017).

In the open ocean, OMZs have increased by 4.5 million square kilometres, an area equivalent in size to Australia (Stramma et al. 2008, 2010). Ocean global climate change models suggest that present rates of oxygen outgassing will double in magnitude by the end of the 21st century (Garcia et al. 1998, Keeling et al. 2010), with roughly a quarter of the predicted outgassing attributable to the direct effect of warming on oxygen solubility and the remainder due to increased stratification, which is also predicted to increase significantly in the 21st century. It is predicted that the oceans will lose 3.5% of their oxygen by 2100, with even greater losses regionally (Long et al. 2016). Regional models that consider local characteristics of an area such as prevailing winds, water circulation patterns, water depths, freshwater and nutrient inputs, and strength of stratification predict even more acute impacts in some cases. For example, along the western coast of California models are predicting that upwelling of water with lower oxygen concentration will increase in frequency north of San Francisco by 2100. The concentration of oxygen off the Californian coast at depths between 200 and 300 m has already declined by 11–33 μmol L^{-1} since 1984 (Bograd et al. 2008). These predicted changes are likely to add further stress to areas that are already experiencing lower than normal oxygen concentrations as well as affecting important biogeochemical cycles that are influenced by oxygen availability.

6. Drivers and Pressures

The single biggest driver of coastal pollution is the transport of nutrients from agricultural activities, with three times as much nitrogen derived from agriculture as from sewage (Bouwman et al. 2005). The Food and Agriculture Organisation of the United Nations estimates that the demand for food will increase by 50% by the end of the 21st century (FAO 2017) as the world's population grows from its current 7.6 billion people to 10 billion people by 2060. To meet this extra demand, food production systems will need to become more productive.

In the 20th century the human population increased from one billion in 1900 to over 6 billion by 2000. This was in part made possible by artificial fertilizers used to enhance the productivity of agricultural lands. The ability to manufacture fertilizer, particularly nitrogen fertilizer, by stripping nitrogen from the air in the Haber-Bosch process allows the world to sustain its current population. Global fertilizer production has increased 10-fold since the 1960s (IFA 2016).

The problem arises when nitrogen and phosphorus fertilizer not taken up by crops is washed off agriculture lands into rivers, lakes and estuaries. The effects in these systems is the same as on land, increased productivity, but over-production can lead to eutrophication. Farmers used 110 million t of nitrogen-rich fertilizer in 2016 to increase the fertility of their soils (up from 10 to 30 million t in the 1960s) (FAOStat 2018). These inputs together with inputs of nitrogen from burning fossil fuels and other human activities are putting more pressure on natural systems. On a global scale it has been estimated that total anthropogenic nitrogen fixation (i.e., artificial fertilizers, fossil fuels and use of legume crops) has doubled the total amount of nitrogen fixation from the atmosphere to the terrestrial environment (Smil 1990, Vitousek et al. 1997). The amount of anthropogenic nitrogen fixation (140 tera grams N yr^{-1}) is now comparable to the amount of "natural" biotic terrestrial fixation (90–130 tg N yr^{-1}) (1 tg = 10^{12} g) (Galloway et al. 1995).

On a regional basis, this increase in nitrogen fixation and the greater availability of biologically active nitrogen pools has increased the flux of nitrogen to the marine environment. Riverine total nitrogen fluxes from most of the temperate regions surrounding the North Atlantic Ocean may have increased by two- to 20-fold. For the North Sea region the increase may have been six- to 20-fold (Howarth et al. 1996). In the Mississippi River, which discharges into the Gulf of Mexico, nitrate loadings increased three-fold from the 1950s to the mid-1990s (Goolsby et al. 1999, Goolsby and Battaglin 2001, Turner et al. 2007). The Baltic Sea saw a four-fold increase in nitrogen load during the 20th century (Larsson 1985).

For the open ocean, the single biggest driver of oxygen loss is climate change (see above) through ocean warming and enhanced stratification. Climate change is also likely to lead to further oxygen loss in coastal waters through increased nutrient runoff under scenarios of increased precipitation and more intense stratification. Thermal warming of coastal waters, which will hold less oxygen, will also contribute to the development and persistence of hypoxia. Increasing sea temperatures will also increase metabolic rates (i.e., respiration) and demand for oxygen. The other effects of climate change, such as ocean acidification and the introduction of new species to areas with changing environmental variables (e.g., sea temperature, salinity, pH), are also likely to add further stresses to resident biological communities already suffering from the effects of hypoxia and anoxia.

The five case studies that follow provide greater detail on the type of damage being caused by the loss of oxygen from the world's oceans and coastal seas.

7. Case Studies

7.1 The Baltic Sea

The Baltic Sea covers an area of 420,000 km^2 and is one of the largest semi-enclosed brackish seas in the world.

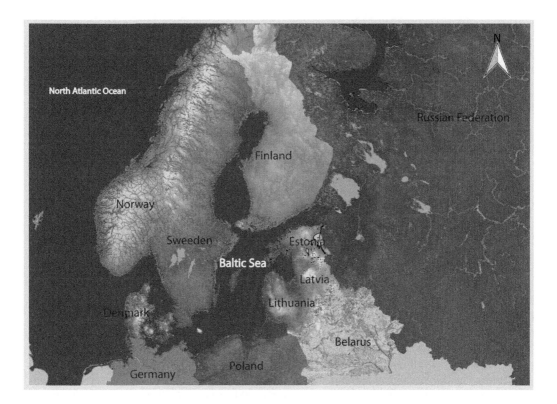

Its catchment area is about four times its surface area and home to approximately 85 million people. It has a narrow connection to the North Sea and this morphological feature reduces the exchange of water between the brackish waters of the Baltic Sea and the more saline waters of the

Fig. 4. A large cyanobacteria bloom in the Baltic Sea in summer 2005. The proliferation of such blooms leads to the consumption of dissolved oxygen in bottom waters following their collapse (Image: Envisat Meris 13 July 2005).

Color version at the end of the book

North Sea. It can take approximately 30 years for the waters of the Baltic Sea to be fully exchanged (Stigebrandt 2001).

The inflow of bottom higher-salinity water rich in oxygen and freshwaters rich in nutrients, together with strong stratification and vertical mixing processes, are the main factors driving the oxygen budget of the Baltic Sea. High-salinity water from the North Sea enters the Baltic Sea through the Danish Straits, mostly in winter. Freshwater reaches the Baltic Sea from numerous rivers, corresponding to about one-fortieth of the total Baltic Sea water volume per year (Bergström et al. 2001). The inflow of high-salinity water through the Straits and inputs of freshwater from the catchment/watershed areas produce a surface brackish water gradient from the higher-salinity water of 15–18 (psu) at the entrance (the Sound) to 7–8 psu in the Baltic Proper and 0–2 psu in the northeast bays. The Baltic Proper is permanently stratified, with a strong permanent halocline forming at depths of 60–80 m. Bottom salinities below the halocline are around 11–13 psu.

Because of its hydrological makeup, the waters of the Baltic Sea have intermittently experienced episodes of hypoxia even dating back to its formation 8000 years ago (Zillen et al. 2008). In terms of anthropogenic influence the issue of nutrient enrichment and eutrophication as a large-scale pressure on the Baltic Sea was first recognized in the 1980s. However, it is now clear from looking at earlier time series that the problem of increasing hypoxia because of human-based nutrient inputs traced back to the 1950s. During the 20th century, the Baltic nitrogen load increased four-fold and phosphorus load increased eight-fold (Larsson 1985). The extent of low-oxygen areas in the Baltic Sea varies greatly as indicated above, but the long-term average of sea bottom covered by hypoxic water in the autumn averaged 49,000 km^2 over the time period 1961–2000. In most years, the deep basins of the Baltic Sea (Gotland Deep, Landsort Deep, northwest Baltic Proper and Gdansk Deep) are permanently hypoxic and in bad hypoxic years low oxygen concentrations migrate higher up the water column and the individual basins become connected to form one large hypoxic area (Conley et al. 2002).

While major high-salinity inflows into the Baltic Sea replenish oxygen levels, they can also lead to significant oxygen depletion in deep waters in subsequent years. The reason is that the increased vertical salinity gradient brought about by these inflows results in stronger vertical stratification and reduced surface-to-bottom ventilation (Gerlach 1994). In 1971, a hypoxia area of 70,000 km^2 formed following large high-salinity inflows in the preceding years. Conversely, one of the smallest areas of hypoxia, 12,000 km^2, occurred in 1993, during the peak of the stagnation period between 1983 to 1993 when higher-salinity inflows from the North Sea were at a minimum (Conley et al. 2002).

The impact of hypoxia on biological communities in the bottom waters and sediments of the Baltic Sea has been significant. Below the halocline, benthic organisms have virtually disappeared from large parts of the Baltic Proper, as indicated by the formation of laminated sediments over large parts of the seafloor (Schaffner et al. 1992). In some areas, hypoxia- tolerant species such as *Scoloplos armiger* and *Capitella* sp. are often the dominant components of remaining benthic assemblages. Above the halocline, the story is quite different. It is estimated that eutrophication has increased pelagic primary production by 30–70% and sedimentation of organic matter by 70–190% (due to greater settlement during the spring bloom), leading to an estimated three- to five-fold increase in the biomass of macrobenthic fauna and an approximate doubling of production (Elmgren 1989). Fish production too seems to have increased with total fish landings, increasing 10-fold between the early part of the 1900s and the 1970s, but it has been speculated that this increase is partly (about half) due to increased fishing effort and the extermination of the fish-eating Baltic seal population rather than extra food availability.

To combat the problem of eutrophication, seven countries bordering the Baltic Sea (Denmark, Finland, Germany (East and West), Poland, USSR and Sweden) agreed to reduce nutrient loadings by about half. This agreement, formalized during a meeting of Baltic Sea environmental ministers in Helsinki in 1988, established a requirement that contracting parties of the HELCOM Convention would report annually on progress in reaching this level of reduction. In 2007, a Baltic Sea unaffected by eutrophication was set as one of the goals of the Baltic Sea Action Plan (HELCOM 2007). Having peaked in the 1980s, loads of nitrogen and phosphorus (about 1 million t of N and 60,000 t of P) declined in the early 2000s to levels seen in the 1960s and 1950s, respectively (Gustafsson et al. 2012). Nevertheless, eutrophication persists across much of the Baltic Sea, with one of the largest areas of hypoxia (65,000 km^2) ever recorded occurring in 2012 (Carstensen et al. 2014).

The continued eutrophic state of the Baltic Sea is mostly blamed on the continued release of phosphate from anoxic sediments during summer (Pitkanen et al. 2001). It is estimated that the amount of phosphate released from sediments during the seasonal expansion of hypoxia in the Baltic Sea is approximately one order of magnitude greater than the total yearly anthropogenic phosphorus loading (Conley et al. 2002). Ironically, the reduction in nitrogen loads from land and increase in phosphate released from sediments have, over time, lowered the N:P ratio in surface waters, favouring the proliferation of nitrogen-fixing cyanobacteria blooms. While other phytoplankton species may be limited by the availability of nitrogen (which also has the effect of lowering the system-wide biological uptake of phosphorus), cyanobacteria, which can fix nitrogen from the atmosphere, benefit from the extra phosphorus not used by nitrogen-limited species. The increased biomass of cyanobacteria adds to the vertical flux of organic material to the seafloor, further reinforcing the occurrence of hypoxia and release of phosphate from the sediments. This feedback loop between hypoxia and the biogeochemical cycling of phosphate that further exacerbates eutrophication has been described as the "vicious cycle of the Baltic Sea" (Vahtera et al. 2007).

While most of the focus is on summer cyanobacteria blooms, the key to solving the problem of eutrophication in the Baltic Sea is reducing the size of the nitrogen-limited spring bloom. This is when most of the organic matter that results in bottom water hypoxia is produced. Reducing its magnitude will reduce the extent of anoxic waters and the release of phosphate from the sediments. Fixing the problem of eutrophication in the Baltic Sea will therefore require a dual nutrient approach, a reduction in nitrogen loads to limit the development of anoxia following the spring bloom and a reduction in phosphorus loads to limit the blooming of nitrogen-fixing cyanobacteria in summer.

The countries of the HELCOM Convention established a nutrient reduction scheme to achieve the goal of no eutrophication in the Baltic Sea. The purpose of the scheme was to share the burden of nutrient reductions amongst the contracting parties of the Convention. The scheme involves first estimating the maximum allowable input of nutrients that would prevent eutrophication in the Baltic Sea and then the allocation of nutrient reduction targets for each country. In 2007, HELCOM estimated that for good environmental status to be achieved, the maximum allowable annual nutrient pollution inputs into the Baltic Sea would be 21,000 t of phosphorus and about 600,000 t of nitrogen. This translates to annual reductions of some 15,000 t of phosphorus and 135,000 t of nitrogen, to achieve the plan's crucial "clear water" objective.

7.2 Gulf of Mexico

The largest area of low bottom water oxygen concentration in the entire western Atlantic, and second largest in the world, is found in the northern Gulf of Mexico.

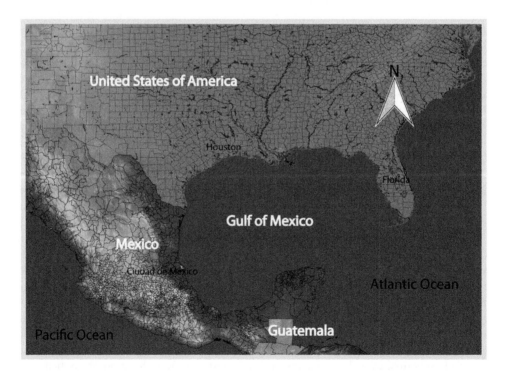

Since 1985 the extent of the hypoxic zone (< 2 mg L^{-1}) has varied from just under 5,000 km^2 in 2000 (the extent of oxygen depletion was negligible in 1988 due to dry weather and a 52-year low river flow) to a maximum of 22,000 km^2 in 2002 (Rabalais et al. 2007). The long-term average over this period is 13,500 km^2. The extent of the low oxygen zone is influenced by the degree of freshwater discharge and nutrient loading from the Mississippi and Atchafalaya rivers. Haline stratification of the water column persists through the year but intensifies in spring and summer due to thermal warming. Taken together, these river systems (hereafter referred to as the Mississippi River System) are the primary sources of freshwater, nitrogen and phosphorus to the northern Gulf of Mexico, delivering 91% of the estimated annual nitrogen load and 88% of the phosphorus load (Dunn 1996). The Mississippi drains 41% of the continental United States.

Hypoxic waters, which are most prevalent from late spring through to late summer, can form close to the shore in shallow water (4–5 m), extending offshore to depths of up to 45 m. Hypoxia is most severe in the summer, but conditions improve in the autumn and return to normal in the winter. Since the 1950s the load of nitrogen, as inorganic nitrate, entering the Gulf from the Mississippi River System has increased by 300% (Goolsby et al. 1999, Goolsby and Battaglin 2001), while freshwater flow increased by 30% (Bratkovich et al. 1994). There is strong evidence from long-term datasets and sediment records that increased primary production in the surface waters of the continental shelf has been fuelled by increases in river nitrate concentrations (Rabalais and Turner 2001). This increased production results in the development of seasonal hypoxia in bottom waters below a strong and persistent saline and thermally stratified water column.

The impact of hypoxia on living resources in the Gulf of Mexico has been extensive. In the hypoxic zone the diversity and abundance of fish, shrimp, benthic invertebrates and other marine organisms have declined. This is due to either the death of sensitive species that are unable to survive or escape the effects of low oxygen conditions (Kidwell et al. 2009) or a shift to more tolerant species that create less diverse benthic communities and poorer food resources for higher trophic levels such as fish (Baustian and Rabalais 2009, Breitburg 2002). Mobile species avoid areas with low oxygen concentrations, with some species being less tolerant than others of these conditions. Elasmobranchs (rays and sharks) are less tolerant of low concentrations and avoid areas with oxygen concentrations less than 3 mg L^{-1}. Many species of demersal species are slightly more tolerant but will leave hypoxic areas when levels fall below 2 mg L^{-1}, although drum, snapper and redfish are occasionally observed when oxygen concentrations are near 1.5 mg L^{-1} (Levin et al. 2009). Mobile species such as fish or shrimp that migrate from the hypoxic zone can suffer indirect effects related to the movement to habitats with poorer food resources or increased predation. In some cases, the aggregation of different species in oxygenated areas at the edge of hypoxic zones can lead to increases in bycatch of species that aggregate with commercially fished species. For example, the aggregation of Atlantic croaker (*Micropogonias undulatus*) with the commercially fished brown shrimp has led to the greater bycatch of the former species (Craig 2012). Even mobile species sometimes fail to escape from low-oxygen waters when landward movements of hypoxic water masses trap them along the coast. Such incidents have been associated with the mass mortality of both demersal fish species and crustaceans.

For benthic organisms with limited or no mobility, options are more limited. Some bottom-dwelling organisms will do what they can to position themselves in water with higher oxygen content. Crabs (e.g., *Libinia* sp., *Persephona* sp.) and sea stars (*Astropecten* sp.) climb to the top of excavated mounds, brittle stars emerge from the sediment and use their arms to raise their disk off the ground, while burrowing shrimp (*Alpheus* sp.) emerge from their burrows, and gastropods (e.g., *Oliva sayana*, *Terebra* sp.) move through the surface sediment with their siphons held directly upwards (Levin et al. 2009). Other organisms lie motionless on the seafloor in a moribund state, probably to suppress their metabolism and the need for oxygen. So many species are able to survive in this motionless state or away from the protection of their burrows because of the absence of bottom-feeding fish that are excluded from these areas. As oxygen falls further from 1 to 0.5 mg L^{-1}, even the most tolerant burrowing organisms, such as polychaetes, will leave their burrows and lie motionless on the seafloor (Rabalais et al. 2001).

In general, commercial and recreational fisheries off the Louisiana coast have suffered because of the presence of a hypoxia zone. The habitat of the brown shrimp, which is one of the most important commercial species fished from these waters, has declined by 25%, with shrimp populations aggregating both inshore and offshore of the hypoxic region (Craig et al. 2005). In the northern Gulf of Mexico, small brown shrimp populations occur in the same years that large hypoxic zones do. There is also evidence to suggest that the Atlantic croaker, an important species for recreational anglers, suffers ecophysiological damage including reproductive impairment due to the presence of hypoxia. A biomarker indicating the presence of ecophysiological stress has been observed in Atlantic croaker when exposed to hypoxia. This biomarker has been seen in other species such as shrimp, indicating that other marine organisms in the Gulf of Mexico are also suffering the impacts of hypoxia (Thomas et al. 2007, Thomas and Rahman 2009, 2010).

A national Action Plan to address the issue of nutrient enrichment and the occurrence of hypoxia in the Gulf of Mexico was established in 2001 (Mississippi River/Gulf of Mexico Watershed Nutrient Task Force 2001). The Plan contained 11 priority actions on funding, governance (i.e., establishment of sub-basin committees), research, monitoring, nutrient reduction strategies, point source pollution, wetland restoration and best management practices for agriculture. These actions, which were voluntary but also incentive-based, were to be reviewed on an on-going basis and the plan was to be adapted if they were not proving effective in reducing the extent of hypoxia. Central to the Plan was the goal to reduce the area of the hypoxic zone to less than 5,000 km² by 2015 and a related second goal to reduce nitrogen discharges to the Gulf, with multistate sub-basin committees responsible for developing nutrient reduction strategies (phosphorus was not viewed as a cause of hypoxia at that time).

The original Plan focused on nitrogen reduction and at the time it was estimated that nitrogen loading to the Gulf would need to be reduced by 30% to achieve the spatial goal set out in the Plan. However, as scientific understanding of the drivers of hypoxia became clearer it was evident that reduction in both nitrogen and phosphorus loads of about 45% would be required. It also became evident that the original target year of 2015 was unrealistic due to a number of factors including the effort and funding required to put actions in place, the time for nutrient reservoirs in soils and sediments to return to equilibrium, the occurrence of extreme rainfall events leading to greater nutrient runoff, and warming of coastal waters, which have less capacity to hold dissolved oxygen. Taking these factors into account, the original target year was extended to 2035, with an interim target of a 20% nutrient load reduction by 2025 as a milestone toward reducing the hypoxic zone to less than 5,000 km² by 2035.

In 2017, the Gulf of Mexico Dead Zone measured 22,720 km² (Fig. 5), the largest size measured since standardized mapping began in 1985 and considerably larger than the long-term average

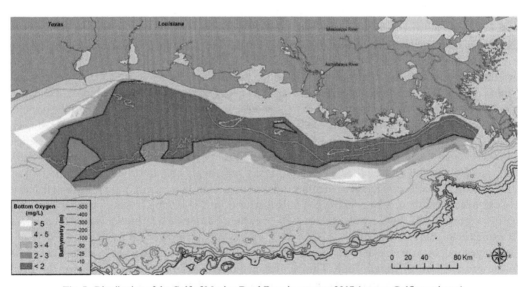

Fig. 5. Distribution of the Gulf of Mexico Dead Zone in summer 2017 (source: Gulfhypoxia.net).

(13,751 km^2) as well as the average over the five years preceding 2017 (14,024 km^2). The Dead Zone is also 4.5 times larger than the coastal goal of 5,000 km^2, indicating that nutrients from the Mississippi River watershed are continuing to seriously affect the nation's coastal resources and habitats in the Gulf.

7.3 Chesapeake Bay

Chesapeake Bay is located on the eastern seaboard of the United States. The Bay is almost 300 km long and covers an area of 11,500 km^2.

A central and relatively deep (20–30 m), narrow channel runs the length of the Bay with either side flanked by shallower waters. The Bay is extensively connected to its watershed/catchment through

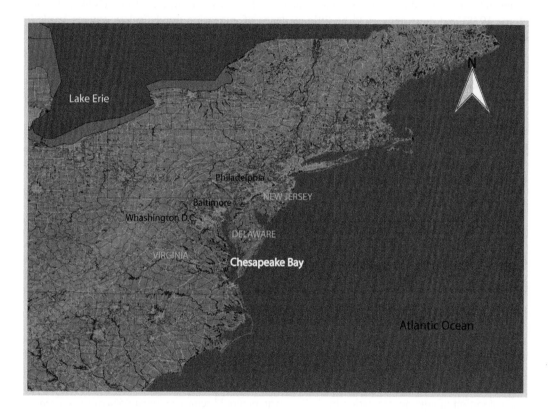

numerous tributaries. Significant freshwater flow into the Bay results in a stratified water column that can isolate the waters of the central channel. The strength of stratification is strongly correlated with river flow, increasing with higher river flows.

The character of the Bay's watershed, which extends to an area of over 164,000 km^2, has changed dramatically since these lands were first colonized by Europeans in the 16th century. Prior to their arrival, the catchment area of Chesapeake Bay comprised temperate woodlands and wetlands along the shoreline. By the 1850s, nearly half of the land area had been converted to farm land and urban developments around smaller settlements and towns started to increase. The human population of the watershed stands at 16 million. The proportion of the watershed committed to agriculture peaked in the 1850s and since then the proportion of land covered by forest has actually increased (Kemp et al. 2005).

While the proportion of the Bay's watershed given over to agriculture was in decline since the 1850s, the use of artificial fertilizer and import of animal feed particularly since the 1950s has meant that nutrient inputs to the Bay increased in the second half of the 20th century. Nutrient concentrations (total nitrogen) in the Susquehanna River, the largest river discharging into the Bay, increased 2.5-fold from 1945 to 1990 (Kemp et al. 2005). Examination of the Bay's sediments has provided a window into the development of hypoxia. Sediment cores have revealed evidence of human disturbance as far back as the 17th and 18th centuries, but nearly all sedimentary indicators (e.g., the proportion of hypoxia-tolerant foraminifera species) indicate that the occurrence of low oxygen increased sharply from the 1950s (Karlsen et al. 2000) with the volume of hypoxic water in summer tripling in the second half of the 20th century (Hagy et al. 2004).

Oxygen depletion in the Bay occurs in spring and is closely linked to the amount of freshwater inflow, which determines the extent of water column stratification and vertical oxygen exchange, and spring water column temperature, which influences respiration rates (Hagy et al. 2004). Low oxygen levels in the main channel of the Bay have been directly observed since the 1930s (Newcombe and Horne 1938) and there appears to be little evidence of increasing trends in low oxygen levels in spring. This suggests that the initial onset of hypoxia is driven by physical factors (i.e., river flow and stratification) rather than anthropogenic nutrient inputs, which are believed to play a more important role in determining the temporal extent of hypoxia in late spring and summer (Hagy et al. 2004).

The sensitivity of the Bay to nutrient inputs appears to be increasing with similar levels of nitrogen inputs resulting in greater volumes of waters suffering from hypoxia. For example, when two periods were compared, one earlier (1950–1979) and one later (1980–2001), it was found that the volume of low-oxygen bottom water was much greater in the later period even under similar levels of nutrient inputs (Hagy et al. 2004). The reason for this increase in sensitivity is thought to be linked to a general deterioration in the ecological health of the Bay and the impacts of low oxygen levels on key biogeochemical processes.

One of the main ecological changes that has occurred in the Bay is the decline of oyster populations (*Crassostrea virginica*) due to overfishing and habitat destruction (Rothschild et al. 1994). Deteriorating water quality in the second half of the 20th century and the occurrence of oyster diseases are also likely to have affected the recovery of depleted stocks. The population has declined 50-fold since the early 20th century and this decline has had a significant impact on the ecology of the Bay and cycling of nutrients. The decrease in oyster biomass has led to a reduction in the amount of nutrients that would have been retained by healthy and growing oyster populations. Furthermore, increasing phytoplankton blooms fuelled by nutrient enrichment have also resulted in declining water clarity in the Bay (Gallegos and Jordan 2002). This has also been exacerbated by the decline in oyster populations, which would have filtered much of this additional algal biomass from the water column. The decline in water quality has led to a decline in bottom-living species of angiosperms such as seagrass and attached macroalgae (Kemp et al. 2004, Newell 1988). Their loss further decreased the natural ability of the Bay to retain nutrients. In fact, it has been estimated that the restoration of aquatic vegetation in shallow waters would remove almost 45% of the current N inputs to the upper Bay through particle trapping and direct uptake (Kemp et al. 2004).

The other main reason for the increased sensitivity of the Bay to nutrient inputs is believed to be the impact of hypoxia on biogeochemical processes that are oxygen-dependent. When sediments underlying hypoxic bottom waters become anoxic, the release of phosphate from the sediments increases. In Chesapeake Bay, it is estimated that the flux of phosphate from sediments increases five-fold when oxygen concentrations in bottom waters fall below 1.5 mg L^{-1} (W. Boynton, unpublished). Similarly, the recycling of ammonium is affected in oxygen-stressed bottom waters. Under normal oxygen conditions, much of the regenerated ammonia would be oxidized to nitrate, which in turn would be denitrified to nitrogen gas and nitrous oxide, which is mostly lost to the atmosphere. However, in oxygen-poor sediments, nearly all the re-mineralized ammonia is released into the overlying water column, where it is available for uptake. This feedback mechanism, in which the release of both ammonium and phosphate becomes more efficient when sediments become oxygen stressed, feeds and reinforces the eutrophication process occurring in the Bay.

The most obvious direct biological effect of hypoxia would appear to be on the abundance of benthic macroinvertebrates. These organisms are mostly absent from the deeper water of the channel along the middle part of the Bay and from analysis of sediment cores it appears that they have been absent from this region for at least 100 years (Schaffner et al. 1992). Although observations on the composition of benthic communities in the Bay are nearly all from a period when eutrophication-driven hypoxia was already established, they generally indicate the replacement of larger longer-lived bivalves by short-lived opportunistic species (Holland et al. 1987). The abundance of these opportunistic species tends to reflect the availability of food, with increased abundance associated with increases in phytoplankton biomass and other inputs of organic matter. This appears to be the case in the upper and lower regions of Chesapeake Bay, but doesn't hold for the middle region of the Bay, where abundance is low even though food availability is high. Here, it would appear that hypoxic stress affects the ability of the benthic community to utilize the available food resource (Hagy 2002). These impacts can include mass mortality events in summer in sediments below the pycnocline that are starved of oxygen. Such events can wipe out the recruitment of opportunistic species in spring and severely limit their re-colonization in autumn, when normal oxygen conditions return (Holland et al. 1987).

The shallower waters, which are normally replete in oxygen, can also be directly and indirectly damaged by the presence of hypoxia in the deeper waters of the main channel. Episodic wind-driven movements and shoaling of the pycnocline can cause deep hypoxic channel water to infringe on the adjacent shoals, resulting in intermittent oxygen shocks. Such short-lived events can result in the mass mortality of species such as fish and crustaceans associated with shallow habitats. On one such occasion, dead and moribund fish and blue crabs were observed along a 20 km stretch of the western shore following the upwelling of anoxic water onto the western flank of the Bay in late summer (Malone et al. 1986). There is also evidence these communities can be impacted indirectly by increased predation from predators displaced from the hypoxic waters of the channel (Kemp and Boynton, 1981).

In terms of impact on fisheries the picture is complex. The presence of hypoxia in the Bay has directly affected the blue crab (*Callinectes sapidus*), which is one of the most economically important species in the Bay. The disappearance of crabs in deep waters below 4–6 m, death of crabs in pots, and the shoaling of crabs in summer have all been observed since the 1950s (Officer et al. 1984). Hypoxia and water quality deterioration affect the crabs in two main ways, first by reducing their food resource and second by altering the quality of habitats used by young crabs for shelter and foraging. It has been estimated that the hypoxic zone in Chesapeake Bay prevents the growth of about 75,000 t of clams and worms every year, which is estimated to be enough to feed about half the annual commercial harvest of blue crabs (Diaz and Rosenberg 2008). The total blue crab population in the Bay plummeted from 791 million in 1990 to 260 million in 2007. In contrast, the catch of pelagic species increased. For example, landings of Atlantic menhaden (*Brevoortia tyrannus*), a surface filter feeder, increased, probably in response to nutrient enrichment and increased primary and secondary productivity (Luo et al. 2001). The increase in pelagic species has, to a certain extent, been at the expense of demersal fish species.

The Chesapeake Bay Watershed Agreement of 2014 (which built upon the earlier Chesapeake Bay Program) contains five themes, consisting of 10 goals, to track progress with the restoration and protection of the Bay's watershed. It builds upon the earlier Chesapeake Bay Program established in 1983 but focuses on the watershed of the Bay rather than just the Bay. The themes cover protection and restoration of species and habitats, clean water, land conservation, community engagement and climate change. The success of each goal is measured using a set of outcomes and indicators that are reported upon on a regular basis. For example, under the theme of abundant life (protection and restoration of species and habitats) there are two goals centred on sustainable fisheries and restoring and protecting habitats to support fish and other wildlife. Under the goal of sustainable fisheries there is a target to increase the number of female blue crabs in the Bay (215 million individuals by 2025) and under the habitats goals there is a target to increase the area of the Bay covered in submerged aquatic vegetation. In 2018, the Chesapeake Bay Program Partners reported positive trends across several of these indicators. This included exceeding the target for adult female blue crabs, with estimates indicating that the Bay was home to 250 million crabs. Indicators also showed an increase in the coverage of submerged aquatic vegetation (mostly seagrass), with nearly 105,000 acres present in 2017, exceeding the 2017 interim target by 15,000 acres and on course to reach the target of 130,000 acres by 2025.

Other positive trends included computer models showing a reduction of 9% in nitrogen and sediment loads and a 20% reduction in phosphorus loads. Measures were put into place to achieve an estimated 33% reduction in nitrogen, 81% reduction in phosphorus and 57% reduction in sediment loads to the Bay. Furthermore, between 2014 and 2016, 40% of the Bay and its tributaries met their required water quality standards, the highest level of compliance achieved since assessments began in 1985. And it appears that reductions in nutrient loading to the Bay are starting to influence the extent of the summer hypoxic zone, particularly in the lower part of the Bay. Analysis of a 30-year time series indicates that the summer dead zone in the lower Bay is breaking up earlier because of the replenishment of oxygen. There are also indications that the improvement in oxygen conditions has led to a reduction in the amount of ammonium in bottom waters. Increases in the rate of nitrification appear to be converting more ammonium into nitrate, which is then removed from the Bay to the atmosphere in the process of denitrification (Testa et al. 2018).

In 2018, however, a wet spring with higher spring flows and nitrogen loading from the Susquehanna and Potomac and prolonged settled weather in summer led to a much higher volume of hypoxic water of just over 4 cubic kilometres in the main channel of the Bay.

7.4 Black Sea

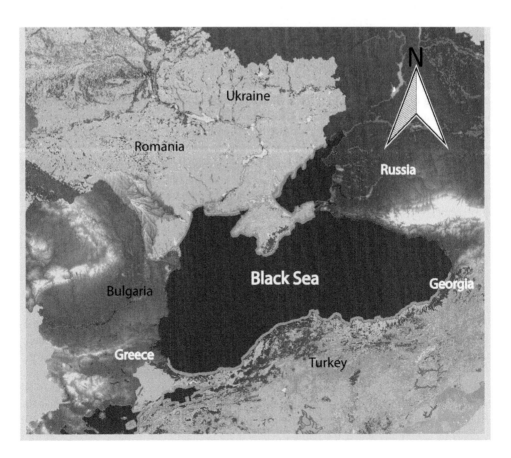

The Black Sea is nearly entirely enclosed, apart from a narrow connection through the Bosphorus straits to the smaller Sea of Marmara, which in turn is connected to the Aegean and Mediterranean Seas. It receives freshwater inputs from two of Europe's largest rivers, the Danube and Dnieper, whose combined watershed covers an area of 1.3 million km^2. The deeper waters of the Black Sea are naturally anoxic below depths of 100 m due to the presence of a permanent pycnocline that separates the higher-salinity waters of Mediterranean origin from a less saline surface layer influenced by freshwater inputs. This represents the world's largest natural anoxic basin and has been in place since the Black Sea was formed 7,500 years ago.

On the northwestern continental shelf, historical oxygen concentrations have been far more variable particularly in recent decades. Bottom water hypoxia covering an area of 3,500 km^2 was first recorded in the summer of 1973 and by 1978 the size of the hypoxic zone had increased by an order of magnitude to 30,000 km^2 (Zaitsev 1992). By 1989 the hypoxic zone covered most of the northwestern shelf (40,000 km^2) and reports of huge quantities of dead and rotting benthic organisms being washed up on the shores of Romania and Ukraine became common.

Prior to this, the bottom waters of the northwestern shelf had been dominated by massive swaths of the red alga *Phyllophora nervosa*, which supported rich communities of invertebrates and fish species (Zaitsev 1992). Characteristic also of this time was the presence of large populations of mussels, such as *Mytilus galloprovincialis* at shallower depths transitioning to *Modiolus phaseolinus* biocoenoses at greater depths. The filtering capacity of these mussel populations played an important role in regulating the biomass of surface phytoplankton blooms.

Intensification of agriculture and increased use of phosphate-based detergents in the countries surrounding the Black Sea led to a significant increase in nutrient inputs. Between 1950 and 1990, nitrogen and phosphorus loads from the Danube catchment more than doubled (from 393,000 t to 923,000 t for N and from 41,000 t to 110,000 t for P (Behrendt et al. 2005). This led to an increase in phytoplankton biomass and a reduction in water clarity and increase in light limitation, which led to the disappearance of the macroalgae mats. This further depressed oxygen conditions as the net oxygen produced by the alga was now lost from the system. As phytoplankton blooms increased, the mussel population was unable to filter the extra organic matter from the water so it sank to the seafloor, creating further oxygen demand and the development of bottom water hypoxia. As conditions worsened, anoxia developed, leading to the release of phosphate and ammonium from the sediments and further reinforcing the eutrophication process. Eventually, conditions became so bad that the mussels stopped filtering material from the water and relied instead on internal nutrition for survival. As hypoxic conditions persisted for longer periods of time the benthic system dominated by mussels declined and eventually collapsed (Mee et al. 2005).

The effect of hypoxia and collapse of the benthic system on the broader Black Sea ecosystem has been more difficult to untangle. In the 1950s the pelagic system was dominated by large predatory fish such as tuna, bonito and horse mackerel. These species exerted a top-down control on smaller planktonivorous fish such as anchovy, which in turn had a cascading effect on the abundance of lower trophic levels such as zooplankton and phytoplankton. Top-down control by plantivorous species reduces zooplankton abundance, which in turn leads to an increase in phytoplankton biomass. In the early 1960s the abundance of top predatory fish declined because of overfishing. This released the pressure of predation on plantivorous species, which rapidly increased during the 1960s. In turn this led to a collapse in zooplankton abundance and increase in phytoplankton biomass from the mid-1970s, and a consequent increase in oxygen demand caused by respiration of the surplus algal biomass. The increase in oxygen demand driven by changes in pelagic-level interactions added to the demand coming from enhanced primary productivity caused by nutrient enrichment in the 1970s and 1980s.

Further fishing down the food chain in the late 1970s and during most of the 1980s reduced the abundance of plantivorous fish. However, this time the response of zooplankton was far more variable. It was thought, in part, to be attributable to the massive increase in the early 1980s in the abundance of the native jellyfish *Aurelia aurita* due to increased biological production and the removal of its main predator (Arai 2001). Also common at this time were massive red-tide blooms of the heterotrophic dinoflagellate *Noctiluca scintillans*, a voracious consumer of phytoplankton, bacteria, detritus, eggs and larval stages of copepods. Further pressure was exerted on zooplankton abundance with the arrival in the 1980s and then prolific expansion in the late 1980s and early 1990s of the non-native introduced jellyfish *Mnemiopsis leidyi* (ctenophore). *Noctiluca* and *Mnemiopsis* are described as dead-end trophic species. Both outcompete native zooplankton species for resources and attain large biomass. But due to the absence of suitable predators most of this biomass is diverted to the microbial loop or to other gelatinous species rather than higher trophic levels. Edible zooplankton biomass on the northwestern shelf, which would have otherwise been utilized by higher trophic levels, declined by three- to four-fold most likely because of predation by *M. leidyi* and other gelatinous species (Oguz 2017).

The explosion in the biomass of *Mnemiopsis* in the late 1980s has also been linked to the collapse of the anchovy fishery in 1989/1990. Prior to this the population of anchovy (*Engraulis encrasicolus*) had experienced a significant increase from a standing stock of approximately 300,000 t in the 1960s to 1.5 million t in the 1970s. This increase is believed to have been influenced by the bottom-up increase in nutrient-driven biological production (Humborg et al. 1997, Oguz and Gilbert 2007) and the reduction in top-down pressures due to overfishing of top predatory fish (Daskalov et al. 2007, Oguz 2007).

There are two main theories proposed to explain the cause of the anchovy *Mnemiopsis* shift. In one, the role of overfishing is invoked to explain the collapse, and the diversion of food resources previously consumed by anchovy to *Mnemiopsis*. In the other, the ability of *Mnemiopsis* to outcompete anchovy for food resources and to directly prey on anchovy eggs and larvae is suggested as the main cause of the fishery collapse and the explosive increase in the population of the ctenophore. More recent ecosystem modelling has suggested that neither overfishing nor competition for food, nor predation on anchovy eggs and larvae can explain the simultaneous collapse of the anchovy fishery and the population explosion in *Mnemiopsis* (Oguz et al. 2008). What appears to have been critical is the ability of the ctenophore to outcompete anchovy larvae, and for that matter its gelatinous competitor *Aurelia*, for food resources. And those food resources, in terms of zooplankton biomass, had increased sharply because of nutrient-fuelled higher primary and secondary production in the 1980s. The greater abundance and predation by *Mnemiopsis* put further pressure on anchovy recruitment, and this together with overfishing led to the collapse of the fishery. Furthermore, favourable climatic conditions at the time promoted the development of the ctenophore's explosive growth in 1989 and 1990 (Oguz et al. 2008). The *Mnemiopsis* population would peak again in 1994/1995 but would decline because of predation by another introduced ctenophore species, *Beroe ovata*, during 1998 (Vinogradov et al. 2000).

The preceding sections have given some indication of the level of complexity that exists in ecosystems and the relative influence of bottom-up (nutrient enrichment) and top-down (predation) controls in determining how ecosystems function. To fully understand the damage caused by nutrient over-enrichment and the development of hypoxia, a comprehensive picture of energy flows through marine food webs is needed. This knowledge can then be used to predict how disturbances caused by nutrient enrichment and other pressure such as overfishing, for example, can interact antagonistically or synergistically to affect the structure and functioning of such systems.

The poor ecological state of the Black Sea prompted its six bordering countries (Bulgaria, Georgia, Romania, Russia, Turkey, and Ukraine) to sign the Bucharest Convention for the Protection of the Black Sea in 1992 (it entered into force in 1994). This was followed by the Black Sea Action Plan, which was launched in 1996. However, it would be the political and economic upheaval following the collapse of the former Soviet Union that would have the most dramatic impacts on hypoxia in the Black Sea. Changes in farm subsidies and supports led to a dramatic decline in the discharge of nutrients to the Black Sea, with phosphate loads almost returning to 1960s levels (Mee 2001). Remarkably, for the first time in decades, hypoxia was almost absent from the northwestern shelf of the Black Sea in 1996 and receded to an area of less than 1000 km² in 1999 (Mee 2001). In 2001, however, late rainfall and higher temperatures triggered a new large-scale hypoxic event (Mee et al. 2005), highlighting the sensitivity of this enclosed sea to variations in nutrient inputs.

The Vertical Structure of Ocean Minimum Zones

The vertical structure of ocean minimum zones can be divided into four layers: a surface well-oxygenated layer; below this a steep, thin oxycline, where oxygen concentrations decline sharply due to increasing biological oxygen consumption and reduction in the level of vertical oxygen exchange; a core where oxygen levels are at their minimum concentration (< 20 µM) due to maximum biological consumption; and finally a lower layer where oxygen concentrations gradually increase as the rate of biological oxygen consumption declines exponentially with depth. The vertical profile of oxygen in the Arabian Sea OMZ is shown in Fig. 6.

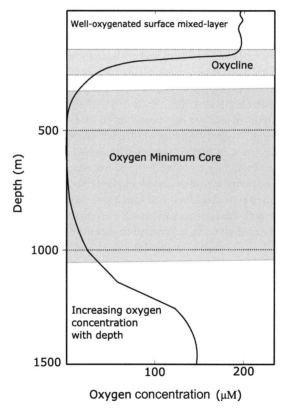

Fig. 6. Vertical profile of oxygen in the Arabian Sea oxygen minimum zone.

Color version at the end of the book

7.5 Arabian Sea oxygen minimum zone and the Pakistan margin

The oxygen minimum zone in the Arabian Sea covers an area of 2.5 million km² and has a vertical extent of 2980 m (± 680 m). The core, where oxygen concentrations are less than 20 µM, has a vertical extent of 760 m (± 340 m) and is the thickest of all the oxygen minimum cores identified to date and second in volume only to the eastern North Pacific OMZ. Much of what is known about the impact of OMZs on biology relates to areas where the OMZ intercepts continental shelves and shelf margins. In the case of the Arabian Sea OMZ this includes the shelf waters and continental margins off Pakistan, western India and Oman. These waters can therefore experience prolonged periods of exposure to waters depleted in oxygen ($< 20 \ \mu M$) and sediments in these regions typically consist of laminated muds rich in organic matter (Schmaljohann et al. 2001).

Along the Pakistan Margin, the distribution of macrofauna above the ocean minimum oxycline (> 270 m) is typical of other comparable shelf regions with very strong bioturbation (burrows a few centimetres apart) and an abundance of benthic macroinvertebrate life (molluscs, bryosoans, pennatulids). In the oxycline and in the upper part of the core (270 to 430 m water depth), the presence of bioturbation and macroinvertebrate life declines and disappears completely in the central part of the core (430 to 940 water depth) where oxygen conditions are at their minimum concentration. This layer is often covered in mat-forming filamentous sulphide-oxidizing bacteria (*Thioploca* and *Beggiatoa*) (Schmaljohann et al. 2001), which can be dense in foraminifera and metazoan meiofauna such as nematodes and polychaetes adapted to survive and indeed thrive in this environment (Levin et al. 2000). Observations indicate that total densities of metazoan meiofauna are never reduced within OMZs. Indeed, meiofaunal densities along the Peru Margin OMZ were shown to reach maximum values at lowest oxygen concentrations, presumably due to abundant particulate food and/or reduced predation intensity (Neira et al. 2001). As oxygen concentrations start to increase in the lower layer of the OMZ, bioturbations increase and molluscs and crustaceans and other macroinvertebrate species begin to appear again. Below the OMZ (below 1100 m water depth) the benthic macroinvertebrate community is rich in species (sponges, gorgonians, crabs, sea urchins, pennatulids, worms, serpulid worms) and abundant in fish (von Rad 1996).

The vertical distribution of marine organisms in an OMZ is determined by their ability to withstand varying degrees of oxygen depletion. Different adaptations allow the most efficient use of low levels of oxygen present in hypoxic water (Childress and Seibel 1998). This can include morphological adaptations including thin and long bodies that improve the efficiency of oxygen transfer to body tissues. And meiofauna communities in OMZ sediments are often dominated by large populations of nematode and polychaete worms. Large gill surfaces allow some organisms to pass large volumes of water through their bodies and extract enough oxygen to sustain aerobic respiration. Other morphological features that increase ventilation include elongated, proliferated and enlarged branchiae making the most efficient use of available oxygen. Such features are commonly observed in polychaete species inhabiting low-oxygen environments (Lamont and Gage 2000). Another adaptation is the presence of blood pigments such as haemoglobin (typically absent from most invertebrate groups) to transport oxygen more efficiently in the blood. Haemoglobin has been observed in a number of bivalves from the Oman margin OMZ (150–1150 m water depth), including the mytilid mussel *Amygdalum anoxicolum* (Oliver 2001), the clams *Pitar sewelli* (Veneroidea), *Indocrassatella indica* (Crassatelloidea) and *Lucinoma* sp. (Lucinoidea), and the scallop *Propeamussium* cf. *alcocki* (Pectinoidea).

Biochemical adaptations to low-oxygen environments include supplementing the aerobic production of ATP with anaerobic production of ATP. For example, some species of polychaetes have an increased number of pyruvate oxidoreductases, enzymes that help to better regulate pyruvate consumption during the transition to anaerobic respiration (González and Quiñones 2000). Some nematodes have developed endosymbiotic relations with sulphide-oxidizing bacteria that reside in the gut of their host (observed in mouthless nematodes). In addition to providing nutrition, the presence of these bacteria is believed to help reduce the toxicity of this environment by removing hydrogen sulphide. In return, the bacteria are thought to benefit from the vertical mobility of the worms, increasing their exposure to sulphide and oxygen (if present) or other compounds such as nitrate (Ott 2004).

One of the main advantages that pelagic species have over bottom-dwelling benthic species is their ability to vertically migrate to shallower, well-oxygenated waters at night. Most of the metazoan species inhabiting OMZs undertake daily vertical migrations. Nevertheless, they still need to use a variety of adaptations, similar in many ways to those used by their benthic counterparts, to survive in low-oxygen environments. Some reduce their energy requirements by suppressing their metabolism; others supplement aerobic metabolic ATP production with anaerobic ATP production; still others have adaptations to enhance the extraction of oxygen from their environment (e.g., increased gill surface area, higher ventilation) and its internal transport (increased heart rate, increased blood pigment levels). Many organisms use a combination of these strategies to survive their daily migration to OMZs. One of the main ways that fish inhabiting these zones reduce their energy consumption is to reduce their motility. Many species of fish have been observed to be "passively drifting" during their day-time visit to the OMZ but to be "actively swimming" when observed at night in shallower well-oxygenated water (Barham 1971). Because locomotion is so closely linked to energy consumption many of the top-performing oceanic predators such as tuna and swordfish avoid low-oxygen environments with concentrations less than 150 mmol L^{-1} (Brill 1984). Similarly, many species of shark such as mako sharks and white sharks avoid areas where the oxygen concentration is less than 50 mmol L^{-1}, and even the great white shark limits the depths to which it descends in hypoxic waters (Vetter et al. 2008, Nasby-Lucas et al. 2009). There are some exceptions to this general observation about the limited vertical migration of top predators in low-oxygen oceanic environments. For example, the big-eye tuna (*Thunnus obesus*) has developed adaptations that allow it to foray to depths with oxygen concentrations of approximately 40 mmol L^{-1}. It is able to do this because of the presence of higher levels of respiratory pigments with a high affinity for oxygen. Other organisms rely on supplementing their aerobic metabolism with the anaerobic production of ATP. However, as mentioned previously, this is an inefficient way to produce energy, which in effect means that energy needs to be consumed sparingly. Another disadvantage of anaerobic metabolism is the production of deleterious end-products such as lactate, octopine and ethanol. Despite these disadvantages, there is evidence for elevated anaerobic metabolic activity as well as the build-up of anaerobic end-products in organisms undertaking daily vertical migrations in pronounced OMZs (Seibel 2011).

A number of very important biogeochemical processes occur in OMZs. The microbial processes of nitrification and denitrification have already been described. Ammamox, the anaerobic oxidation of ammonium to nitrogen gas, is also an important process in OMZ environments (Lam et al. 2009, Kuypers et al. 2005). The suboxic (almost oxygen-free) waters of the Arabian Sea OMZ core provide ideal conditions for the denitrification of nitrate into gaseous nitrogen (e.g., nitrogen gas and nitrous oxide). When fixed nitrogen (i.e., nitrate) is denitrified into gaseous nitrogen it can be lost to the atmosphere, leading to nitrogen deficits and ultimately constraining primary production in the sunlit surface layers. In the Arabian Sea, denitrification is a huge sink of fixed nitrogen, removing an estimated 60 tg N yr^{-1}. This represents over a third of water column denitrification for the global ocean estimated at 150 tg N yr^{-1}, most of which occurs in ocean OMZs (135 tg N yr^{-1}) (Codispoti et al. 2001). It also represents about 10% of total (sediments and water column) oceanic denitrification, which has been estimated at approximately 450 tg N yr^{-1}.

This loss in fixed nitrogen equates to a potential reduction in oceanic new primary production of about 1.3 billion t of carbon per year (about 2% of total global ocean net primary production). This represents a significant reduction in the oceanic biological pump's ability to sequester atmospheric carbon dioxide. During periods when denitrification is removing fixed N from the ocean, the pump can be reversed, and ocean biology can add rather than subtract carbon dioxide to the atmosphere (Codispoti et al. 1989). What is most notable is that these globally important biogeochemical processes occur in relatively small volumes of ocean water. For example, the volume of the Arabian Sea OMZ where denitrification happens represents about 0.1% of the global ocean volume and yet accounts for about 10% of oceanic denitrification (Codispoti et al. 2001). Any change to the oxygen content of these zones including an increase in their volume could have very significant effects on oceanic denitrification and nitrogen loss and the ability of the ocean to sequester atmospheric carbon dioxide.

Another concern is the greater production of nitrous oxide, which is a potent greenhouse gas. The Arabian Sea is known to have some of the highest subsurface nitrous oxide concentrations (90 nmol kg^{-1}) of any of the world's ocean basins (Freing et al. 2012). Most of the nitrous oxide produced in the suboxic waters of the OMZ is produced and consumed by denitrification; nitrous oxide is an intermediary product of denitrification but is also converted to nitrogen gas if the denitrification pathway is completed. However, in some cases, excess amounts of nitrous oxide can be released to the atmosphere before being converted to nitrogen gas. For example, in the eastern South Pacific, OMZ supersaturated levels of nitrous oxide drive a massive efflux of 0.2–0.9 tg of nitrogen emitted as nitrous oxide per year, equivalent to 5–22% of previous estimates of global marine nitrous oxide emissions (Arevalo-Martínez et al. 2015). Data collected from vertical profiles suggests that coupled nitrification-denitrification in the upper OMZ led to the production of nitrous oxide that was subsequently brought to the surface by upwelling. These observations suggest there may be hotspots of nitrous oxide production associated with OMZs and that further ocean warming that will increase the volume of OMZs, as well as nitrous oxide supersaturation, could lead to a substantial increase in the flux of this potent greenhouse gas to the atmosphere.

8. Conclusions

The insights from the five case studies presented in this chapter have shown how deoxygenation can damage not only the abundance of marine species but the composition and distribution of biological communities, the structure of marine food webs and the biogeochemical cycling of key nutrients and greenhouse gases. From a human perspective, the damage caused by the spread of coastal and oceanic low-oxygen zones is degrading the ecosystem services and regulatory functions that these environments provide, from the cycling and retention of nutrients to the provision of food and the removal of atmospheric carbon dioxide. The continued loss of oxygen from the world's oceans and coastal areas is putting further stress on ecosystems that are already stressed from rising sea temperatures, ocean acidification, overfishing and habitat loss. Even in oxygen minimum zones, species that have evolved biological and behavioural mechanisms to deal with low-oxygen environments may struggle to tolerate the even lower oxygen concentrations predicted to occur under different climate change scenarios.

In many parts of the world there is a trend of increasing low-oxygen dead zones in our coastal and ocean environments. The rate is also alarming, with the number of dead zones increasing 10-fold since the 1950s. And it seems likely that the greater use of artificial and organic fertilizers to feed a human population approaching 10 billion people by the end of the century will continue to put even more pressure on the marine environment. There is also the added complication of climate change. As we have seen from projections in the Gulf of Mexico, higher water temperatures, reduced oxygen solubility and greater nutrient loadings due to extreme and episodic weather events will delay recovery times. In the case of the Gulf, recovery may be delayed by many decades. Another study from the central North Sea predicts that dissolved oxygen concentrations in bottom waters of the oyster grounds may decrease by 11.5% by 2100 because of climate-induced changes in the stratification and oxygen solubility.

The threat of coastal and ocean hypoxia must be addressed on two fronts. First, the use and subsequent loss of nutrients from food production systems and from the burning of fossil fuels must be reduced, and second, the economies of the world must reduce their emissions of greenhouse gases and their dependency on fossil fuels.

We have seen how important regional action plans have been in trying to solve the issue of nutrient over-enrichment. And we can point to examples where areas have recovered, or are recovering (e.g., Chesapeake Bay), from hypoxia due to the implementation of regional action plans. The Thames estuary is often cited as an example of the benefits of improved water quality. Having been labelled "biologically dead" in the 1950s, the estuary is now home to over 100 species of fish and 400 species of invertebrates. Most of these plans set targets to reduce nutrient inputs from both point sources and more widely from diffuse catchment sources. In formulating these plans, it is critical that the nature of the problem is fully understood and that the contribution from both point and diffuse sources is fully integrated in any plan that seeks to reduce nutrient loss from a catchment. The success of such plans will ultimately depend on changes in agricultural practices, the use of new technologies (e.g., improved fertilizers, more targeted nutrient delivery systems) and the cultivation of crops and grasses that have lower demands for fertilization. The environmental pressures associated with modern food systems may also need to be addressed by changing diets and reducing food waste.

Ocean deoxygenation is just one of the many environmental issues that is either indirectly or directly affected by climate change. A global effort will be required to reduce the widespread impact of climate change on the natural environment, food production and human safety. That global effort most recently manifested itself in the form of the Paris Agreement, a commitment (signed by 195 countries) to keep global temperature rise this century to well below 2 degrees Celsius and to pursue efforts to limit the temperature increase to 1.5 degrees Celsius. Countries also committed to setting aside $100 billion annually to fund climate change mitigation efforts in developing countries. Each country committed to produce plans every five years with stepped-up efforts to further reduce greenhouse gas emissions. To sufficiently decarbonize their economies, countries will need to put in place new technologies that will become central in the context of carbon pricing, which will incentivize countries, businesses and individuals to evolve in a way that reduces the amount of carbon released into the atmosphere. Putting a price on the damage caused by carbon emissions will encourage industries to reduce their emissions, encourage innovators to invent new technologies and ensure fairer comparisons between energy produced from traditional hydrocarbon-based sources and renewable energies such as wind, wave and solar power.

Many of the measures put in place to reduce nutrient enrichment may have joint benefits that address both issues. For example, restoration of coastal wetlands could help with nutrient retention as well as act as a sink for carbon. Improvements in oxygen conditions also improve the resilience of ecosystems to the additional stresses caused by climate change, such as rising sea temperatures and ocean acidification, and by other human activities that damage the marine environment.

The benefits of dealing with the problem of oxygen loss from the marine environment are obvious but a more integrated approach is needed to reverse current trends.

References

Arai, M.N. 2001. Pelagic coelenterates and eutrophication: a review. Hydrobiologia 451: 69–87.

Arevalo-Martínez, D.L., A. Kock, C.R. Löscher, R.A. Schmitz and H.W. Bange. 2015. Massive nitrous oxide emissions from the tropical South Pacific Ocean. Nature Geoscience 8(7): 530.

Baden, S.P., L. Pihl and R. Rosenburg. 1990. Effects of oxygen depletion on the ecology, blood physiology and fishery of the Norway lobster, *Nephrops norvegicus*. Marine Ecology Progress Series 67: 151–155.

Barham, E.G. 1971. Deep-sea fishes. Lethargy and vertical orientation. *In*: Proceedings of an International Symposium on Biological Sound Scattering in the Ocean (pp. 100–118). US GOVT. PRINT OFFICE.

Baustian, M.M. and N.N. Rabalais. 2009. Seasonal composition of benthic macroinfauna exposed to hypoxia in the northern Gulf of Mexico. Estuaries and Coasts 32(5): 975–983.

Bergström, S., H. Alexandersson, B. Carlsson, W. Josefsson, K.G. Karlsson and G. Westring. 2001. Climate and hydrology of the Baltic Basin. *In*: A Systems Analysis of the Baltic Sea. Springer, Berlin, Heidelberg, pp. 75–112.

Bouwman, A.F., G. Van Drecht, J.M. Knoop, A.H.W. Beusen and C.R. Meinardi. 2005. Exploring changes in river nitrogen export to the world's oceans. Global Biogeochemical Cycles 19(1): 1–14.

Behrendt, H., J. van Gils, H. Schreiber and M. Zessnex. 2005. Arch. Hydrobiol. Suppl. 158/1–2, Large Rivers 16(1-2): 221–247.

Bratkovich, A., S.P. Dinnel and D.A. Goolsby. 1994. Variability and prediction of freshwater and nitrate fluxes for the Louisiana-Texas shelf: Mississippi and Atchafalaya River source functions. Estuaries 17(4): 766–778.

Breitburg, D.L. 1992. Episodic hypoxia in Chesapeake Bay: Interacting effects of recruitment, behaviour, and physical disturbance. Ecological Monograph 62: 525–546.

Breitburg, D.L., T. Loher, C.A. Pacey and A. Gerstein. 1997. Varying effects of low dissolved oxygen on trophic interactions in an estuarine food web. Ecological Monographs 67(4): 489–507.

Breitburg, D.L. 2002. Effects of hypoxia, and the balance between hypoxia and enrichment, on coastal fishes and fisheries. Estuaries 25: 767–781.

Breitburg, D., L.A. Levin, A. Oschlies, M. Grégoire, F.P. Chavez, D.J. Conley et al. 2018. Declining oxygen in the global ocean and coastal waters. Science 359(6371): 7240.

Brill, R.W. 1994. A review of temperature and oxygen tolerance studies of tunas pertinent to fisheries oceanography, movement models and stock assessments. Fisheries Oceanography 3(3): 204–216.

Bograd, S.J., C.G. Castro, E. Di Lorenzo, D.M. Palacios, H. Bailey, W. Gilly et al. 2008. Oxygen declines and the shoaling of the hypoxic boundary in the California Current. Geophysical Research Letters 35: L12607.

Carstensen, J., J.H. Andersen, B.G. Gustafsson and D.J. Conley. 2014. Deoxygenation of the Baltic Sea during the last century. Proceedings of the National Academy of Sciences 111(15): 5628–5633.

Catling, D.C. and M.W. Claire. 2005. How Earth's atmosphere evolved to an oxic state: A status report. Earth and Planetary Science Letters 237(1-2): 1–20.

Childress, J.J. and B.A. Seibel. 1998. Life at stable low oxygen levels: Adaptations of animals to oceanic oxygen minimum layers. Journal of Experimental Biology 201(8): 1223–1232.

Codispoti, L.A., J.A. Brandes, J.P. Christensen, A.H. Devol, S.W.A. Naqvi, H.W. Paerl et al. 2001. The oceanic fixed nitrogen and nitrous oxide budgets: Moving targets as we enter the anthropocene? Science Marine 65(2): 85–105.

Conley, D.J., C. Humborg, L. Rahm, O.P. Savchuk and F. Wulff. 2002. Hypoxia in the Baltic Sea and basin-scale changes in phosphorus biogeochemistry. Environmental Science & Technology 36(24): 5315–5320.

Craig, K. 2012. Aggregation on the edge: Effects of hypoxia avoidance on the spatial distribution of brown shrimp and demersal fishes in the Northern Gulf of Mexico. Marine Ecology Progress Series 445: 75–95.

Craig, J.K. and L.B. Crowder. 2005. Hypoxia-induced habitat shifts and energetic consequences in Atlantic croaker and brown shrimp on the Gulf of Mexico shelf. Marine Ecology Progress Series 294: 79–94.

Daskalov, G.M., A.N. Grishin, S. Rodianov and V. Mihneva. 2007. Trophic cascades triggered by overfishing reveal possible mechanisms of ecosystem regime shifts. Proc. Natl Acad. Sci. 104(25): 10518–10523.

Diaz, R.J. and R. Rosenberg. 1995. Marine benthic hypoxia: A review of its ecological effects and the behavioural responses of benthic macrofauna. Oceanography and Marine Biology: An Annual Review 33: 245–303.

Diaz, R.J. and R. Rosenberg. 2008. Spreading dead zones and consequences for marine ecosystems. Science 321: 926–929.

Diaz, R.J., H. Eriksson-Hägg and R. Rosenberg. 2013. Hypoxia. pp. 67–97. *In*: Noone, K.J., U.R. Sumaila and R.J. Diaz (eds.). Managing Ocean Environments in a Changing Climate. Elsevier, Amsterdam, Netherlands.

Dunn, D.D. 1996. Trends in Nutrient Inflows to the Gulf of Mexico from Streams Draining the Conterminous United States 1972–1993. U.S. Geological Survey, Water-Resources Investigations Report 96-4113. Prepared in cooperation with the U.S. Environmental Protection Agency, Gulf of Mexico Program, Nutrient Enrichment Issue Committee, U.S. Geological Survey, Austin, Texas.

Elmgren, R. 1989. Man's impact on the ecosystem of the Baltic Sea: Energy flows today and at the turn of the century. Ambio 18: 326–332.

FAO. 2017. The future of food and agriculture – Trends and challenges. Rome, Italy.

FAOStat. 2018—http://www.fao.org/faostat/en/#home.

Freing, A., D.W. Wallace and H.W. Bange. 2012. Global oceanic production of nitrous oxide. Philosophical Transactions of the Royal Society of London B: Biological Sciences 367(1593): 1245–1255.

Galloway, J.N., W.H. Schlesinger, H. Levy, A. Michaels and J.L. Schnoor. 1995. Nitrogen-fixation—Anthropogenic enhancement-environmental response. Glob. Biogeochem. Cycle 9: 235–252.

Gallegos, C.L. and T.E. Jordan. 2002. Impact of the spring 2000 phytoplankton bloom in Chesapeake Bay on optical properties and light penetration in the Rhode River, Maryland. Estuaries 25: 508–518.

Garcia, H., A. Cruzado, L. Gordon and J. Escanez. 1998. Decadal-scale chemical variability in the subtropical North Atlantic deduced from nutrient and oxygen data. J. Geophys. Res. 103: 2817–2830.

Gerlach, S.A. 1994. Oxygen conditions improve when the salinity in the Baltic Sea decreases. Marine Pollution Bulletin 28(7): 413–416.

González, R.R. and R.A. Quiñones. 2000. Pyruvate oxidoreductases involved in glycolytic anaerobic metabolism of polychaetes from the continental shelf off central-south Chile. Estuarine, Coastal and Shelf Science 51(4): 507–519.

Goolsby, D.A., W.A. Battaglin, G.B. Lawrence, R.S. Artz, B.T. Aulenbach, R.P. Hooper et al. 1999. Flux and Sources of Nutrients in the Mississippi-Atchafalaya River Basin, Topic 3 Report for the Integrated Assessment of Hypoxia in the Gulf of Mexico. National Oceanic and Atmospheric Administration Coastal Ocean Program Decision Analysis Series No. 17. National Oceanic and Atmospheric Administration Coastal Ocean Program, Silver Spring, Maryland.

Goolsby, D.A. and W.A. Battaglin. 2001. Long-term changes in concentrations and flux of nitrogen in the Mississippi River Basin, USA. Hydrologic Processes 15: 1209–1226.

Gray, J.S., R. Shiu-sun Wu and Or. Ying Ying. 2002. Effects of hypoxia and organic enrichment on the coastal marine environment. Marine Ecology Progress Series 238: 249–279.

Gustafsson, B.G., F. Schenk, T. Blenckner, K. Eilola, H.M. Meier, B. Müller-Karulis et al. 2012. Reconstructing the development of Baltic Sea eutrophication 1850–2006. Ambio 41(6): 534–548.

Hagy, J.D. 2002. Eutrophication, hypoxia and trophic transfer efficiency in Chesapeake Bay. PhD dissertation, University of Maryland, College Park, MD.

Hagy, J.D., W.R. Boynton, C.W. Keefe and K.V. Wood. 2004. Hypoxia in Chesapeake Bay, 1950–2001: long-term change in relation to nutrient loading and river flow. Estuaries 27(4): 634–658.

Helly, J.J. and L.A. Levin. 2004. Global distribution of naturally occurring marine hypoxia on continental margins. Deep Sea Research Part I: Oceanographic Research Papers 51(9): 1159–1168.

Helm, K.P., N.L. Bindoff and J.A. Church. 2011. Observed decreases in oxygen content of the global ocean. Geophys. Res. Lett. 38: L23602. Doi: 10.1029/2011GL049513.

Helsinki Commission for Protection of the Baltic Sea. 2007. HELCOM Baltic Sea Action Plan. Available at www. helcom.fi/Documents/Baltic%20sea%20action%20plan/BSAP_Final.pdf. Accessed March 21, 2014.

Hochachka, P.W. and G.N. Somero. 2002. Biochemical adaptation: mechanism and process in physiological evolution. Oxford University Press.

Holland, A.F., A.T. Shaughnessy and M.H. Hiegel. 1987. Long-term variation in mesohaline Chesapeake Bay macrobenthos: Spatial and temporal patterns. Estuaries 10: 227–245.

Howarth, R.W., G. Billen, D. Swaney, A. Townsend, N. Jaworski, K. Lajtha et al. 1996. Regional nitrogen budgets and riverine N & P fluxes for the drainages to the North Atlantic Ocean: Natural and human influences. *In*: Nitrogen cycling in the North Atlantic Ocean and its watersheds. Springer, Dordrecht, pp. 75–139.

Humborg, C., V. Ittekkot, A. Cociasu and B.V. Bodungen. 1997. Effect of Danube River dam on Black Sea biogeochemistry and ecosystem structure. Nature 386(6623): 385.

IFA International Fertilizer Association, IFADATA. 2016. http://ifadata.fertilizer.org/ucSearch.aspx.

Isensee, K., L.A. Levin, D.L. Breitburg, M. Gregoire and V. Garcon. 2015. The Ocean is losing its breath. Ocean and Climate, Scientific notes. www.ocean-climate.org:20-28.

Justic, D., T. Legovic and L. Rottini-Sandrini. 1987. Trends in oxygen content 1911–1984 and occurrence of benthic mortality in the northern Adriatic Sea. Estuarine and Coastal Shelf Science 26: 184–189.

Karlsen, A.W., T.M. Cronin, E.S. Ishman, D.A. Willard, C.W. Holmes, M. Marot et al. 2000. Historical trends in Chesapeake Bay dissolved oxygen based on benthic Foraminifera from sediment cores. Estuaries 23: 488–508.

Kemp, W.M. and W.R. Boynton. 1981. External and internal factors regulating metabolic rates of an estuarine benthic community. Oecologia 51: 19–27.

Kemp, W.M., R. Batiuk, R. Bartleson, P. Bergstrom and 12 others. 2004. Habitat requirements for submerged aquatic vegetation in Chesapeake Bay: Water quality, light regime, and physical-chemical factors. Estuaries 27: 263–377.

Kemp, W.M., W.R. Boynton, J.E. Adolf, D.F. Boesch, W.C. Boicourt, G. Brush et al. 2005. Eutrophication of Chesapeake Bay: Historical trends and ecological interactions. Marine Ecology Progress Series 303: 1–29.

Keeling, R.F., A. Körtzinger and N. Gruber. 2010. Ocean deoxygenation in a warming world. Annu. Rev. Mar. Sci. 2: 199–229.

Kidwell, D.M., A.J. Lewitus, S. Brandt, E.B. Jewett and D.M. Mason. 2009. Ecological impacts of hypoxia on living resources. Journal of Experimental Marine Biology and Ecology 381: S1–S3.

Kuypers, M.M., G. Lavik, D. Woebken, M. Schmid, B.M. Fuchs, R. Amann et al. 2005. Massive nitrogen loss from the Benguela upwelling system through anaerobic ammonium oxidation. Proceedings of the National Academy of Sciences 102(18): 6478–6483.

Lam, P., G. Lavik, M.M. Jensen, J. van de Vossenberg, M. Schmid, D. Woebken et al. 2009. Revising the nitrogen cycle in the Peruvian oxygen minimum zone. Proceedings of the National Academy of Sciences 106(12): 4752–4757.

Lamont, P.A. and J.D. Gage. 2000. Morphological responses of macrobenthic polychaetes to low oxygen on the Oman continental slope, NW Arabian Sea. Deep Sea Research Part II: Topical Studies in Oceanography 47(1-2): 9–24.

Larsson, U. 1985. Eutrophication and the Baltic Sea: Causes and consequences. Ambio 14: 9–14.

Levin, L.A., J.D. Gage, C. Martin and P.A. Lamont. 2000. Macrobenthic community structure within and beneath the oxygen minimum zone, NW Arabian Sea. Deep Sea Research Part II: Topical Studies in Oceanography 47(1-2): 189–226.

Levin, L.A., W. Ekau, A.J. Gooday, F. Jorissen, J.J. Middelburg, S.W.A. Naqvi et al. 2009. Effects of natural and human-induced hypoxia on coastal benthos.

Long, M.C., C.A. Deutsch and T. Ito. 2016. Finding forced trends in oceanic oxygen. Glob. Biogeochem. Cycles 30: 381–397.

Luo, J., K.J. Hartman, S.B. Brandt and C.F. Cerco. 2001. A spatially explicit approach for estimating carrying capacity: an application for the Atlantic menhaden (Brevortia tyrannus) in Chesapeake Bay. Estuaries 24: 545–556.

Malone, T., W.M. Kemp, H. Ducklow, W. Boynton, J. Tuttle and R. Jonas. 1986. Lateral variation in the production and fate of phytoplankton in a partially stratified estuary. Mar. Ecol. Prog. Ser. 32: 149–160.

Mee, L.D. 2001. Eutrophication in the Black Sea and a basin-wide approach to its control. pp. 71–94. *In*: Bodungen, B. and K. Turner (eds.). Science and Integrated Coastal Management, Dahlem Workshop, Berlin.

Mee, D., J. Friedrich and M. Gomoiu. 2005. Restoring the Black Sea in times of uncertainty. Oceanography 18: 100–112.

Mississippi River/ Gulf of Mexico Watershed Nutrient Task Force. 2001. Action Plan for Reducing, Mitigating, and Controlling Hypoxia in the Northern Gulf of Mexico.

Nasby-Lucas, N., H. Dewar, C.H. Lam, K.J. Goldman and M.L. Domeier. 2009. White shark offshore habitat: A behavioral and environmental characterization of the eastern Pacific shared offshore foraging area. PloS One 4(12): e8163.

Neira, C., J. Sellanes, L.A. Levin and W.E. Arntz. 2001. Meiofaunal distributions on the Peru margin: Relationship to oxygen and organic matter availability. Deep Sea Research Part I: Oceanographic Research Papers 48(11): 2453–2472.

Newcombe, C.L. and W.A. Horne. 1938. Oxygen-poor waters in the Chesapeake Bay. Science 88: 80–81.

Newell, R.I.E. 1988. Ecological changes in Chesapeake Bay: Are they the result of overharvesting the eastern oyster (*Crassostrea virginica*)? pp. 536–546. *In*: Lynch, M.P. and E.C. Krome (eds.). Understanding the Estuary: Advances in Chesapeake Bay Research, Chesapeake Research Consortium Publication, 129.

Officer, C.B., R.B. Biggs, J.L. Taft, L.E. Cronin, M.A. Tyler and W.R. Boynton. 1984. Chesapeake Bay anoxia: Origin, development, and significance. Science 223: 22–27.

Oguz, T. 2007. Nonlinear response of Black Sea pelagic fish stocks to over-exploitation. Mar. Ecol. Prog. Ser. 345: 211–228.

Oguz, T. and D. Gilbert. 2007. Abrupt transitions of the top-down controlled Black Sea pelagic ecosystem during 1960–2000: Evidence for regime shifts under strong fishery exploitation and nutrient enrichment modulated by climate-induced variations. Deep-Sea Res. I 54: 220–242.

Oguz, T., B. Fach and B. Salihoglu. 2008. Invasion dynamics of the alien ctenophore Mnemiopsis leidyi and its impact on anchovy collapse in the Black Sea. Journal of Plankton Research 30(12): 1385–1397.

Oguz, T. 2017. Controls of multiple stressors on the black Sea fishery. Frontiers in Marine Science 4: 110. Doi: 10.3389/fmars.2017.00110.

Oliver, P.G. 2001. Functional morphology and description of a new species of Amygdalum (Mytiloidea) from the oxygen minimum zone of the Arabian Sea. Journal of Molluscan Studies 67(2): 225–241.

Ott, J. 2004. Symbioses between marine nematodes and sulfur-oxidizing chemoautotrophic bacteria. Symbiosis 36: 103–126.

Paulmier, A. and D. Ruiz-Pino. 2009. Oxygen minimum zones (OMZs) in the modern ocean. Progress in Oceanography 80(3-4): 113–128.

Pavela, J.S., J.L. Ross and M.E. Chittendun. 1983. Sharp reductions in abundance of fishes and benthic macroinvertebrates in the Gulf of Mexico off Texas associated with hypoxia. Northeast Gulf Science 6: 167–173.

Pitkänen, H., J. Lehtoranta and A. Räike. 2001. Internal nutrient fluxes counteract decreases in external load: The case of the estuarial eastern Gulf of Finland, Baltic Sea. AMBIO: A Journal of the Human Environment 30(4): 195–201.

Rabalais, N.N. and R.E. Turner. 2001. Hypoxia in the northern Gulf of Mexico: description, causes and change. pp. 1–36. *In*: Rabalais, N.N. and R.E. Turner (eds.). Coastal and Estuarine Studies, Coastal Hypoxia: Consequences for Living Resources and Ecosystems, American Geophysical Union, Washington, D.C.

Rabalais, N.N., R.E. Turner, B.K. Sen Gupta, D.F. Boesch, P. Chapman and M.C. Murrell. 2007. Characterization and long-term trends of hypoxia in the northern Gulf of Mexico: Does the science support the Action Plan? Estuaries and Coasts 30(5): 753–772.

Rothschild, B.J., J.S. Ault, P. Goulletquer and M. Heral. 1994. Decline of the Chesapeake Bay oyster population: A century of habitat destruction and overfishing. Mar. Ecol. Prog. Ser. 111: 29–39.

Schaffner, L.C., P. Jonsson, R.J. Diaz, R. Rosenberg and P. Gapcynski. 1992. Benthic communities and bioturbation history of estuarine and coastal systems: Effects of hypoxia and anoxia. Sci. Total Environ. (Suppl.): 1001–1016.

Schmaljohann, R., M. Drews, S. Walter, P. Linke, U. von Rad and J.F. Imhoff. 2001. Oxygen-minimum zone sediments in the northeastern Arabian Sea off Pakistan: a habitat for the bacterium Thioploca. Marine Ecology Progress Series 211: 27–42.

Schmidtko, S., L. Stramma and M. Visbeck. 2017. Decline in global oceanic oxygen content during the past five decades. Nature 542: 335–339. Doi: 10.1038/nature21399; pmid: 28202958; https://www.nature.com/articles/nature21399.

Seibel, B.A. 2011. Critical oxygen levels and metabolic suppression in oceanic oxygen minimum zones. Journal of Experimental Biology 214(2): 326–336.

Smil, V. 1990. Nitrogen and phosphorus. pp. 423–436. *In*: Turner, II, B.L., W.C. Clark, R.W. Kates, J.F. Richards, J.T. Mathews and W.B. Meyer (eds.). The Earth as Transformed by Human Action: Global and Regional Changes in the Biosphere Over the Past 300 Years. Cambridge University Press, Cambridge, U.K.

Stigebrandt, A. 2001. Physical oceanography of the Baltic Sea. *In*: A systems analysis of the Baltic Sea. Springer, Berlin, Heidelberg, pp. 19–74.

Stramma, L., G.C. Johnson, J. Sprintall and V. Mohrholz. 2008. Expanding oxygen-minimum zones in the tropical oceans. Science 320: 655–658.

Stramma, L., S. Schmidtko, L.A. Levin and G.C. Johnson. 2010. Ocean oxygen minima expansions and their biological impacts. Deep-Sea Res. Part I 57: 587–595. Doi: 10.1016/ j.dsr.2010.01.005.

Testa, J.M., W.M. Kemp and W.R. Boynton. 2018. Season-specific trends and linkages of nitrogen and oxygen cycles in Chesapeake Bay. Limnology and Oceanography 63(5): 2045–2064.

Thomas, P., M.S. Rahman, I.A. Khan and J.A. Kummer. 2007. Widespread endocrine disruption and reproductive impairment in an estuarine fish population exposed to seasonal hypoxia. Proceedings of the Royal Society of London B: Biological Sciences 274(1626): 2693–2702.

Thomas, P. and M.S. Rahman. 2009. Biomarkers of hypoxia exposure and reproductive function in Atlantic croaker: a review with some preliminary findings from the northern Gulf of Mexico hypoxic zone. Journal of Experimental Marine Biology and Ecology 381: S38–S50.

Thomas, P. and M.S. Rahman. 2010. Region-wide impairment of Atlantic croaker testicular development and sperm production in the northern Gulf of Mexico hypoxic dead zone. Marine Environmental Research 69: S59–S62.

Tyrrell, T. 1999. The relative influences of nitrogen and phosphorus on oceanic primary production. Nature 400(6744): 525.

Vahtera, E., D.J. Conley, B.G. Gustafsson, H. Kuosa, H. Pitkänen, O.P. Savchuk et al. 2007. Internal ecosystem feedbacks enhance nitrogen-fixing cyanobacteria blooms and complicate management in the Baltic Sea. AMBIO: A Journal of the Human Environment 36(2): 186–194.

Vaquer-Sunyer, R. and C.M. Duarte. 2008. Thresholds of hypoxia for marine biodiversity. Proceedings of the National Academy of Sciences 105(40): 15452–15457.

Vetter, R., S. Kohin, A.N.T.O.N.E.L.L.A. Preti, S.A.M. Mcclatchie and H. Dewar. 2008. Predatory interactions and niche overlap between mako shark, Isurus oxyrinchus, and jumbo squid, Dosidicus gigas, in the California Current. CalCOFI Report 49: 142–156.

Vinogradov, M.E., E.A. Shushkina, L.L. Anokhina, S.V. Vostokov, N.V. Kucheruk and T.A. Lukashova. 2000. Mass development of the ctenophore Beroe ovata Eschscoltz near the northeastern coast of the Black Sea. Oceanology (Eng. Transl.) 40: 52–55.

Vitousek, P.M., J. Aber, R.W. Howarth, G.E. Likens, P.A. Matson, D.W. Schindler et al. 1997. Human alteration of the global nitrogen cycle: Sources and consequences. Ecological Applications 7: 737–750.

von Rad, U., H. Rösch, U. Berner, M. Geyh, V. Marchig and H. Schulz. 1996. Authigenic carbonates derived from oxidized methane vented from the Makran accretionary prism off Pakistan. Marine Geology 136(1-2): 55–77.

Zaitsev, Y.P. 1992. Recent changes in the trophic structure of the Black Sea. Fisheries Oceanography 1(2): 180–189.

Zillén, L., D.J. Conley, T. Andrén, E. Andrén and S. Björck. 2008. Past occurrences of hypoxia in the Baltic Sea and the role of climate variability, environmental change and human impact. Earth-Science Reviews 91(1-4): 77–92.

3

Harmful Algal Blooms in Coastal, Estuarine, and Offshore Waters

Jun Zhao,[1,2,*] *Bin Ai,*[1,2,*] *Lin Qi,*[1,2,3] *Wei Huang,*[1,2] *Chunlei Ma,*[1,2] *Xiaoping Xu*[1,2] and *Jiahui Liu*[1,2]

1. Introduction

Harmful algal blooms (HABs), also referred to as red tide, are caused by the proliferation of phytoplankton. In addition to their negative effects on aquaculture, fisheries, and tourism operations, HABs have major impacts on the aquatic environment and human health (Hallegraeff 1993). Because of these effects, HABs have been under scrutiny for decades.

The first paper about HAB recorded in Web of Science was published in 1947 (Cornman 1947), although there were earlier recorded reports of human poisoning correlated with toxins produced by bloomed phytoplankton. Comparisons of paralytic shellfish poison distribution for 1970 and 2015 indicate significant increase of HAB occurrence across the world (https://www.whoi.edu/redtide/regions/world-distribution). As shown in Fig. 1, statistics of papers about HABs published in Web of Science showed a significant increasing trend from 1947 to the present. For example, there were 629 papers about HABs published in 2018.

HABs can be dominated by different bloom species. As reported by Hallegraeff (1993), 300 out of 5000 species of extant marine phytoplankton can result in discoloration of water surface and 40 or so species can produce toxins. For example, the periodical *Karenia brevis* bloom on the west Florida shelf produces polyether neurotoxins (Pierce and Henry 2008). Cyanobacterial toxins such as microcystin and nodularin have been frequently present in Taihu Lake (Ye et al. 2009) and Lake Erie (Miller et al. 2017). Other HABs may be non-toxic but can cause kills of fish and invertebrates by damaging or clogging their gills and by depleting the oxygen in the water.

[1] School of Marine Sciences, Sun Yat-sen University, 135 Xingang Xi Road, Haizhu District, Guangzhou, Guangdong, 510310, China.
[2] Southern Laboratory of Ocean Science and Engineering, Zhuhai, Guangdong, 519000, China.
[3] College of Marine Science, University of South Florida, 140 Seventh Avenue, South, St. Petersburg, FL 33701, USA.
* Corresponding authors: zhaojun28@mail.sysu.edu.cn; abin@mail.sysu.edu.cn

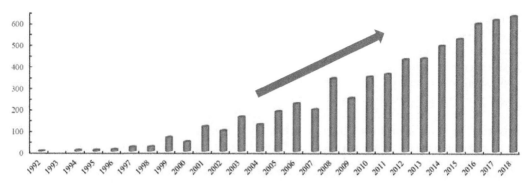

Fig. 1. Statistics for papers published in Web of Science using the key word of "harmful algal blooms". The arrow demonstrates the increasing trend of published papers about harmful algal bloom (HAB).

The effects of HABs have been well recognized and can be grouped into four types (Isabella et al. 2016): (1) human health impact; (2) fishery impact; (3) tourism and recreation impact; (4) monitoring and management impact. In the United States of America, for example, the estimated economic loss caused by HAB was approximately $75 million per year over the period 1987–2000 (Anderson 2009). With HAB outbreaks over global waters, it can be inferred that the economic losses due to HABs become larger and larger. According to Isabella et al. (2016), the estimated annual HAB-caused economic impacts on human health is $670,000 in Canada and $20 million in the US.

In this chapter, we review the major impacts of HABs (including fish and shellfish poisoning, and bird mortality) and mass mortality events, illustrate HAB events with different dominant species in different areas, and predict and forecast HAB occurrence.

2. Data and Method

MODIS/Aqua data used in this work were downloaded from ocean color data archive (https://oceancolor.gsfc.nasa.gov). Satellite data were processed with SeaWiFS Data Analysis System (SeaDAS, version 7.5). Normalized fluorescence line height (nflh) and remote sensing reflectance (R_{rs}) were generated. R_{rs} at 547, 488, and 443 nm were used to composite enhanced red-green-blue (ERGB) images. The ERGB images are very helpful in differentiating dark features caused by chlorophyll and/or colored dissolved organic matter (CDOM) from bright features due to high contents of suspended sediment and/or shallow bottom. Landsat 8 Operational Land Imager (OLI) data were downloaded from the United States Geological Survey and processed with SeaDAS.

3. Review of Mass Mortality Events and Impacts of HABs

Human-related activities such as pollution, climatic extremes, and other environmental influences play main roles in mass mortality events, which have become more common for bird, fish and marine invertebrates. Reports of mass mortality events contribute much to demonstrate their ecological and evolutionary importance. According to the analysis of 727 published mass mortality events from across the globe, it was found that the magnitude of mass mortality events has been intensifying for birds, fishes, and marine invertebrates; has remained steady for mammals; and has been decreasing for reptiles and amphibians (Fey et al. 2015).

In terms of mass mortality in marine ecosystems, researchers discovered that climatic anomalies, especially sea surface temperature (SST) warming, coupled with reduced food resources, resulted in mass mortality events such as pufferfish mortality in Hawaii, mass coral mortality on the southeastern coast of Ishigaki Island, Japan, and unusual cetacean mortality events in the northern Gulf of Mexico (Carlson et al. 2018, Litz et al. 2014, Work et al. 2017). Mass mortality events of coral reefs, seagrass, and invertebrates have presented the most significant responses to climatic anomalies (Sasano et al. 2016). For instance, more than 90% of shallow corals were killed on most Indian Ocean reefs in 1998 due to the high sea surface temperature. A Pacific oyster mortality rate of 77.4% resulted from a rise in seawater temperature from 20°C to 25°C (Green et al. 2018). A rapid and severe mass mortality of rocky shore invertebrates occurred along the north-central California coast of the northeastern Pacific Ocean (Jurgens et al. 2015). The relationship between climatic anomalies and occurrences of mass mortality events is a popular topic. It was predicted that the probable repeat occurrence patterns of mass coral mortality in the Indian Ocean as result of rise in mean SST from 2010 to 2025 would show typically geographical differences (Sheppard 2003). It was also discovered that mass coral bleaching quickly ensued on Dongsha Atoll, in the northern South China Sea, and 40% of the resident coral community was killed because the SST increased by 2°C in response to the developing Pacific El Niño. This mass coral mortality event in June 2015 was unprecedented in at least the past 40 years and related to the unusually weak winds amplifying the 2°C basin-scale anomaly. This phenomenon resulted in water temperatures on the reef flat exceeding 6°C above normal summertime levels (DeCarlo et al. 2017). Coma et al. (2009) found that many biological processes in the Mediterranean Sea are expected to be affected by the lengthening of warmer summer conditions, culminating in such events as mass mortality of invertebrates.

HABs, which are extreme biological events, have direct impacts on human health and negative influences on human well being, mainly through their consequences for coastal ecosystem services (fisheries, tourism, and recreation) and on other marine organisms and environments (Berdalet et al. 2016). One of the typical impacts HABs have on the environment is frequent mass mortalities of fish, plankton and benthos. Studies have shown that algae toxins that are toxic to fish include paralytic shellfish poison (*Alexandrium* spp. and *Gymnodinium* spp.), brevetoxin (also known as neurotoxic shellfish poison, *Gynodinium breve*), hemolytic toxin (*A. tamarense* and *Klebsiella kawaii*), and cytolytic toxin (*G. mikimotoi* and *A. catenella*) (Yin et al., 2000). For instance, a mass mortality of pufferfish in Hawaii (USA) erupted in 2010, which was associated with HABs caused by anomalous westerly winds (Work et al. 2017). A mass mortality of small pelagic fish species due to the dinoflagellate blooms in southern Benguela was detected by the analysis of *in situ* and remotely sensed data (van der Lingen et al. 2016). Brevetoxin-associated mass mortality event of bottlenose dolphins and manatees along the east coast of Florida, USA, occurred between October 2007 and January 2008 (Fire et al. 2015). Differential mortality of North Atlantic bivalve molluscs was well described and it was discovered that blooms of *Cochlodinium polykrikoides* pose significant age- and species-specific threats to native and cultured bivalve shellfish (Griffith et al. 2018). Hundreds of sea turtles were found dead along the Pacific coastline of El Salvador, killed by the anomalous HABs off the coast (Amaya et al. 2018). HABs also cause organ failure in marine mammals and have increased the occurrences of marine mammal mass strandings with broad geographical and widespread taxonomic impact (Pyenson et al. 2014). Correlation analysis between historical monthly stranding event numbers and mean monthly chlorophyll concentration discovered that the periodic trend of cetacean mass stranding events in the Australian island state of Tasmania was dominated by domoic acid poisoning (Nash et al. 2017). Over 2000 dead seabirds were stranded on the coast of Washington in 1942 coincident with an *Alexandrium catenella* bloom event; similarly, 636 dead seabirds were found near Farne Island in 1968, about 620 birds were killed from Maine to Massachusetts in 1972, and hundreds of birds were found stranded in Oregon and Washington because of *Akashiwo sanguinea* bloom in 2009 (Gaydos 2012). In Monterey Bay, California, 207 fresh dead birds and 550 stranded live birds were collected during the bloom of dinoflagellate *Akashiwo sanguinea* in 2007 (Jessup et al. 2009). Bird mortality events along the central west Florida coast due to dinoflagellate *Karenia brevis* occurred from February 2005 through November 2006 (Fauquier et al. 2013). Many bird group mortality events due to HABs have been documented, including cormorants, waterfowl, terns, pelicans, alcids and shearwaters (Jones et al. 2017).

It should be noted that the damage mechanism of different bloom species of phytoplankton is different. Dinoflagellate *Karenia brevis* can produce a powerful neurotoxin named brevetoxin, which is released into water when the algal cells lyse and can remain in the environment for up to one year (Fauquier 2014). It is transferred through the food chain and thus is also lethal to marine organisms, particularly those that eat shellfish and fish (Landsberg et al. 2009). Even several months after the end of an HAB, birds may still be at risk of brevetoxin poisoning, because shellfish and fish can accumulate brevetoxin continually (Fauquier et al. 2013). Brevetoxin acts by depolarizing nerve cells by binding to voltage-gated sodium channels, thus leading to interference with nerve transmission (Pierce and Henry 2008). Another neurotoxin named domoic acid (DA), produced by *Pseudo-nitzschia*, shows high affinity for glutamate receptors; it is transmitted to predators through trophic transfer like brevetoxin by fish feeding mainly on plankton. This has led to mass death of many kinds of birds in California, including Brandt's cormorants, brown pelicans and marbled murrelets; clinical symptoms include ataxia, seizures, and coma (Gaydos 2012). Potent neurotoxins named saxitoxins can be produced by numerous dinoflagellates including *Gymnodinium catenatum*, *Pyrodinium bahamense var. compressum*, several cyanobacteria and about 11 species of the genus *Alexandrium* (Gaydos 2012). Saxitoxins have also been documented to be related to many sporadic mortalities (Landsberg 2002); clinical symptoms include vomiting, pupil constriction, paralysis and loss of equilibrium (Gaydos 2012). Dinoflagellate *Akashiwo sanguinea* can also result in mortality because of surfactant, which is in itself minimally toxic or nontoxic (Jessup et al. 2009). The proteinaceous foam formed after the breakdown of the *Akashiwo sanguinea* cells can destroy the waterproofing ability of birds if it coats their feathers, and finally can cause hypothermia, hypoglycemia, and even death (Gaydos 2012).

Because of the impacts and unselective nature of mass mortality events caused by HABs, some researchers seized the opportunity to investigate demographic parameters of marine life (Humple et al. 2011). However, it is still difficult to link the mortality of marine life directly to HABs and there were few studies about changes in populations or community structure caused by HABs (Fauquier 2014). Furthermore, long-term monitoring of many species of birds is crucial for evaluating the effects of HABs on marine ecosystems.

4. Case Studies of HABs across the World

4.1 Case study of HAB in the Arabian/Persian Gulf

The Arabian/Persian Gulf is a shallow, semi-enclosed marginal sea with an average depth of 35 m. The area is characterized by semi-arid weather conditions. The precipitation over the region is very low with a rate of 0.07–0.1 m per year (Marcella and Eltahir 2008). On the other hand, the evaporation rate over the area is high and exceeds rainfall and freshwater discharge, resulting in high salinity over 45‰ (Zhao et al. 2017b). In recent years, HAB outbreaks were recorded almost every year over the Arabian/Persian Gulf (Glibert et al. 2002, Zhao and Ghedira 2014).

In 2008, there was an extensive HAB event in the Arabian/Persian Gulf. The dominant bloom species was dinoflagellate *Cochlodinium polykrikoides*. The event lasted almost a year from August 2008 to August 2009. Field sampling results demonstrated that the cell density amounted to $1.1–2.1 \times 10^7$ cells L^{-1} in October in the Sea of Oman and the chlorophyll-a concentration reached 40 mg m^{-3} (Moradi and Kabiri 2012, Richlen et al. 2010). Zhao and Ghedira (2014) successfully used multi-sensor satellite imagery to detect and track the HAB event. They found that the HAB event was triggered by upwelling in the Sea of Oman that transported nutrient-rich bottom water to the surface through Ekman pumping and supported the growth of phytoplankton.

Figure 2 shows satellite imagery for December 23, 2008, over the Arabian/Persian Gulf. The dark color in the ERGB image indicates high contents of phytoplankton and/or CDOM, while the bright color is caused by high concentrations of suspended sediment and/or shallow bottom. Satellite-derived nflh has been widely used as a proxy of phytoplankton biomass. In Fig. 2b, fronts in the middle of the Arabian Gulf can be clearly seen. The highest nflh reached 2 W m^{-2}μm^{-1} sr^{-1}. Combined with the ERGB image, the high nflh can be used for detection of HAB. Figure 2c shows

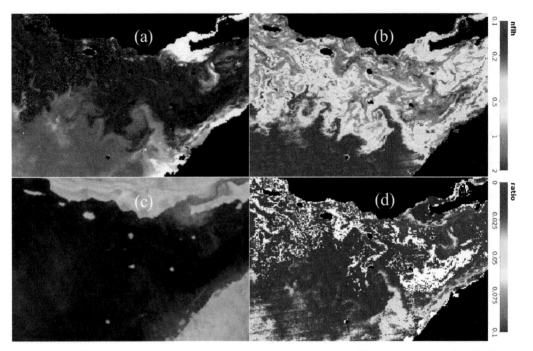

Fig. 2. Satellite-derived enhanced red-green-blue composite (ERGB) (a), normalized fluorescence line height (nflh) (b), true color image (c), the ratio of particulate backscattering (b_{bp}) to nflh for December 23, 2008, over the Arabian Gulf when an extensive HAB event occurred.

Color version at the end of the book

the true color image. Over the bloom area, the water surface shows reddish-brown color. Further, the ratio of particulate backscattering to nflh is generally < 0.02 W^{-1} m µmsr in the bloom area, which is significantly smaller than in the non-bloom area. According to the approach proposed by Zhao et al. (2015), it can be inferred that the dominant species was *Cochlodinium polykrikoides*, which agrees well with *in situ* measurements.

In recent years, the bloom species in the Arabian Sea was found to be dominated by dinoflagellate *Noctiluca scintillans*. Previously, the bloom species comprised mainly diatoms. Do Rosário Gomes et al. (2014) reported that the massive outbreaks of *Noctiluca scintillans* in the Arabian Sea were facilitated by an unprecedented influx of oxygen-deficient waters into the euphotic zone and the efficient carbon fixation ability of the species under hypoxic conditions. This phase shift of dominant bloom species is consistent with the finding by National Aeronautics and Space Administration (NASA), which reported that populations of diatom have declined more than 1% per year between 1998 and 2012 (https://nasaviz.gsfc.nasa.gov/11934). They also mentioned that the phenomenon is related to the shoaling of mixed layer depth.

4.2 Case study of HAB in the Gulf of Mexico

In the eastern GOM, *Karenia brevis* blooms occur on an annual basis. The dinoflagellate *Karenia brevis* can produce brevetoxin, causing fish kills and other environmental and economic problems. Compared with traditional field surveys with research vessels or buoys, satellite observations have the advantages of synoptic view over the sea surface over large temporal scales. Various remote-sensing detection methods have been proposed to detect and quantify these blooms (see Soto et al. 2015 and references therein). A fundamental principle in all these methods is to estimate

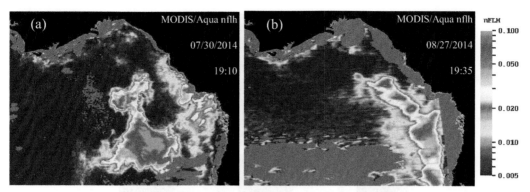

Fig. 3. MODIS/Aqua derived nflh imagery for July 30, 2014 (a), and August 27, 2014 (b).

Color version at the end of the book

chlorophyll-a concentration (Chl-a) or its equivalent from satellite measurements. An alternative way is through the solar-stimulated chlorophyll-a fluorescence because it does not suffer from CDOM perturbations (Hu et al. 2005). It has been successfully used to track HAB on the west Florida shelf (Zhao et al. 2013).

MODIS/Aqua-derived nflh images for July 30 and August 27, 2014, are displayed in Fig. 3. MODIS/Aqua nflh images were used as the "truth" for detecting bloom patterns, where the offshore bloom patch associated with high nflh values (> 0.015 mW cm^{-2} μm^{-1} sr^1) was confirmed to contain elevated concentrations (> 10^3 cells L^{-1}) of the toxic *Karenia brevis* (Hu et al. 2015). From the images, it can be seen that the bloom patches were advected from offshore regions to coastal waters and extended over a large area on the west Florida shelf.

Le et al. (2013) proposed a red-green-chlorophyll-a index (RGCI) to estimate chlorophyll-a concentration in optically complex coastal waters. Likewise, the success of the VIIRS RGCI algorithm in estimating chlorophyll-a is due to its avoidance of blue bands to minimize CDOM perturbation. In water rich in suspended sediments, the perturbation mainly comes from particle scattering, which influences not only the green band but also the red tide. In such environments, the RGCI may be a useful index for the combined effects of phytoplankton and suspended particles, where separating these two effects is still a challenging task. Nevertheless, as long as the fundamental principles are well defined and the optical complexity of the inland or coastal systems is well understood, locally tuned algorithms should be able to estimate chlorophyll-a and other phytoplankton pigments, which enable the studies of phytoplankton dynamics in complex environments.

4.3 Case study of HAB in Southern Benguela, South Africa

The Benguela region in South Africa is well known as one of the four major eastern boundary current regions of the world ocean (Jury 2014). It is characterized by wind-driven upwelling along the whole western coast of southern Africa (Pitcher et al. 1998). The Benguela upwelling system has been subject to HABs for decades. The HABs were attributed to one or another dinoflagellate species (Pitcher and Weeks 2006). HABs mainly happen between winter and spring during the latter months of the upwelling season (Pitcher and Calder 2000).

Bernard et al. (2014) applied various algorithms to MERIS data and investigated the evolution of several bloom events in the Benguela system. They also made recommendations with respect to optimal ocean color use for high biomass HABs in coastal waters. Pitcher et al. (2008) reported HABs in the later summer and autumn of 2007 in two bays of the southern Benguela upwelling system. Their *in situ* measurements showed high concentrations of the dinoflagellate *Gonyaulax polygramma*

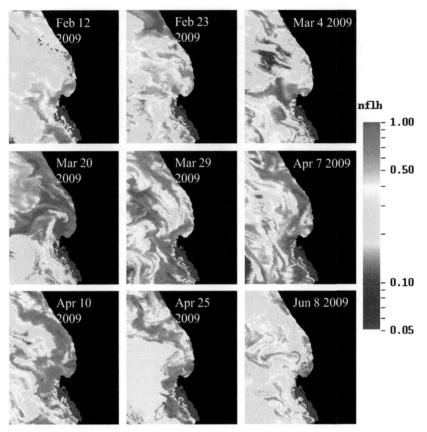

Fig. 4. Satellite-observed nflh between February 12 and June 8 2009 over the southern Benguela region, South Africa.

reaching 2×10^7 cells L^{-1}. They used remote sensing data to identify the scale and physical forcing of the HAB.

In the autumn of 2009, there was an exceptional HAB event dominated by dinoflagellate *Ceratium belechii* in the St Helena Bay, which induced mass mortality of macrofauna due to anoxia. *In situ* sampling analysis by Pitcher and Probyn (2011)showed that the maximum phytoplankton cell density reached 7.3×10^6 cells L^{-1} on March 25, 2009. Figure 4 shows MODIS/Aqua observed normalized chlorophyll-a fluorescence (nflh) between February 12 and June 8, 2009, in the southern Benguela upwelling system. The bloom was first detected in mid-February in the St Helena Bay. Induced by diminished upwelling activity in late summer and inshore counter ocean circulation in early fall, the bloom patches gradually moved southward. In late April and early May, the bloom accumulated in the shallow water of the St Helena Bay. In early winter, the bloom degraded.

In the southern Benguela upwelling system, oxygen deficiency has been frequently observed. It has pronounced negative impact on the local marine ecosystems. Large episodic mortalities of the rock lobster *Jasus lalandii* were reported (Bernard et al. 2014). The environmental conditions led to dinoflagellate HABs. The upwelling-downwelling cycle determined the nutrient concentrations in the water column. This, in turn, affected the fate of the HABs in the region.

4.4 Case study of HAB in the Daya Bay, South China Sea

During the last two decades, HABs have been frequently recorded in coastal waters of China, dominated by different types of phytoplankton, such as dinoflagellate, haptophyte, and pelagophyte species (Yu et al. 2018). As one of the well-developed economic zones in China, the Pearl River Delta has also been subject to the effects of HABs. Wei et al. (2012) investigated the occurrence of HABs between 2000 and 2009 in the Pearl River estuary and found that the bloom species included *Skeletonema costatum* and *Phaeocystis globosa*. Daya Bay is a semi-enclosed bay. It covers an area of about 550 km². It is shallow, with an average depth of 11 m. The hydrodynamic conditions are dominated by tides. There is no major river discharge into the bay area. In recent years, HABs were also reported in Daya Bay.

In August 2017, there was a HAB event in Daya Bay (http://www.twoeggz.com/news/3263036. html). The bloom species was dinoflagellate *Scrippsiella trochoidea*. Satellite data from Landsat 8 OLI are presented in Fig. 5. The satellite sensor overpassed the region around 2:45 GMT August 13, 2017. The satellite-derived true color imagery is shown in Fig. 5a. Close to the northeastern coastal area, there was an anomalous region demonstrating a dark brown color, which is magnified in Fig. 5c. The floating algae index is useful in detecting the occurrence of extensive HAB events (Hu 2009). For example, it has been successfully used to track the notorious green tide event in the

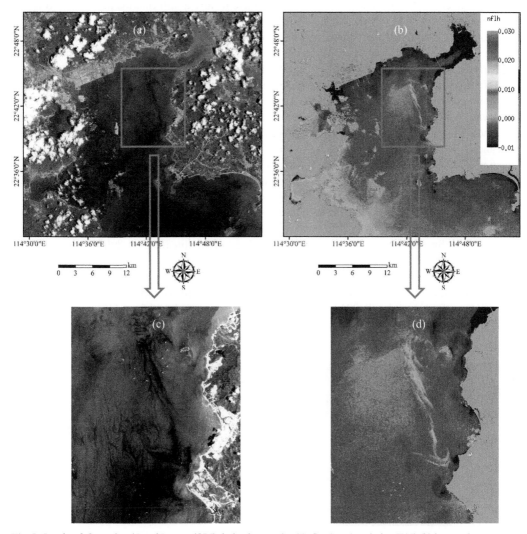

Fig. 5. Landsat 8 Operational Land Imager (OLI) derived true color (a), floating algae index (FAI) (b) images for August 13 2017 over the Daya Bay, China. The area outlined with a red box is zoomed in (c) and (d).

Yellow Sea and East China Sea (Hu et al. 2010). The corresponding floating algae index values in the bloom region (Figs. 5b and 5d) are apparently larger than in surrounding waters.

HABs were also reported in other areas of the South China Sea. Yin et al. (1999) studied the spring 1998 HAB event in Hong Kong and examined its potential relationship with El Niño. Tang et al. (2004) reported an extensive HAB off southeastern Vietnam from late June to July 2002. Zhao et al. (2018) investigated the summer 2007 HAB event southeast of Vietnam and found that the HAB event was related to upwelling caused by Ekman pumping.

It can be concluded that HABs in the South China Sea may be triggered by different factors, including terrestrial runoffs, upwelling, and climate change. More localized studies are required to improve our understanding of HAB occurrence in the region. Further, HAB-related information in nearly real time is urgently required for monitoring variations of the marine ecosystems in the Guangdong-Hong Kong-Macau Bay area.

4.5 Modeling and forecast of HABs

In situ and satellite observations can provide present and historical information of HABs. However, forecast and prediction of HABs are of great importance to minimize the potential adverse effects of HABs on aquaculture, marine animals, and human health.

Karki et al. (2018) developed data-driven models to identify factors that control HAB propagation in Charlotte County, southwestern Florida, which used spatiotemporal remote sensing and filed data. Their results provide nowcast and forecast of HAB occurrences up to three days in advance. California-Harmful Algae Risk Mapping (C-HARM) model (https://www.cencoos.org/data/models/ habs) was created as a combination of sophisticated circulation model, ocean color remote-sensing data, and statistical models for bloom and toxin likelihoods to predict HAB conditions for one to three days ahead. The National Ocean and Atmosphere Administration (NOAA) developed an ensemble bloom prediction model by relating bloom size to spring total phosphorus load in Lake Erie (http:// scavia.seas.umich.edu/hypoxia-forecasts/). They aimed to alert water resources managers and the public to avoid the potential severity of HABs in Lake Erie (Bertani et al. 2016). The European Union funded the ASIMUTH project (Applied simulations and integrated modeling for the understanding of toxic and harmful algal blooms) to generate short-term forecasts of HABs along the European Atlantic coasts by using a combination of both modeling and satellite imagery (https://cordis.europa. eu/project/rcn/96871/factsheet/en).

Zhao et al. (2017a) used satellite-derived information for HABs to initiate a numerical model and track the movements of HABs in the Arabian Gulf. Figure 6a shows the nflh observed by MODIS/Aqua for November 7, 2013, in the Arabian Gulf, as reproduced from J Zhao et al. (2017a). The ROMS-simulated HAB patches are shown in Fig. 6b. It can be seen that the model-simulated

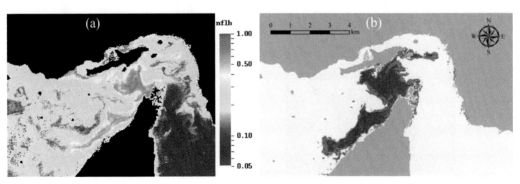

Fig. 6. MODIS/Aqua observed nflh for November 7 2013 over the Hormuz Strait (a). Satellite-observed and ROMS-simulated bloom patches for November 7 2013. Reproduced from Zhao et al. (2017).

results agree well with satellite observations. On the other hand, it should also be noted that most recent satellite observations should be integrated into the numerical model to improve the simulation accuracy of HAB occurrence.

5. Conclusion

Because of climate change and/or anthropogenic activities, the frequency of HAB occurrence exhibits a significant increasing trend globally. To avoid potential losses, timely and efficient observations and detection of HABs are required. By combining all appropriate approaches for HAB monitoring, including *in situ* measurements and satellite remote sensing, we can better understand the entire initiation and degradation procedure of HABs. With lessons learned from historical events, we can predict and forecast HAB occurrence with accuracy. Thus, the fate of HABs can be modeled.

It should also be noted that an individual HAB event has its own specific features in terms of trigger, spatiotemporal distribution, and dominant bloom species. Localized biogeochemical studies are needed to improve our understanding of HABs, on the basis of which we can combat HABs more efficiently.

Acknowledgements

We thank the Ocean Biology Processing Group of NASA for providing MODIS/Aqua data. We also express our great appreciation to the US Geological Survey for providing Landsat 8 OLI data. We are grateful to Sun Yat-sen University for their funding support.

References

Amaya, O., R. Quintanilla, B.A. Stacy, M.-Y. Dechraoui Bottein, L. Flewelling, R. Hardy et al. 2018. Large-scale sea turtle mortality events in El salvador attributed to paralytic shellfish toxin-producing algae blooms. Front. Mar. Sci. 5(411). Doi:10.3389/fmars.2018.00411.

Anderson, D.M. 2009. Approaches to monitoring, control and management of harmful algal blooms (HABs). Ocean Coast. Manage. 52(7): 342–342. Doi: 10.1016/j.ocecoaman.2009.04.006.

Berdalet, E., L.E. Fleming, R. Gowen, K. Davidson, P. Hess, L.C. Backer et al. 2016. Marine harmful algal blooms, human health and wellbeing: challenges and opportunities in the 21st century. J. Mar. Biol. Assoc. UK 96(1): 61–91. Doi: 10.1017/s0025315415001733.

Bernard, S., G. Pitcher, H. Evers-King, L. Robertson, M. Matthews, A. Rabagliati et al. 2014. Ocean colour remote sensing of harmful algal blooms in the Benguela system. pp. 185–203. *In*: Barale, V. and M. Gade (eds.). Remote Sensing of the African Seas, Springer Netherlands, Dordrecht. Doi: 10.1007/978-94-017-8008-7_10.

Bertani, I., D.R. Obenour, C.E. Steger, C.A. Stow, A.D. Gronewold and D. Scavia. 2016. Probabilistically assessing the role of nutrient loading in harmful algal bloom formation in western Lake Erie. J. Great Lakes Res. 42(6): 1184–1192. Doi: https://doi.org/10.1016/j.jglr.2016.04.002.

Carlson, D.F., L.A. Yarbro, S. Scolaro, M. Poniatowski, V. McGee-Absten and P.R. Carlson, Jr. 2018. Sea surface temperatures and seagrass mortality in Florida Bay: Spatial and temporal patterns discerned from MODIS and AVHRR data. Remote Sens. Environ. 208: 171–188. Doi: 10.1016/j.rse.2018.02.014.

Coma, R., M. Ribes, E. Serrano, E. Jimenez, J. Salat and J. Pascual. 2009. Global warming-enhanced stratification and mass mortality events in the Mediterranean. P. Natl. Acad. Sci. USA 106(15): 6176–6181 Doi: 10.1073/pnas.0805801106.

Cornman, I. 1947. Retardation of Arbacia egg cleavage by dinoflagellate-contaminated sea water (red tide). Biol. Bull. 93(2): 205.

DeCarlo, T.M., A.L. Cohen, G.T.F. Wong, K.A. Davis, P. Lohmann and K. Soong. 2017. Mass coral mortality under local amplification of 2 degrees C ocean warming. Sci. Rep. 7: Doi: 10.1038/srep44586.

do Rosário Gomes, H., J.I. Goes, S.G.P. Matondkar, E.J. Buskey, S. Basu, S. Parab et al. 2014. Massive outbreaks of Noctiluca scintillans blooms in the Arabian Sea due to spread of hypoxia. Nat. Commun. 5: 4862. Doi: 10.1038/ncomms5862, https://www.nature.com/articles/ncomms5862#supplementary-information.

Fauquier, D. 2014. Impacts of red tide toxins on seabirds. *In*: Annual Summary of the Activities and Findings of the Chicago Zoological Society's Sarasota Dolphin Research Program, p. 33, Chicago Zoological Society.

Fauquier, D.A., L.J. Flewelling, J.M. Maucher, M. Keller, M.J. Kinsel, C.K. Johnson et al. 2013. Brevetoxicosis in Seabirds Naturally Exposed to Karenia Brevis Blooms along the Central West Coast of Florida. J. Wildlife Dis. 49(2): 246–260. Doi: 10.7589/2011-09-270.

Fey, S.B., A.M. Siepielski, S. Nussle, K. Cervantes-Yoshida, J.L. Hwan, E.R. Huber et al. 2015. Recent shifts in the occurrence, cause, and magnitude of animal mass mortality events. P. Natl. Acad. Sci. USA 112(4): 1083–1088. Doi: 10.1073/pnas.1414894112.

Fire, S.E., L.J. Flewelling, M. Stolen, W.N. Durden, M. de Wit, A.C. Spellman et al. 2015. Brevetoxin-associated mass mortality event of bottlenose dolphins and manatees along the east coast of Florida, USA. Mar. Ecol. Prog. Ser. 526: 241–251. Doi: 10.3354/meps11225.

Gaydos, J.K. 2012. Deadly Diatoms: The Latest on Harmful Algal Blooms, Paper Presented at North American Veterinary Conference, Orlando, Florida.

Glibert, P.M., J.H. Landsberg, J.J. Evans, M.A. Al-Sarawi, M. Faraj, M.A. Al-Jarallah et al. 2002. A fish kill of massive proportion in Kuwait Bay, Arabian Gulf, 2001: The roles of bacterial disease, harmful algae, and eutrophication. Harmful Algae 1(2): 215–231.

Green, T.J., N. Siboni, W.L. King, M. Labbate, J.R. Seymour and D. Raftos. 2018. Simulated marine heat wave alters abundance and structure of vibrio populations associated with the pacific oyster resulting in a mass mortality event. Microb. Ecol. Doi: 10.1007/s00248-018-1242-9.

Griffith, A.W., S.E. Shumway and C.J. Gobler. 2018. Differential mortality of North Atlantic bivalve molluscs during harmful algal blooms caused by the Dinoflagellate, Cochlodinium (a.k.a. Margalefidinium) polykrikoides. Estuar. Coast. Doi: 10.1007/s12237-018-0445-0.

Hallegraeff, G.M. 1993. A review of harmful algal blooms and their apparent global increase. Phycologia 32(2): 79–99.

Hu, C., F.E. Muller-Karger, C. Taylor, K.L. Carder, C. Kelble, E. Johns et al. 2005. Red tide detection and tracing using MODIS fluorescence data: A regional example in SW Florida coastal waters. Remote Sens. Environ. 97(3): 311–321. Doi: https://doi.org/10.1016/j.rse.2005.05.013.

Hu, C. 2009. A novel ocean color index to detect floating algae in the global oceans. Remote Sens. Environ. 113(10): 2118–2129. Doi: https://doi.org/10.1016/j.rse.2009.05.012.

Hu, C., D. Li, C. Chen, J. Ge, F.E. Muller-Karger, J. Liu et al. 2010. On the recurrent Ulva prolifera blooms in the Yellow Sea and East China Sea. J. Geophys. Res.: Oceans 115(C5). Doi: 10.1029/2009JC005561.

Hu, C., B. Barnes, L. Qi and A. Corcoran. 2015. A harmful algal bloom of Karenia brevis in the Northeastern Gulf of Mexico as revealed by MODIS and VIIRS: A comparison. Sensors 15(2): 2873.

Humple, D.L., H.M. Nevins, E.M. Phillips, C. Gibble, L.A. Henkel, K. Boylan et al. 2011. Demographics of Aechmophorus Grebes killed in three mortality events in California. Mar. Ornithol. 39(2): 235–242.

Isabella, S., C. Diana, P. Luca, D. Srdan and L. Teresa. 2016. Algal Bloom and its Economic Impact Rep., Joint Research Center, the European Commission.

Jessup, D.A., M.A. Miller, J.P. Ryan, H.M. Nevins, H.A. Kerkering, A. Mekebri et al. 2009. Mass stranding of marine birds caused by a surfactant-producing red tide. Plos One 4(2). Doi: 10.1371/journal.pone.0004550.

Jones, T., J.K. Parrish, A.E. Punt, V.L. Trainer, R. Kudela, J. Lang et al. 2017. Mass mortality of marine birds in the Northeast Pacific caused by Akashiwo sanguinea. Mar. Ecol. Prog. Ser. 579: 111–127. Doi: 10.3354/meps12253.

Jurgens, L.J., R.B. Laura, P.T. Raimondi, L.M. Schiebelhut, M.N. Dawson, R.K. Grosberg et al. 2015. Patterns of mass mortality among rocky shore invertebrates across 100 km of Northeastern Pacific Coastline. Plos One 10(6): e0126280.

Jury, M.R. 2014. Environmental forcing of red tides in the southern Benguela. Int. J. Oceanog. 2014: 1–16.

Karki, S., M. Sultan, R. Elkadiri and T. Elbayoumi. 2018. Mapping and forecasting onsets of harmful algal blooms using MODIS data over coastal waters surrounding Charlotte County, Florida. Remote Sens. 10(10): 1656.

Landsberg, J.H. 2002. The effects of harmful algal blooms on aquatic organisms. Rev. Fish. Sci. 10(2): 113–390. Doi: 10.1080/20026491051695.

Landsberg, J.H., L.J. Flewelling and J. Naar. 2009. Karenia brevis red tides, brevetoxins in the food web, and impacts on natural resources: Decadal advancements. Harmful Algae 8(4): 598–607. Doi: 10.1016/j.hal.2008.11.010.

Le, C., C. Hu, D. English, J. Cannizzaro and C. Kovach. 2013. Climate-driven chlorophyll-a changes in a turbid estuary: Observations from satellites and implications for management. Remote Sens. Environ. 130: 11–24. Doi: https://doi.org/10.1016/j.rse.2012.11.011.

Litz, J.A., M.A. Baran, S.R. Bowen-Stevens, R.H. Carmichael, K.M. Colegrove and L.P. Garrison. 2014. Review of historical unusual mortality events (UMEs) in the Gulf of Mexico (1990–2009): Providing context for the multi-year northern Gulf of Mexico cetacean UME declared in 2010. Dis. Aquat. Organ. 112(2): 161–175. Doi: 10.3354/dao02807.

Marcella, M.P. and E.A.B. Eltahir. 2008. The hydroclimatology of Kuwait: Explaining the variability of rainfall at seasonal and interannual time scales. J. Hydrometeorol. 9(5): 1095–1105. Doi: 10.1175/2008jhm952.1.

Miller, T.R., L.J. Beversdorf, C.A. Weirich and S.L. Bartlett. 2017. Cyanobacterial toxins of the laurentian great lakes, their toxicological effects, and numerical limits in drinking water. Mar. Drugs 15(6): 160. Doi: 10.3390/md15060160.

Moradi, M. and K. Kabiri. 2012. Red tide detection in the Strait of Hormuz (east of the Persian Gulf) using MODIS fluorescence data. Int. J. Remote Sens. 33(4): 1015–1028. Doi: 10.1080/01431161.2010.545449.

Nash, S.M.B., M.C. Baddock, E. Takahashi, A. Dawson and R. Cropp. 2017. Domoic acid poisoning as a possible cause of seasonal cetacean mass stranding events in Tasmania, Australia. Bull. Environ. Contam. Toxicol. 98(1): 8.

Pierce, R.H. and M.S. Henry. 2008a. Harmful algal toxins of the Florida red tide (Karenia brevis): natural chemical stressors in South Florida coastal ecosystems. Ecotoxicology 17(7): 623–631. Doi: 10.1007/s10646-008-0241-x.

Pitcher, G.C., A.J. Boyd, D.A. Horstman and B.A. Mitchell-Innes. 1998. Subsurface dinoflagellate populations, frontal blooms and the formation of red tide in the southern Benguela upwelling system. Mar. Ecol. Prog. Ser. 172: 253–264. Doi: 10.3354/meps172253.

Pitcher, G.C. and D. Calder. 2000. Harmful algal blooms of the Southern Benguela current: A review and appraisal of monitoring from 1989 to 1997. Afr. J. Mar. Sci. 22: 255–271.

Pitcher, G.C. and S.J. Weeks. 2006. The variability and potential for prediction of harmful algal blooms in the southern Benguela ecosystem. pp. 125–146. *In*: Shannon, V., G. Hempel, P. Malanotte-Rizzoli, C. Moloney and J. Woods (eds.). Large Marine Ecosystems, Elsevier. Doi: https://doi.org/10.1016/S1570-0461(06)80012-1.

Pitcher, G.C., S. Bernard and J. Ntuli. 2008. Contrasting bays and red tides in the southern Benguela upwelling system. Oceanography 21(3): 82–91.

Pitcher, G.C. and T.A. Probyn. 2011. Anoxia in southern Benguela during the autumn of 2009 and its linkage to a bloom of the dinoflagellate Ceratium balechii. Harmful Algae 11: 23–32. Doi: https://doi.org/10.1016/j.hal.2011.07.001.

Pyenson, N.D., C.S. Gutstein, J.F. Parham, J.P. Le Roux, C. Chavarría, L. Holly et al. 2014. Repeated mass strandings of Miocene marine mammals from Atacama Region of Chile point to sudden death at sea. Proc. Biol. Sci. 281(1781): 20133316.

Richlen, M.L., S.L. Morton, E.A. Jamali, A. Rajan and D.M. Anderson. 2010. The catastrophic 2008–2009 red tide in the Arabian gulf region, with observations on the identification and phylogeny of the fish-killing dinoflagellate Cochlodinium polykrikoides. Harmful Algae 9(2): 163–172. Doi: 10.1016/j.hal.2009.08.013.

Sasano, M., M. Imasato, H. Yamano and H. Oguma. 2016. Development of a regional coral observation method by a fluorescence imaging LIDAR installed in a Towable Buoy. Remote Sens. 8(1). Doi: 10.3390/rs8010048.

Sheppard, C.R.C. 2003. Predicted recurrences of mass coral mortality in the Indian Ocean. Nature 425(6955): 294–297. Doi: 10.1038/nature01987.

Soto, I., J. Cannizaro, F.E. Muller-Karger, C. Hu, J. Wolny and D. Goldgof. 2015. Evaluation and optimization of remote sensing techniques for detection of Karenia brevis blooms on the West Florida Shelf. Remote Sens. Environ. 170: 239–254.

Tang, D., H. Kawamura, H. Doan-Nhu and W. Takahashi. 2004. Remote sensing oceanography of a harmful algal bloom off the coast of southeastern Vietnam. J. Geophys. Res.: Oceans 109(C3). Doi: 10.1029/2003JC002045.

van der Lingen, C.D., L. Hutchings, T. Lamont and G.C. Pitcher. 2016. Climate change, dinoflagellate blooms and sardine in the southern Benguela Current Large Marine Ecosystem. Environ. Dev. 17: 230–243. Doi: 10.1016/j.envdev.2015.09.004.

Wei, G., H. Wang, W. Cai and B. Yi. 2012. 10-year retrospective analysis on the harmful algal blooms in the Pearl River estuary. Mar. Sci. Bull. 31(4): 466–474.

Work, T.M., P.D. Moeller, K.R. Beauchesne, J. Dagenais, R. Breeden, R. Rameyer et al. 2017. Pufferfish mortality associated with novel polar marine toxins in Hawaii. Dis. Aquat. Organ. 123(2): 87.

Ye, W., X. Liu, J. Tan, D. Li and H. Yang. 2009. Diversity and dynamics of microcystin—Producing cyanobacteria in China's third largest lake, Lake Taihu. Harmful Algae 8(5): 637–644. Doi: https://doi.org/10.1016/j.hal.2008.10.010.

Yin, K., J.H. Paul, C. Jay, H. Wei and Q. Pei-Yuan. 1999. Red tides during spring 1998 in Hong Kong: is El NiÃƒÂ±o responsible? Mar. Ecol. Prog. Ser. 187: 289–294.

Yin, Y.W., Z.H. Wang, T.J. Jiang, Y.M. Luo and L.C. Xie. 2000. Toxic effects of red tide toxins on fishes. Mar. Environ. Sci. 19(04): 62–65.

Yu, R.-C., S.-H. Lü and Y.-B. Liang. 2018. Harmful algal blooms in the coastal waters of China. pp. 309–316. *In*: Glibert, P.M., E. Berdalet, M.A. Burford, G.C. Pitcher and M. Zhou (eds.). Global Ecology and Oceanography of Harmful Algal Blooms, Springer International Publishing, Cham. Doi: 10.1007/978-3-319-70069-4_15.

Zhao, H., J. Zhao, X. Sun, F. Chen and G. Han. 2018. A strong summer phytoplankton bloom southeast of Vietnam in 2007, a transitional year from El Niño to La Niña. PLOS ONE 13(1): e0189926. Doi: 10.1371/journal. pone.0189926.

Zhao, J., C. Hu, J.M. Lenes, R.H. Weisberg, C. Lembke, D. English et al. 2013. Three-dimensional structure of a Karenia brevis bloom: Observations from gliders, satellites, and field measurements. Harmful Algae 29: 22–30. Doi: https://doi.org/10.1016/j.hal.2013.07.004.

Zhao, J. and H. Ghedira. 2014. Monitoring red tide with satellite imagery and numerical models: A case study in the Arabian Gulf. Mar. Pollut. Bull. 79(1): 305–313. Doi: https://doi.org/10.1016/j.marpolbul.2013.10.057.

Zhao, J., M. Temimi and H. Ghedira. 2015. Characterization of harmful algal blooms (HABs) in the Arabian Gulf and the Sea of Oman using MERIS fluorescence data. ISPRS J. Photogramm. 101: 125–136. Doi: https://doi.org/10.1016/j.isprsjprs.2014.12.010.

Zhao, J., M. Temimi, M. Al Azhar and H. Ghedira. 2017a. Analysis of bloom conditions in fall 2013 in the Strait of Hormuz using satellite observations and model simulations. Mar. Pollut. Bull. 115(1): 315–323. Doi: https://doi.org/10.1016/j.marpolbul.2016.12.024.

Zhao, J., M. Temimi and H. Ghedira. 2017b. Remotely sensed sea surface salinity in the hyper-saline Arabian Gulf: Application to landsat 8 OLI data. Estuar. Coast. Mar. Sci. 187: 168–177. Doi: https://doi.org/10.1016/j.ecss.2017.01.008.

Section II

Inorganic Pollutants and Associated Effects

4

Mercury Cycling in the Coastal and Open Ocean and Associated Health Risks

A Global Overview

Noelia S. La Colla,[1,*] *Sandra E. Botté,*[1,2]
Carlos E.S.S. Monteiro,[3,4] *Rute I. Talhadas Cesario,*[3,4]
Marcos A.L. Franco[5] and *Jorge E. Marcovecchio*[1,6,7]

1. Introduction

The estuaries and oceans of the world are the ultimate repository for every pollutant deliberately or accidentally discharged, with land acting as the main contributor for water pollution, either through runoff and discharges (44%) or through the atmosphere (33%). Maritime activity and shipping accidents contribute 12% to the total amount of pollution in the seas. Sewage discharges, dumping of garbage, offshore drilling and mining make up the remaining contributions (11%) (Potters 2013).

Estuaries and coastal marine waters are acknowledged as the aquatic systems with the highest pollution impact, since the vast majority of countries have 80% to 100% of their population living

[1] Instituto Argentino de Oceanografia (CONICET-UNS), Carrindanga km 7.5, Bahía Blanca, 8000, Argentina.
[2] Departamento de Biología, Bioquímica y Farmacia, UNS (8000) Bahía Blanca, Buenos Aires, Argentina.
[3] Instituto Português do Mar e da Atmosfera, Av. Alfredo Magalhães Ramalho, 6, 1495-165 Lisboa, Portugal.
[4] Environmental Biogeochemistry, Centro de Química Estrutural, Instituto Superior Técnico, Universidade de Lisboa, Av. Rovisco Pais, 1; 1049-001 Lisboa, Portugal.
[5] Universidade Estadual do Norte Fluminense Darcy Ribeiro. Departamento de Biociências e Biotecnologia, Laboratório de Ciências Ambientais. Av. Alberto Lamego, 2000, Campos dos Goytacazes, 28013-602, RJ, Brasil.
[6] Universidad de la Fraternidad de Agrupaciones Santo Tomás de Aquino, Gascón 3145, Mar del Plata, 7600, Argentina.
[7] Universidad Tecnológica Nacional – FRBB, 11 de Abril 445, Bahía Blanca, 8000, Argentina.
   Emails: sbotte@iado-conicet.gob.ar; carlos.monteiro@ipma.pt; rcesario@ipma.pt; malfranco@yahoo.com.br; jorgemar@iado-conicet.gob.ar
* Corresponding author: nlacolla@iado-conciet.gob.ar

close to the coastline, within the first 100 km, and coastal regions have average population densities of over three times the global mean (Deycard et al. 2014). The human burden enhances pressure on these coastal ecosystems through habitat conversion, infrastructure for manufacturing, waste products disposal, and other activities (Martínez et al. 2007). Rapid population growth without proper planning creates significant increase in waste produced. As the discharge and dumping of waste in coastal environments is nowadays being closely monitored, increasing pressure is expected on the deep open ocean as a final repository of waste and unwanted substances from a variety of human activities (Kennish 1997, Mengerink et al. 2014).

The deep ocean comprises waters deeper than 200 m, at the shelf break, and visualized by a clear change of fauna from shallow to deep waters (Ramirez-Llodra et al. 2011). The deep ocean forms the largest ecosystem on Earth, covering an area of 360 million km^2, equivalent to about 50% of the surface of the Earth, and with an average depth of 3800 m (Ramirez-Llodra et al. 2011). The waters in the deep sea have an average residence time of up to 1000 years and the paths back to the surface are diffuse and very remote (Tyler 2003). The deep ocean, one of the least contaminated environments on Earth, is the deliberately planned dumping ground for radioactive waste, sewage, litter, unused munitions, toxic chemicals, and terrestrial mine tailings and is also the unintended final resting place for a variety of land-based anthropogenic debris and pollutants (Ramirez-Llodra et al. 2011, Cau et al. 2018, Czub et al. 2018).

Among pollutants, metals are elements of interest because of their non-biodegradability, environmental persistence, biogeochemical recycling and biomagnification in the food chain (e.g., Papagiannis et al. 2004, Duarte et al. 2014, De Souza Machado et al. 2016). Metals occur in the aquatic environment both as a result of natural sources, mainly the weathering of soil and rocks, erosion, forest fires, and volcanic eruptions, and from a huge variety of human activities (e.g., FranÇa et al. 2005, Rahman et al. 2012, Yao et al. 2016). Within the aquatic system, the mobility and partitioning of metals depends on factors such as the pH, salinity, redox conditions, background levels, resuspension of sediment due to dredging, tidal action, winds and storms, flocculation and coagulation of colloidal material and adsorption onto suspended particles (Zwolsman and Van Eck 1999, Oursel et al. 2014).

Metal accumulation in coastal systems is a complex, worldwide problem since these environments are commonly used for disposal and dilution of terrestrial wastes (Spencer et al. 2006) and thereby continuously degraded. Anthropogenic metal sources range from industrial activities and manufacturing processes to poorly treat domestic and agricultural wastewaters containing high metal concentrations, often discharged into the environment in developing countries (Gupta 2008, Bhattacharya et al. 2015). Even though metal emissions have been greatly reduced in industrialized countries, the legacy of such indiscriminate emissions over the past 150 years remains as a potential source of metals available even after changes in land use (Lacerda 2007).

Among the metal elements, mercury (Hg), in its elemental form, exists as liquid at room temperature. It is easily subjected to liquid–vapor changes because of its low latent heat of evaporation (295 kJ/kg) and relative absence from ambient air (Estrade et al. 2009, Rice et al. 2014). Hg is rarely found in the reduced form and it is generally associated with minerals such as cinnabar (HgS), the most common ore, as well as living stonite ($HgSb_4S_8$) and corderoite ($Hg_3S_2Cl_2$) (Ariya et al. 2015). Hg has long been recognized as a nonessential and naturally occurring trace element, with background environmental concentrations present since long before humans appeared (UNEP 2013). Hg is toxic to humans and animals and listed as a high-priority environmental pollutant within the Convention for the Protection of the Marine Environment of the North-East Atlantic (OSPAR Convention 1992) and the United States Environmental Protection Agency (US EPA) (Lillebø et al. 2011). Environmental bodies such as the World Health Organization (WHO) and the United Nations Environment Programme (UNEP) have called for reduction or, wherever possible, elimination of its use.

The ocean has a main role in controlling the fate and transport of Hg at the Earth's surface. Moreover, the pollution of the ocean with high Hg concentrations is of worldwide concern due to its capacity to transform into a highly bio-accumulative organic form for organisms and humans. The aim of this chapter is therefore to provide a global and up-to-date overview of our understanding of Hg sources and cycling in the coastal and open oceans, affecting aquatic organisms and human populations.

2. Mercury Sources in the Aquatic Environment

Mercury in the environment exists in three main forms: elemental Hg (Hg^0) as metallic Hg and Hg vapor, inorganic Hg compounds (mercurous Hg_2^{2+} and mercuric Hg^{2+}), and organic Hg, which includes Hg bonded to a structure containing carbon atoms (methyl-, ethyl- and phenyl-Hg or similar groups). The different Hg species result in specific physicochemical properties and their deposition rates vary and change according to their transformation (Ariya et al. 2015). Hg species are subject to complex inter-conversions, principally through oxidation–reduction and methylation–demethylation reactions, and transportation on a worldwide scale; processes that together are termed the "Hg cycle" (Holmes et al. 2009) (Fig. 1).

The ocean plays a crucial role in the Hg global cycle as it is a source of Hg to the atmosphere, as well as a sink for Hg from the terrestrial and atmospheric environments with toxic effects on aquatic organisms (Mason and Fitzgerald 1996). It acts both as a dispersion medium and as an exposure pathway by containing a substantial fraction of the global Hg reservoir, strongly affecting atmospheric concentrations through air–sea exchange (Soerensen et al. 2010, Zhang et al. 2015b). Various studies have established significant differences concerning the quantity of Hg presently circulating in

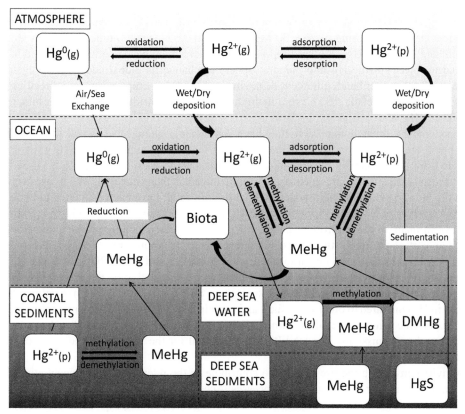

Fig. 1. Overview of global mercury cycling in the environment. (g): gaseous form; (p): particulate form; MeHg: methylmercury; DMHg: dimethylmercury.

seawater. Lamborg et al. (2014) estimated that the ocean contains 278 Gg of dissolved total Hg, taking the North and South Atlantic concentration profiles each to represent a quarter of the whole ocean and the Pacific profiles to represent the other half. Amos et al. (2013) obtained another estimate of 222 Gg of Hg. Other reports estimate Hg concentrations in the oceans of only 66 Gg (Zhang et al. 2014) or highest concentrations of approximately 350 Gg of Hg (Sunderland and Mason 2007).

Mercury cycling between the ocean, atmosphere, and terrestrial ecosystems undergoes a series of complex chemical and physical processes. Among them, the atmosphere, through wet and dry deposition, generally dominates the majority of Hg inputs to open ocean (Mason and Sheu 2002, Sunderland and Mason 2007, Soerensen et al. 2010). Various global Hg assessments (e.g., Cohen et al. 2004, Selin 2009, Pacyna et al. 2010, Driscoll et al. 2013, Gustin et al. 2016) have identified the atmosphere as the dominant pathway for the distribution of Hg on a global scale. Hg is predominantly in its gaseous elemental form (Hg^0) and is oxidized to Hg^{2+}, which is then rapidly deposited. Indeed, Hg cycling models assumed that ocean water achieves a rapid equilibrium with atmospheric inputs (Strode et al. 2007, Sunderland and Mason 2007). According to the UNEP (2013), Hg atmospheric depositions to the open ocean are about 3.7 Gg of Hg yr^{-1}, and come from a myriad of natural and anthropogenic sources. In aquatic ecosystem from remote areas, the primary Hg sources are long-range atmospheric transport and the subsequent Hg deposition, directly to water or indirectly via deposition to terrestrial watersheds (Wang et al. 2009).

Regarding initial Hg sources to the atmosphere, natural Hg arises from the degassing of the Earth's crust through volcanic eruption, geothermal activity and marine hydrothermal vents from the oceans (e.g., Ferrara et al. 2000, Nriagu and Becker 2003, Pyle and Mather 2003, Bagnato et al. 2015). It is speculated that there are about 5 to 6 Gg of Hg in the atmosphere (Amos et al. 2013, UNEP 2013, Ariya et al. 2015, Obrist et al. 2018). The UNEP (2013) estimated emissions of around 80–600 Mg Hg yr^{-1} from volcanoes, whereas Bagnato et al. (2015) estimated volcanic flux from continuous passive degassing of 76 ± 30 Mg Hg yr^{-1}. In the volcanic plume, more than 90% is in the dominant species of Hg^0. Studies proposed that cataclysmic volcanic eruptions would have the potential to inject volatile Hg into the stratosphere to change its global and regional cycle for a few years. On the other hand, quiescent degassing and moderate eruptions exhale directly into the troposphere (Bagnato et al. 2015). From hydrothermal vents, Outridge et al. (2018) estimated less than 600 Mg Hg yr^{-1}.

Since ancient times, human activities have significantly altered the natural global Hg cycle through emissions to the atmosphere (Streets et al. 2011). Global anthropogenic Hg emissions can originate from a great number of economically important activities such as mining, the use of Hg in precious metal extraction, its presence as a trace contaminant in many materials, and its use in products and in the industry. Current anthropogenic Hg emissions to air have been estimated to be between 2 Gg yr^{-1} (Pacyna et al. 2010, Pirrone et al. 2010, UNEP 2013) and 2.5 ± 0.5 Gg yr^{-1} (Outridge et al. 2018). In many of the anthropogenic activities, Hg is released as a by-product, unintentionally emitted as part of the impurities in fuels and raw materials. There are several studies regarding unintentional Hg emissions to the atmosphere (e.g., Yudovich and Ketris 2005, Streets et al. 2011, Giang et al. 2015, Wang et al. 2016). Coal combustion is estimated to be the first global source of unintentional anthropogenic Hg emissions to air (e.g., Streets et al. 2017, 2018). Coal is relatively enriched in Hg compared to other fossil fuels and contains 0.01–1 mg kg^{-1} Hg by mass (Mukherjee et al. 2008). As large volumes of coal are burned, all the Hg within is vaporized as Hg^0 to the atmosphere, contributing large emissions. The UNEP (2013) reported for the year 2010 that coal combustion released 474 Mg of Hg to the atmosphere.

Activities involving the use of oil also contribute to the release of low concentrations of Hg to the environment, during the refining processes. According to the UNEP (2017), global average Hg concentrations in crude oil were between 3.4 mg Mg^{-1} and 5.7 mg Mg^{-1} or about 15–20 Mg of Hg in the year 2015. The raw materials and the burning of fossil fuels involved in cement production are another source of anthropogenic Hg. They were estimated to contribute 10% to the global Hg annual emission inventory estimated by the UNEP in 2010 (UNEP 2013). China, as the largest cement producer in the world, accounting for over half of the world's production, contributed with a total of 93.5 Mg Hg yr[-1] in 2010 (Zhang et al. 2015a). Non-ferrous metal production is among the top emitters and has become the forefront of efforts to combat global atmospheric Hg pollution (Li et al. 2017).

On the other hand, Hg emissions may also be the result of an intentional use. Artisanal and small-scale gold mining (ASGM) uses Hg to create an amalgam separating gold from other materials. Artisanal gold miners have been reported in over 50 countries, where amalgamation with Hg added to the whole ore to extract gold is the most popular method used. According to Veiga et al. (2006), miners lose 1 to 2 g of Hg per gram of gold produced, amounting to an estimated annual loss, with a concomitant release to the environment, of between 800 and 1000 Mg of Hg. The amounts of Hg emitted are often difficult to estimate, since it is a typically unregulated activity. There are several reports regarding ASGM, world emissions and human health risks (e.g., Telmer and Veiga 2009, Gibb and O'Leary 2014).

Another major industrial use of Hg is found in the chlor-alkali plants (CAPs), where Hg-cell technology is used in the production of chlorine and caustic soda. According to the UNEP (2017), of the three main chloride production methods, Hg-cell accounts for about 8% of the global chlorine production capacity of about 60 million Mg. While part of the Hg waste generated during the normal operation of Hg-cell chlor-alkali facilities may sometimes be sent to recycling, the rest goes to disposal. Moreover, according to Biester et al. (2002), most Hg emitted from CAPs is dispersed over long distances, and only small amounts are deposited in the vicinity of the plants. The environmental impact of Hg emissions from CAPs has already been dealt with in several articles (e.g., Biester et al. 2002, Ullrich et al. 2007, Reis et al. 2009) and direct releases were estimated to be 2.8 Mg yr[-1] (UNEP 2013).

The production of polyvinyl chloride (PVC) from acetylene (UNEP 2013) is another source of Hg to the environment. Hg is part of a catalyst used to ensure the reactivity and selectivity of vinyl chloride monomer (VCM) formation, which is afterwards transformed into PVC. The catalyst is activated carbon impregnated with Hg chloride ($HgCl_2$) and normally contains 8 to 15% of it. It has been estimated that 1 to 2 kg of Hg-impregnated catalyst is consumed to produce 1 Mg of PVC throughout the process. According to Ren et al. (2014) the process for producing PVC uses the largest amount of Hg in China. Besides these major contributors, there are many other Hg-containing consumer products such as batteries, fluorescent bulbs, thermostats, thermometers, and car switches that incorporate different amounts of Hg into the environment.

Hg can also be released to the environment from re-emission processes. Re-emission is the result of natural processes that convert inorganic and organic forms of Hg that were previously deposited on soils, surface waters, and vegetation, to its elemental form, which is volatile and therefore able to return to the air. It is difficult or impossible to identify the specific origin (natural or anthropogenic) of the Hg by the time it is re-emitted (UNEP 2013), and thus it should not be considered as a natural source. While Hg emission sources increase the global pool of Hg in surface reservoirs, Hg re-emission sources redistribute it among and within ecosystems (Driscoll et al. 2013). The re-emission of previously deposited Hg is associated with land use changes, forest fires, biomass burning, meteorological conditions and exchange mechanisms of gaseous Hg at air–water/topsoil/snow–ice pack interfaces (Selin 2009, Pirrone et al. 2010).

Like the atmosphere, the terrestrial system is a source of the previously atmospherically deposited Hg to aquatic systems (Grigal et al. 2000), terrestrial runoff being the main Hg source to streams and lakes (Obrist et al. 2016). Hg sequestered in terrestrial ecosystems represents a legacy of past natural and anthropogenic emissions that may have been deposited and accumulated throughout time (Grigal et al. 2000, Amos et al. 2013). The soils are one of the largest actively cycling reservoirs in the biogeochemical cycle of Hg and they generally account for over 90% of the Hg stored at the whole ecosystem level in terrestrial ecosystems (Obrist et al. 2016). Soil Hg pool sizes were estimated globally to be between 150 Gg (Outridge et al. 2018) and 300 Gg (Amos et al. 2013). Nevertheless, other simulations using additional observational constraints estimated higher worldwide soil Hg pools of 250–1000 Gg with a best estimate of 500 Gg (Streets et al. 2011, Obrist et al. 2018).

Vegetation plays an important role in the Hg cycle that ultimately reaches the coastal area through runoff. Hg accumulates on leaves through precipitation and dry deposition and can be incorporated into leaves via three processes: oxidation and adsorption of Hg^0, Hg^0 uptake by stomata or adsorption of Hg^{2+}, and particulate Hg (Hg-P) (Schroeder and Munthe 1998, Teixeira et al. 2012). Hg in aboveground biomass is taken mainly from the atmosphere, whereas Hg in the roots comes from the soil (Selin 2009). For example, in estuaries, metals like Hg are bound to sulfides in anoxic sediments. However, plants can oxidize the sediments in the root zone through the movement of oxygen downwards through aerenchyma tissue (Moorhead and Reddy 1988), and this oxidation can remobilize the metal, thus increasing the otherwise low availability of Hg in wetland sediments. The Hg deposited to vegetation is then incorporated into the soil pool via throughfall and litterfall (Grigal et al. 2000). Upon deposition to the ground via throughfall or contained in senesced leaves, needles, bark and dead wood, Hg is sequestered by reduced sulfur groups in the humic matter of the organic soil (Friedli et al. 2007), then reaches the aquatic system through runoff.

The impact of rivers as carriers of Hg to the marine environment has been estimated to be around 2,800 Mg of Hg yr^{-1}, and may have been understudied. Predominant Hg sources include direct release from Hg-containing effluents, atmospheric Hg deposition and river runoff containing atmospheric Hg deposits that accumulated in terrestrial environments (Obrist et al. 2018). Though previous studies suggested that 1,000–2,000 Mg of Hg yr^{-1} were discharged from rivers to ocean margins, the global present-day estimates are 5,400 Mg of Hg yr^{-1} (Amos et al. 2014), of which 28% reaches the open ocean and the rest is buried in ocean margin sediments by particle settling on short time scales (Zhang et al. 2015b). The estimated discharges are larger than the previous ones due to recently published data suggesting greater contamination in Asian rivers (e.g., Guo et al. 2008, Lin et al. 2012), even though there is uncertainty driven by the scarcity and large variability of measurements of Hg concentrations in rivers (Amos et al. 2014).

3. Mercury Contributions from a Global Point of View

Mercury cycling among various reservoirs (atmosphere, lithosphere, biosphere, and hydrosphere) results both from anthropogenic and natural Hg emission sources. Sedimentary and ice core records provided evidence of a two- to five-fold enrichment in present-day atmospheric Hg deposition relative to preindustrial levels (Amos et al. 2013, Obrist et al. 2018). Hg emissions from different sources have been estimated in a number of global inventories (e.g., Pirrone et al. 2010, Sprovieri et al. 2010, UNEP 2013, Pacyna et al. 2016, Obrist et al. 2018, Outridge et al. 2018). Europe and North America showed growing emissions over the 19th and 20th centuries, with the addition of emissions from Asian territories over the past decades (Pacyna et al. 2010, Streets et al. 2011, Amos et al. 2013). Despite significant emission reductions in Europe and North America over the past 15 years, global Hg emissions have not changed significantly because of emissions growth in other parts of the world (McHugh et al. 2016). According to the latest reports from the UNEP emission inventory (UNEP 2013, 2017), the greatest proportion of anthropogenic Hg emissions to the atmosphere comes from Southeast Asia, contributing about 40% of the global total. China is considered to be the largest anthropogenic Hg source region in the world (Fu et al. 2012, Zhang et al. 2015a). Indeed, China, Mexico and Indonesia are the major sources of Hg primary production, with mining as the most significant source of Hg used in products and processes and responsible, in 2015, for 1,630 to 2,150 Mg of Hg. The second largest source of Hg supply, with an emission of 1,040 to 1,410 Mg, is due to product and waste recycling, dominated by recycling of the catalysts used in the production of VCM in China (UNEP 2017). Energy consumption, especially fossil fuel consumption, as well as the production of cement, non-ferrous metals, iron and steel, are other main sources of anthropogenic Hg emissions in China (Wang et al. 2014).

The UNEP (2017) estimated that, between 2005 and 2015, there was an increase in the use of Hg in sectors such as ASGM, VCM and measuring and control devices, whereas in other Hg uses such as in CAPs, batteries and lamps, a general decline was apparent. For instance, Hg-cell CAPs are gradually being phased out in Europe and North America, being replaced by cleaner technologies. In less developed countries, where stringent environmental controls are lacking, there are still significant remaining emissions from existing Hg-cell plants (Ullrich et al. 2007).

In view of these global Hg emissions, 128 countries, including China and Mexico, signed an international agreement in 2013 called the Minamata Convention in order to control the supply, trade and use of Hg, reduce its atmospheric emissions from point source categories (e.g., coal-fired power plants and smelters), and develop additional Hg research (Ariya et al. 2015). This agreement entered into force in August 2017 and, thus, more accurate emission inventories are expected in the future. The Convention on the Prevention of Marine Pollution by Dumping of Wastes and Other Matter is also intended to protect the marine environment from human activities. Its objective is to promote the effective control of all sources of marine pollution and to take all practicable steps to prevent pollution of the sea by dumping of wastes and other matter. Under the protocol all dumping is prohibited, except for possibly acceptable wastes including sewage sludge, dredged material, and industrial fish processing waste, suggesting that by-products from human activities at sea, such as oil and gas, sediment and tailings from mining, as well as organic matter from fish processing, can be disposed of at sea (Tyler 2003).

4. Mercury Cycling from the Atmosphere to the Ocean

Understanding atmospheric Hg redox chemistry is critical for determining whether Hg will be deposited in waters or not, and for understanding its biogeochemical cycling. Hg has been regarded as a "global pollutant" because of the long residence time of certain Hg species with the potential to travel long distances within air and water masses. Hg emitted in one part of the world may be transported through the atmosphere to another part of the world, conditioned by its chemical behavior and by the wind direction and speed (Sundseth et al. 2017); it can therefore have an impact on soil, water and vegetation in remote and pristine areas (Sprovieri et al. 2010, De Simone et al. 2014). Evidence of the "global transport" of Hg includes its presence in the Arctic and Antarctic (Pirrone et al. 2008, Kim et al. 2016), regions with no significant emissions. Indeed, Hg accumulation in marine biota of the Arctic and Antarctic coastal areas has been extensively reviewed (e.g., Rigét et al. 2011, Dietz et al. 2013, Braune et al. 2015, McKinney et al. 2015).

In the atmosphere, Hg exists mainly as gaseous Hg^0, representing 90% to 99% of the total Hg amounts (Gustin et al. 2013), and as inorganic Hg compounds, but only in trace amounts. Hg^0 is the least chemically reactive form with low water solubility; it takes part in the volatilization process at the earth surface leading to an atmospheric lifetime between 0.8 to 1.7 years (Ariya et al. 2015). This makes it a global pollutant (Dastoor and Larocque 2004). Deposition of atmospheric Hg^0 involves its becoming charged and precipitating in its inorganic forms (Berry and Ralston 2008) (Fig. 1). Even though it was first thought that the oxidants OH and O_3 had a main role in the oxidation process of Hg^0, the Br radical may play the predominant role as oxidant (Shah et al. 2016, Horowitz et al. 2017). The mechanism might involve a two-step process, where Hg^0 reacts with the Br radical to form HgBr, and this compound then reacts with multiple potential oxidants (e.g., I, OH) to form different inorganic Hg species (Obrist et al. 2018).

Inorganic Hg species are, on the other hand, highly water-soluble and reactive species, where Hg^{2+} compounds may occur as gases, most probably $HgCl_2$, $Hg(OH)_2$, or compounds of other halides (Lindberg and Stratton 1998, Sheu and Mason 2001, Gustin et al. 2013). Also, they may be associated as Hg-P with airborne particles such as dust, sea-salt aerosols, or ice crystal or may be the result of the adsorption of Hg^{2+} compounds onto atmospheric particles (Poissant et al. 2005). Hg^{2+} species and Hg-P have atmospheric residence times that range from hours to up to a few weeks, and as a result are generally deposited locally or regionally (Driscoll et al. 2013, Marusczak et al. 2016, Horowitz et al. 2017). They are important Hg species with respect to atmospheric deposition due to their large dry deposition velocities and scavenging coefficients. Indeed, wet removal of water-soluble Hg^{2+} compounds and Hg-P are more efficient than gaseous Hg^0 (Shannon and Voldner 1995), and 60% of the global Hg deposition is estimated to occur by wet and dry deposition of these forms of Hg (Selin et al. 2008).

Once in the aquatic system, the main chemical forms of Hg are Hg^0, inorganic Hg complexes and organic Hg forms (Fig. 1), and are strongly influenced by the redox and pH conditions, temperature and biological interactions, inorganic and organic complexing agents concentrations, and other factors (Horvat et al. 2003, UNEP 2013). Most surface waters are supersaturated with Hg^0 relative to the atmosphere, and therefore elemental Hg is readily lost from the water to the atmosphere (Horvat et al. 2003, Soerensen et al. 2010). Indeed, about 70% of the Hg deposited into the ocean is re-emitted to the atmosphere (UNEP 2013) after being reduced to Hg^0. Newly deposited Hg has been shown to preferentially revolatilize, in a phenomenon that has been termed "prompt recycling" (Selin 2009). Hg^0 evasion depends mostly on water temperature, salinity, wind speed and available water-borne binding sites (Loux 2004) and is estimated to be 30% of the total annual flux of Hg to the atmosphere (Mason et al. 1995). Hg^0 production by the reduction of inorganic Hg compounds is mainly mediated by aquatic microorganisms (Ullrich et al. 2001, Kotnik et al. 2007, Bratkic et al. 2018). The second important source of elemental Hg is by photochemical transformations of organic matter that may result in abiotic Hg^{2+} reduction, mediated by UV light and dissolved organic matter (DOM) (Ullrich et al. 2001, Gu et al. 2011). Moreover, intense geotectonic activity may be another source of Hg^0 (Horvat et al. 2003), especially in the Mediterranean basin, which is tectonically very active (Kotnik et al. 2007). Hg^0 exchange at the atmosphere–surface water is then bidirectional: both dry deposition and evasion from the ocean occur at the interface, whereas wet deposition of Hg^0 is insignificant due to its low solubility (Song et al. 2016).

Besides Hg evasion, Hg deposited in surface waters is assimilated and available as a substrate for microbially mediated methylation, following incorporation of methyl Hg (MeHg) in the marine food webs (Horvat et al. 2003). Once inorganic Hg is available in the water fraction it is transformed into MeHg, the Hg species that ultimately controls burdens in upper-trophic level biota (Fig. 1). Possible biotic sources of MeHg in the aquatic environment may include its production in coastal and shelf sediments (Mason and Sheu 2002, Mason et al. 2012), and from *in situ* water column methylation processes (Lehnherr et al. 2011, Mason et al. 2012). Moreover, as abiotic MeHg sources, hydrothermal vents and deep-sea sediments (Mason et al. 2012) may release MeHg directly into the aquatic system. In freshwater environments, the most stable MeHg species is found as methylmercuric hydroxide (CH_3Hg-OH), whereas in seawater MeHg is present mainly as the chloride form (CH_3Hg-Cl) (Ullrich et al. 2001).

In coastal environments, biotic methylation of inorganic Hg into MeHg is also done by microorganisms, and Hg methylation occurs in both anaerobic and aerobic conditions (Batrakova et al. 2014). Sulfate and iron-reducing bacteria are the main agents responsible for MeHg production in sediments and estuaries (King et al. 1999, Kerin et al. 2006, Batrakova et al. 2014, Schartup et al. 2014). The rate of MeHg formation is mediated by an array of *in situ* biogeochemical factors, classified as factors that influence bacterial activity and those that can alter inorganic Hg bioavailability (Horvat et al. 2003, Schartup et al. 2014). Environmental characteristics such as the sulfur concentrations, total organic carbon, the sediment structure and composition, temperature, salinity and pH are known to affect bacterial activity. Moreover, the bioavailability of inorganic Hg in sediments is associated with the partitioning of Hg between sediment–porewater fractions (Schartup et al. 2014). This sediment–porewater partitioning coefficient is controlled by the amounts of Hg-binding ligands, such as sulfide and thiols, in both solid and dissolved phases (Cesário et al. 2017a). In addition, the rate of methylation decreases with increasing salinity, most probably because of the inhibitory influence of chlorine complexes (Batrakova et al. 2014). Within estuaries, Hg methylation is facilitated by the presence of wetlands, which act as sources of DOM, reducing oxygen levels and thereby increasing the activity of sulfate-reducing bacteria (De Marco et al. 2006). The methylation of Hg^{2+} in the water column of coastal areas is less clear compared to the MeHg production in the open waters (Bratkic et al. 2018) with imprecision in the signals of the water column methylation due to the sedimentary sources and the watershed inputs (Balcom et al. 2015).

In the open ocean, MeHg production occurs mainly in the water column (Lehnherr et al. 2011, Mason 2012) (Fig. 1). Hg conversion into its organic form takes place largely at intermediate depths, between 200 and 1000 m in the water column, and mainly through biotic pathways (Ullrich et al. 2001). For instance, two main zones of MeHg productivity have been reported in the Mediterranean Sea water column: one at the bottom of the euphotic layer and the other at the oxygen minimum in the thermocline (Cossa et al. 2009, Zagar et al. 2014). Biotic Hg methylation occurs through natural bacterial processes and the methylation activity is usually lower than in freshwater and coastal environments because of the high salinity and the presence of charged sulfide and chloride complexes that promote demethylation processes (Kotnik et al. 2007). Hg methylation in the water column is correlated with the decomposition of organic matter (Batrakova et al. 2014) and information on the microorganisms involved in the MeHg production in the water column is still limited (Mason 2012). According to current modeling, MeHg stays in the upper ocean for about 11 years (UNEP 2013). Even though abiotic methylation is not generally important in the open ocean, one of the most substantial abiotic sources of MeHg is the activity of hydrothermal vents and submarine volcanoes. MeHg produced by hydrothermal fluids may deposit in sediments from the deep ocean. The contribution of abiotic sources to MeHg concentrations is only significant locally and may average 1–2% of all sources of MeHg in the ocean (Batrakova et al. 2014). Moreover, MeHg in the ocean may also be produced from dimethyl Hg (DMHg) (Lehnherr et al. 2011).

In the deep ocean waters, Hg enrichment is smaller than in surface and subsurface waters due to the long time scales for lateral and vertical Hg transport to the deep ocean (Sunderland and Mason 2007). The mobilized Hg is biologically methylated to DMHg in the low oxygen region, thus the dominant methylated species is DMHg (Cossa et al. 1994, Mason et al. 1995) (Fig. 1). Moreover, the bottom sediments of oceans are thought to be the ultimate Hg sink, where it is deposited as highly insoluble HgS (Baeyens et al. 2012). On the other hand, volatile and unstable DMHg diffuses in the mixed layer and/or decomposes into MeHg and subsequently Hg^0, which tends to diffuse back into the mixed layer and eventually into the atmosphere (Cossa et al. 1994). No DMHg has been detected in surface oceanic waters (Horvat et al. 2003) and it is not considered to be available for accumulation in aquatic organisms (Horvat et al. 2003, Cossa et al. 2009). Its lack of presence in surface oceanic waters is likely due to the presence of light and the additional potential loss of DMHg via gas exchange at the water surface (Mason et al. 1995). Many studies have already discussed MeHg and DMHg distribution and transformations in the aquatic systems (Louis et al. 2007, Ogrinc et al. 2007, Selin 2009).

5. Mercury Interactions along the Aquatic Food Chain

In the aquatic environment, Hg methylation into MeHg is the fundamental process by which Hg is bioaccumulated and biomagnified throughout the food chain. MeHg shows high stability and low water solubility and is considered to be relatively lipid-soluble (Ullrich et al. 2001). Hg is taken up by organisms that incorporate MeHg much more efficiently than its inorganic form, resulting in Hg plankton concentrations that are as high as 10,000 times the levels found in seawater (UNEP 2013). Both $HgCl_2$ and CH_3HgCl diffuse through membranes at about the same rate and they are both reactive with cellular components and are efficiently retained by microorganisms. Nevertheless, the efficiency of transference between phytoplankton and zooplankton is four times as high for MeHg as for Hg^{2+} (Mason et al. 1996). According to Morel et al. (1998), the difference between bioaccumulation of Hg^{2+} and MeHg relies on the fact that Hg^{2+} becomes bound mainly to the particulate cellular material (membranes) of phytoplankton that are excreted rather than absorbed by zooplankton. MeHg, in turn, is associated with the soluble fraction of the phytoplankton cell and is efficiently assimilated by the zooplankton species (Mason et al. 1996). This difference in the efficiency between Hg^{2+} and MeHg is applicable to other unicellular microorganisms and their predators (Morel et al. 1998). Moreover, DMHg and Hg^0 are not even bioaccumulated (Ullrich et al. 2001) simply because they are not reactive and are not retained in plankton species; they diffuse out as readily as they diffuse in (Morel et al. 1998).

Among aquatic organisms, fishes pose serious concern about risk to Hg exposure largely as MeHg, as it can be found in nearly all fish species, although MeHg levels are different according to many fish characteristics (Canli and Atli 2003, Akan et al. 2012, Kim et al. 2016). Hg in fish is mainly accumulated through the diet, accounting for 32% to 92% of the total Hg accumulation (Kim et al. 2016, Bradley et al. 2017). Direct uptake from the water across the gills is of minor importance and is dependent on water chemistry; while an increase in DOM values may decrease Hg uptake, the presence of chloride concentrations favors the production of $HgCl_2$ and thus increases Hg absorption (Wang and Wang 2010). Once Hg is assimilated in fishes, it accumulates in the protein fraction of the muscle as it binds to thiol groups (Clarkson and Magos 2006, Bosch et al. 2016). MeHg can be easily accumulated by fish because it is efficiently assimilated and difficult to eliminate (Wang and Wang 2017, Chouvelon et al. 2018). Indeed, MeHg contributes over 95% of total Hg in marine organisms on top of the food chain (Fitzgerald et al. 2007), except in pilot whales and probably other sea mammals where inorganic Hg may account for 50% of the total Hg values (Clarkson and Magos 2006). Other variables affecting Hg pollution of aquatic food webs include the landscape methylation potential, food chain length, and food web complexity (Chasar et al. 2009, Lavoie et al. 2013).

In general, Hg concentrations in fish tend to increase with age and/or size. MeHg concentrations undergo a remarkable biomagnification process to achieve their highest values in the muscle tissues of the long-lived predatory fish, particularly piscivorous fish such as tuna and shark in ocean waters (Clarkson and Magos 2006, Storelli et al. 2007, Carroll and Warwick 2017). Large predatory fish may contain 100,000 times as much MeHg as the surrounding water environment (Kim et al. 2016). Top predator fish species are usually used as bioindicators of Hg pollution and biomagnification as Hg is assimilated very efficiently by higher trophic levels (Zamani-Ahmadmahmoodi et al. 2014, Chouvelon et al. 2018). For instance, muscles of four shark species and their main prey were analyzed to determine Hg contents in Baja California peninsula, Mexico (Maz-Courrau et al. 2012). Linear regression showed a certain tendency for Hg concentration to augment at larger sizes, showing that younger sharks have been exposed for less time to the metal than the adults. Some other reasons for these differences may be that larger individuals have more efficient predation mechanisms (and consequently accumulate more) or different trophic habits (feeding on pelagic organisms, piscivorous), that some species (i.e., blue shark) have more efficient Hg elimination mechanisms (higher synthesis of metallothionein, a detoxification protein), and ontogenic trophic changes (Marcovecchio et al. 2015). This bioaccumulation property making an organism more prone to Hg biomagnification in food chains has been reviewed (e.g., Bosch et al. 2016, Lyons et al. 2017, Makedonski et al. 2017).

6. Mercury and Impacts to Public Health

Understanding Hg cycling is of particular relevance for worldwide public health since dietary consumption of marine fish and other seafood is a major route of Hg exposure in humans (Baeyens et al. 2003, Malczyk and Branfireun 2015, Sundseth et al. 2017, Liu et al. 2018), with MeHg as the fraction that poses risks to both marine animals and humans. Legislation of the Food and Agriculture Organization (FAO 2003) has stipulated a maximum Hg limit of 0.5 mg kg^{-1} (wet weight (w.w.)) in muscle tissues of fish and seafood with the exception of predatory fish (e.g., shark, tuna and swordfish), for which the upper limit is 1.0 mg of Hg kg^{-1} (w.w.). The WHO (2011) indicated that billions of people worldwide rely on fish as the major source of protein in their diet. World per capita apparent fish consumption increased from an average of 9.9 kg in the 1960s to 20.2 kg in 2015, at an average rate of about 1.5% a year. Of the global total consumption of 149 Tg of fishes in 2015, Asia consumed more than two-thirds (106 Tg at 24.0 kg per capita), whereas Oceania and Africa consumed the lowest share (FAO 2018). Globally, fish accounts for about 17% of animal protein intake. This share, however, exceeds 50% in many countries (Thilsted et al. 2014).

In humans, MeHg is widely distributed throughout the body (Bernhoft 2012) and 95% of the ingested MeHg is absorbed throughout the gastrointestinal tract. It may then bind covalently to glutathione and cysteine protein groups and enter the red blood cells and the brain (Spiller 2017). Its target tissues are mainly the neuroendocrine and nervous systems (Zhang et al. 2013). The biological half-life of MeHg is 39 to 70 d depending on body burden (Rice et al. 2014) and, since urinary excretion is negligible, MeHg is primarily eliminated from the body in an inorganic form through the action of the biliary system at the rate of 1% of the body burden per day. MeHg may be demethylated in the liver, where MeHg degradation may occur via interaction with hydroxyl radicals produced by cytochrome P-450 reductase. Demethylation may also take place in the mitochondria via interaction with a superoxide anion produced by the electron transfer system (Spiller 2017). Owing to the long biological half-life of MeHg, tremendous and undesirable effects have been observed on human health, particularly on the nervous system, resulting in problems including psychopathy, hearing and vision loss, loss of control over the body, general weakness, and anxiety attack, as well as effects on the fetal nervous system (Manavi and Mazumder 2018).

High-profile Hg poisoning incidents through fish consumption have drawn scientific and policy attention to the risks of Hg exposure to the general population. The first known event with a large number of human casualties was the Minamata incident in Japan. From 1932 and for approximately 40 years, over 150 Mg of Hg, mixed in with effluents, were discharged into the Minamata Bay from an acetaldehyde-producing factory (Minamata City 2007) that used Hg as a catalyst in the synthesizing process. Total Hg sediment concentrations near the overflow of the plant were as high as 2000 mg kg^{-1} (Tomiyasu et al. 2008). The disease was first acknowledged in 1956 and was recognized as being related to pollution 12 years later. Minamata disease is a disease of the central nervous system caused by MeHg and affects people who eat large amounts of fish and shellfish contaminated with MeHg compounds. The disease had devastating health effects on thousands of people who consumed the fish as their main food source (Ha et al. 2017). It is well documented that prenatal or postnatal exposure to MeHg produces adverse neurological impacts in both adults and children. Patients with chronic Hg poisoning complain of distal paresthesias of the extremities and the lips even 30 years after cessation of exposure to MeHg (Ha et al. 2017).

Although fish and seafood are the predominant sources of MeHg in the diets of humans and wildlife, there are other MeHg sources associated with health risks. Rice cultivated in areas contaminated with Hg can contain relatively high levels of MeHg (e.g., Horvat et al. 2003, Hong et al. 2016). MeHg has also been reported in different food groups such as meats of terrestrial animals (Ysart et al. 2000), as well as in chicken and pork, probably as a result of the use of fish meal as livestock feed (Lindberg et al. 2004). The agricultural use of MeHg fungicides has also caused great damage to human health. In 1971 in Iraq, farmers and their families, instead of using a fungicide-treated grain for planting, used it for homemade bread and some of the barley was fed to domestic animals, mainly sheep. This outbreak resulted in more than 6,000 cases of severe poisoning and more than 600 deaths throughout the country (Clarkson 1993).

Other Hg sources to global populations include dental amalgams. Dentists and dental personnel working with amalgam are chronically exposed to Hg vapors, which may enter the body through the respiratory system, pass readily into the circulation (Bernhoft 2012), and accumulate in their bodies at much higher levels than in those with other occupations (e.g., Gul et al. 2016, Kim et al. 2016, Bengtsson and Hylander 2017). In many countries, babies are also exposed to ethyl Hg through vaccination, since this form is the active ingredient of the preservative thimerosal used in vaccines (Clarkson et al. 2003).

7. Mercury–Selenium Interactions

MeHg toxicity in wildlife and human beings is affected by its interaction with other micronutrients. Selenium (Se) is considered an essential micronutrient for most organisms, including humans (Eisler 2000, Hamilton 2004), and has long been known to offer some protection against Hg toxicity (Berry and Ralston 2008, Zhang et al. 2013, Lino et al. 2018). Se is a metalloid that occurs naturally in the terrestrial and marine environment. It originates from natural sources like volcanic eruptions, wildfires, volatilization from water bodies and plants, and weathering of soils and rocks, being especially abundant with sulfide minerals of various metals, such as iron, lead, and copper (Lantzy and Mackenzie 1979, Eisler 2000, Arcagni et al. 2017). Average Se concentrations in the earth's crust are approximately 0.09 mg g^{-1} (Ralston et al. 2008). Anthropogenic Se entering the atmosphere is estimated to be 3,500 Mg yr^{-1}, and the major source is attributed to the combustion of coal and the irrigation of high Se soils for crop production (Eisler 2000). Anthropogenic Se pollution sources to the aquatic environment include agricultural drain water, sewage sludge, oil refineries, and mining of phosphates and metal ores (Hamilton 2004, Arcagni et al. 2017).

In the aquatic environment Se is found as dissolved selenite (SeO$_3^{2-}$), selenate (SeO$_4^{2-}$), and dissolved organic selenides (organic-Se^{2+}), even though only selenate is thought to be present in oxic seawater (Cutter and Cutter 2001, Ralston et al. 2008). Background Se values in marine waters are approximately 0.02–0.04 μg L^{-1} and under the impact of geogenic or anthropogenic Se emissions, Se concentrations in water are typically in the range of 1–10 μg L^{-1}, and may exceptionally exceed 100 μg L^{-1} (Ralston et al. 2008). Both selenite and selenate can be further reduced to insoluble elemental Se due to microbial activity; thus, in strongly reducing waters it is not expected that either selenite or selenate will be found.

Se species, such as selenite, selenate or organic-Se, are effectively taken up by organisms after the consumption of food and water. The incorporation occurs through passive diffusion and active uptake according to the specific molecular form, since it is an essential element for many biological functions (Gailer 2007). Se exists in selenoproteins as selenocysteine (Sec) and selenomethionine (SeMet) and is incorporated into the active sites of antioxidant selenoenzymes, where Se displays main biological functions that affect processes such as free radical metabolism, immune function, reproductive function, and apoptosis (Zhang et al. 2013). Nonetheless, Se may be harmful to humans and animals at high exposures because of the narrow margins between the amount that is essential and the levels associated with toxicity. Data available regarding the effects of high doses of Se to the aquatic biota have been reviewed in Hamilton (2004) and Ralston et al. (2008).

The Hg–Se interaction and the antagonistic phenomenon due to the protective effects of Se on Hg toxicity was first reported by Parizek and Ostadalova (1967), who demonstrated that selenite greatly reduced the mortality of rats treated with inorganic Hg (HgCl$_2$). In the following decades, Se has been shown to interact with Hg in a variety of organisms (Berry and Ralston 2008, Yang et al. 2008, Truong et al. 2014, Bjorklund et al. 2017, Wang and Wang 2017). The liver of aquatic organisms may be involved in the mechanisms for demethylation and/or sequestration of both organic and inorganic forms of Hg, and Se may play a role in Hg toxicity reduction with Hg redistribution to less sensitive target organs, and suggesting the formation of an insoluble, stable and inert Hg:Se complex (Kehrig et al. 2009). MeHg may predominantly bind to cysteine thiols to form a MeHg–cysteine complex (MeHg–Cys). Then MeHg–Cys reaches the active sites of selenoenzyme, where the sulfur (S) atom of MeHg–Cys can be directly replaced by the ionized Se of Sec to form the unavailable MeHg–Sec complex. This is due to the fact that the binding affinity between Hg and Se is several orders of magnitude greater than the affinity of Hg and S (Spiller 2017). Se may either reduce MeHg bioavailability (thus decrease assimilation) or increase the elimination of MeHg, leading to an overall decreased MeHg bioaccumulation (Wang and Wang 2017). On the other hand, the same Hg–Se interaction may also be responsible for the MeHg toxicity. The formation of the unavailable MeHg–Sec complex inhibits the bioavailability of MeHg but also results in an efficient sequestration of the biologically required Se in intracellular cycles of Sec synthesis, impairing selenoprotein form and function (Burger and Gochfeld 2013). Therefore, the irreversible MeHg–Sec complex is one of the possible mechanisms for MeHg toxicity, especially when the organism is in a Se-deficient state (Zhang et al. 2013).

8. Mercury in the Tagus River Estuary: A Case Study

For more than a century, the Tagus River Estuary (Fig. 2) has traditionally been a region of industry, agriculture, fisheries, salt extraction, and other economic activities. Consequently, it represents a strategic region of economical relevance. With increase in population and economic activities, the estuary banks have been overloaded with pollutants over the years (Dias and Marques 1999), resulting in the presence of inorganic and organic forms of Hg, such as MeHg. The Tagus River Estuary is a mesotidal system with approximately 320 km^2 of total area, from which about 120 km^2 correspond to intertidal areas, making it one of the largest estuaries in Europe. The estuary has relatively high hydrodynamic conditions and is vertically well mixed because of spring tides and low river flows (Fortunato et al. 1999, Alvera-Azcárate et al. 2003). However, stratification varies widely under extreme conditions (Fortunato et al. 1999).

The estuary has been contaminated by Hg since the mid-20th century, from two major industrial sources first identified by Figuères et al. (1985) and considered "hotspots" of Hg environmental pollution: Cala do Norte and Barreiro (CN and BRR in Fig. 2). Environmental concerns were raised

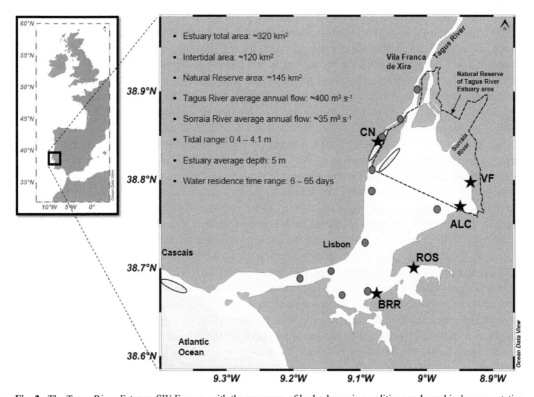

Fig. 2. The Tagus River Estuary, SW Europe, with the summary of hydrodynamic conditions and graphical representation of the most important sites studied over the past years: CN, Cala do Norte; BRR, Barreiro; ROS, Rosário salt marsh; ALC, Alcochete; VF, Vale de Frades. Black stars indicate the sites sampled for sediments, pore waters, microorganisms and salt marsh plants, depending on the study; grey circles indicate the sites sampled for waters and SPM; and ellipses indicate the sites sampled for benthic and pelagic organisms. Base maps were created using Ocean Data View software (Schlitzer 2017).

in this Iberian estuary and Hg research was actively pursued (e.g., Canário 2004, Monteiro 2010, Cesário 2017). Despite the decommissioning of most industries in Barreiro by the end of the 20th century, Hg has persisted in sediments and pore waters over the years (Canário et al. 2003a), spreading within the estuary and reaching protected areas without direct anthropogenic pressure (Cesário et al. 2016). Nowadays, the river still receives discharges from industrial and agricultural effluents, mainly incoming from the northern part of the estuary.

Total Hg concentrations in superficial sediments reported by Canário et al. (2005) varied within a broad spatial range up to 50 µg Hg g^{-1}, similar to those previously reported by Figuères et al. (1985), with the highest concentrations found in the vicinity of industrial areas (CN and BRR). Recently, Cesário et al. (2017b) reported higher Hg concentrations than those in deeper sediment layers of the same sites. At BRR, a core collected during summer presented concentrations up to 126 µg Hg g^{-1}, while in the winter core the concentrations were lower, up to 41 µg Hg g^{-1}. Concentrations in pore waters showed the same seasonal variation (Cesário et al. 2016, 2017b), with total dissolved concentrations reaching up to 365 ng Hg L^{-1}. These variations in Hg amounts within the same site could be explained by some spatial variation, particle heterogeneity of the sediments and simultaneous occurrence of different biogeochemical processes. In fact, Canário et al. (2003b) and Cesário et al. (2017b) evidenced the incorporation of Hg in the abundant Fe-oxides present in CN sediments, which are extremely enriched in sulfate due to discharges of a chlor-alkali industrial effluent. In BRR, sorption and/or co-precipitation with Fe and Mn oxyhydroxides are also key in the retention of Hg in sediments. Moreover, organic matter plays an important role in controlling Hg bioavailability and cannot be disregarded (Cesário et al. 2017b).

Another important aspect of Hg cycling in the Tagus River Estuary has been the evaluation of MeHg distribution in the different environmental compartments. The work from Canário et al. (2005) regarding total Hg and MeHg concentrations in the sediments reported that the proportion of MeHg to the total Hg varied in the entire estuary from 0.02% to 0.4%. In the same work, the authors estimated that 23 Mg of Hg could be stored in the first 5 cm of sediments, 24 kg as MeHg. Later, an increase of 37% in MeHg concentrations over total Hg was found during the summer, reflecting the warmer and reducing properties of surface sediments (Canário et al. 2007). These results highlighted the seasonal variability of MeHg, reported also by Monteiro et al. (2016) and Cesário et al. (2016, 2017b). Furthermore, the latter authors found that in more pristine areas of the estuary (ALC, in Fig. 2) the percentage of MeHg could increase up to 95% in pore waters, despite the lower Hg concentrations found when compared to the contaminated sites. In general, sediments tend to act as a source of dissolved Hg to the water column. However, diffusion fluxes of dissolved MeHg at the sediment–water interface tend to vary seasonally, i.e., sediments retain MeHg during summer and only act as a source to the water column in colder periods. The estimated annual budgets were larger in the most contaminated sites than in the pristine areas (Cesário et al. 2016, Cesário 2017). In addition, the observed exchanges across the sediment–water interface point to a major influence of advective transport, by a factor of 5, compared to the predicted diffusive fluxes.

During a tidal inundation experiment, Cesário et al. (2017c) observed dramatically fast changes in the sediment–water interface in the first 10 min, reporting a decrease of dissolved Hg and MeHg concentrations in pore waters. On the other hand, in the overlying waters an increase was observed on both Hg species concentrations during that period. After 10 min, concentrations decreased sharply and remained constant. The authors concluded that the release of dissolved Hg species from the sediments during tidal flushing is significant enough to affect the water column concentrations. Furthermore, water-to-air Hg volatilization was also measured in this study and it was suggested that the effect of tidal flushing may also control this process. A similar pattern was observed during the flooding period for the Hg volatilization flux in overlying water and dissolved gaseous Hg (DGM) concentrations in the water column (Cesário et al. 2017c). The fast release of the produced DGM suggests that kinetics of DGM production overtakes the advective transport of Hg from water to atmosphere. Nonetheless, intertidal sediments of Tagus River Estuary are repeatedly exposed to the atmosphere and therefore chemical reactions occur directly between the sediment surface layer and atmosphere, as previously reported by Canário and Vale (2004). Thus, intertidal areas may act as a potential source of elemental Hg to the atmosphere during the tidal cycles. This mechanism may be enhanced by sediment resuspension and bioturbation, increasing the photochemical reduction of Hg released in pore waters in a short period of time and escaping rapidly to the atmosphere (Canário and Vale 2004).

Cesário et al. (2018) studied the temporal variation of Hg and MeHg in the water column at CN site (Fig. 2), along a tidal cycle and during one year. Concentrations found in suspended particulate matter varied between 0.6–6.6 µg Hg g^{-1} and 6.9–690 ng MeHg g^{-1}. In the dissolved fraction, concentrations varied between 10–49 ng Hg L^{-1} and 0.069–6.9 ng MeHg L^{-1}. As previously reported by Cesário et al. (2017b), sediments may act as a source of dissolved Hg and MeHg to the water column. However, in the particular case of CN the exchange of Hg species with the estuary is highly dependent on the hydrodynamic conditions. Despite the possible Hg export through both ends of the CN channel towards the rest of the Tagus River Estuary, this mechanism is preferential during the ebb through the downstream end.

Since microbial activity is key to MeHg formation, Figueiredo et al. (2014) isolated Hg-resistant microbial communities from the Tagus River Estuary sediments at CN, BRR and ALC sites, identifying aerobic and anaerobic bacteria, namely, sulfate-reducing bacteria (SRB). The distribution of these bacteria and the resistance levels were consistent with Hg contamination along the depth of the sediments. Recently, Figueiredo et al. (2018) used isotope-enriched Hg incubation experiments to assess the effects of microbial communities on methylation and demethylation processes. Their findings showed a positive correlation between SRB and the methylation process, which occurs in anoxic environments. Demethylation was also shown for both aerobic and anaerobic microbial communities in anoxic and oxic environments. Furthermore, the use of Hg stable isotopic techniques recently allowed the measurement of methylation and demethylation rates in two salt marsh areas of the Tagus River Estuary (ROS and ALC, Fig. 2), using non-colonized sediments and sediments colonized by the species *Spartina maritima* and *Sarcoconia fruticosa*. Cesário et al. (2017a) estimated Hg methylation rates (K_M) ranging from 0.009 to 0.043 d^{-1}, pointing out the favorable environmental conditions for MeHg production in colonized sediments of the most contaminated area. Demethylation rates (K_D) varied between 3.2 and 5.7 d^{-1}, contrasting with K_M despite the site-specific differences. The authors considered it reasonable that the true behavior of MeHg in sediments was represented by the spiked tracer of MeHg, since it has no chance to "age" in a sediment, unlike the behavior of inorganic Hg. In another study using a salt marsh plant, *Halimione portucaloides*, collected at ALC, Cabrita et al. (2019) used Hg isotopes in exposure experiments to assess the uptake of Hg and MeHg, translocation and Hg released by volatilization through leaves, as well as plant toxicity and tolerance responses. Their findings showed that the plant incorporated both Hg and MeHg, with roots concentrating higher levels of both Hg isotopes than the other tissues. After long-term exposure, the relatively low levels of both Hg species in the aerial parts suggested potential Hg release by stems and leaves, as a mechanism to eliminate the metal despite the high tolerance of this plant species.

Canário (2004) quantified organic and inorganic Hg concentrations in different organisms of CN. It was reported that top predators are affected by Hg contamination. Organic Hg levels indicated that fish and cephalopod species, such as *Pomatoschistus minutus* (0.13–0.35 µg MeHg g^{-1}), *Liza ramada* (0.11–0.72 µg MeHg g^{-1}) and *Sepia officinalis* (0.50 µg MeHg g^{-1}), were the most contaminated, showing MeHg concentrations over 95%. The species in the lower levels of the food web presented MeHg concentrations below 95%, such as shrimps (0.06–0.22 µg MeHg g^{-1}) and crabs (0.19–0.35 µg MeHg g^{-1}). By that time, the species *Liza ramada* and *Sepia officinalis* presented Hg concentrations exceeding the threshold levels (0.5 µg Hg g^{-1}) established in Portuguese law for human consumption (Canário 2004). Raimundo et al. (2010) also reported Hg contamination and bioaccumulation in *Octopus vulgaris* specimens off the Tagus Estuary, in Cascais. Despite the site's being less Hg-contaminated than the inner estuary, the analyzed tissues of octopus specimens showed an elevated percentage of MeHg. The mantle exhibited significant enhancements of Hg and MeHg compared to the digestive gland, with tendency to a proportional increase in the two organs. According to the authors, this proportionality suggests that MeHg enters the digestive gland via dietary intake, being mobilized and partially stored in mantle. Moreover, Pereira et al. (2013) reported Hg concentrations in fish eyes from *Liza aurata* (median concentration of 0.05 μg Hg g^{-1}) that faithfully reflected the Hg contamination of the water column, supporting the use of this organ in environmental health assessment.

From all the research performed in the Tagus River Estuary over the past years concerning Hg and its different chemical species, some aspects still need additional work to fully characterize this system and to understand Hg behavior under varying hydrodynamic conditions. For instance, speciation and hydrodynamic modelling should address the latest findings with respect to the Hg and MeHg budgets recently reported, in order to predict accurately potential hazardous impacts. With increasing knowledge over the last few years, it is evident that the persistence of Hg, mostly in sediments from industrialized areas after decades of anthropogenic emissions, still poses a major environmental concern. Thus, further monitoring is needed, as well as the implementation of stricter mitigation procedures and remediation mechanisms, in view of the environmental and economic importance of the Tagus River Estuary.

Acknowledgements

N.L.C., S.E.B. and J.E.M. would like to acknowledge support from the National Council of Scientific and Technological Research (CONICET-Argentina). C.M. and R.C. would like to gratefully acknowledge the support of Portuguese Foundation for Science and Technology (FCT) projects: UID/QUI/001002013; PTDC/MAR/102748/2008 PROFLUX – Processes and Fluxes of Mercury and Methylmercury in a Contaminated Coastal Ecosystem, Tagus Estuary, Portugal; and PTDC/AAC-AMB/115798/2009 PLANTA – Effect of salt marsh plants on mercury methylation, transport and volatilization to the atmosphere. R.C. would also like to acknowledge FCT for the grant funding of her PhD (SFRH/BD/86441/2012).

References

Akan, J.C., S. Mohmoud, B.S. Yikala and V.O. Ogugbuaja. 2012. Bioaccumulation of some heavy metals in fish samples from River Benue in Vinikilang, Adamawa State, Nigeria. Am. J. Analyt. Chem. 3(11): 727.

Alvera-Azcárate, A., J.G. Ferreira and J.P. Nunes. 2003. Modelling eutrophication in mesotidal and macrotidal estuaries. The role of intertidal seaweeds. Estuar. Coast. Shelf Sci. 57: 715–724. Doi: 10.1016/S0272-7714(02)00413-4.

Amos, H.M., D.J. Jacob, D.G. Streets and E.M. Sunderland. 2013. Legacy impacts of all time anthropogenic emissions on the global mercury cycle. Global Biogeochem. Cycles 27(2): 410–421.

Amos, H.M., D.J. Jacob, D. Kocman, H.M. Horowitz, Y. Zhang, S. Dutkiewicz et al. 2014. Global biogeochemical implications of mercury discharges from rivers and sediment burial. Environ. Sci. Technol. 48(16): 9514–9522.

Arcagni, M., A. Rizzo, R. Juncos, M. Pavlin, L.M. Campbell, M.A. Arribére et al. 2017. Mercury and selenium in the food web of Lake Nahuel Huapi, Patagonia, Argentina. Chemosphere 166: 163–173.

Ariya, P.A., M. Amyot, A. Dastoor, D. Deeds, A. Feinberg, G. Kos et al. 2015. Mercury physicochemical and biogeochemical transformation in the atmosphere and at atmospheric interfaces: A review and future directions. Chem. Rev. 115(10): 3760–3802.

Baeyens, W., M. Leermakers, T. Papina, A. Saprykin, N. Brion, J. Noyen et al. 2003. Bioconcentration and biomagnification of mercury and methylmercury in North Sea and Scheldt Estuary fish. Arch. Environ. Contam. Toxicol. 45(4): 498–508.

Baeyens, W., R. Ebinghaus and O. Vasiliev. 2012. Global and Regional Mercury Cycles: Sources, Fluxes and Mass Balances (Vol. 21). Springer Science & Business Media.

Bagnato, E., G. Tamburello, G. Avard, M. Martinez-Cruz, M. Enrico, X. Fu et al. 2015. Mercury fluxes from volcanic and geothermal sources: An update. pp. 263–285. *In*: Zellmer, G.F., M. Edmonds and S.M. Straub (eds.). The

Role of Volatiles in the Genesis, Evolution and Eruption of Arc Magmas. Geological Society, London, Special Publications, 410(1).

Balcom, P.H., A.T. Schartup, R.P. Mason and C.Y. Chen. 2015. Sources of water column methylmercury across multiple estuaries in the Northeast US. Mar. Chem. 177: 721–730.

Batrakova, N., O. Travnikov and O. Rozovskaya. 2014. Chemical and physical transformations of mercury in the ocean: a review. Ocean Sci. 10(6): 1047–1063.

Bengtsson, U.G. and L.D. Hylander. 2017. Increased mercury emissions from modern dental amalgams. Biometals 30(2): 277–283.

Bernhoft, R.A. 2012. Mercury toxicity and treatment: A review of the literature. J. Environ. Public Health 2012. Doi: 10.1155/2012/460508.

Berry, M.J. and N.V. Ralston. 2008. Mercury toxicity and the mitigating role of selenium. EcoHealth 5(4): 456–459.

Bhattacharya, B.D., D.C. Nayak, S.K. Sarkar, S.N. Biswas, D. Rakshit and M.K. Ahmed. 2015. Distribution of dissolved trace metals in coastal regions of Indian Sundarban mangrove wetland: A multivariate approach. J. Clean Prod. 96: 233–243.

Biester, H., G. Müller and H.F. Schöler. 2002. Estimating distribution and retention of mercury in three different soils contaminated by emissions from chlor-alkali plants: Part I. Sci. Total Environ. 284(1-3): 177–189.

Bjørklund, G., J. Aaseth, O.P. Ajsuvakova, A.A. Nikonorov, A.V. Skalny, M.G. Skalnaya et al. 2017. Molecular interaction between mercury and selenium in neurotoxicity. Coord. Chem. Rev. 332: 30–37.

Bosch, A.C., B. O'Neill, G.O. Sigge, S.E. Kerwath and L.C. Hoffman. 2016. Mercury accumulation in Yellowfin tuna (Thunnus albacares) with regards to muscle type, muscle position and fish size. Food Chem. 190: 351–356.

Bradley, M.A., B.D. Barst and N. Basu. 2017. A review of mercury bioavailability in humans and fish. Int. J. Environ. Res. Public Health 14(2): 169.

Bratkič, A., T. Tinta, N. Koron, S.R. Guevara, E. Begu, T. Barkay et al. 2018. Mercury transformations in a coastal water column (Gulf of Trieste, northern Adriatic Sea). Mar. Chem. 200: 57–67.

Braune, B., J. Chételat, M. Amyot, T. Brown, M. Clayden, M. Evans et al. 2015. Mercury in the marine environment of the Canadian Arctic: Review of recent findings. Sci. Total Environ. 509: 67–90.

Burger, J. and M. Gochfeld. 2013. Selenium and mercury molar ratios in commercial fish from New Jersey and Illinois: variation within species and relevance to risk communication. Food Chem. Toxicol. 57: 235–245.

Cabrita, M.T., B. Duarte, R. Cesário, R. Mendes, H. Hintelmann, K. Eckey et al. 2019. Mercury mobility and effects in the salt-marsh plant Halimione portulacoides: Uptake, transport, and toxicity and tolerance mechanisms. Sci. Total Environ. 650: 111–120.

Canário, J., C. Vale and M. Caetano. 2003a. Mercury in sediments and pore waters at a contaminated site in the Tagus estuary. Cienc. Mar. 29: 535–545.

Canário, J., C. Vale, M. Caetano and M. Madureira. 2003b. Mercury in contaminated sediments and pore waters enriched in sulphate (Tagus Estuary, Portugal). Environ. Pollut. 126: 425–433.

Canário, J. 2004. Mercúrio e Monometilmercúrio na Cala do Norte do Estuário do Tejo. Ph.D. Thesis. Faculdade de Ciências e Tecnologia, Universidade Nova de Lisboa, Portugal.

Canário, J. and C. Vale. 2004. Rapid release of mercury from intertidal sediments exposed to solar radiation: A field experiment. Environ. Sci. Technol. 38: 3901–3907.

Canário, J., C. Vale and M. Caetano. 2005. Distribution of monomethylmercury and mercury in surface sediments of the Tagus Estuary (Portugal). Mar. Pollut. Bull. 50: 1142–5.

Canário, J., V. Branco and C. Vale. 2007. Seasonal variation of monomethylmercury concentrations in surface sediments of the Tagus Estuary (Portugal). Environ. Pollut. 148: 380–383.

Canli, M. and G. Atli. 2003. The relationships between heavy metal (Cd, Cr, Cu, Fe, Pb, Zn) levels and the size of six Mediterranean fish species. Environ. Pollut. 121(1): 129–136.

Carroll, R.W. and J.J. Warwick. 2017. The importance of dynamic mercury water column concentrations on body burdens in a planktivorous fish: A bioenergetic and mercury mass balance perspective. Ecol. Modell. 364: 66–76.

Cau, A., A. Bellodi, D. Moccia, A. Mulas, A.P. Pesci, R. Cannas et al. 2018. Dumping to the abyss: single-use marine litter invading bathyal plains of the Sardinian margin (Tyrrhenian Sea). Mar. Pollut. Bull. 135: 845–851.

Cesário, R., C.E. Monteiro, M. Nogueira, N.J. O'Driscoll, M. Caetano, H. Hintelmann et al. 2016. Mercury and methylmercury dynamics in sediments on a protected area of Tagus Estuary (Portugal). Water Air Soil Pollut. 227(12): 475.

Cesário, R. 2017. Processes and Fluxes of Mercury and Methylmercury in a contaminated ecosystem (Portugal). Ph.D. Thesis. Instituto Superior Técnico, Universidade de Lisboa, Portugal.

Cesário, R., H. Hintelmann, N.J. O'Driscoll, C.E. Monteiro, M. Caetano, M. Nogueira et al. 2017a. Biogeochemical cycle of mercury and methylmercury in two highly contaminated areas of Tagus estuary (Portugal). Water Air Soil Pollut. 228(7): 257.

Cesário, R., H. Hintelmann, R. Mendes, K. Eckey, B. Dimock, B. Araújo et al. 2017b. Evaluation of mercury methylation and methylmercury demethylation rates in vegetated and non-vegetated saltmarsh sediments from two Portuguese estuaries. Environ. Pollut. 226: 297–307.

Cesário, R., L. Poissant, M. Pilote, N.J. O'Driscoll, A.M. Mota and J. Canário. 2017c. Dissolved gaseous mercury formation and mercury volatilization in intertidal sediments. Sci. Total Environ. 603-604: 279–289.

Cesário, R., A.M. Mota, M. Caetano, M. Nogueira and J. Canário. 2018. Mercury and methylmercury transport and fate in the water column of Tagus estuary (Portugal). Mar. Pollut. Bull. 127: 235–250.

Chasar, L.C., B.C. Scudder, A.R. Stewart, A.H. Bell and G.R. Aiken. 2009. Mercury cycling in stream ecosystems–3. Trophic dynamics and methylmercury bioaccumulation. Environ. Sci. Technol. 43: 2733–2739.

Chouvelon, T., P. Cresson, M. Bouchoucha, C. Brach-Papa, P. Bustamante, S. Crochet et al. 2018. Oligotrophy as a major driver of mercury bioaccumulation in medium-to high-trophic level consumers: A marine ecosystem-comparative study. Environ. Pollut. 233: 844–854.

Clarkson, T.W. 1993. Mercury: major issues in environmental health. Environ. Health Perspect. 100: 31.

Clarkson, T.W., L. Magos and G.J. Myers. 2003. The toxicology of mercury—current exposures and clinical manifestations. N. Eng. J. Med. 349(18): 1731–1737.

Clarkson, T.W. and L. Magos. 2006. The toxicology of mercury and its chemical compounds. Crit. Rev. Toxicol. 36(8): 609–662.

Cohen, M., R. Artz, R. Draxler, P. Miller, L. Poissant, D. Niemi et al. 2004. Modeling the atmospheric transport and deposition of mercury to the Great Lakes. Environ. Res. 95(3): 247–265.

Cossa, D., J.M. Martin and J. Sanjuan. 1994. Dimethylmercury formation in the Alboran Sea. Mar. Pollut. Bull. 28(6): 381–384.

Cossa, D., B. Averty and N. Pirrone. 2009. The origin of methylmercury in open Mediterranean waters. Limnol. Oceanogr. 54(3): 837–844.

Cutter, G.A. and L.S Cutter. 2001. Sources and cycling of selenium in the western and equatorial Atlantic Ocean. Deep Sea Res. Part 2 Top. Stud. Oceanogr. 48(13): 2917–2931.

Czub, M., L. Kotwicki, T. Lang, H. Sanderson, Z. Klusek, M. Grabowski et al. 2018. Deep sea habitats in the chemical warfare dumping areas of the Baltic Sea. Sci. Total Environ. 616: 1485–1497.

Dastoor, A.P. and Y. Larocque. 2004. Global circulation of atmospheric mercury: a modelling study. Atmos. Environ. 38(1): 147–161.

De Marco, S.G., S.E. Botté and J.E. Marcovecchio. 2006. Mercury distribution in abiotic and biological compartments within several estuarine systems from Argentina: 1980–2005 period. Chemosphere 65(2): 213–223.

De Simone, F., C.N. Gencarelli, I.M. Hedgecock and N. Pirrone. 2014. Global atmospheric cycle of mercury: a model study on the impact of oxidation mechanisms. Environ. Sci. Pollut. Res. 21(6): 4110–4123.

De Souza Machado, A.A., K. Spencer, W. Kloas, M. Toffolon and C. Zarfl. 2016. Metal fate and effects in estuaries: a review and conceptual model for better understanding of toxicity. Sci. Total Environ. 541: 268–281.

Deycard, V.N., J. Schäfer, G. Blanc, A. Coynel, J.C. Petit, L. Lanceleur et al. 2014. Contributions and potential impacts of seven priority substances (As, Cd, Cu, Cr, Ni, Pb, and Zn) to a major European Estuary (Gironde Estuary, France) from urban wastewater. Mar. Chem. 167: 123–134.

Dias, M. and J. Marques. 1999. Estuários. Estuário do Tejo: O seu valor e um pouco da sua história. Reserva Natural do Estuário do Tejo, Instituto da Conservação da Natureza. Lisboa, Portugal.

Dietz, R., C. Sonne, N. Basu, B. Braune, T. O'Hara, R.J. Letcher et al. 2013. What are the toxicological effects of mercury in Arctic biota? Sci. Total Environ. 443: 775–790.

Driscoll, C.T., R.P. Mason, H.M. Chan, D.J. Jacob and N. Pirrone. 2013. Mercury as a global pollutant: sources, pathways, and effects. Environ. Sci. Technol. 47(10): 4967–4983.

Duarte, B., G. Silva, J.L. Costa, J.P. Medeiros, C. Azeda, E. Sá et al. 2014. Heavy metal distribution and partitioning in the vicinity of the discharge areas of Lisbon drainage basins (Tagus Estuary, Portugal). J. Sea Res. 93: 101–111.

Eisler, R. 2000. Handbook of Chemical Risk Assessment: Health Hazards to Humans, Plants, and Animals, Three Volume Set. CRC press.

Estrade, N., J. Carignan, J.E. Sonke and O.F. Donard. 2009. Mercury isotope fractionation during liquid–vapor evaporation experiments. Geochim. Cosmochim. Acta 73(10): 2693–2711.

FAO (Food and Agriculture Organization). 2003. Heavy Metals Regulations Legal Notice No 66/2003. FAO, Rome.

FAO (Food and Agriculture Organization). 2018. The State of World Fisheries and Aquaculture 2018—Meeting the sustainable development goals. Rome. Licence: CC BY-NC-SA 3.0 IGO.

Ferrara, R., B. Mazzolai, E. Lanzillotta, E. Nucaro and N. Pirrone. 2000. Volcanoes as emission sources of atmospheric mercury in the Mediterranean basin. Sci. Total Environ. 259(1-3): 115–121.

Figueiredo, N., M.L. Serralheiro, J. Canário, A. Duarte, H. Hintelmann and C. Carvalho. 2018. Evidence of mercury methylation and demethylation by the estuarine microbial communities obtained in stable Hg isotope studies. Int. J. Environ. Res. Public Heal. Doi: 10.3390/ijerph15102141.

Figueiredo, N.L.L., J. Canário, A. Duarte, M.L. Serralheiro and C. Carvalho. 2014. Isolation and characterization of mercury-resistant bacteria from sediments of Tagus Estuary (Portugal): implications for environmental and human health risk assessment. J. Toxicol. Environ. Health. A 77: 155–68.

Figuères, G., J.M. Martin, M. Meybeck and P. Seyler. 1985. A comparative study of mercury contamination in the Tagus estuary (Portugal) and major French estuaries (Gironde, Loire, Rhône). Estuar. Coast. Shelf Sci. 20: 183–203.

Fitzgerald, W.F., C.H. Lamborg and C.R. Hammerschmidt. 2007. Marine biogeochemical cycling of mercury. Chem. Rev. 107(2): 641–662.

Fortunato, A., A. Oliveira and A.M. Baptista. 1999. On the effect of tidal flats on the hydrodynamics of the Tagus estuary. Oceanol. Acta 22: 31–44.

FranÇa, S., C. Vinagre, I. CaÇador and H.N. Cabral. 2005. Heavy metal concentration in sediment benthic invertebrates and fish in three salt marsh areas subjected to different pollution loads in the Tagus estuary (Portugal). Mar. Pollut. Bull. 50: 993–1018.

Friedli, H.R., L.F. Radke, N.J. Payne, D.J. McRae, T.J. Lynham and T.W. Blake. 2007. Mercury in vegetation and organic soil at an upland boreal forest site in Prince Albert National Park, Saskatchewan, Canada. J. Geophys. Res. Biogeosci. 112(G1).

Fu, X., X. Feng, J. Sommar and S. Wang. 2012. A review of studies on atmospheric mercury in China. Sci. Total Environ. 421: 73–81.

Gailer, J. 2007. Arsenic–selenium and mercury–selenium bonds in biology. Coord. Chem. Rev. 251(1-2): 234–254.

Giang, A., L.C. Stokes, D.G. Streets, E.S. Corbitt and N.E. Selin. 2015. Impacts of the minamata convention on mercury emissions and global deposition from coal-fired power generation in Asia. Environ. Sci. Technol. 49(9): 5326–5335.

Gibb, H. and K.G. O'Leary. 2014. Mercury exposure and health impacts among individuals in the artisanal and small-scale gold mining community: a comprehensive review. Environ. Health Perspect. 122(7): 667–672.

Grigal, D.F., R.K. Kolka, J.A. Fleck and E.A. Nater. 2000. Mercury budget of an upland-peatland watershed. Biogeochemistry 50(1): 95–109.

Gu, B., Y. Bian, C.L. Miller, W. Dong, X. Jiang and L. Liang. 2011. Mercury reduction and complexation by natural organic matter in anoxic environments. Proc. Natl. Acad. Sci. Doi: 10.1073/pnas.1008747108.

Gul, N., S. Khan, A. Khan, J. Nawab, I. Shamshad and X. Yu. 2016. Quantification of Hg excretion and distribution in biological samples of mercury-dental-amalgam users and its correlation with biological variables. Environ. Sci. Pollut. Res. 23(20): 20580–20590.

Guo, Y., X. Feng, Z. Li, T. He, H. Yan, B. Meng et al. 2008. Distribution and wet deposition fluxes of total and methyl mercury in Wujiang River Basin, Guizhou, China. Atmos. Environ. 42(30): 7096–7103.

Gupta, D.A. 2008. Implication of environmental flows in river basin management. Phys. Chem. Earth 33(5): 298–303.

Gustin, M.S., J. Huang, M.B. Miller, C. Peterson, D.A. Jaffe, J. Ambrose et al. 2013. Do we understand what the mercury speciation instruments are actually measuring? Results of RAMIX. Environ. Sci. Technol. 47(13): 7295–7306.

Gustin, M.S., D.C. Evers, M.S. Bank, C.R. Hammerschmidt, A. Pierce, N. Basu et al. 2016. Importance of integration and implementation of emerging and future mercury research into the Minamata Convention. Environ. Sci. Technol. 50(6): 2767–2770.

Ha, E., N. Basu, S. Bose-O'Reilly, J.G. Dórea, E. McSorley, M. Sakamoto et al. 2017. Current progress on understanding the impact of mercury on human health. Environ. Res. 152: 419–433.

Hamilton, S.J. 2004. Review of selenium toxicity in the aquatic food chain. Sci. Total Environ. 326(1-3): 1–31.

Holmes, P., K.A.F. James and L.S. Levy. 2009. Is low-level environmental mercury exposure of concern to human health?. Sci. Total Environ. 408(2): 171–182.

Hong, C., X. Yu, J. Liu, Y. Cheng and S.E. Rothenberg. 2016. Low-level methylmercury exposure through rice ingestion in a cohort of pregnant mothers in rural China. Environ. Res. 150: 519–527.

Horowitz, H.M., D.J. Jacob, Y. Zhang, T.S. Dibble, F. Slemr, H.M. Amos et al. 2017. A new mechanism for atmospheric mercury redox chemistry: Implications for the global mercury budget. Atmos. Chem. Phys. 17(10): 6353–6371.

Horvat, M., J. Kotnik, M. Logar, V. Fajon, T. Zvonarić and N. Pirrone. 2003. Speciation of mercury in surface and deep-sea waters in the Mediterranean Sea. Atmos. Environ. 37: 93–108.

Kehrig, H.D.A., T.G. Seixas, E.A. Palermo, A.P. Baêta, C.W. Castelo-Branco, O. Malm et al. 2009. The relationships between mercury and selenium in plankton and fish from a tropical food web. Environ. Sci. Pollut. Res. 16(1): 10–24.

Kennish, M.J. 1997. Pollution Impacts on Marine Biotic Communities (Vol. 14). CRC Press.

Kerin, E.J., C.C. Gilmour, E. Roden, M.T. Suzuki, J.D. Coates and R.P. Mason. 2006. Mercury methylation by dissimilatory iron-reducing bacteria. Appl. Environ. Microbiol. 72(12): 7919–7921.

Kim, K.H., E. Kabir and S.A. Jahan. 2016. A review on the distribution of Hg in the environment and its human health impacts. J. Hazard. Mater. 306: 376–385.

King, J.K., F.M. Saunders, R.F. Lee and R.A. Jahnke. 1999. Coupling mercury methylation rates to sulfate reduction rates in marine sediments. Environ. Toxicol. Chem. 18(7): 1362–1369.

Kotnik, J., M. Horvat, E. Tessier, N. Ogrinc, M. Monperrus, D. Amouroux et al. 2007. Mercury speciation in surface and deep waters of the Mediterranean Sea. Mar. Chem. 107(1): 13–30.

Lacerda, L.D. 2007. Biogeoquímica de contaminantes no Antropoceno. Oecol. Bras. 11(2): 139–144.

Lamborg, C.H., C.R. Hammerschmidt, K.L. Bowman, G.J. Swarr, K.M. Munson, D.C. Ohnemus et al. 2014. A global ocean inventory of anthropogenic mercury based on water column measurements. Nature 512(7512): 65.

Lantzy, R.J. and F.T. Mackenzie. 1979. Atmospheric trace metals: global cycles and assessment of man's impact. Geochim. Cosmochim. Acta 43(4): 511–525.

Lavoie, R.A., T.D. Jardine, M.M. Chumchal, K.A. Kidd and L.M. Campbell. 2013. Biomagnification of mercury in aquatic food webs: a worldwide meta-analysis. Environ. Sci. Technol. 47: 13385–13394.

Lehnherr, I., V.L. St Louis, H. Hintelmann and J.L. Kirk. 2011. Methylation of inorganic mercury in polar marine waters. Nat. Geosci. 4: 298–302.

Li, J.S., B. Chen, G.Q. Chen, W.D. Wei, X.B. Wang, J.P. Ge et al. 2017. Tracking mercury emission flows in the global supply chains: A multi-regional input-output analysis. J. Clean. Prod. 140: 1470–1492.

Lillebø, A.I., P.J. Coelho, P. Pato, M. Válega, R. Margalho, M. Reis et al. 2011. Assessment of mercury in water, sediments and biota of a southern European estuary (Sado estuary, Portugal). Water Air Soil Pollut. 214(1-4): 667–680.

Lin, Y., R. Vogt and T. Larssen. 2012. Environmental mercury in China: a review. Environ. Toxicol. Chem. 31(11): 2431–2444.

Lindberg, A., K.A. Björnberg, M. Vahter and M. Berglund. 2004. Exposure to methylmercury in non-fish-eating people in Sweden. Environ. Res. 96(1): 28–33.

Lindberg, S.A. and W.J. Stratton. 1998. Atmospheric mercury speciation: concentrations and behavior of reactive gaseous mercury in ambient air. Environ. Sci. Technol. 32(1): 49–57.

Lino, A.S., D. Kasper, Y.S. Guida, J.R. Thomaz and O. Malm. 2018. Mercury and selenium in fishes from the Tapajós River in the Brazilian Amazon: An evaluation of human exposure. J. Trace Elem. Med. Biol. 48: 196–201.

Liu, C.B., X.B. Hua, H.W. Liu, B. Yu, Y.X. Mao, D.Y. Wang et al. 2018. Tracing aquatic bioavailable Hg in three different regions of China using fish Hg isotopes. Ecotoxicol. Environ. Saf. 150: 327–334.

Louis, V.L.S., H. Hintelmann, J.A. Graydon, J.L. Kirk, J. Barker, B. Dimock et al. 2007. Methylated mercury species in Canadian high Arctic marine surface waters and snowpacks. Environ. Sci. Technol. 41(18): 6433–6441.

Loux, N.T. 2004. A critical assessment of elemental mercury air/water exchange parameters. Chem. Speciation Bioavailability 16(4): 127–138.

Lyons, K., A.B. Carlisle and C.G. Lowe. 2017. Influence of ontogeny and environmental exposure on mercury accumulation in muscle and liver of male Round Stingrays. Mar. Environ. Res. 130: 30–37.

Makedonski, L., K. Peycheva and M. Stancheva. 2017. Determination of heavy metals in selected black sea fish species. Food Control 72: 313–318.

Malczyk, E.A. and B.A. Branfireun. 2015. Mercury in sediment, water, and fish in a managed tropical wetland-lake ecosystem. Sci. Total Environ. 524: 260–268.

Manavi, P.N. and A. Mazumder. 2018. Potential risk of mercury to human health in three species of fish from the southern Caspian Sea. Mar. Pollut. Bull. 130: 1–5.

Marcovecchio, J.E., S.G. De Marco, N.S. Buzzi, S.E. Botté, A.C. Labudia, N.S. La Colla et al. 2015. Fish and seafood. pp. 621–643. *In*: Miguel de la Guardia and Salvador Garrigues (eds.). Handbook of Mineral Elements in Food.

Martínez, M.L., A. Intralawan, G. Vazquez, O. Perez-Maqueo, P. Sutton and R. Landgrave. 2007. The coasts of our world: ecological, economic and social importance. Ecol. Econ. 63: 254–272.

Marusczak, N., J.E. Sonke, X. Fu and M. Jiskra. 2016. Tropospheric GOM at the Pic du Midi observatory-correcting bias in denuder based observations. Environ. Sci. Technol. 51(2): 863–869.

Mason, R.P., K.R. Rolfhus and W.F. Fitzgerald. 1995. Methylated and elemental mercury cycling in surface and Deep Ocean waters of the North Atlantic. Water Air Soil Pollut. 80(1-4): 665–677.

Mason, R.P. and W.F. Fitzgerald. 1996. Sources, sinks and biogeochemical cycling of mercury in the ocean. *In*: Global and regional mercury cycles: sources, fluxes and mass balances. Springer, Dordrecht, pp. 249–272.

Mason, R.P., J.R. Reinfelder and F.M. Morel. 1996. Uptake, toxicity, and trophic transfer of mercury in a coastal diatom. Environ. Sci. Technol. 30(6): 1835–1845.

Mason, R.P. and G.R. Sheu. 2002. Role of the ocean in the global mercury cycle. Global Biogeochem. Cycles 16(4): 1–14.

Mason, R.P. 2012. The methylation of metals and metalloids in aquatic systems. In Methylation-From DNA, RNA and Histones to Diseases and Treatment. InTech.

Mason, R.P., A.L. Choi, W.F. Fitzgerald, C.R. Hammerschmidt, C.H. Lamborg, A.L. Soerensen et al. 2012. Mercury biogeochemical cycling in the ocean and policy implications. Environ. Res. 119: 101–117.

Maz-Courrau, A., C. López-Vera, F. Galvan-Magaña, O. Escobar-Sánchez, R. Rosíles-Martínez and A. Sanjuan-Munoz. 2012. Bioaccumulation and biomagnification of total mercury in four exploited shark species in the Baja California Peninsula, Mexico. Bull. Environ. Contam. Toxicol. 88(2): 129–134.

McHugh, B., R. Berbee, E. Farmer, R. Fryer, N. Green, M.M. Larsen et al. 2016. Mercury assessment in the marine environment: assessment criteria comparison (EAC/EQS) for mercury. OSPAR Commission.

McKinney, M.A., S. Pedro, R. Dietz, C. Sonne, A.T. Fisk, D. Roy et al. 2015. A review of ecological impacts of global climate change on persistent organic pollutant and mercury pathways and exposures in arctic marine ecosystems. Curr. Zool. 61(4): 617–628.

Mengerink, K.J., C.L. Van Dover, J. Ardron, M. Baker, E. Escobar-Briones, K. Gjerde et al. 2014. A call for deep-ocean stewardship. Science 344(6185): 696–698.

Minamata City. 2007. Minamata disease: Its history and lessons. Kumamoto: Kumamoto Prefecture: Minamata City Planning Division.

Monteiro, C.E. 2010. Mercury and Methylmercury in sediment cores from the Tagus estuary. MSc Thesis. Universidade de Aveiro, Portugal.

Monteiro, C.E., R. Cesário, N.J. O'Driscoll, M. Nogueira, M. Válega, M. Caetano et al. 2016. Seasonal variation of methylmercury in sediment cores from the Tagus Estuary (Portugal). Mar. Pollut. Bull. 104: 162–170.

Morel, F.M., A.M. Kraepiel and M. Amyot. 1998. The chemical cycle and bioaccumulation of mercury. Annu. Rev. Ecol. Syst. 29(1): 543–566.

Mukherjee, A.B., R. Zevenhoven, P. Bhattacharya, K.S. Sajwan and R. Kikuchi. 2008. Mercury flow via coal and coal utilization by-products: a global perspective. Resour. Conserv. Recycl. 52: 571–591.

Nriagu, J. and C. Becker. 2003. Volcanic emissions of mercury to the atmosphere: global and regional inventories. Sci. Total Environ. 304(1-3): 3–12.

Obrist, D., C. Pearson, J. Webster, T. Kane, C.J. Lin, G.R. Aiken et al. 2016. A synthesis of terrestrial mercury in the western United States: Spatial distribution defined by land cover and plant productivity. Sci. Total Environ. 568: 522–535.

Obrist, D., J.L. Kirk, L. Zhang, E.M. Sunderland, M. Jiskra and N.E. Selin. 2018. A review of global environmental mercury processes in response to human and natural perturbations: Changes of emissions, climate, and land use. Ambio 47(2): 116–140.

Ogrinc, N., M. Monperrus, J. Kotnik, V. Fajon, K. Vidimova, D. Amouroux et al. 2007. Distribution of mercury and methylmercury in deep-sea surficial sediments of the Mediterranean Sea. Mar. Chem. 107(1): 31–48.

Oursel, B., C. Garnier, I. Pairaud, D. Omanović, G. Durrieu, A.D. Syakti et al. 2014. Behaviour and fate of urban particles in coastal waters: Settling rate, size distribution and metals contamination characterization. Estuar. Coast. Shelf Sci. 138: 14–26.

Outridge, P.M., R.P. Mason, F. Wang, S. Guerrero and L.E. Heimbürger-Boavida. 2018. Updated global and oceanic mercury budgets for the United Nations Global Mercury Assessment 2018. Environ. Sci. Technol. 52(20): 11466–11477.

Pacyna, E.G., J.M. Pacyna, K. Sundseth, J. Munthe, K. Kindbom, S. Wilson et al. 2010. Global emission of mercury to the atmosphere from anthropogenic sources in 2005 and projections to 2020. Atmos. Environ. 44(20): 2487–2499.

Pacyna, J.M., O. Travnikov, F.D. Simone, I.M. Hedgecock, K. Sundseth, E.G. Pacyna et al. 2016. Current and future levels of mercury atmospheric pollution on a global scale. Atmos. Chem. Phys. 16: 12495–12511.

Papagiannis, I., I. Kagalou, J. Leonardos, D. Petridis and V. Kalfakakou. 2004. Copper and zinc in four freshwater fish species from Lake Pamvotis (Greece). Environ. Int. 30(3): 357–362.

Pařízek, J. and I. Ošťádalová. 1967. The protective effect of small amounts of selenite in sublimate intoxication. Experientia 23(2): 142–143.

Pereira, P., J. Raimundo, J. Canário, A. Almeida and M. Pacheco. 2013. Looking at the aquatic contamination through fish eyes—A faithful picture based on metals burden. Mar. Pollut. Bull. 77: 375–379.

Pirrone, N., I.M. Hedgecock and F. Sprovieri. 2008. New Directions: Atmospheric mercury, easy to spot and hard to pin down: impasse? Atmos. Environ. 42: 8549–8551.

Pirrone, N., S. Cinnirella, X. Feng, R.B. Finkelman, H.R. Friedli, J. Leaner et al. 2010. Global mercury emissions to the atmosphere from anthropogenic and natural sources. Atmos. Chem. Phys. 10(13): 5951–5964.

Poissant, L., M. Pilote, C. Beauvais, P. Constant and H.H. Zhang. 2005. A year of continuous measurements of three atmospheric mercury species (GEM, RGM and Hgp) in southern Quebec, Canada. Atmos. Environ. 39(7): 1275–1287.

Potters, G. 2013. Marine Pollution, Bookboon.com.

Pyle, D.M. and T.A. Mather. 2003. The importance of volcanic emissions for the global atmospheric mercury cycle. Atmos. Environ. 37(36): 5115–5124.

Rahman, M.S., A.H. Molla, N. Saha and A. Rahman. 2012. Study on heavy metals levels and its risk assessment in some edible fishes from Bangshi River, Savar, Dhaka, Bangladesh. Food Chem. 134(4): 1847–1854.

Raimundo, J., C. Vale, J. Canário, V. Branco and I. Moura. 2010. Relations between mercury, methyl-mercury and selenium in tissues of Octopus vulgaris from the Portuguese Coast. Environ. Pollut. 158(6): 2094–2100.

Ralston, N.V., J. Unrine and D. Wallschläger. 2008. Biogeochemistry and analysis of selenium and its species. North American Metals Council, Washington, DC.

Ramirez-Llodra, E., P.A. Tyler, M.C. Baker, O.A. Bergstad, M.R. Clark, E. Escobar et al. 2011. Man and the last great wilderness: human impact on the deep sea. PLoS One 6(8): e22588.

Reis, A.T., S.M. Rodrigues, C. Araújo, J.P. Coelho E. Pereira and A.C. Duarte. 2009. Mercury contamination in the vicinity of a chlor-alkali plant and potential risks to local population. Sci. Total Environ. 407(8): 2689–2700.

Ren, W., L. Duan, Z. Zhu, W. Du, Z. An, L. Xu et al. 2014. Mercury transformation and distribution across a polyvinyl chloride (PVC) production line in China. Environ. Sci. Technol. 48(4): 2321–2327.

Rice, K.M., E.M. Walker Jr, M. Wu, C. Gillette and E.R. Blough. 2014. Environmental mercury and its toxic effects. J. Prev. Med. Public Health 47(2): 74.

Rigét, F., B. Braune, A. Bignert, S. Wilson, J. Aars, E. Born et al. 2011. Temporal trends of Hg in Arctic biota, an update. Sci. Total Environ. 409(18): 3520–3526.

Schartup, A.T., P.H. Balcom and R.P. Mason. 2014. Sediment-porewater partitioning, total sulfur, and methylmercury production in estuaries. Environ. Sci. Technol. 48(2): 954–960.

Schlitzer, R. 2017. Ocean Data View, odv.awi.de.

Schroeder, W.H. and J. Munthe. 1998. Atmospheric mercury—an overview. Atmos. Environ. 32(5): 809–822.

Selin, N.E., D.J. Jacob, R.M. Yantosca, S. Strode, L. Jaeglé and E.M. Sunderland. 2008. Global 3-D land-ocean-atmosphere model for mercury: Present-day versus preindustrial cycles and anthropogenic enrichment factors for deposition. Global Biogeochem. Cycles 22(2).

Selin, N.E. 2009. Global biogeochemical cycling of mercury: a review. Annu. Rev. Environ. Resour. 34: 43–63.

Shah, V., L. Jaeglé, L.E. Gratz, J.L. Ambrose, D.A. Jaffe, N.E. Selin et al. 2016. Origin of oxidized mercury in the summertime free troposphere over the southeastern US. Atmos. Chem. Phys. 16(3): 1511–1530.

Shannon, J.D. and E.C. Voldner. 1995. Modeling atmospheric concentrations of mercury and deposition to the Great Lakes. Atmos. Environ. 29(14): 1649–1661.

Sheu, G.R. and R.P. Mason. 2001. An examination of methods for the measurements of reactive gaseous mercury in the atmosphere. Environ. Sci. Technol. 35(6): 1209–1216.

Soerensen, A.L., E.M. Sunderland, C.D. Holmes, D.J. Jacob, R.M. Yantosca, H. Skov et al. 2010. An improved global model for air-sea exchange of mercury: High concentrations over the North Atlantic. Environ. Sci. Technol. 44(22): 8574–8580.

Song, S., N.E. Selin, L.E. Gratz, J.L. Ambrose, D.A. Jaffe, V. Shah et al. 2016. Constraints from observations and modeling on atmosphere–surface exchange of mercury in eastern North America. http://dx.doi.org/10.12952/journal.elementa.000100.

Spencer, K.L., C.L. MacLeod, A. Tuckett and S.M. Johnson. 2006. Source and distribution of trace metals in the Medway and Swale estuaries, Kent. Mar. Pollut. Bull. 52: 226–231.

Spiller, H.A. 2017. Rethinking mercury: the role of selenium in the pathophysiology of mercury toxicity. Clin. Toxicol. 56(5): 313–326.

Sprovieri, F., N. Pirrone, R. Ebinghaus, H. Kock and A. Dommergue. 2010. A review of worldwide atmospheric mercury measurements. Atmos. Chem. Phys. 10(17): 8245–8265.

Storelli, M.M., G. Barone, G. Piscitelli and G.O. Marcotrigiano. 2007. Mercury in fish: concentration vs. fish size and estimates of mercury intake. Food Addit. Contam. 24(12): 1353–1357.

Streets, D.G., M.K. Devane, Z. Lu, T.C. Bond, E.M. Sunderland and D.J. Jacob. 2011. All-time releases of mercury to the atmosphere from human activities. Environ. Sci. Technol. 45(24): 10485–10491.

Streets, D.G., H.M. Horowitz, D.J. Jacob, Z. Lu, L. Levin, A.F. Ter Schure et al. 2017. Total mercury released to the environment by human activities. Environ. Sci. Technol. 51(11): 5969–5977.

Streets, D.G., Z. Lu, L. Levin, A.F. Ter Schure and E.M. Sunderland. 2018. Historical releases of mercury to air, land, and water from coal combustion. Sci. Total Environ. 615: 131–140.

Strode, S.A., L. Jaeglé, N.E. Selin, D.J. Jacob, R.J. Park, R.M. Yantosca et al. 2007. Air-sea exchange in the global mercury cycle. Global Biogeochem. Cycles 21(1).

Sunderland, E.M. and R.P. Mason. 2007. Human impacts on open ocean mercury concentrations. Global Biogeochem. Cycles 21(4).

Sundseth, K., J.M. Pacyna, E.G. Pacyna, N. Pirrone and R.J. Thorne. 2017. Global sources and pathways of mercury in the context of human health. Int. J. Environ. Res. Public Health 14(1): 105.

Teixeira, D.C., R.C. Montezuma, R.R. Oliveira and E.V. Silva-Filho. 2012. Litterfall mercury deposition in Atlantic forest ecosystem from SE–Brazil. Environ. Pollut. 164: 11–15.

Telmer, K.H. and M.M. Veiga. 2009. World emissions of mercury from artisanal and small scale gold mining. *In*: Mercury fate and transport in the global atmosphere. Springer, Boston, MA, pp. 131–172.

Thilsted, S.H., D. James, J. Toppe, R. Subasinghe and I. Karunasagar. 2014. Maximizing the contribution of fish to human nutrition. In ICN2 Second International Conference on Nutrition. FAO and World Health Organisation.

Tomiyasu, T., A. Matsuyama, T. Eguchi, K. Marumoto, K. Oki and H. Akagi. 2008. Speciation of mercury in water at the bottom of Minamata Bay, Japan. Mar. Chem. 112(1-2): 102–106.

Truong, H.Y.T., Y.W. Chen, M. Saleh, S. Nehzati, G.N. George, I.J. Pickering et al. 2014. Proteomics of Desulfovibrio desulfuricans and X-ray absorption spectroscopy to investigate mercury methylation in the presence of selenium. Metallomics 6(3): 465–475.

Tyler, P.A. 2003. Disposal in the deep sea: analogue of nature or faux ami? Environ. Conserv. 30(1): 26–39.

Ullrich, S.M., M.A. Ilyushchenko, I.M. Kamberov and T.W. Tanton. 2007. Mercury contamination in the vicinity of a derelict chlor-alkali plant. Part I: Sediment and water contamination of Lake Balkyldak and the River Irtysh. Sci. Total Environ. 381(1-3): 1–16.

Ullrich, S.M., T.W. Tanton and S.A. Abdrashitova. 2001. Mercury in the aquatic environment: a review of factors affecting methylation. Crit. Rev. Environ. Sci. Technol. 31(3): 241–293.

UNEP. 2013. Global Mercury Assessment, Sources, emissions, releases and environmental transport. UNEP Chemicals Branch, Geneva, Switzerland, Vol. 42.

UNEP. 2017. Global mercury supply, trade and demand. United Nations Environment Programme, Chemicals and Health Branch. Geneva, Switzerland.

Veiga, M.M., P.A. Maxson and L.D. Hylander. 2006. Origin and consumption of mercury in small-scale gold mining. J. Clean. Prod. 14(3-4): 436–447.

Wang, F., S. Wang, L. Zhang, H. Yang, Q. Wu and J. Hao. 2016. Characteristics of mercury cycling in the cement production process. J. Hazard. Mater. 302: 27–35.

Wang, R. and W.X. Wang. 2010. Importance of speciation in understanding mercury bioaccumulation in tilapia controlled by salinity and dissolved organic matter. Environ. Sci. Technol. 44(20): 7964–7969.

Wang, S., L. Zhang, L. Wang, Q. Wu, F. Wang and J. Hao. 2014. A review of atmospheric mercury emissions, pollution and control in China. Front. Environ. Sci. Eng. 8(5): 631–649.

Wang, X. and W.X. Wang. 2017. Selenium induces the demethylation of mercury in marine fish. Environ. Pollut. 231: 1543–1551.

Wang, Z., X. Zhang, J. Xiao, C. Zhijia and P. Yu. 2009. Mercury fluxes and pools in three subtropical forested catchments, southwest China. Environ. Pollut. 157(3): 801–808.

World Health Organization (WHO). 2011. Report of the joint FAO/WHO expert consultation on the risks and benefits of fish consumption, 25–29 January 2010, Rome, Italy (No. FIPM/R978 (En)). Geneva: World Health Organization.

Yang, D.Y., Y.W. Chen, J.M. Gunn and N. Belzile. 2008. Selenium and mercury in organisms: interactions and mechanisms. Environ. Rev. 16(NA): 71–92.

Yao, Q., X. Wang, H. Jian, H. Chen and Z. Yu. 2016. Behavior of suspended particles in the Changjiang estuary: size distribution and trace metal contamination. Mar. Pollut. Bull. 103(1): 159–167.

Ysart, G., P. Miller, M. Croasdale, H. Crews, P. Robb, M. Baxter et al. 2000. 1997 UK total diet study dietary exposures to aluminium, arsenic, cadmium, chromium, copper, lead, mercury, nickel, selenium, tin and zinc. Food. Addit. Contam. 17(9): 775–786.

Yudovich, Y.E. and M.P. Ketris. 2005. Mercury in coal: A review: Part 1. Geochemistry. Int. J. Coal Geol. 62(3): 107–134.

Zagar, D., N. Sirnik, M. Četina, M. Horvat, J. Kotnik, N. Ogrinc et al. 2014. Mercury in the Mediterranean. Part 2: Processes and mass balance. Environ. Sci. Pollut. Res. 21(6): 4081–4094.

Zamani-Ahmadmahmoodi, R., A.R. Bakhtiari and J.A.R. Martín. 2014. Spatial relations of mercury contents in Pike (Esox lucius) and sediments concentration of the Anzali wetland, along the southern shores of the Caspian Sea, Iran. Mar. Pollut. Bull. 84(1-2): 97–103.

Zhang, H., X. Feng, H.M. Chan and T. Larssen. 2013. New insights into traditional health risk assessments of mercury exposure: implications of selenium. Environ. Sci. Technol. 48(2): 1206–1212.

Zhang, L., S. Wang, L. Wang, Y. Wu, L. Duan, Q. Wu et al. 2015a. Updated emission inventories for speciated atmospheric mercury from anthropogenic sources in China. Environ. Sci. Technol. 49(5): 3185–3194.

Zhang, Y., D.J. Jacob, S. Dutkiewicz, H.M. Amos, M.S. Long and E.M. Sunderland. 2015b. Biogeochemical drivers of the fate of riverine mercury discharged to the global and Arctic oceans. Global Biogeochem. Cycles 29(6): 854–864.

Zhang, Y., L. Jaeglé, L. Thompson and D.G. Streets. 2014. Six centuries of changing oceanic mercury. Global Biogeochem. Cycles 28(11): 1251–1261.

Zwolsman, J.J.G. and G.T.M. Van Eck. 1999. Geochemistry of major elements and trace metals in suspended matter of the Scheldt estuary, southwest Netherlands. Mar. Chem. 66(1e2): 91–111.

5

Copper
Essential and Noxious to Aquatic Organisms

William Senior,[1,]* *Ricardo de La Cruz*[1] and *Luis Troccoli*[2]

1. Introduction

Aquatic chemistry has become a rewarding and substantial area of research that is attracting many scientists. Its literature has changed from a compilation of composition tables to studies of chemical reactions that take place within aquatic environments. Given that the rivers deliver to the world's oceans most of their dissolved and particulate components, the interactions of these two sets of waters determine the vitality of our coastal waters. This chapter not only provides an introduction to the dynamics of aquatic chemicals, but also identifies materials that endanger marine and fluvial resources. The information presented here will be of great value to environmental scientists dedicated to maintaining renewable hydrosphere resources. As the size of the world population will increase in the near future and the uses of materials and energy show parallel increases, rivers and oceans should be considered as a resource to accept some of society's waste. The capacity of these waters and the sediments to accommodate the waste must be evaluated continuously. The information presented in this chapter is based on the review and evaluation of scientific publications and technical reports from various sources.

Global copper production during the past 60 centuries is estimated at 307 million metric tons (million mt), most of which (79%) was produced since 1900; annual global production of copper is now estimated at 13.6 million tons (Nriagu 1979b). Copper occurs naturally in many minerals and as uncombined metal (Eisler 1997). The three most important sources of copper are chalcocite (Cu_2S), chalcopyrite ($CuFeS_2$), and malachite ($CuCO/Cu(OH)_2$) (ATSDR 1990, 2004, Ndoro and Witik 2017).

The United States is a major world producer and consumer of copper and its compounds. Most of the copper produced is used to manufacture electrical equipment, pipe, and machinery. Copper

---

[1] Departamento de Oceanografía, Instituto Oceanográfico de Venezuela, Universidad de Oriente, Núcleo de Sucre, Venezuela.
[2] Escuela de Ciencias Aplicadas al Mar. Universidad de Oriente, Núcleo Nueva Esparta, Venezuela, State University of the Santa Elena Peninsula, Ecuador.
    Emails: ricaedela@gmail.com; ltroccoli@gmail.com
* Corresponding author: senior.william@gmail.com

releases to the global biosphere—which may approach 1.8 million mt per year—come mostly from anthropogenic activities such as mining and smelting, industrial emissions and effluents, and municipal wastes and sewage sludge. Copper compounds are widely used as biocides to control nuisance algae and macrophytes, freshwater snails that may harbor schistosomiasis and other diseases, ectoparasites of fish and mammals, marine fouling organisms, and mildew and other diseases of terrestrial crop plants. Copper compounds are also used in agricultural fertilizers, in veterinary and medical products, in the food industry, and as a preservative of wood and other materials (Eisler 1997, Okocha and Adedeji 2012, Wells and Trainer 2016).

State-owned Codelco (Corporación Nacional Del Cobre de Chile) is the world's largest copper producer, producing 1.79 million tons in 2013. According to the company, it controls approximately 9% of the global copper reserves. In 2014, Chile as a whole accounted for more than one-third of global copper production, with total mine output of 5.8 million tons. According to the Central Intelligence Agency, copper production accounts for approximately 19% of government revenue, making the Chilean copper industry one of the most important both domestically and globally (PlusMining 2017).

China's copper consumption expanded markedly since 2001. Between 2001 and 2011, China's copper use increased by 5.1 million mt, or 215%. By 2002, China surpassed the United States to become the world's largest copper consumer (Mills 2015). In 2011, China consumed 7.9 million mt of refined copper products, accounting for 40% of the global demand. China's copper consumption is projected to rise to 9.7 million mt in 2014, accounting for 84% of global copper demand growth between 2011 and 2014 (Hammer and Jones 2012, Daly 2018).

The amount of copper entering the global ecosystem annually is unknown, but estimates range from 211,000 mt to 1.8 million mt (NAS 1977, Nriagu 1979b, Eisler 2007). About 80.7% of this copper is deposited in terrestrial compartments, 15.7% is deposited in the hydrosphere, and 3.6% is released into the atmosphere (Nriagu 1979b, Haywood 2015). The residence time for copper in the deep ocean is 1,500 years; in soils it may be retained for as long as 1,000 years; in air, copper persists for about 13 days (Nriagu 1979b, Eisler 2007). Copper in the atmosphere results mainly (73%) from human activities such as copper production and combustion of fossil fuels; the remainder is from natural sources that include sea salt sprays, windblown dusts, volcanogenic particles, and decaying vegetation (Nriagu 1979b, 1979d, ATSDR 2004, Sen and Peucker-Ehrenbrink 2012).

Inputs of copper into the aquatic ecosystems increased sharply during the 20th century and include inputs from waste discharges into saline waters, industrial discharges into freshwater, and leaching of antifouling marine paints and wood preservatives (Rodney et al. 2007, Bighiu et al. 2017). Present anthropogenic inputs of copper are two to five times higher than natural loadings; the atmosphere is a primary recipient of these inputs (Nriagu 1979c, Eisler 2007, Mohamed et al. 2015). In mining and industrial areas, precipitation of atmospheric fallout is a significant source of copper to the aquatic environment (USEPA 1980, Rodney et al. 2007).

More than 99.9% of oceanic copper fell as clay and manganese oxide particles in precipitation (NAS 1977, LaBarre 2014). In the lower Great Lakes, direct atmospheric inputs of copper—in mt per year—range from 55 to 2,300 for Lake Michigan, 120 to 330 for Lake Erie, and 72 to 123 for Lake Ontario; regional disparities in atmospheric deposition of copper are related to the intensity of industrial activity and to the regional wind systems (Nriagu 1979c, Robertson and Saad 2011, Sensarma et al. 2016).

Copper in soils may come from a variety of anthropogenic sources: mining and smelting activities; other industrial emissions and effluents; traffic; fly ash; dumped waste materials; contaminated dusts and rainfall; sewage and sludge; pig slurry; composted refuse; and agricultural fertilizers, pesticides, and fungicides (ATSDR 1990, 2004, Chauhan 2016, Kumar et al. 2016). In the case of Florida citrus groves, copper-containing fertilizers applied during the early 1900s accounted for as much as 34 kg Cu ha^{-1} annually, and routine fungicidal sprays contributed another 10 kg Cu ha^{-1} annually. Surface soils (0–15 cm) from some mature citrus groves contained as much as 540 kg Cu ha^{-1} (Eisler

2007, Obreza and Morgan 2017). Copper deposition rates in soils are usually higher in cities and near highways, railroads, power plants, and industrial activities (Nriagu 1979c, Su et al. 2014, Yan et al. 2018).

The top domestic markets for copper and its alloys in 1986 were, in order of importance, plumbing, building wire, telecommunications, power utilities, in-plant equipment, air conditioning, automotive electrical, automotive non-electrical, business electronics, and industrial valves and fittings (ATSDR 1990, 2004, Ciacci et al. 2017). A small percentage of copper production is used to manufacture chemicals, mainly copper sulfate (ATSDR 1990, 2004). Of the copper sulfate used domestically, 65% is used in agriculture for fungicides, algicides, nutritional supplements, insecticides, and repellents; 28% is used industrially in froth flotation production of chromated copper arsenate wood preservatives, in electroplating, and in the manufacture of azo dyes; and 7% is used in water treatment to control nuisance algae (ATSDR 1990, 2004, Eisler 2007).

Copper is widely used to control unwanted species of freshwater algae and macrophytes (NAS 1977, USEPA 1980, Eisler 2007, Wagner et al. 2017). Chelated copper products are claimed to be effective algicides in hard water; the chelation of copper by organic compounds, such as ethanolamines or ethanolamine complexes, protects copper from precipitation and complexation (Eisler 2007, Bishop and Rodgers 2012). Copper sulfate is approved by the US Environmental Protection Agency (USEPA) as an algicide in waters used to raise fish for human consumption (Viriyatum 2013). In algae, copper inhibits photosynthesis, nitrogen fixation, and phosphorus uptake; it selectively eliminates cryptophytes but spares diatoms (Rastogi et al. 2015).

Copper sulfate at low concentrations has been used to control freshwater algae in Wisconsin since 1918 without any conclusively proven effect on diversity or abundance of non-target species (Eisler 2007, Watson and Yanong 2011). But reduced abundance of freshwater benthos was noted in Lake Monova, Wisconsin, which received 771 mt of copper to control algae over a 26-year period and had sediment levels as high as 1,093 mg Cu kg^{-1} DW (Han et al. 2001, Okocha and Adedeji 2012). Copper sulfate used to control algal blooms in Wisconsin lakes at 1.25 mg Cu L^{-1} killed non-target fishes, crustaceans, snails, and amphibians in 14 d or less; however, 0.25 mg Cu L^{-1} was not fatal to these species in 20 d (Eisler 2007, Beaver et al. 2018). Concentrations as low as 0.03 mg Cu L^{-1} inhibited growth in two of four species of nuisance aquatic weeds in Lake Mendota, Wisconsin, and 0.3 mg Cu L^{-1} was fatal to all four species (Anderson et al. 2009, Lewtas et al. 2015). Copper sulfate controlled algae in cranberry bogs at 0.4 mg Cu L^{-1} but this concentration also killed resident fishes (Eisler 2007). Copper was not measurable in the surface waters of cranberry bogs within 10 d of treatment, regardless of initial copper concentration; it is probable that copper was adsorbed onto bog soils (Eisler 2007, USEPA 2004). Copper salts are intentionally added to drinking water supplies of some municipalities to control growth of algae; concentrations as high as 59 µg Cu L^{-1} are maintained in New York City (USEPA 1980).

Copper compounds are used routinely and widely to control freshwater snails that serve as intermediate vectors of schistosomiasis and other diseases that afflict humans (NAS 1977, Al-Sabri et al. 1993, Colley et al. 2014). These compounds include copper sulfate, copper pentachlorophenate, copper carbonate, copper-tartaric acid, Paris green (copper arsenite acetate), copper oxide, copper chloride, copper acetyl acetonate, copper dimethyl dithiocarbamate, copper ricinoleate, and copper rosinate (Schmitt et al. 1999, Hoang et al. 2008). Also, many species of oyster enemies are controlled by copper sulfate dips (Pérez and Pérez 2012). All tested species of marine gastropods, tunicates, echinoderms, and crabs that had been dipped for 5 sec in a saturated solution of copper sulfate died if held in air for as little as a few seconds to 8 h; mussels, however, were resistant (Matheson et al. 2007).

Copper sulfate is used to control protozoan fish ectoparasites including *Ichthyopthirius*, *Trichodina,* and *Costia*; the effectiveness of the treatment diminishes with increasing total alkalinity and total hardness of the water (Athanassopoulou et al. 2009, Bowker et al. 2014). Copper compounds now used to control protozoan parasites of cultured red drum (*Sciaenops ocellatus*) include copper sulfate, copper sulfate plus citric acid, and chelated copper compounds (forms of copper bound by sequestering agents, such as ethanolamine); chelated copper compounds are considered less toxic to fish than copper sulfate and at least as effective in controlling parasites (Mitchell and Hobbs 2003, Athanassopoulou et al. 2009).

Copper is the active agent in many antifouling paints applied to watercraft (Ytreberg et al. 2016, Telegdi et al. 2016) and is associated with elevated copper concentrations in the Pacific (Ryder et al. 2014). These paints are associated with elevated copper concentrations found in Pacific oysters (*Crassostrea gigas*) farmed in the Bay of Arcachon, France (Gamain et al. 2016).

Copper compounds are used in agriculture to treat mildew and other plant diseases; in the food industry as preservatives, additives, or coloring agents; in preservatives of wood, leather, and fabrics; in coin manufacture; and in water treatment (ATSDR 1990, 2004, Hand 2004). The use of copper-containing pesticides is traditional along the Mediterranean coast, especially the use of Bordeaux mixture, a copper sulfate-based fungicide that has been widely used for more than a century to prevent mildew on grape vines (Van Zwieten et al. 2007). However, at current application rates of about 0.8 mg Cu cm^{-2}, Bordeaux mixture significantly reduces the life span and breeding rate of the fruit fly (*Drosophila melanogaster*) (Bretman et al. 2013).

Copper is widely used in veterinary clinics in medical products (Szymański et al. 2012, Humann-Ziehank 2016). Copper sulfate is used by veterinarians to treat cattle and sheep for helminthiasis and infectious pododermatitis (NAS 1977). Cuprol (a 1% solution of cupric oleinate) is used to control lice (Eisler 2007). Copper is routinely used as a growth supplement in the diets of swine (*Sus* sp.) in the United Kingdom and elsewhere; diets may contain as much as 250 mg Cu kg^{-1} ration (USEPA 1980, Veum et al. 2004). The intensity of pig farming within about 10 km from the coast may influence copper content in estuarine sediments (Ismail and Ramli 1997, Eisler 2007). For example, intensive pig farming in coastal Brittany, France, increased soil copper concentrations by 0.6 kg ha^{-1} annually and increased coastal sediment copper concentrations to as much as 49.6 mg kg^{-1} DW (Van der Werf et al. 2005).

In human medicine, metallic copper is used in some intrauterine devices, and various copper compounds are used as emetics and to treat rheumatoid arthritis (USEPA 1980, Szymański et al. 2012, RSA 2014). Some individuals wear copper bracelets as treatment for arthritis, although its therapeutic value has little support (USEPA 1980, MacGill 2018).

2. Copper Is Both Essential and Toxic to Life

The use of heavy metals is intimately related to human history (Jan et al. 2015). After discovering the elements called metals, humans began to use them in industrial and agricultural activities (Moss 2008, Jaramillo and Restrepo 2017). However, information about metals not only describes their benefits but also their negative influence on the environment (Dechezleprêtre and Sato 2017). Today, industrial activities and agriculture generate a wide variety of chemical species that contain lead, mercury, and cadmium, which are generally associated with environmental pollution due to their toxicity and bioaccumulation properties (Fashola et al. 2016). Metals are also used in medicine as ingredients in various compounds for treatments, preservation of vaccines (WHO 2004, Baker 2008) or contrast medium (Klein et al. 2008, Dórea 2015).

Copper is a very important mineral for the development of human life and the contributions of this element must be noted, but in its proper measure. High doses of copper cause serious health disorders, such as anemia, stomach problems, and damage to the kidneys and liver (IMSP 2017, López 2018). The symptoms of excess copper can include headaches, dizziness, and irritation of the nose, mouth, and eyes (IMSP 2017).

Copper is present in our homes in pipes and in kitchen containers and therefore gets into our water and/or cooked food (USEPA 1999, EPA 2013a). It is used massively in industry and that is what has made our exposure to copper much higher than normal (WHO 2004, Pappas 2014).

With industrialization increasing in coastal cities, metal pollution of the marine environment, especially in estuaries and along coasts, has become a worldwide problem (He et al. 2014). To strengthen environmental management and to minimize impacts of metals on marine organisms, the USEPA has recommended criteria continuous concentrations (CCCs) for nine metals or metalloids in saltwater: arsenic (As), cadmium (Cd), hexavalent chromium (Cr(VI)), copper (Cu), lead (Pb), mercury (Hg), nickel (Ni), selenium (Se) and zinc (Zn) since 1986 (Qie et al. 2017). However, for other metals, CCCs are still lacking, which limits the power of government regulators to assess water quality and to make sound environmental management decisions. Thus, there is a need to derive CCCs for additional metals (Qie et al. 2017).

Copper was one of the first metals used by humans; its use dates back approximately to 5000 years ago in the Aegean region, where it was used for the creation of valuable objects (Stern et al. 2007). The exposure of humans to copper is mainly due to the consumption of food and drinking water, and the intake of copper in relation to food and water depends on the geographical location; generally, 20–25% of the copper intake comes from drinking water (Stern et al. 2007, Abdolmaleki et al. 2013). The deficits and excessive increases of copper have been of great importance in human health, because it is an element whose deficiency has been associated with altered production of energy, abnormalities in the metabolism of glucose and cholesterol, increase in iron ions, structural and physiological alteration of the blood circulation, cardiac affections (Medeiros et al. 2002), and alterations at the level of the cells of the immune system (Keen et al. 2003). Other effects identified are bone alterations in children with low birth weight (Negrato and Gomes 2013). With regard to genetics, it has been reported that the existence of mutations at the level of copper-binding proteins can favor the development of diabetes (Flores et al. 2004) and neurological damage (Xu et al. 2004).

On the other hand, the toxicity caused by this element is attributed to accidental intake (Stern 2007, Gamakaranage et al. 2011). Acute toxicity may be associated with the existence of oxidative stress in various areas of the body or endocrine disorders (Jaishankar et al. 2014). Other mechanisms reported are exposures through the skin or respiratory tract (Hostynek et al. 2003). Chronic toxicity arises as a consequence of prolonged exposure in small quantities (Amador et al. 2015). In dental care, the measurement of copper through saliva as a biological matrix can be useful, since it is possible to find traces of this element associated with dental caries, a widely prevalent disease (Bhattacharya et al. 2016).

Metal contamination is a global environmental challenge because metals are neither chemically nor biologically degradable, so they persist once they have entered the environment (Tchounwou et al. 2012). In addition to natural sources of metals, anthropogenic sources, consumer products containing metals and metal nanoparticles are added to the load (Hill and Julang 2017). Millions of tons of electronic waste containing metals such as cadmium, copper, chromium, silver, nickel and cobalt end up in the environment every year (Needhidasan et al. 2014, Chen et al. 2018). Metals are filtered from landfills and, to a much greater extent, they are re-mobilized from sediments, for example, during flood events (Redelstein et al. 2015) and enter the aquatic environment, where they harm aquatic organisms, causing lethal effects on activity, growth, metabolism, and reproduction (Solomon 2008, Kibria 2016). Especially fish are important indicators of the contamination of metals in fresh water, since they can accumulate metals through respiration and absorption of food (Tchounwou et al. 2012, Rajeshkuma and Li 2018). The main reason for the high sensitivity of fish to metal toxicity is the constant exposure of the large respiratory epithelium of the gills, where the metal ions cause topical cellular toxicity and interfere with the ion exchange and transport processes (Thangam et al. 2014). However, the early stages of development of fish, which have not yet developed gills, are also prone to metal toxicity, causing mortality or sublethal development deficiencies (Hernández et al. 2011, Witeska et al. 2014). The molecular basis of these adverse effects is still less understood than metal toxicity in adult fish (Tchounwou et al. 2012, Andreoli and Sprovieri 2017).

Environmentally important metals, such as copper, have been extensively studied because of their high toxicity to aquatic organisms such as daphnia and fish (Hoppe et al. 2015, Nikitin et al.

2018). Once they have entered the aquatic ecosystems, they are transformed through biogeochemical processes and distributed among several species with different physical and chemical characteristics, for example, particulate (> 0.45 μm), colloidal (1 nm–0.45 μm) and dissolved species (< 1 nm) (Martorell 2010, Reyes et al. 2016). Particulate and colloidal matter, both organic and inorganic, plays a key role in coagulation, sedimentation and adsorption processes, which influence the residence times and transport of trace metals from the water column to the sediments and other matrices (Turner and Millward 2002, Reyes et al. 2016).

Copper naturally occurs in the aquatic environment in low concentrations. Major aquifers of the USA have Cu concentrations less than 10 ppb (Tchounwou et al. 2012), while Canadian freshwaters have 1–8 ppb Cu (ATSDR 1990, 2004), and streams in Bristol Bay (Alaska, USA) have 0.04–5.60 ppb Cu (Woody and O'Neal 2012). Seawater copper concentrations are generally less than 1 ppb (Lee et al. 2011).

Elevated aquatic copper concentrations primarily occur near copper mining and smelting facilities and in urbanized areas (Davis et al. 2000, Eisler 2000). Aquatic habitats are susceptible to copper pollution because they are the ultimate receptor of industrial and urban wastewater, storm water run-off, and atmospheric deposition (Nriagu 1979a, Davis et al. 2000). Copper is acutely toxic (lethal) to freshwater fish in soft water at low concentrations ranging from 10 to 20 ppb (Woody and O'Neal 2012). Elevated copper concentrations observed in mine-impacted Mineral Creek, Colorado, were as high as 410 ppb (Runkel et al. 2009, Kimball et al. 2009, Woody and O'Neal 2012, Hua et al. 2018) and in mine-impacted Copperas Brook in Vermont they were 4600 ppb (Hammarstrom et al. 1999, Woody and O'Neal 2012). Sansalone et al. (1997) and Woody and O'Neal (2012) documented urban storm water run-off copper concentrations of 325 ppb. Such copper concentrations are lethal to fish and aquatic life (Eisler 2000, Solomon 2009, Kiaune and Singhasemanon 2011).

About 15 million mt of copper is used worldwide yearly for construction, electrical conduit, agriculture, manufacturing and other uses; one third of this copper is from recycling, while the rest is from mining (ATSDR 1990, 2004, Woody and O'Neal 2012). Because copper is a non-degradable potentially toxic pollutant that builds up in the environment, continued releases are of global concern (Woody and O'Neal 2012, Tchounwou et al. 2012).

3. Impact on Seagrass

Copper is an essential trace element for all biota (Hothem et al. 2015). It is incorporated into a variety of enzymes that play important roles in physiological processes (e.g., enzymes involved in cellular respiration, free radical defense, neurotransmitter function, connective tissue biosynthesis and other functions), as well as into some structural proteins (WHO 1998, Gaetke et al. 2014). Nonetheless, when organisms are excessively exposed, homeostatic control mechanisms become overwhelmed, and toxicity arises owing to adverse effects of copper on the structure and function of macromolecules such as DNA and proteins (WHO 1998, Emamverdian et al. 2015).

Since copper from anthropogenic sources eventually contaminates water bodies, the toxicity of this metal to aquatic organisms has been intensively studied over the past two decades (WHO 1998, Paul 2017). Comparative toxicity of different freshwater species, however, has been much less studied (Oliveira-Filho et al. 2004, Katagi 2008, Shuhaimi-Othman et al. 2012, Katagi and Tanaka 2016). For comparative purposes it is desirable to conduct toxicity assays with different species in parallel and at the same laboratory, because toxicity indices such as LC_{50} and EC_{50} are known to be subject to a wide interlaboratory variability (Chinedu et al. 2013, Dong et al. 2016). Comparative toxicity studies are thus important to establish more precise margins of safety (magnitude of differences between toxic concentrations to target and to non-target species) when copper products are intended to be used as algicides, aquatic herbicides or molluscicides (Geer et al. 2016, Wagner et al. 2017).

Toxicity of copper to aquatic species depends, on one side, on organism sensitivity and, on the other side, on the concentration of copper and its bioavailability (Lage et al. 1996, Casares et al. 2012, Hoppe et al. 2015). In natural settings, copper bioavailability in water bodies depends on a variety of factors such as adsorption to particles, complexation by organic matter (e.g., humic and fulvic acids), presence of other cations and pH (WHO 1998, Fonseca et al. 2013, Olaniran et al. 2013). Owing to this variability, copper bioavailability in a certain natural water body may be quite different from that in the standardized assay water and thus laboratory bioassay data should be used with caution by risk assessors (Pereira et al. 2000, Hook et al. 2014). Growth inhibitions of different species of freshwater algae have been noted at comparable concentrations of copper or copper salts in several studies. Copper sulfate, for instance, has been reported to inhibit (72-h IC_{50}, static) *Selenastrum capricornutum* (*Raphidocelis subcapitata*) growth at concentrations as low as 0.047 mg L^{-1}, *Chlamydomonas reinhardtii* at 0.079 mg L^{-1}, and *Scenedesmus subspicatus* at 0.120 mg L^{-1} (WHO 1998, Wang et al. 2017, Wu et al. 2017).

The toxicity data provided by the studies suggest that copper sulfate is more readily bioavailable in the assay soft water than the other two copper-based pesticides. In the alga growth inhibition test, on the other hand, 96-h IC_{50}—expressed in terms of Cu—for copper sulfate was slightly higher than those calculated for copper oxychloride and cuprous oxide (Oliveira-Filho et al. 2004, Silva et al. 2017). Contrasting with the assay soft water, in the algal assay medium, bioavailability of copper from copper sulfate may have been lower than those of Cu derived from the two other copper-based pesticides (Oliveira-Filho et al. 2004).

In conclusion, toxicity data provided by the studies reviewed confirm that planktonic crustaceae and algae are extremely susceptible to increases in free copper levels in water bodies (Oliveira-Filho et al. 2004, Grimm and Gerhardt 2018). Since phyto- and zooplanktonic organisms form the basis of aquatic food webs, increased levels of bioavailable copper are likely to dramatically affect freshwater ecosystems (Wang 2002, Rostern 2017). Furthermore, our results also indicate that the use of copper compounds, either as algicides and aquatic plant herbicides, or as molluscicides, is likely to adversely affect a variety of non-target aquatic species (Oliveira-Filho et al. 2004, Siemering et al. 2008, Viriyatum 2013, Carvalho and Caramujo 2017).

Seagrasses, marine flowering plants, have a long evolutionary history but are now challenged with rapid environmental changes as a result of coastal human population pressures (Brodie and N'Yeurt 2018). Seagrasses provide key ecological services, including organic carbon production and export, nutrient cycling, sediment stabilization, enhanced biodiversity, and trophic transfers to adjacent habitats in tropical and temperate regions (Santos et al. 2011, Brankovits 2017). They also serve as "coastal canaries," global biological sentinels of increasing anthropogenic influences in coastal ecosystems, with large-scale losses reported worldwide (IUCN 2012, LWLI 2011, 2013). Multiple stressors, including sediment and nutrient run-off, physical disturbance, invasive species, disease, commercial fishing practices, aquaculture, overgrazing, algal blooms, and global warming, cause seagrass declines at scales of square meters to hundreds of square kilometers (Berdalet et al. 2016). Reported seagrass losses have led to increased awareness of the need for seagrass protection, monitoring, management, and restoration (De Los Santos et al. 2014, Johnson et al. 2017, Nordlund 2017). However, seagrass science, which has rapidly grown, is disconnected from public awareness of seagrasses, which has lagged behind awareness of other coastal ecosystems (Hale et al. 2013, Hedberg 2017). There is a critical need for a targeted global conservation effort that includes a reduction of watershed nutrient and sediment inputs to seagrass habitats and a targeted educational program informing regulators and the public of the value of seagrass meadows (Hale et al. 2013).

Seagrass meadows form some of the most productive ecosystems in the world (Johnson et al. 2017, Nordlund et al. 2017) and provide high-value ecological services (Halpern 2008, Nordlund et al. 2017, Mehvar et al. 2018). Seagrass debris provides abundant food to epiphytes, which are fed upon by epifaunal organisms, which then provide food to fish foraging in seagrass beds. Seagrass thereby provides an important link between primary producers, e.g., microalgae, and higher-level consumers (Arnaud et al. 2018) and is positioned in a crucial ecological niche between infauna and pelagic species (da Silva et al. 2015).

Seagrass also support important economic services, for example by providing a habitat for the echinoid *Paracentrotus lividus*, which in many parts of the Mediterranean Sea is commercially important for its gonads, considered a seafood delicacy in European countries (Halpern 2008, Darius et al. 2018). Seagrass communities also supply significant biogeochemical functions, e.g., nutrient cycling (Allgeier et al. 2016, Liu et al. 2018) and carbon sequestration (Chen et al. 2017).

An important feature of seagrass is the formation of a dense composite of leaves, rhizomes and roots, which acts to significantly reduce ambient energy resulting in sediment stabilization (Deyanova et al. 2017) and an increase in the fine fraction in bottom sediment (Deyanova et al. 2017). Elevated fine material provides additional adsorption surfaces, increasing the abundance of metals in seagrass substrate (Jeyasanta et al. 2018), making metals a significant environmental companion of seagrass systems (Hosokawa et al. 2016). Some metals may be incorporated into seagrass tissue from sediments resulting in inhibited growth (Brodersen et al. 2017) and adverse effects to biochemical pathways, e.g., photosynthesis (Fraser and Kendrick 2017). Sedimentary metals thus constitute a significant threat to seagrass functioning in estuarine environments (Ahmad 2015, Copertino et al. 2016). Given that roots are more closely aligned to bottom sediments, a closer relationship between root and sediment than between leaves and sediment may be expected; however, the relationship between metals in ambient surficial sediment and seagrass tissue is inconsistent and not well understood (Howley 2001, Amado-Filho et al. 2008).

Because seagrass often occupies intensely urbanized sheltered coastal areas, these epiflora are frequently exposed and vulnerable to human disturbance (Lane et al. 2013). Direct and indirect reasons have been recognized for loss of seagrass coverage (Manikandan et al. 2011). Direct influences include disease and over-exploitation, including fishing, boating, coastal engineering (Howarth et al. 2000) and natural causes, e.g., cyclones and tsunamis (Polidoro et al. 2010). Indirect impacts are considered more damaging and include declining water quality resulting from increased nutrient and contaminant inputs and sediment run-off (Erftemeijer et al. 2012). Other indirect effects are from aquaculture, invasive species, overfishing causing loss of predators (herbivores), and climate change (FAO 2008, Llope et al. 2011).

Historic declines in seagrass distribution worldwide have been associated with increased industrial and urban development (Larkum and West 1982, Telesca et al. 2015). Since the earliest records of seagrass meadows in 1879, in all areas of the world for which data are available, seagrass meadows have declined (Röhr et al. 2016). Globally, seagrass areas have disappeared at a rate of 110 km^2 yr^{-1} since 1980 and 29% of the known area of these meadows has disappeared since records were kept (Telesca et al. 2015, Gundersen 2017). Moreover, the rate of decline has increased from a median of 0.9% yr^{-1} before 1940 to 7% yr^{-1} since 1990 (Traganos and Reinartz 2018). As in the rest of the world, declining seagrass stocks are evident in Australian waters (Lavery 2013, Nordlund et al. 2017). The coastal environments of Australia support the highest number of seagrass species and the largest seagrass beds in the world (Walker and McComb 1992, Short et al. 2007, Lin and Zhang 2015). In recent years concern has been raised over the extensive loss in seagrass communities in Australian coastal environments, where a loss of 45,000 ha of seagrass has been recorded since the 1960s (Walker and McComb 1992, Clark and Johnston 2017, Romañach et al. 2018). In Sydney estuary, the current area of seagrass coverage is approximately 50 ha, representing only 15% of the estimated pre-European, pristine distribution (West and Williams 2008, Clark and Johnston 2017), whereas mean sedimentary metal enrichment has increased to > 10 times pre-anthropogenic concentrations (maximum is > 100 times) and adverse effects on benthic populations are estimated in 2% to 36% of the waterway, depending on the metal (McCready et al. 2004, 2006a, b, c, Birch 2017).

Despite many recent studies of seagrass, there has been little attention to the response of seagrass metal tissue content to high sedimentary metal mixtures in the estuarine environment (Howley 2001, Dowsett and Rayburg 2011, Ahmad et al. 2015, Lin et al. 2016). Research into the relationship between sediment and seagrass tissue metals has been undertaken in Sydney estuary in recognition of the role contamination plays in threats to seagrass health and functioning in estuarine ecosystems (Serrano et al. 2016). These studies are essential to understand the potential for adaptability and tolerance of seagrasses in declining coastal environments with varying degrees of environmental stability and suitability (Serrano et al. 2016, Birch et al. 2018).

The results then presented were part of earlier research on seagrass-sediment relationships for another epifloral species (*Zostera capricorni*), which shares a habitat with *Halophila ovalis* in Sydney estuaries (Birch et al. 2018). Because the concentration and bioavailability of metals in sediment is confounded by variable grain size (Forstner and Calmano 1998, Paller and Knox 2013), tissue-sediment relationships were investigated for both fine (< 62.5 μm) and total sediment (Huang et al. 2018). Seagrass tissue metal characteristics influence patterns of growth and morphology (Baby et al. 2017) and possible seagrass loss (Soissons et al. 2018). Dissimilar life histories for different seagrass species may have ramifications for biomonitoring and formulation of preventive measures (Ray et al. 2014, McMahon et al. 2017).

4. Alaska Copper Water Quality Standards

Copper and other metals released from mining and urban sites can contaminate water sources and affect fish through water or food (Younger et al. 2002, Clearwater et al. 2002, Lapointe et al. 2011, Woody and O'Neal 2012, Reyes et al. 2016). Fish and aquatic organisms are very sensitive to increased Cu concentrations in water. However, Cu toxicity depends, at least in part, on water quality (Solomon 2009, Kiaune and Singhasemanon 2011, Casares et al. 2012, Woody and O'Neal 2012, Jaishankar et al. 2014).

Water-borne copper exists in a variety of forms, with the dissolved form dCu (cupric ion Cu^{2+}) considered the most toxic to aquatic life (Eisler 2000, USEPA 2007, Woody and O'Neal 2012). Some water parameters affecting Cu toxicity include water hardness, pH, anions and dissolved organic carbon (DOC) (Russell et al. 2015, Zhang et al. 2017). Toxicity of Cu to aquatic life varies with:

1) **Water hardness:** dCu is more lethal in soft waters than in hard waters rich in cations (e.g., Ca_2^+ and Mg_2^+), as cations reduce bioavailability of dCu and thus toxic effects (Woody and O'Neal 2012, Croteau et al. 2014, Ying et al. 2018).

2) **pH:** Cu is more toxic under acidic conditions (pH < 6; Gaetke et al. 2014, Lewis et al. 2016).

3) **Anions and DOC** bind to dCu, creating compounds that reduce dCu concentrations and toxic effects (USEPA 2007, Woody and O'Neal 2012).

The Alaska Department of Environmental Conservation uses the Aquatic Water Quality (AWQ) criteria to protect freshwater species from increased copper inputs (ADEC 2010, Woody and O'Neal 2012, Zhang et al. 2017). Acute AWQ copper criteria address lethal effects of copper using a regression model of lethal dCu concentrations as a function of water hardness; the model also takes into account pH and alkalinity (USEPA 2007, Woody and O'Neal 2012).

Freshwaters in mine leases near Iliamna Lake are "soft" (low hardness of 1 to 31 mg L^{-1}; EPA 2012, Woody and O'Neal 2012); corresponding lethal dCu concentrations at this hardness range from 0.18 to 4.46 ppb dCu respectively; chronic dCu toxicity ranges from 0.18 to 3.29 ppb (Woody and O'Neal 2012, Kiyani et al. 2013). This is because soft waters are limited in their ability to ameliorate toxic effects of increased Cu inputs (Al-Reasi et al. 2011, Casares et al. 2012, Woody and O'Neal 2012). Cusimano et al. (1986) and Woody and O'Neal (2012) found that 50% of exposed rainbow trout died in 96 h at a concentration of 2.8 ppb Cu in water of 9.2 mg L^{-1} hardness. Taylor et al. (2000) found that copper was approximately 20 times more toxic to rainbow trout weighing 1 to 2 g in soft water (20 mg L^{-1}) than in hard water (120 mg L^{-1} as $CaCO_3$). Increases in dCu concentrations can be lethal at very low concentrations in soft waters (NAS 1986, Taylor et al. 2000, Kiaune and Singhasemanon 2008, Woody and O'Neal 2012).

Nevertheless, the Alaska hardness-based acute (lethal) AWQ is under-protective for numerous species comprising aquatic food chains (Scannell 2009, Woody and O'Neal 2012). A review of 75 published reports by Scannell (2009) for the Alaska Department of Fish and Game indicated the hardness-based AWQ was underprotective in acute (lethal) Cu toxicity studies for:

- 5% of reviewed fish test values (Scannell 2009),
- 14% of reviewed aquatic invertebrate test values (Fair et al. 2011, Woody and O'Neal 2012),
- 40% of freshwater mussel test values (Fair et al. 2008, Elison et al. 2015), and
- 38% of zooplankton test values (Fair et al. 2008, 2011, Elison et al. 2015).

The USEPA (2007) recommended the use of a more complex AWQ model for copper because the hardness model may sometimes be underprotective or overly protective and site-specific studies required to fine-tune models can be expensive (Woody and O'Neal 2012, Simons 2015). The Biotic Ligand Model (BLM) uses 10 parameters to calculate AWQ for aquatic species: temperature, pH, DOC, calcium, magnesium, sodium, potassium, sulfate, chloride and alkalinity (USEPA 2007, Lynch 2014). The BLM can be useful in predicting acute or lethal effects to aquatic life that occur at fish gills in waters of different chemistries (Hoppe et al. 2015, Zhang et al. 2017). But BLM assumptions have been questioned and failures of the model to protect aquatic life are documented (McConaghie and Matzke 2016). For example, the BLM did not adequately predict Cu toxicity to trout in soft water and underestimated acute toxicity at higher pH and chronic toxicity at lower pH (Scannnell 2009, Woody and O'Neal 2012, Crémazy et al. 2017). Another issue with the BLM is that DOC varies in form and consequently can limit accurate Cu toxicity predictions of the BLM (Wood et al. 2011, Casares et al. 2012, Tait 2013). Further research is needed to resolve inadequacies of the BLM to ensure salmon and aquatic species are protected from increased Cu inputs (Woody and O'Neal 2012).

5. Copper Toxicity to Aquatic Organisms

Copper is an essential trace metal for growth and metabolism of all living organisms. In vertebrates, including fish, copper forms part of many enzymes and glycoprotein, is important for nervous system function and is necessary for hemoglobin synthesis (Woody and O'Neal 2012, Fresco 2016). Deficiencies are rare as copper is plentiful in the environment; however, deficiencies in mammals are linked to anemia, gastrointestinal disturbances, aortic aneurisms, abnormal bone development and death (Woody and O'Neal 2012). Precise copper dietary needs and deficiency effects in wild fish and aquatic species are unclear and the subject of ongoing research (Weltje et al. 2013).

Copper is one of the most toxic elements to aquatic species; at levels just above that needed for growth and reproduction it can accumulate and cause irreversible harm to some species (Eisler 2000, Tierney et al. 2010). Copper is acutely toxic (lethal) to freshwater fish via their gills in soft water at concentrations ranging from 10 to 20 ppb (NAS 1977, Spokas et al. 2008).

The aquatic environment and especially the marine environment is one of the most exposed to pollutants, because the discharges, whether by land, aquatic-terrestrial or atmosphere, have as a final receptacle the marine environment (Márquez et al. 2012, 2016). In these systems microalgae constitute the main component of the phytoplankton that supports the trophic chain; in this way, a qualitative or quantitative change produced by a contaminant could have a drastic impact on the ecosystem (Franklin et al. 2000, Lu et al. 2015).

Toxicity of Cu to aquatic organisms depends on its "bioavailability" or its potential to transfer from water or food to a receptor (e.g., gills, olfactory neurons) on an organism where toxic effects can occur (Scannell 2009, Tchounwou et al. 2012, Woody and O'Neal 2012, Islam et al. 2017). Toxic effects of Cu are classified as "acute" (lethal) and "chronic" (where sublethal exposures result in reduced growth, immune response, reproduction and/or survival) (Li et al. 2016).

Adverse effects have been demonstrated on various fish "receptors" including gills, olfactory receptors, and lateral line cilia and scientists are now learning more about how Cu affects fish DNA and molecules (Woody and O'Neal 2012, Li et al. 2016). Copper is known to reduce fish resistance to diseases; disrupts migration (i.e., fishes avoid copper-contaminated spawning grounds); alters swimming; causes oxidative damage; impairs respiration; disrupts osmoregulation structure and pathology of kidneys, liver, gills, and other stem cells; impacts mechanoreceptors of lateral line canals; impairs functions of olfactory organs and brain; and is associated with changes in behavior, blood chemistry, enzyme activities, corticosteroid metabolism and gene transcription and expression (Eisler 2000, Tierney et al. 2010).

Copper is acutely toxic to freshwater fish via the gills in soft water at concentrations ranging from 10 to 20 ppb (NAS 1977). Marr et al. (1998) demonstrated that 50% of rainbow trout died when exposed to 14 ppb Cu and 20% died when exposed to 8 ppb Cu in waters of about 25 mg L^{-1} hardness. Playle et al. (1993) demonstrated a dose-response relationship between 120 h mortality of rainbow trout juveniles and Cu gill accumulation after 24 h. Playle et al. (1993) showed that Cu binding to fish gills and the quantity of metal bound to gill was related to dCu concentrations. Waters in mine claims near Iliamna Lake are very soft and have low DOC concentrations (Levit 2010, Woody and O'Neal 2012).

In the evaluation of ecological risk in aquatic ecosystems, the evaluation of exposure and effect is considered (Hook 2014). For the determination of the effect, biological methods known as acute and chronic toxicity bioassays have been proposed, with representative organisms of different trophic levels, which give a global response to the dissolved pollutants in the surface waters that receive the discharge of chemical contaminants (O'Farrel et al. 2002, Chinedu et al. 2013).

In aquatic ecosystems, the organisms that can be affected by potentially toxic chemical agents are the primary producers, which are key in their structure and functioning, so any negative effect on them will affect higher trophic levels (Cid et al. 2012, Brodin et al. 2014). Some microalgae are considered study models for the performance of toxicity bioassays, because they have high sensitivity to various chemical substances, their nutritional requirements are known, they have a high growth rate that allows researchers to know in a few days the density and the effect caused by the toxic agent, and their manipulation is relatively simple in the laboratory (Show et al. 2017, Zhao et al. 2018).

Some microalgae of temperate zones have been considered as model species to study the effects of contamination, for their sensitivity to various test materials (Subirats 2015) and their high growth rate and easy cultivation in the laboratory (Show et al. 2017). However, few tropical species have been studied (Tacon 1987, Sevcikova et al. 2016, Sipaúba-Tavares et al. 2017).

Although the majority of studies of acute toxic effects in aquatic organisms are carried out as a response to periods of 96 h (Canesi and Fabbri 2015), the condition of rapid reproduction of microalgae and the greater effect of heavy metal lethality shown before the 48 h period observed in the present study suggest the reduction of the exposure time of the metals in microalgae to 48 h for a protocol of acute effects (Wang et al. 2017). This reduction of evaluation time would reduce the cost of the test and allow for a faster evaluation, avoiding possible microalgal adaptation to the metals, which could mask the toxic effect of the xenobiotic (Altenburger et al. 2013).

6. Copper and Fish Behavior

Effects of contaminants on fish behavior are now a topic of intensive research (Tierney et al. 2010, Vilches et al. 2014). Fish behavior is linked to individual survival and reproduction: if a salmon returns to spawn but instead gets eaten by a bear then that salmon's genes are out of the gene pool (Gende et al. 2002, Wong and Candolin 2015). Because copper impairs key senses such as smell it has the potential to impair complex fish behaviors important to survival, such as predator avoidance, social interactions, and reproductive behavior (Woody and O'Neal 2012, Lin et al. 2016). Copper can impair olfaction in fish and hardness-based dCu AWQ values are less protective at the fish nose than at the fish gill (Calfee et al. 2016). Fish are highly sensitive to odors in their environment and can detect natural chemical cues over long distances, such as mating pheromones, at ppb or ppt concentrations (Laberge and Hara 2003, Nielsen 2015). Approximately one hundred different olfactory receptors receive and trigger critical physiological and/or behavioral responses in fish (Tierney et al. 2010, Leduc et al. 2013), such as sperm production (Sorensen and Stacey 2004, Mostafa et al. 2012), predator recognition and avoidance, food location (Block et al. 2017), kin recognition (Heuschele and Selander 2014), recognition of conspecifics (Derby and Sorensen 2008), migration (Heuschele and Selander 2014), homing and reproduction (Derby and Sorensen 2008). Alteration of olfactory ability can impair behaviors important to survival and has the potential to affect population biodiversity if predator avoidance, homing, migration and spawning are affected (Woody and O'Neal 2012, Lennox et al. 2016).

Hatchery and wild salmon and steelhead exposed to 5 ppb and 20 ppb for 3 h at 58 mg L^{-1} hardness showed impaired olfaction (Baldwin et al. 2011). Sandahl et al. (2007) demonstrated that coho salmon exposed to just 2 ppb increases in Cu for 3 h at 120 mg L^{-1} hardness showed significantly impaired olfactory detection of predator alarm cues and a 50% decline in normal predator avoidance response; impairment in ability to detect and avoid predators can be lethal. Salmonids avoid waters with low dCu contamination, which disrupts their normal migration patterns (Meador 2014, Thomas et al. 2016). For example, coho salmon yearlings held in 5–30 ppb Cu for as little as 6 d showed altered downstream migration patterns (EPA 1977, Sundström et al. 2010, Woody and O'Neal 2012). Chinook avoided at least 0.7 ppb Cu, whereas rainbow trout avoided at least 1.6 ppb dissolved Cu (Mohammadbakir 2016). Laboratory avoidance of Cu by rainbow trout was observed at 0.1, 1.0 and 10 ppb Cu (Wang et al. 2013, Sappal et al. 2014). Birge et al. (1993) and Wang et al. (2013) demonstrated that salmon and other fish can be attracted to very high concentrations of dissolved Cu (4,560 ppb), which is lethal. However, studies on salmon olfaction and lateral line systems indicate hardness plays a lesser role in reducing toxic effects of Cu (Woody and O'Neal 2012).

Fish exposed to sublethal or chronic Cu concentrations can potentially suffer the following direct and indirect effects and further study is needed:

- Impaired neurological and brain function (Baldwin et al. 2003, Tierney et al. 2010),
- Impaired reproduction (Woody and O'Neal 2012, Moe et al. 2013),
- Impaired predator detection and avoidance (Baldwin et al. 2003, Tierney et al. 2010),
- Impaired ability to find food (Fleeger et al. 2003, Gibbons et al. 2015),
- Impaired ability to recognize members of their own species (Gibbons et al. 2015, Parent and Burd 2016),
- Impaired ability to recognize siblings (Parent and Burd 2016),
- Impaired homing ability (Baldwin et al. 2003, Parent and Burd 2016),
- Impaired migration behaviors (Gibbons et al. 2015),
- Impaired growth (Fleeger et al. 2003, Gibbons et al. 2015), and
- Depressed immune response (Filipuci 2011).

7. Diet-borne Copper

Studies on the toxic effects of Cu on fish have primarily focused on water-borne Cu toxicity but food-borne Cu paths in and through aquatic food chains are of increasing interest (Jaishankar et al. 2014).

Studies indicate that Cu uptake efficiency is similar for diet and water (Clearwater et al. 2002). Diet-borne Cu caused a quantitatively more important effect on gene transcription levels for proteins involved in energy metabolism, metal detoxification and protein protection compared to fish exposed to water-borne Cu (Hogstrand 2001, Lapointe et al. 2011). Like other heavy metals, copper can be accumulated in fish tissues from ingestion (Clearwater et al. 2002, Vineeta et al. 2007). Vineeta et al. (2007) and Bashir and Alhemmali (2015) studied fish fed a diet with high Cu concentration and observed Cu accumulation in descending order from gill > kidney > liver > muscle. Toxic concentrations of diet-borne Cu have been described for rainbow trout, carp and channel catfish, but data appear contradictory (see review by Clearwater et al. 2002). In one study, rainbow trout dietary copper toxicity occurred at 730 mg Cu kg^{-1} and maximum tolerable level was 665 mg Cu kg^{-1}; adverse effects included reduced growth, increased feed:gain ratios, food refusal and elevated liver copper levels (Daglish et al. 2004, FEEDAP 2016). Clearwater et al. (2002) determined Cu toxicity occurred at daily intake levels of 1–15 mg kg^{-1} body weight per day (depending on life stage) for Atlantic salmon and at 35–45 mg kg^{-1} body weight per day for rainbow trout (*Oncorhynchus mykiss*). Further studies are needed to improve understanding of potential effects of ingested Cu because once released into the environment Cu can accumulate in aquatic sediments and continue to recycle into aquatic food webs (Rhind 2009, Woody and O'Neal 2012).

8. Copper and the Freshwater Food Chain

Aquatic food chains and energy pathways are organized in a hierarchical way and Cu can be transferred through aquatic food chains (Winemiller 1990, McGreer et al. 2003, Nielsen 2015). Copper can affect salmonid ecosystems from the bottom of the food chain to top predators (see reviews by Eisler 2000, Solomon 2009). Studies on cumulative adverse effects of Cu on productivity of aquatic food chains are lacking; however, numerous studies have documented adverse effects on freshwater algae, zooplankton, and mussels at levels below Aquatic Water Standards (Scannell 2009, Martorell 2010, Woody and O'Neal 2012), which could result in reduced prey abundance and quality to support fish growth and reproduction (James et al. 2018).

Copper is one of the most toxic metals to unicellular algae, which form the base of the salmonid food chain (Solomon 2009, Rostern 2017). Photosynthetic algae production (*Chlorella* spp.) can decline at just 1.0 to 2 ppb Cu and photosynthesis can be inhibited at 5.0 to 6.3 ppb (USEPA 1980, Woody and O'Neal 2012, Rivas et al. 2016).

Zooplankton feed on algae and their growth and reproduction are affected by food availability; declines in algae production can cause declines in zooplankton production (Juneja et al. 2013), which implies reduced food for fish that feed on zooplankton.

Zooplanktons are the preferred food of juvenile sockeye salmon, which rear in lakes one to two years prior to seaward migration (Scannell 2009, Haskell et al. 2017). Zooplankton are highly sensitive to acute Cu effects and studies in waters of high hardness show *Cladocera* may not be adequately protected by current Alaska AWQ criteria (Woody and O'Neal 2012, Knillmann 2013), particularly because freshwaters in and near Bristol Bay mine claims are very soft and have low levels of DOC (also see review by Scannell 2009, Williams 2018).

Freshwater mussels and gastropods (snails) live in sediments and are filter feeders (Harrold and Guralnick 2010, Gordon et al. 2016). They recycle dead organic matter in lake and river bottoms and they in turn are prey for fish (Woody and O'Neal 2012). For example, freshwater mussels and snails are a primary food of humpback whitefish (Brown 2007), which in turn are prey for larger fish and are a preferred subsistence species for residents of the Kvichak River watershed (Williams 2018).

9. Use of Antifouling Paints

Since our ancestors first boarded a boat, sailors have tried every conceivable mix to keep creatures and floods in check, from blends of tar and sulfur to paints with copper, tin, arsenic, pesticides and even the occasional flask of chili powder (IMO 2001, 2002, 2009, 2011). Although environmental concerns have eliminated the most harmful mixtures of yesteryear, background painting continues to be a complicated issue (IMO 2001, 2002, 2009, 2011). Antifouling paint is a specialized coating that normally contains a formulation of bioactive ingredients and organostannic compounds soluble in water, which is then applied to the hull of the boat to slow the growth of barnacles, algae and marine organisms that adhere to the hull at sea (Etzkorn and Allan 2007, IMO 2011).

The antifouling acts as a barrier against corrosion in metal helmets (Omae 2003, Etzkorn and Allan 2007); improves the water flow around the hull and, therefore, the speed and overall performance of the boats (Ramesh et al. 2014, Penttila 2017); reduces fuel consumption and damage to the propellers; and prevents blockage of the entrances and exits of water of the motor and damages to the surface of the helmet from marine growth (MEP 2006, Willsher 2007).

To avoid incrustations in submerged structures such as helmets, antifouling paints are designed to kill or deter the organism from settling through the slow release of toxins in the aquatic environment (Amara et al. 2018). Several different biocides have been used for this purpose throughout history, but the main biocide used at present is a cuprous oxide (Cu_2O) (Lindgren et al. 2018). The release of biocides from paint directly into the aquatic environment leads to an increase in concentrations in water and/or sediments and, consequently, negatively affects non-target organisms (Lagerström et al. 2018). Zinc oxide (ZnO) is also added to most paints as a means to control the rate of erosion (Lagerström et al. 2018). Although zinc oxide is not classified as an active substance in the EU Biocidal Products Regulation (BPR, Regulation (EU) 528/2012), studies have shown that the release of Zn from antifouling paints can have toxic effects on organisms (Karlsson et al. 2010, Lagerström et al. 2018). Within the EU, antifouling paints are regulated through the BPR, which requires all biocides to obtain authorization before they can be made available on the market (Nilsson and Gipperth 2015, EC 2016). In most member countries, the authorization process includes a risk assessment, which requires manufacturers to present the product's biocidal product and its release rates (Altenburger et al. 2013, Lagerström et al. 2018). Regulatory bodies in the EU member states (like those in, for example, the United States and New Zealand) accept leaching rate data generated from either of the two standardized methods available (EPA New Zealand 2011): a rotating cylinder method (ASTM D6442-06/ISO 15181: 2007) and a mass balance method (ISO 10890: 2010).

Biofouling can be defined as "the undesirable phenomenon of adhesion and accumulation of biotic deposits on an artificial surface submerged or in contact with seawater" (Nguyen et al. 2012). This accumulation or embedding consists of an organic film composed of microorganisms embedded in a polymer matrix created by themselves (biofilm), where they can reach and retain inorganic particles (salts and/or corrosion products) as a result of other types of fouling developed in the process (Mojica 2008, Kulaga 2014). Antifouling paints are applied to the live surface of the boat to prevent the fixation of marine organisms (Sylvander 2007, Xiao 2014). They act by releasing bioactive substances that wrap the treated surface in a biocidal mantle that prevents the fixation of the organisms, so the concentration of toxins in the water layer is what determines the antifouling power of the paint (Joshi et al. 2013, Chen and Qian 2017). There is continual research on new formulations of antifouling, or antifouling paints that can be applied together with anticorrosive coatings and cathodic protection, because of the environmental problems generated by the main chemicals used for this purpose (Canosa et al. 2012).

In the 1960s the chemical industry created the organotin compound tributyltin (TBT), the most effective biocide and antifouling agent ever known, to prevent the development of algae, molluscs and other organisms that slow down boats (Guerreiro et al. 2009). Although its toxic effect is devastating on marine flora and fauna (Acevedo-Whitehouse and Duffus 2009) it is used by 70–80% of the world merchant fleet because it reduces fuel consumption and can be reapplied easily (IMO 2002, Morrisey et al. 2013). By 2003, antifouling paints without TBT were manufactured, and the International Convention on the Control of Systems of Antifouling Paints on Ships of September 17, 2008, prohibited the use of antifouling paints with TBT (IMO 2002, Eklund and Eklund 2014). In recent years, cuprous oxide has replaced TBT in Galician recreational vessels (Abdulla and Linden 2008). This compound is much less harmful than tin derivatives, but it is still a biocidal compound (Baldissera et al. 2015). Copper acts as an algistatic agent or algaecide in two microalgae: *Tetraselmis suecica* and *Dunaliella salina* (Guerreiro et al. 2009). Among the toxic effects it has on crustaceans, its effects on the nauplii larvae of *Artemesia longinaris* can be cited, causing mortality and alterations of swimming movement, growth and development in the survivors (Scelzo 1997). The lethal effects on the population of certain molluscs have also been studied; in the study carried out by Acosta and Lodeiros (2001) and Acosta et al. (2002), copper at high concentrations is discovered to be a stressor of the green mussel, *Perna viridis*. Another fact that contributes to the destructive character of copper is that it has become the most frequently used chemical to control the "plague" of mussels in shrimp farming ponds (Rodero 2011).

The contribution of cuprous oxide to the environment by antifouling paints in the recreational boats of Galicia during the last few years can be calculated at 27 t assuming an occupation of 100%; it was 24 t with the actual occupation in 2008 (Guerreiro et al. 2009, Nogués 2017).

Antifouling paints are evaluated for environmental risk according to their biocidal release rates to the aqueous phase (Felipe 2011). *In situ* release rates of Cu and Zn were derived for five commercial paints in two marine recreational sites with different salinities (5 and 14 psu) (Ytreberg et al. 2016, Lagerström et al. 2018). Salinity was found to significantly affect the release of Cu, with twice the Cu released at higher salinity, while the influence on Zn release was specific to the paint (Kiaune and Singhasemanon 2011, Lagerström et al. 2018). Site-specific release rates for water bodies with salinity gradients, e.g., the Baltic Sea, are therefore necessary for a more realistic assessment of the risks of antifouling paints (Lagerström et al. 2018). In addition, *in situ* release rates were up to 8 times higher than those generated using a laboratory or standardized calculation methods (Lagerström et al. 2018). The assessment of repeated environmental risk with field release rates concludes that it is questionable whether the products studied should be allowed in the Swedish market (Lagerström et al. 2018).

10. The Role of Sediments

Sediments are important in the study of the pollution of aquatic environments and are known to transport different pollutants; they also constitute sinks or sources of some pollutants to the water column (Herrero et al. 2018). For a better evaluation of the pollution process in the marine coastal environment, several authors propose an analysis based on the joint determination of the chemical composition and toxicity tests in the laboratory (Tiam et al. 2016). Bioaccumulation studies of different pollutants can help in the identification of the bioavailability of chemical compounds in sediments (Azizi et al. 2018) and seawater since, unlike chemical analysis, bioaccumulation provides a measure of bioavailable contaminants (Olaniran et al. 2013).

The bioavailability of chemical compounds, which depends on biogeochemical and physiological processes, is an important factor, often neglected in ecotoxicological and risk assessment (Márquez et al. 2016). The bioavailable fraction is critical in the incorporation and, finally, concentration in the organisms at the site in question (Petänen 2001, Copaja et al. 2016). The bioaccumulation factor makes it possible to predict whether the bioavailable chemical compound is bioaccumulated (Urien et al. 2015). The presence of contamination does not indicate the potential of the adverse effects (Odonkor and Ampofo 2013, Pandey et al. 2014). A contaminant can present toxic effects only if it is in a bioavailable form (Mandal 2017) and the compound can be incorporated into the animal's cell or influence the normal process of the organism (Jaishankar et al. 2014).

The bioaccumulation of contaminants in tissues of organisms can be useful for the evaluation of the potential trophic transfer of contaminants (Landers 2016). Studies based on the exposure of organisms to sediment samples under laboratory conditions can provide an approximation of the origins of metal bioaccumulation (Hsu et al. 2016).

After 21 d of exposure, metal concentrations (Cu, Zn, Pb, Hg and As) in gill tissues of *Carcinus maenas* showed significantly higher values for all the sediments analyzed compared with the control, except for GL2 (estuary of the Guadalquivir), the most distant area of the Aznalcóllar mining spill (Tornero et al. 2014).

Copper significantly bioaccumulated ($P < 0.001$) in organisms exposed to sediments from the Ría de Huelva area, especially in Hu1 (548.90 mg kg^{-1}, Ría de Huelva) (Riba et al. 2011). On the other hand, the metals Zn, Pb and As, originating in the Aznalcóllar mining accident, showed greater bioaccumulation in individuals exposed to sediments of the Guadalquivir area, specifically in sediments of GR2, the area closest to the Aznalcóllar mining spill (Garcia-Ordiales et al. 2018). These results confirm those obtained by Riba et al. (2005) and Aouini et al. (2018), who, after exposing the same sediments to the fish *Solea senegalensis* and the clam *Ruditapes philippinarum* in the laboratory, found bioaccumulation of Zn in organisms exposed to sediments of the Guadalquivir Estuary (GR2), associated with the enrichment of this metal by mining discharge. The bioaccumulation of Zn associated with sediments historically affected by mining activities has already been mentioned by other authors such as Marín-Guirao et al. (2005) in the alga *Cymodocea nodosa* (Mar Menor, SE Spain) and by Gutiérrez-Galindo et al. (1999) in the mussel *Modiolus capax* (Santa Rosalia, Baja California, Mexico).

During the first days after the Aznalcóllar mining spill, there was a significant discharge of metals into the Guadalquivir River, some of which showed very high concentrations that were toxic to autochthonous species such as the crab *Carcinus maenas* and the clam *R. philippinarum*, in simulations in the laboratory (Guerrero et al. 2008). In these studies a significant bioaccumulation was observed with respect to the control in the organisms exposed to different concentrations of Cd, Cu and Zn determined in the Guadalquivir Estuary after the mining accident of Aznalcóllar (Aouini et al. 2018). The bioaccumulation of these metals in tissues of the gill, digestive gland and gonad was associated with alterations and cell damage.

The contribution of the concentration of metals associated with sediments to the total bioaccumulation determined in exposed organisms is a complex process and encompasses homeostasis, natural variability and the extent of bioavailability (Hook et al. 2014, Tornero et al. 2014).

Biota-sediment accumulation factors were calculated to determine the bioavailability of the metals associated with the sediments analyzed (Sarmiento et al. 2016). The metals present in the sediment of GR2 were the most bioavailable of all the sediments analyzed (Tornero et al. 2014). The sediments of GR2 are characterized by the highest percentage of fines (91.00%), which could influence the increase of the bioavailability of metals for *C. maenas* (Tornero et al. 2014). When determining the contribution of sediments as sources of metals, the bioavailability of these metals is an important tool to understand the bioaccumulation that can occur in exposed organisms (Olaniran et al. 2013).

However, there are several abiotic and biotic parameters that affect the bioavailability and accumulation in the tissues of organisms. Of the abiotic parameters, the most important are the speciation of metals, salinity, temperature, pH, height in the water column, microhabitat, and interaction with other metals. Of the biotic parameters, the most important are size and/or body weight, phenotypic differences, sex, physiological conditions and reproductive state (Piculell et al. 2008). Therefore, it is more appropriate to assess bioavailability by direct measures of bioaccumulation of specific compounds in organisms (Borgå 2013). These bioavailability studies have been carried out in several countries for sediment evaluation (Selck et al. 2011, Egorova and Ananikov 2017). Many authors have studied the relationships between bioavailability and bioaccumulation for the evaluation of sediment toxicity (Otero et al. 2005).

11. CO_2-Induced Ocean Acidification Modulates Copper Toxicity in Marine Organisms

Copper is considered to be toxic to macroalgae at higher levels (Huang et al. 2010, Duinker et al. 2016). Ocean acidification can also alter the physiological performances of macroalgae (Liu et al. 2014, Gao et al. 2017). However, little is known regarding the interactive effects of Cu and ocean acidification on macroalgae (Gao et al. 2017). Photosynthetic apparatus is considered to be the primary target of damage caused by Cu in plants (Kumar et al. 2014). It can damage both electron donors and acceptors of photosynthetic electron transport and thus inhibit the primary photochemical reaction

(Küpper et al. 2002). Therefore, copper can generally reduce algal growth (Moenne et al. 2016). In addition, copper can inhibit the synthesis of D1 protein in PSII, thereby hindering recovery from photoinhibition (Komenda 2000).

The atmospheric concentration of carbon dioxide increased by 40% to 391 ppm between 1750 and 2011 due to human activity (Butler and Montzka 2018, EEA 2018); a rate of increase that is unprecedented within at least the last 800,000 years (IPCC 2013). When CO_2 dissolves in seawater, it forms carbonic acid and as more CO_2 is taken up by the ocean's surface, the pH decreases, moving towards a less alkaline and therefore more acidic state, termed ocean acidification (WHOI 2012, Ceceña 2014). The mean surface ocean pH has already decreased by 0.1 units since the beginning of the industrial era, corresponding to a 26% increase in hydrogen ion concentration (IPCC 2013). Ocean uptake of anthropogenic CO_2 will continue, leading to the increase of ocean acidification (Raven et al. 2005, IPCC 2013, Bennett 2018). Maroalgae have developed multiple strategies in inorganic carbon (Ci) acquisition, with different carboxylation efficiencies associated with different photosynthetic affinities for Ci (Giordano et al. 2005, Zou and Gao 2010). Most macroalgae can take up HCO_3^- and/or CO_2 by active transport (termed carbon concentrating mechanisms), while a few red and green macroalgae acquire Ci solely by diffusion of dissolved CO_2 (Gao et al. 2016, Cornwall et al. 2018).

Neither ocean acidification nor heavy metal pollution occurs in isolation (Breitburg et al. 2015); rather, they proceed simultaneously, especially in coastal areas (Breitburg et al. 2015, Schönberg et al. 2017). Furthermore, ocean acidification alters not only the seawater carbon chemistry but also the bioavailability of water-borne metals (Lewis et al. 2016). For instance, the inorganic speciation of copper is dominated by the complexation to CO_3^{2-} and OH^-, which will be reduced by decreased pH and ocean acidification (Richards et al. 2011). Consequently, the toxic free-ion concentration of copper is predicted to increase by as much as 115% in coastal waters over the next 100 years because of the projected decline in pH (Richards et al. 2011). Therefore, greater metal toxicity in organisms has been predicted in the context of ocean acidification and this hypothesis is supported by the findings of increased bioaccumulation of metals in the squid *Loligo vulgari* and increased toxicity of metals to amphipod *Corophium volutator* (Roberts et al. 2013) and the green algae *Chlorella vulgaris* (Král'Ová et al. 2004) under conditions of elevated pCO_2 (Flynn 2015). On the other hand, increased pCO_2 and decreased pH could also alleviate the toxicity of heavy metals to algae (Franklin et al. 2000). For instance, the toxicity of Cd to phosphate uptake in *Scenedesmus quadricaerda* decreased strongly with declining pH over a range of 5.5–8.5 and Cu toxicity also decreased when the pH changed from 6.5 to 5.8 (Gao et al. 2016). The concentrations of copper and uranium required to inhibit growth rate of *Chlorella* sp. by 50% increased from 1.5 to 35 mg L^{-1} and 44 to 78 mg L^{-1} respectively as the pH decreased from 6.5 to 5.7 (Franklin et al. 2000, Liu et al. 2017). The different effects of ocean acidification on the toxicology of heavy metal to organisms remain unknown (Jaishankar et al. 2014, Li et al. 2016).

Marine macroalgae, mainly inhabiting the intertidal and subtidal zones of coastal waters, are economically significant in the food, pharmaceutical, chemical and other industries (Feroz 2018). As an important part of global primary producers, they also play a key role in the marine biological CO_2 pump and the coastal carbon cycle (Ji et al. 2016). Among them, *Ulva* is cosmopolitan and the only genus that causes green tides (Smetacek and Zingone 2013). Therefore, the effects of copper on *Ulva* have been widely studied (Wichard et al. 2015). Maximal chlorophyll fluorescence (F_v/F_m) and maximum electron transport rate (ETR_{max}) in *U. pertusa* decreased with the increase of Cu concentration (0.125–1 mg L^{-1}) (Gao et al. 2016). The gross photosynthetic rate in *U. flexuosa* also decreased from 23.68 +/– 4.95 to ‾0.99 +/– 0.25 mg O_2 mg FW^{-1} h^{-1} when the Cu concentration increased from 50 to 250 mg L^{-1} (Andrade et al. 2004). Han et al. (2008) compared copper (Cu) toxicity to *U. pertusa* and *U. armoricana* and found the maximum efficiency of photosystem II, ETR_{max} and non-photochemical quenching in *U. armoricana* were not affected by Cu except at the highest concentration (250 mg L^{-1}), while *U. pertusa* showed a noticeable decrease in those parameters at much lower Cu concentrations. Baumann et al. (2009) and Arantes et al. (2016) investigated the effects of Cu, Cr, Zn, Cd and Pb on photosynthetic activity in seven species of green, red and brown macroalgae, in which *U. intestinalis* accumulated the highest amounts of all metals, and the metals were accumulated in the order of Cu > Pb > Zn > Cr > Cd.

12. Effect of Physicochemical Form on Copper Availability to Aquatic Organisms

An important goal of biogeochemistry is to understand the transport, fate, and effects of metals introduced into natural ecosystems (Fashola et al. 2016). Copper is of special interest—its partitioning among abiotic compartments and availability to biota in aquatic ecosystems is dependent on its physicochemical form (Ho et al. 2012).

Not all physicochemical forms of copper are equally available to aquatic organisms. Soluble copper is accumulated by most, if not all organisms, but the quantities accumulated are related to the chemical form (species) of copper in the water (Jaishankar et al. 2014). Particulate copper in the water column is known to be effectively concentrated by some filter feeders and copper in bottom sediments is accumulated in varying amounts by deposit feeders (Ho et al. 2018).

Numerous attempts have been made to quantify the physicochemical forms of copper in water systems (Benson et al. 2013). Field measurements of forms of copper in water systems have been made using a variety of techniques (Camacho-Flores et al. 2015). The validity of some measurements is in question because of the sampling and analytical methods used (Gawande et al. 2016). Furthermore, comparison of results from different sites is difficult because the techniques used by different investigators are rarely the same (Bengtsson 2016).

Studies have been carried out to assess the physicochemical forms and effects of copper in effluents from power stations adjacent to aquatic ecosystems with water that differed in salinity, pH, and concentrations of organic and inorganic constituents (Mkadmi et al. 2018). In addition, responses of ecologically and economically important marine and freshwater organisms to increased concentrations of soluble copper were evaluated (Basaling and Praveen 2011).

It is well documented that bioavailability and, in turn, the toxicity of copper in aquatic ecosystems are related to the concentrations of inorganic and organic ligands in the soluble phase and the affinity for copper of metal binding sites in the particulate phase (Solomon 2009). Most data available is from controlled laboratory experiments in which the test organisms were exposed to acute levels of copper and mortality monitored (Calfee et al. 2016). However, many exposures to copper in the field are chronic and low in concentration (Tchounwou et al. 2012, Li et al. 2016).

Chronic exposure to low concentrations of copper may have sublethal effects on an organism's potential for growth, its reproductive capacity, its ability to resist further change, or its effectiveness in competing with other species in the ecosystem (Gharedaashi et al. 2013). Evidence for these kinds of effects due to increased availability of copper in a field situation was obtained from studies of fish in the H.B. Robinson Impoundment (FCSAP 2013). Monitoring of the abundance of adult and larval fish populations showed smaller fish populations and higher incidence of structural deformities in bluegills collected from 1976 to 1978 than those found in previous years (Pope et al. 2009). Considerable effort was expended investigating possible causes and the results eliminated included temperature, selenium, pesticides, common water-quality properties, and vitamin C storage as causes (Findlay et al. 2006). Because total copper and labile copper concentrations were much higher in the effluent from this power station than in those from other stations, studies were initiated to determine whether copper was responsible for the adverse effects on fish populations (Luan 2016).

Bluegills collected from the intake and discharge zones of the impoundment and from a control pond were examined for tissue copper concentrations and for the metals associated with metalloproteins in the livers. Separation of liver metalloproteins was performed because research on mammalian livers indicates that more information about potential adverse effects of metals can be obtained from analysis of the kinds and quantities of metals associated with the different metalloprotein pools than of the total quantities of metals in the tissues (Jan et al. 2015). Large differences were found in the quantities of copper associated with the three categories of metalloproteins resolved in the soluble fraction of liver homogenates.

The low molecular weight (LMW) fraction (6,000 to 40,000 daltons) contains the metallothionein-like proteins considered to be important in the homeostasis and detoxification of copper (Isani and Carpenè 2014); the intermediate molecular weight (IMW) fraction (40,000 to 126,000 daltons) and the high molecular weight (HMW) fraction (126,000 to > 670,000 daltons) contain metalloenzymes required for metabolic processes (Lee et al. 2011).

In all samples of fish, the LMW fraction contained by far the largest amount of copper. Comparison of the quantities of copper in the LMW, IMW, and HMW fractions shows that copper concentrations were much higher in fish from the intake and discharge sites than from the control site (Thomas 2014). The mean of the copper concentrations in the LMW fraction from the 10 fish from the discharge site was 20 times greater than in those from the control site; the mean from the intake site was six times greater (Boag 2017). The quantities of copper bound to proteins in the IMW and HMW fractions were small except in the homogenates from four of the five pairs of fish sampled from the discharge site (Lorenzo et al. 2017). Also in these same four pairs of fish, copper appeared to be displacing zinc from metalloenzymes in the HMW and LMW fractions (Jitar et al. 2015). High copper concentrations in the HMW fraction and displacement of zinc are evidence that the metallothionein-like protein detoxification system was becoming overloaded (Calvo et al. 2017).

13. Copper and Aquaculture

Aquaculture has increased worldwide and is considered economically important for rural coastal communities (Tull et al. 2016). While aquatic farming has been projected to provide 93 million t of the world's seafood supply by 2030 (FAO 2016), its rapid expansion has in some cases been to the detriment of the environment (De Silva 2001). Aquaculture can contribute to environmental degradation in a number of ways depending on location, cultivated species, and method of farming (FAO 2006, Martinez-Porchas and Martinez-Cordova 2012). However, one factor that is ubiquitous, with the single exception of farming aquatic plants, is the contribution of aquaculture to the widespread nutrification of coastal waters. At times this exceeds the assimilative capacity of the ecosystem, resulting in changes to trophic structure or function or triggering of eutrophic or hypertrophic events. Intensive monocultures, particularly those using feed species of a high trophic level, can contribute biogenic wastes (dissolved organic matter, particulate organic matter, dissolved inorganic nitrogen, dissolved inorganic phosphate) and non-biogenic wastes into the environment directly (open sea, flow-through net pen systems) or indirectly (land-based, particularly flow-through systems) (Lazzari et al. 2016).

Copper sulfate is used in commercial freshwater shrimp farms as algaecide for the control of cyanobacteria that decrease the dissolved oxygen content in water and produce bad taste and odor in meat and of others that produce toxic compounds (Rodgers 2008, Rastogi et al. 2015). It is also recommended for the treatment of parasitic ctenophores and protozoa that cause the decline of shrimp and shrimp farm production, with approximately 2 kg ha^{-1} of copper sulfate being dumped (Osunde et al. 2004). According to Massaut (1999), the toxicity of the copper ion in freshwater organisms, given the high solubility of copper sulfate in water, causes interruption in the osmoregulation of the gills and causes mechanical and physiological damage to organisms. In sublethal doses it brings about diverse problems in the cultivated organisms, such as increase in the respiratory rate, decrease of hemocytes, deficiency in phagocytosis (Sang 2010, Ellis 2012), problems of behavior (Burba 1999), inhibition of the metamorphosis, mutation inhibition, malformations (Scelzo 1997), decrease in growth (Méndez and Green 2005), decrease in food intake (Boock and Neto 2000), bioaccumulation, inhibition of enzyme systems related to the formation of ammonium, and mortality (Ramírez et al. 2002), all of them depending on concentration, species, stage and physiology.

Because copper does not discolor the water, it is a preferred treatment for use in display aquaria (Yanong 2013). Water chemistry and other environmental factors determine how much copper will be biologically available and for how long (Tchounwou et al. 2012, Wang et al. 2013). However, the copper concentrations required for effective treatment may be acutely toxic for some species of finfish and are lethal for most invertebrates (Solomon 2009, Yanong 2013). Chronic copper exposure will also adversely affect fish health (Jaishankar et al. 2014). Sublethal and toxic levels of copper damage gills and other tissues of fish, and also are known to depress the immune system (El-Gazzar et al. 2014). Because of all these concerns, it is important to understand how copper works and how copper availability is affected by the environment in which it is used (Yanong 2013). Calculations and follow-up procedures required for the use of copper in marine systems are different from those used for copper in freshwater (Tait 2013). Factors including parasite life cycle, susceptibility and non-target species sensitivities will also factor in the determination of whether or not to use copper and, if it is used, how long to continue the treatment to ensure it is both effective and safe (Yanong 2013, Beyth et al. 2015).

For aquaculture to continue to expand in an environmentally sound manner the issue of nutrient enrichment, among others, needs to be addressed; integrated multi-trophic aquaculture (IMTA) is one potential method by which this could be achieved (Buck et al. 2018). It is attractive because it can lead not only to a higher ecological stability (FAO 2010, Buck et al. 2018), but also to increased economic potential (FAO 2010), economic resilience and social acceptance (Chopin et al. 2012). IMTA is a growing sub-sector within aquaculture, either intentionally for increased profit (FAO 2009) or bio-mitigation (Chopin et al. 2012, Waite et al. 2014), or unintentionally by consequence of pressure on limited coastal zone space by diverse stakeholders (e.g., Icely et al. 2018). The IMTA concept aims to recreate a simplified and balanced ecosystem, taking account of and buffered by the natural host ecosystem, where unused inputs to and from the fed species (feed and waste) are recycled in a planned manner (Chopin et al. 2012). This is achieved by the use of deposit and filter feeders for the removal of particulate organic matter and seaweeds for the extraction of dissolved inorganic nutrients (Shah et al. 2017). The culture of these additional organisms thus realizes a saleable crop from an otherwise lost resource and simultaneously reduces the ecological footprint of the cultivation unit and/or increases the aquacultural carrying capacity of the system (EC 2011a).

There are many possible species combinations in IMTA systems and intemperate, open, marine waters. The predominant model to have received attention is the salmon-mussel-kelp system (Gvozdenović et al. 2017). Since the 1990s there has been a steady expansion in intensively farmed salmonid monocultures (FAO 2018), which currently results in an estimated 49 kg N and 9 kg P per t of salmon produced lost to the environment (Gvozdenović et al. 2017). However, the infrastructure of these farms is such that mussel and kelp longlines can be incorporated easily (Lal 2015). Kelps have been favored because of their fast growth rate and biomass potential, because markets already exist and additional markets are growing in new and expanding sectors (e.g., in Europe), and because the basic methods for their cultivation are relatively simple and well established (Bañeras 2014). It has been well established that under certain circumstances cage aquaculture contributes to the metal loading of sediments surrounding cage sites, although with little evidence of effects beyond several hundred meters (Environment Canada 2009, Jaysankar et al. 2009). Increased sediment levels of Cu, Zn, Fe and Cd have been attributed to waste (uneaten or excreted) feed (Environment Canada 2009, Jaysankar et al. 2009) and a number of other elements are also known to be contained in manufactured feeds (Mn, Mg, Se, Co, Ni, Pb, Hg) (Squadrone et al. 2016, Munshi et al. 2018). Presence of trace metals is reported in aquaculture marine ecosystems of the northwestern Mediterranean Sea (Italy) (Squadrone et al. 2016). Furthermore, increased sediment copper has been found to be associated with the use of copper-based antifoulants on the moored cage structures, net pens and farm boats (Yiğit et al. 2018). Most research to date has focused on exploring the dissolved nutrient and particulate dynamics between the species of IMTA systems (Kinney 2017), changes to the benthos below and surrounding finfish cage sites (Wartenberg et al. 2017). By comparison, there are few investigations on the dynamics of trace metals, including priority metal pollutants, between species in the system (Jiang et al. 2014).

Given that certain metals associated with salmon production (As, Cd, Cu, Zn) are priority pollutants (Evans and Edwards 2011, Fu et al. 2016, Su et al. 2017) and that kelps possess a strong affinity for their uptake and/or adsorption (Evans and Edwards 2011, Tett et al. 2018), there is a need to address the potential transfer of metals from finfish farming activities to cultivated macroalgae where these cultivation units occur in close proximity to one another, as in IMTA situations (Evans and Edwards 2011). In the current situation of increasing interest around seaweed as a human food source (Gadberry et al. 2018), it is of interest not only to investigate whether aquaculture practices are contributing metals in bioavailable forms to the water column, but also to evaluate the potential contribution of *Laminaria digitata*, positive or negative, to metals in the human food chain (Evans and Edwards 2011, Buck et al. 2018). This study will examine metal concentrations from cultivated and wild kelp populations in Ireland with the following specific aim: to determine metal concentrations of cultivated and wild *L. digitata* and assessments in metal content between undisturbed wild populations, seaweed-only populations (monocultivation), and populations cultured alongside organic salmon (IMTA), particularly with regard to priority metal pollutants (As, Cd, Cu, Sb, Zn) (Ronan et al. 2017). The discussion is extended to include consideration of the contribution of *L. digitata* to metals in animal feeds and in the human food chain (Zaharudin 2017).

14. Conclusions

During the 20th century, the discharge of untreated or partially treated wastewater was probably the most widespread, most documented and undoubtedly the best known form of pollution entering the aquatic environment. In the past two decades, however, there has been growing public awareness of the risks from contamination of the water environment by toxic substances associated with the mining industry. The world demand for minerals has intensified the exploitation of natural resources. In most of the developed and newly developed countries, the important mining proposals are now strictly regulated to protect the environment. This implies economic and legislative measures and the use of appropriate control technologies. Undoubtedly, this concern will continue to spread throughout the world and will require a program of environmental protection management policies and practices. This book has been prepared as a synthesis of our current understanding of the effects of various heavy metals and acid discharges that can contaminate the environment as a direct result of mining activities. The review is based on providing a better scientific understanding of the causes of environmental problems associated with this industry. We are pleased to publish this information for use and application by a wider audience.

Human societies have begun to realize the unsustainability of the idea of "dilution". Many national laws and international protocols currently prohibit the dumping of harmful substances into the oceans, although their application is often uncertain. Marine sanctuaries are being created to maintain virgin marine ecosystems. Likewise, isolated initiatives are being carried out that have achieved some success in the restoration of estuaries and bays.

Copper can enter the environment through releases from copper mines and other metals, and from factories that manufacture or use metallic copper or copper compounds. Copper can also enter the environment from landfills, domestic wastewater, burning waste and fossil fuels, wood production, production of phosphate fertilizers, and natural sources (for example, dust in the air and from the ground, volcanoes, decaying vegetation, forest fires and sea foam). Therefore, copper is widely distributed in the environment. Approximately 640,000,000,000 g of copper were released into the environment by industries in the year 2000. Copper is often found near mines, smelters, industrial plants, landfills and waste sites. When copper is released into the soil, it can adhere strongly to organic matter and other components (e.g., clay, sand) in the topsoil and may not move very far when it is released. When copper and copper compounds are released into the water, the copper that dissolves can be transported in the surface water either in the form of copper or free copper compounds or, more likely, as copper bound to suspended particles in water. Even when copper adheres strongly to suspended particles or sediments, there is evidence to suggest that some of the soluble copper compounds enter groundwater. The copper that enters the water is eventually deposited in the sediments of rivers, lakes and estuaries. Copper is transported in particles emitted by foundries and plants that process minerals, and returns to the earth by gravity or in rain or snow. Copper is also transported by wind to air in metal powders. The release of copper in closed areas comes mainly from combustion processes (for example, kerosene heaters). Elemental copper does not degrade in the environment. Copper can be found in plants and animals, and in high concentrations in organisms that filter their food, such as mussels and oysters. Copper is also found in a variety of concentrations in many beverages and foods, even in drinking water.

Metal pollution from the ocean or seas is less visible and direct than other types of marine pollution, but its effects on marine ecosystems and humans are extensive. The presence of metals and metal ions varies among fish species; this depends on age, stage of development and other physiological factors. Fish accumulate a substantial concentration of metals in their tissues and, therefore, may represent an important dietary source of this element for humans. Much research has been done on the effect of heavy metals on human health. In this chapter we try to explain situations that we currently find in the coastal zone and rivers and we realize how the acidification of the oceans, seas and rivers exacerbates the mobilization of metals in bodies of water and how acid waters (low pH) have contributed to change metal states to more lethal forms of metals: metal ions. Very low levels of contaminant may not have an apparent effect on the fish, as it would not show external signs of disease, but it could decrease the fecundity of the fish, causing a long-term decline and punctual extinction of fish populations. The contamination can affect reproduction directly (affecting the free gametes (sperm or ovules) that are released into the water) or indirectly (accumulating in the reproductive organs, which can interrupt the normal production of steroids, causing defective production in both males and females, decreasing the quality and quantity of production of sperm and ovules).

References

Abdolmaleki, A.S., A.G. Ahangar and J. Soltani. 2013. Artificial Neural Network (ANN) approach for predicting Cu concentration in drinking water of chahnimeh1 reservoir in Sistan-Balochistan, Iran. J. Health Scope. 2(1). http://dx.doi.org/10.17795/jhealthscope-9828 accessed 23 September 2018.

Abdulla, A. and O. Linden. 2008. Maritime traffic effects on biodiversity in the Mediterranean Sea. Volume 1— Review of impacts, priority areas and mitigation measures, 170 p. https://cmsdata.iucn.org/downloads/maritime_v1_lr.pdf accessed 13 October 2018.

Acevedo-Whitehouse, K. and A.L.J. Duffus. 2009. Philos. Trans. R. Soc. Lond. B. Biol. Sci. 364(1534): 3429–3438. https://dx.doi.org/10.1098%2Frstb.2009.0128 accessed 09 September 2018.

Acosta, V. and C. Lodeiros 2001. Evaluación del efecto del cobre sobre juveniles del Mejillón Verde Pernas Viridis mediante la concenrtación de ADN y la relación ARN/ADN en el músculo abductor. FCV-LUZ. Vol. XI. Nº 6: 485–490. http://produccioncientificaluz.org/index.php/boletin/article/download/17/17 http://www.produccioncientifica.luz.edu.ve/index.php/cientifica/article/viewFile/14804/14781 accessed 03 August 2018.

Acosta, V., C. Lodeiros, W. senior and G. Martínez. 2002. Niveles de metales pesados en sedimentos superficiales de los litorales de Playa Güiria (Estado Sucre), Boca de Paparo y Río Chico (Estado Miranda), Venezuela. Interciencia 27(12): 686–690. www.redalyc.org/html/339/33907507/ accessed 03 August 2018.

ADEC. 2010. Copper and aquatic life citeria. Power Point presentation available from Alaska Water Quality Standards and parameters of concern. Water Quality Standards Division of Water Alaska Dept. of Environmental Conservation Jim Powell. Presentation. Overview of Water Quality Standards Brief history and statutory basis. https://www.slideserve.com/manny/alaska-water-quality-standards-and-parameters-of-concern.

Ahmad, F., S. Azman, M.I.M. Said and L. Baloo. 2015. Biomonitoring of metal contamination in estuarine ecosystem using seagrass. J. Environ. Health Sci. Eng. 13: 41. https://dx.doi.org/10.1186%2Fs40201-015-0198-7 accessed on 26 August 2018.

Al-Reasi, H.A., C.M. Wood and D.S. Smith. 2011. Physicochemical and spectroscopic properties of natural organic matter (NOM) from various sources and implications for ameliorative effects on metal toxicity to aquatic biota. Aq. Tox. 103: 79–190. http://dx.doi.org/10.1016/j.aquatox.2011.02.015.

Al-Sabri, I.Y.N., J.D. Smith and J.D. Thomas. 1993. Copper molluscicides for controls of schistosomiasis. 3. Absorption by clay suspensions. Environ. Sci. Technol. 27: 299–303. http://citeseerx.ist.psu.edu/viewdoc/download?doi=10.1.1.850.7220&rep=rep1&type=pdf accessed 23 October 20118.

Allgeier, J.E., A. Valdivia, C. Cox and C.A. Layman. 2016. Fishing down nutrients on coral reefs. Nature Communications 7, Article number: 12461 http://dx.doi.org/10.1038/ncomms12461.

Altenburger, R., Å. Arrhenius, T. Backhaus, A. Coors, M. Faust and D. Zitzkat. 2013. Ecotoxicological combined effects from chemical mixtures. Part 1: Relevance and adequate consideration in environmental risk assessment of plant protection products and biocides. 217 p. https://www.umweltbundesamt.de/sites/default/files/medien/378/publikationen/texte_92-2013_ecotoxicological_combined_effects_from_chemical_mixtures_part_1.pdf accessed 01 September 2018.

Amado-Filho, G.M., L.T. Salgado, M.F. Rebelo, C.E. Rezende, C.S. Karez and W.C. Pfeiffer. 2008. Heavy metals in benthic organisms from Todos os Santos Bay, Brazil. Braz. J. Biol. 68(1): 95–100. http://dx.doi.org/10.1590/S1519-69842008000100013 accessed on 26 August 2018.

Amador, L.R.T., F.D.G. Martínez, L.J.M. Hernández, L.A.W. Vergara and J.N.C. Suárez. 2015. Niveles de metales pesados en muestras biológicas y su importancia en salud. Rev. Nac. Odonto. 11(21): 83–99. http://dx.doi.org/10.16925/od.v11i21.895 acessed 23 September 2018.

Amara, I., W. Miled, R. BenSlam and N. Ladhar. 2018. Antifouling processes and toxicity effects of antifouling paints on marine environment. A review. Environ. Toxicol. Pharmacol. 57: 115–130. https://doi.org/10.1016/j.etap.2017.12.001 accessed 10 August 2018.

Anderson, S., B. Johnson, S. Kishbaugh, M. Balyszak, M. Wunderlich and B. Zhu. 2009. Aquatic Weeds: Nuisance and Necessity. Managing Waterweeds in Cayuga, Owasco and Seneca Lakes, 28 p. https://www.cayugacounty.us/portals/0/wqma/projects/Aquatic_Weeds_Nuisance_and_Necessity.pdf.

Andrade, L.R., M. Farina and G.M.A. Filho. 2004. Effects of copper on Enteromorpha flexuosa (*Chlorophyta*) *in vitro*. Ecotoxicol. Environ. Saf. 58: 117–125. https://doi.org/10.1016/S0147-6513(03)00106-4.

Andreoli, V. and F. Sprovieri. 2017. Genetic aspects of susceptibility to mercury toxicity: An overview. Int. J. Environ. Res. Public Health 14: 93. https://doi.org/10.3390/ijerph14010093.

Aouini, F., C. Trombini, M. Volland, M. Elcafsi and J. Blasco. 2018. Assessing lead toxicity in the clam *Ruditapes philippinarum*: Bioaccumulation and biochemical responses. Ecotox. Environ. Safety 158: 193–203. https://doi.org/10.1016/j.ecoenv.2018.04.033 accessed 20 October 2018.

Arantes, F.P., L.A. Savassi, H.B. Santos, M.V.T. Gomes and N. Bazzol. 2016. Bioaccumulation of mercury, cadmium, zinc, chromium, and lead in muscle, liver, and spleen tissues of a large commercially valuable catfish species from Brazil. An. Acad. Bras. Cienc., 12 p. http://dx.doi.org/10.1590/0001-3765201620140434 accessed 06 October 2018.

Arnaud, A., P. Matthew, G. Sylvie and B. Joseph. 2018. Seascape ecology in Posidonia oceanica seagrass meadows: Linking structure and ecological processes for management. Ecological Indicators 87: 1–13. http://dx.doi.org/10.1016/j.ecolind.2017.12.029 accessed on 24 August 2018.

Athanassopoulou, F., I.S. Pappas and K. Bitchava. 2009. An overview of the treatments for parasitic disease in Mediterranean aquaculture. pp. 65–83. *In*: Rogers, C. and B. Basurco (eds.). The Use of Veterinary Drugs and Vaccines in Mediterranean Aquaculture. Zaragoza: CIHEAM (Options Méditerranéennes: Série A. Séminaires Méditerranéens; n. 86). https://om.ciheam.fr/om/pdf/a86/00801063.pdf accessed 29 August 2018.

ATSDR (Agency for Toxic Substances and Disease Registry). 1990. Toxicological Profile for Copper. US Public Health Service, Atlanta, Georgia. TP-90-08. 143 pp.

ATSDR (Agency for Toxic Substances and Disease Registry). 2004. Toxicological profile for Copper. U.S. Department of Health and Human Services, Public Health Service, Atlanta, 314 p. https://www.atsdr.cdc.gov/toxprofiles/tp132.pdf accessed 24 October 2018.

Azizi, G., M. Akodad, M. Baghour, M. Layachi and A. Moumen. 2018. The use of Mytilus spp. mussels as bioindicators of heavy metal pollution in the coastal environment. A review. J. Mater. Environ. Sci. 9(4): 1170–1181. https://doi.org/10.26872/jmes.2018.9.4.129.

Baby, L., G.T.R. Kumar, K.R. Remyakumari, J. Varkey, T.V. Sankar and N. Chandramohanakumar. 2017. Comparison of hydrographic and sediment characteristics of seagrass meadows of Gulf of Mannar and Palk Bay, South West Coast of India. Int. J. Fish. Aqua. Stud. 5(2): 80–84. http://www.fisheriesjournal.com/archives/2017/vol5issue2/PartB/4-6-91-582.pdf.

Baker, J.P. 2008. Mercury, vaccines, and autism. Am. J. Public Health. 98(2): 244–253. https://dx.doi. org/10.2105%2FAJPH.2007.113159 accessed 23 September 2018.

Baldissera, A.F., K.L. Miranda, C. Bressy, C. Martin, A. Margaillan and C.A. Ferreira. 2015. Using conducting polymers as active agents for marine antifouling paints. Mat. Res. 18(6): 1129–1139. http://dx.doi.org/10.1590/1516-1439.261414 accessed 15 October2018.

Baldwin, D.H., J.F. Sandahl, J.S. Labenia and N.L. Scholz. 2003. Sublethal effects of copper on coho. Environmental Toxicology and Chemistry 22(10): 2266–2274. https://doi.org/10.1897/02-428 accessed 01 October 2018.

Baldwin, D.H., C.P. Tatara and N.L. Scholz. 2011. Copper-induced olfactory toxicity in salmon and steelhead: Extrapoloation across species and rearing environments. Aquatic Toxicology 101(1): 295–297. https://doi. org/10.1016/j.aquatox.2010.08.011.

Bañeras, S.B. 2014. Análisis del sector de producción de algas con fines alimentarios. Percepción del consumidor frente al consumo de algas. Trabajo de final de grado en Ingeniería de Sistemas Biológicos. Escola Superior d'Agricultura de Barcelona. UPC–BarcelonaTech, 158 p. https://upcommons.upc.edu/bitstream/handle/2099.1/22271/memoria.pdf?sequence=4&isAllowed=y.

Basaling, H. and D. Praveen. 2011. Evaluation of acute toxicity of copper cyanide to freshwater fish, *Catla catla* (Hamilton). Journal of Central European Agriculture 12(1): 135–144. https://doi.org/10.5513/JCEA01/12.1.890.

Bashir, F.A. and E.M. Alhemmali. 2015. Analysis of some heavy metal in marine fish in muscle, liver and gill tissue in two marine fish spices from Kapar coastal waters, Malaysia. *In*: The Second Symposium on Theories and Applications of Basic and Biosciences 2(1): 16 p. https://www.researchgate.net/publication/301554192_Analysis_of_some_heavy_metal_in_marine_fish_in_muscle_liver_and_gill_tissue_in_two_marine_fish_spices_from_Kapar_coastal_waters_Malaysia accessed 17 October 2018.

Baumann, H.A., L. Morrison and D.B. Stengel. 2009. Metal accumulation and toxicity measured by PAM-chlorophyll fluorescence in seven species of marine macroalgae. Ecotoxicol. Environ. Saf. 72: 1063–1075. https://doi. org/10.1016/j.ecoenv.2008.10.010 accessed on 28 August 2018.

Beaver, J.R., T.R. Renicker, C.E. Tausz, J.L. Young, J.C. Thomason, Z.L. Wolf et al. 2018. Winter swarming behavior by the exotic cladoceran *Daphnia lumholtzi* Sars, 1885 in a Kentucky (USA) reservoir. BioInvasions Records 7(1): 43–50. https://doi.org/10.3391/bir.2018.7.1.06.

Bengtsson, M. 2016. How to plan and perform a qualitative study using content analysis. NursingPlus Open. 2: 8–14. https://doi.org/10.1016/j.npls.2016.01.001.

Bennett, J. 2018. Ocean acidification. https://ocean.si.edu/ocean-life/invertebrates/ocean-acidification accessed 24 October 2018 .

Benson, N.U., W.U. Anake and I.O. Olanrewaju. 2013. Analytical relevance of trace metal speciation in environmental and biophysicochemical systems. American J. Anal. Chem. 4: 633–641. http://dx.doi.org/10.4236/ajac.2013.411075.

Berdalet, E., L.E. Fleming, R. Gowen, K. Davidson, P. Hess, L.C. Backer et al. 2016. Marine harmful algal blooms, human health and wellbeing: challenges and opportunities in the 21st century. J. Mar. Biol. Asso. U.K. 96(1): 61–91. https://doi.org/10.1017/S0025315415001733.

Beyth, N., Y. Houri-Haddad, A. Domb, W. Khan and R. Hazan. 2015. Alternative antimicrobial approach: Nano-antimicrobial materials. Evidence-Based Complementary and Alternative Medicine 2015: 16. http://dx.doi. org/10.1155/2015/246012 accessed 01 September 2018.

Bhattacharya, P.T., S.R. Misra and M. Hussain. 2016. Nutritional aspects of essential trace elements in oral health and disease: An extensive review. Scientifica, pp: 1–12. http://dx.doi.org/10.1155/2016/5464373.

Bighiu, M.A., A.K. Eriksson-Wiklund and B. Eklund. 2017. Biofouling of leisure boats as a source of metal pollution. Environ. Sci. Pollut. Res. Int. 24(1): 997–1006. https://dx.doi.org/10.1007%2Fs11356-016-7883-7 accessed on 28 August 2018.

Birch, G.F. 2017. Assessment of human-induced change and biological risk posed by contaminants in estuarine/harbour sediments: Sydney Harbour/estuary (Australia). Mar. Pollut. Bull. 116: 234–248. http://dx.doi. org/10.1016/j.scitotenv.2016.12.028.

Birch, G.F., B.M. Cox and C.H. Besley. 2018. Metal concentrations in seagrass (*Halophila ovalis*) tissue and ambient sediment in a highly modified estuarine environment (Sydney estuary, Australia). Marine Pollution Bulletin 131(2018): 130–141. https://doi.org/10.1016/j.marpolbul.2018.04.010.

Birge, W.J., R.D. Shoyt, J.A. Black, M.D. Kercher and W.A. Robison. 1993. Effects of chemical stresses on behavior of larval and juvenile fishes and amphibians. Am. Fish. Soc. Symp. 14, Water Quality and the Early Life Stages of fishes (Fuiman, L., Ed.). American Fisheries Society, Bethesad, MD, 55.

Bishop, W.M. and J.H. Jr. Rodgers. 2012. Responses of Lyngbya wollei to exposures of copper-based algaecides: the critical burden concept. Arch. Environ. Contam. Toxicol. 62(3): 403–10. https://doi.org/10.1007/s00244-011-9711-x accessed on 24 August 2018.

Block, E., V.S. Batista, H. Matsunami, H. Zhuang and Lucky Ahmed. 2017.The role of metals in mammalian olfaction of low molecular weight organosulfur compounds. Nat. Prod. Rep. 34(5): 529–557. https://dx.doi.org/10.1039%2Fc7np00016b accessed 03 October 2018.

Boag, M.D. 2017. The Immobilization of Copper in Peatlands: Characterizing the Interactions Between Copper and Natural Organic Matter. Thesis. Master of Science in Chemistry. The University of Guelph. Canada,158 p. https://atrium.lib.uoguelph.ca/xmlui/bitstream/handle/10214/10858/Boag_Matt_201706_MSc.pdf?sequence=3 accessed 09 October 2018.

Boock, M. and M. Neto. 2000. Estudos toxicológicos do oxicloreto de cobre para tilápia vermelha (*Oreochromis* sp.) Arq. Inst. Biol. 67(2): 215–221. http://www.biologico.sp.gov.br/uploads/docs/arq/V67_2/13.pdf accessed on 24 August 2018.

Borgå, K. 2013. Ecotoxicology: Bioaccumulation. Reference Module in Earth Systems and Environmental Sciences. https://doi.org/10.1016/b978-0-12-409548-9.00765-x accessed 20 October 2018.

Bowker, J.D., J.T. Trushenski, M.P. Gaikowski and D.L. Straus (eds.). 2014. Guide to using drugs, biologics, and other chemicals in aquaculture. American Fisheries Society Fish Culture Section. https://www.fws.gov/fisheries/aadap/PDF/guidetousingdrugs.pdf.

Brankovits, D., J.W. Pohlman, H. Niemann, M.B. Leigh, M.C. Leewis, K.W. Becker et al. 2017. Methane- and dissolved organic carbon-fueled microbial loop supports a tropical subterranean estuary ecosystem. Nature Communications 8(1) 1835: pp. 1–12. https://doi.org/10.1038/s41467-017-01776-x.

Breitburg, D.L., J. Salisbury, J.M. Bernhard, W.J. Cai, S. Dupont, S.C. Doney et al. 2015. And on top of all that…: coping with ocean acidification in the midst of many stressors. Oceanography, Special Issue on Emerging Themes in Ocean Acidification Science 28(2): 48–61. https://www.jstor.org/stable/24861870 accessed on 24 August 2018.

Brodersen, K.E., K.J. Hammer, V. Schrameyer, A. Floytrup, M.A. Rasheed, P.J. Ralph et al. 2017. Sediment resuspension and deposition on seagrass leaves impedes internal plant aeration and promotes phytotoxic H_2S intrusion. Front. Plant. Sci. 8: 657. https://dx.doi.org/10.3389%2Ffpls.2017.00657 accessed 26 August 2018.

Brodin, T., S. Piovano, J. Fick, J. Klaminder, M. Heynen and M. Jonsson. 2014. Ecological effects of pharmaceuticals in aquatic systems—impacts through behavioural alterations. Philos. Trans. R. Soc. Lond. B. Biol. Sci. 369(1656): 20130580. https://dx.doi.org/10.1098%2Frstb.2013.0580 accessed 01 October 2018.

Bretman, A., J.D. Westmancoat and T. Chapman. 2013. Male control of mating duration following exposure to rivals in fruitflies. J. Insect. Physiol. 59(8): 824–827. https://doi.org/10.1016/j.jinsphys.2013.05.011 accessed 22 September 2018.

Brodie, G. and A.D.R. N'Yeurt. 2018. Effects of climate change on seagrasses and seagrass habitats relevant to the Pacific Islands. Pacific Mar. Climate Change Rep. Card. Science Review, pp. 112–131.

Brown, R. 2007. Freshwater mollusks survive fish gut passage. Arctic 60(2): 124–128. https://www.jstor.org/stable/40513128.

Buck, B.H., M.F. Troell, G. Krause, D.L. Angel, B. Grote and T. Chopin. 2018. State of the art and challenges for offshore Integrated Multi-Trophic Aquaculture (IMTA). Front. Mar. Sci. https://doi.org/10.3389/fmars.2018.00165.

Burba, A. 1999. The effect of copper on behavioural reactions of noble crayfish *Astacus astacus* L. Act. Zoo. Lit. Hydrobiologia 9: 30–36. https://doi.org/10.1080/13921657.1999.10512284.

Butler, J.H. and S.A. Montzka. 2018. The NOAA Annual Greenhouse Gas Index (AGGI). https://www.esrl.noaa.gov/gmd/aggi/aggi.html accessed 05 October 2018.

Calfee, R.D., H.J. Puglis, E.E. Little, W.G. Brumbaugh and C.A. Mebane. 2016. Quantifying fish swimming behavior in response to acute exposure of aqueous copper using computer assisted video and digital image analysis. J. Vis. Exp. 108: 53477. https://doi.org/10.3791/53477 accessed 24 October 2018.

Calvo, J., H. Jung and G. Meloni. 2017. Copper metallothioneins. IUBMB Life. 69(4): 236–245. https://doi.org/10.1002/iub.1618.

Camacho-Flores, B.A., O. Martínez-Álvarez, M.C. Arenas-Arrocena, R. Garcia-Contreras, L. Argueta-Figueroa, J. de la Fuente-Hernández et al. 2015. Copper: Synthesis techniques in nanoscale and powerful application as an antimicrobial agent. J. Nanomaterials 2015. Art. ID 415238, 10 p. http://dx.doi.org/10.1155/2015/415238 accessed 07 October 2018.

Canesi, L. and E. Fabbri. 2015. Environmental effects of BPA. Dose Response 13(3): 1559325815598304. https://dx.doi.org/10.1177%2F1559325815598304 accessed on 27 August 2018.

Canosa, G., P.V. Alfieri and C.A. Giudice. 2012. Environmentally friendly, nano lithium silicate anticorrosive coatings. Progress in Organic Coatings 73(2-3): 178–185. https://doi.org/10.1016/j.porgcoat.2011.10.013 accessed 15 October 2018.

Casares, M.V., L.I. de Cabo, R.S. Seoane, O.E. Natale, M.C. Ríos, C. Weigandt et al. 2012. Measured copper toxicity to *Cnesterodon decemmaculatus* (Pisces: Poeciliidae) and predicted by biotic ligand model in pilcomayo river water: A step for a cross-fish-species extrapolation. J. Toxi. 849315. http://doi.org/10.1155/2012/849315.

Carvalho, C.C.C.R.D. and M.J. Caramujo. 2017. Carotenoids in aquatic ecosystems and aquaculture: A colorful business with implications for human health. Front. Mar. Sci. 4. https://doi.org/10.3389/fmars.2017.00093.

Ceceña, L.A. 2014. The effect of ocean acidification on the organic complexation of iron and copper. Thesis for Doctor of Philosophy. University of Southampton. Faculty of Natural and Environmental Sciences. Ocean and Earth Sciences, 191 p. https://eprints.soton.ac.uk/377295/1/Avendano%252C%2520Lizeth_PhD_Mat_15.pdf accessed 05 October 2018.

Chauhan, S. 2016. Assessment in water related to public, Bilaua, Gwalior M.P. India. Inter. J. Scient. Res. Growth (ijirg) 1(5): 156–160. http://www.ijsrg.com/ijsrg/wp-content/uploads/2017/02/Savita_Chauhan-1.pdf.

Chen, L. and P.Y. Qian. 2017. Review on molecular mechanisms of antifouling compounds: An update since 2012. Mar. Drugs. 15(9): 264. 20 p. https://dx.doi.org/10.3390%2Fmd15090264 accessed 15 October 2018.

Chen, G., M.H. Azkab, G.L. Chmura, S. Chen, P. Sastrosuwondo, Z. Ma et al. 2017. Mangroves as a major source of soil carbon storage in adjacent seagrass meadows. Scient. Rep. 7, Article number: 42406. https://doi.org/10.1038/srep42406.

Chen, Y., M. Chen, Y. Li, B. Wang, S. Chen and Z. Xu. 2018. Impact of technological innovation and regulation development on e-waste toxicity: A case study of waste mobile phones. Sci. Rep. 8, Article number: 7100. https://doi.org/10.1038/s41598-018-25400-0.

Chinedu, E., D. Arome and F.S. Ameh. 2013. A new method for determining acute toxicity in animal models toxicology international 20(3): 224–226. http://doi.org/10.4103/0971-6580.121674.

Chopin, T., J.A. Cooper, G. Reid, S. Cross and C. Moore. 2012. Open-water integrated multi-trophic aquaculture: environmental biomitigation and economic diversification of fed aquaculture by extractive aquaculture. Rev. Aquacul. 4: 209–220. http://doi.org/10.1111/j.1753-5131.2012.01074.x accessed 09 October 2018.

Ciacci, L., I. Vassura and F. Passarini. 2017. Urban mines of copper: Size and potential for recycling in the EU. Resources 6(1): 6. https://doi.org/10.3390/resources6010006 accessed on 24 August 2018.

Cid, A., R. Prado, C. Rioboo, P. Suarez-Bregua and C. Herrero. 2012. Use of microalgae as biological indicators of pollution: Looking for new relevant cytotoxicity endpoints. pp. 311–32. *In*: Johnsen, M.N. (ed.). Microalgae: Biotechnology, Microbiology and Energy. Nova Science Publishers, New York. https://core.ac.uk/download/pdf/61918530.pdf accessed 01 October 2018.

Clark, G.F. and E.L. Johnston. 2017. Australia state of the environment 2016: coasts, independent report to the Australian Government Minister for Environment and Energy, Australian Government Department of the Environment and Energy, Canberra, 167 p. https://soe.environment.gov.au/sites/g/files/net806/f/soe2016-coasts-launch-17feb.pdf?v=1488793015.

Clearwater, S.J., A.M Farag and J.S. Meyer. 2002. Bioavailability and toxicity of dietborne copper and zinc to fish. Comparative Biochemistry and Physiology Part C: Toxicology & Pharmacology 132: 269–313. https://doi.org/10.1016/S1532-0456(02)00078-9.

Colley, D.G., A.L Bustinduy, W.E. Secor and C.H King. 2014. Human schistosomiasis. Lancet 383(9936): 2253–2264. https://dx.doi.org/10.1016%2FS0140-6736(13)61949-2 accessed 29 Aout 2018.

Copaja, S.V., V.R. Nuñez, G.S. Muñoz, G.L. González, I. Vila and D. Véliz. 2016. Heavy metal concentrations in water and sediments from affluents and effluents of mediterranean chilean reservoirs. J. Chil. Chem. Soc. 61(1): 2797–2804. http://dx.doi.org/10.4067/S0717-97072016000100011 accessed 02 October 2018.

Copertino, M.S., J.C. Creed, M.O. Lanari, K. Magalhães, K. Barros, P.C. Lana et al. 2016. Seagrass and Submerged Aquatic Vegetation (VAS) Habitats off the Coast of Brazil: state of knowledge, conservation and main threats. Braz. J. Oceanogr. 64 no.spe2 São Paulo 2016. http://dx.doi.org/10.1590/S1679-875920161036064sp2.

Cornwall, C.E., A.T. Revill, J.M. Hall-Spencer, M. Milazzo, J.A. Raven and C.L. Hurd. 2018. Inorganic carbon physiology underpins macroalgal responses to elevated CO_2. Sci. Rep. 7: 46297. http://dx.doi.org/10.1038/srep46297.

Crémazy, A., C.M. Wood, T.Y.T. Ng, D.S. Smith and M.J. Chowdhury. 2017. Experimentally derived acute and chronic copper biotic ligand models for rainbow trout. Aqua. Toxicol. 192: 224–240. http://dx.doi.org/10.1016/j.aquatox.2017.07.013 accessed 01 October 2018.

Croteau, M.N., S.K. Misra, S.N. Luoma and E. Valsami-Jones. 2014. Bioaccumulation and toxicity of CuO nanoparticles by a freshwater invertebrate after waterborne and dietborne exposures. Environ. Sci. Technol. 48(18): 10929–10937. https://doi.org/10.1021/es5018703 accessed on 25 September 2018.

Cusimano, R.F., D.F. Brakke and G.A. Chapman. 1986. Effects of pH on the toxicities of cadmium, copper, and zinc to steelhead trout (*Salmo gairdneri*). Canadian Journal of Fisheries and Aquatic Sciences 43(8): 1497–1503. https://doi.org/10.1139/f86-187 accessed on 25 September 2018.

Daglish, R.W., B.F. Nowak and T.W. Lewis. 2004. Copper/metal ratios in the gills of rainbow trout (*Oncorhynchus mykiss*) provide evidence of copper exposure under conditions of mixed-metal exposure. Arch. Environ. Contam. Toxicol. 47(1): 110–116.

Daly, T. 2018. China's refined copper imports to fall 7.5 pct in 2018—Antaike. https://www.reuters.com/article/china-metals-copper/chinas-refined-copper-imports-to-fall-7-5-pct-in-2018-antaike-idUSL3N1R43VN.

Darius, H.T., M. Roué, M. Sibat, J. Viallon, C. Mahanaitiatti, Mark W. Vandersea et al. 2018. Toxicological investigations on the sea Urchin *Tripneustes gratilla* (*Toxopneustidae*, Echinoid) from Anaho Bay (Nuku Hiva, French Polynesia): Evidence for the presence of Pacific Ciguatoxins. Mar. Drugs 16(4): 122. https://doi.org/10.3390/md16040122.

Da Silva, A.F., N.S. Sales., R.E.C.C. Oliveira and A.L.M. Pessanha. 2015. Trophic relationships among fish assemblages on a mudflat within a Brazilian Marine protected area. Braz. J. Oceanogr. 63(4) São Paulo Oct./Dec. 2015. http://dx.doi.org/10.1590/S1679-87592015091306304.

Davis, R.A., A.T. Welty, J. Borrego, J.A. Morales, J.G. Pendon and J.G. Ryan. 2000. Rio Tinto estuary (Spain): 5000 years of pollution. Environ. Geol. 39(10): 1107–1116. https://doi.org/10.1007/s002549900096 accessed 23 September 2018.

Dechezleprêtre, A. and M. Sato. 2017. The impacts of environmental regulations on competitiveness. Review of Environmental Economics and Policy 11(2): 183–206. https://doi.org/10.1093/reep/rex013.

De Los Santos, C.B., R. Sigurðardóttir, A. Cunha, K. Cook, J.M. Wiktor, A. Tatarek et al. 2014. A survey-based assessment of seagrass status, management and legislation in Europe. https://doi.org/10.3389/conf.FMARS.2014.02.00027.

Derby, C. and P.W. Sorensen. 2008. Neural processing, perception, and behavioral responses to natural chemical stimuli by fish and crustaceans. J. Chem. Ecol. 34(7): 898–914. https://doi.org/10.1007/s10886-008-9489-0 accessed 03 October 2018.

De Silva, S.S. 2001. A global perspective of aquaculture in the new millennium. pp. 431–459. In: Subasinghe, R.P., P. Bueno, M.J. Phillips, C. Hough, S.E. McGladdery and J.R. Arthur (eds.). Aquaculture in the Third Millennium. Technical Proceedings of the Conference on Aquaculture in the Third Millennium, Bangkok, Thailand, 20–25 February 2000. NACA, Bangkok and FAO, Rome. http://www.fao.org/docrep/003/ab412e/ab412e27.htm.

Deyanova, D., M. Gullström, L.D. Lyimo, M. Dahl, M.I. Hamisi, M.S.P. Mtolera et al. 2017. Contribution of seagrass plants to CO_2 capture in a tropical seagrass meadow under experimental disturbance. PLoS One 12(7): e0181386. https://dx.doi.org/10.1371%2Fjournal.pone.0181386.

Dong, Y., R.K. Rosenbaum and M.Z. Hauschild. 2016. Assessment of metal toxicity in marine ecosystems: comparative toxicity potentials for nine cationic metals in coastal seawater. Environ. Sci. Techno., American Chemical Society 50(1): 269–278. https://doi.org/10.1021/acs.est.5b01625.

Dórea, J.G. 2015. Exposure to mercury and aluminum in early life: Developmental vulnerability as a modifying factor in neurologic and immunologic effects. Int. J. Environ. Res. Public. Health 12(2): 1295–1313. https://dx.doi.org/10.3390%2Fijerph120201295 accessed 23 September 2018.

Dowsett, N.S. and S. Rayburg. 2011. Heavy metal distribution in estuarine sediments: a comparison of a seagrass bed and adjacent bare sediment. Proceedings of the 34th World Congress of the International Association for Hydro-environment Research and Engineering; 33rd Hydrology and Water Resources Symposium and. In: 10th Conference on Hydraulics in Water Engineering. Barton, ACT, Australia. Engineers Australia, pp. 1047–1053. https://search.informit.com.au/documentSummary;dn=335239714301421;res=IELENG.

Duinker, A., I.S. Roiha, H. Amlund, L. Dahl, E.J. Lock, T. Kögel et al. 2016. Potential risks posed by macroalgae for application as feed and food—a Norwegian perspective. NIFES (National Institute of Nutrition and Seafood Research), 25 p. https://www.mattilsynet.no/mat_og_vann/produksjon_av_mat/fisk_og_sjomat/rapport_makroalger_2016_nifes.23097/binary/Rapport%20makroalger%202016%20Nifes accessed 29 September 2018.

EC (European Commission). 2011a. Food, Agriculture and Fisheries, and Biotechnology. FP7 Cooperation Work Programme, 80 p. https://ec.europa.eu/research/participants/data/ref/fp7/89419/b-wp-201201_en.pdf.

EC (European Commission). 2016. Report from the Commission to the European Parliament and the Council on the sustainable use of biocides pursuant to Article 18 of Regulation (EU) No 528/2012 of the European Parliament and of the Council concerning the making available on the market and use of biocidal products, 13 p. https://ec.europa.eu/health/sites/health/files/biocides/docs/2016_report_sustainableuse_biocides_en.pdf accessed 22 August 2018.

EEA (European Environment Agency). 2018. Atmospheric greenhouse gas concentrations. https://www.eea.europa.eu/downloads/2544a644eb9a4648b40c998b9bd3148b/1517395177/assessment.pdf accessed 05 October 2018.

Egorova, K.S. and V.P. Ananikov. 2017. Toxicity of metal compounds: Knowledge and myths. Organomet 36: 4071–4090. http://dx.doi.org/10.1021/acs.organomet.7b00605 accessed 08 October 2018.

Eisler, R. 1997. Copper hazards to fish, wildlife, and invertebrates: A synoptic review. U.S. Geological Survey, Biological Resources Division, Biological Science Report USGS/BRD/BSR-1997-0002 98 pp.

Eisler, R. 2000. Handbook of chemical risk assessment: health hazards to humans, plants and animals. Vol. 1: Metals. Lewis Publishers, New York. https://epdf.tips/handbook-of-chemical-risk-assessment.html.

Eisler, R. 2007. Copper. *In*: Eisler's Encyclopedia of Environmentally Hazardous Priority Chemicals Escrito por Ronald Eisler. Cap. 9: 161–189. https://epdf.tips/eislers-encyclopedia-of-environmentally-hazardous-priority-chemicals.html.

Eklund, B. and D. Eklund. 2014. Pleasure boatyard soils are often highly contaminated. Environ. Manage. 53(5): 930–946. https://dx.doi.org/10.1007%2Fs00267-014-0249-3 accessed 15 October 2018.

EL-Gazzar, A.M., K.E. Ashry and Y.S. El-Sayed. 2014. Physiological and oxidative stress biomarkers in the freshwater Nile Tilapia, Oreochromis Niloticus L., exposed to sublethal doses of cadmium. Alexan. J. Veter. Sci. 40: 29–43. https://dx.doi.org/10.5455/ajvs.48333 accessed 09 October 2018.

Elison, T., A. Tiernan and D. Taylor. 2015. 2012 Kuskokwim area management report. Alaska Department of Fish and Game, Fishery Management Report No. 15–29, Anchorage, 127 p. http://www.adfg.alaska.gov/FedAidPDFs/FMR15-29.pdf accessed 01 October 2018.

Ellis, R.P. 2012. The impact of ocean acidification, increased seawater temperature and a bacterial challenge on the immune response and physiology of the blue mussel, Mytilus edulis, 273 p. Thesis presented to the University of Plymouth in part fulfilment for the degree of Doctor of Philosophy. School of Marine Science & Engineering. Faculty of Science & Technology. https://pdfs.semanticscholar.org/4363/1b568e4bb4846d6fdb09e4426952116619f9.pdf accessed 02 September 2018.

Emamverdian, A., Y. Ding, F. Mokhberdoran and Y. Xie. 2015. Heavy metal stress and some mechanisms of plant defense response. Sci. World J. 2015. Article ID 756120, 18 pages. http://dx.doi.org/10.1155/2015/756120.

Environment Canada. 2009. Organic Waste and Feed Deposits on Bottom Sediments from Aquaculture Operations: Scientific Assessment and Guidance. Ecosystem Health: Science-bassed Solutions. Report No 1–14. National Guidelines and Standard Office, Environment Canada. p 68. http://publications.gc.ca/collection_2009/ec/En13-1-14-2009E.pdf.

EPA. 1977. Effects of Copper and Zinc on Smoltification of Coho Salmon, 83 p. https://nepis.epa.gov/Exe/ZyPDF.cgi/9101EXTD.PDF?Dockey=9101EXTD.PDF accessed 03 October 2018.

EPA. 2012. An Assessment of Potential Mining Impacts on Salmon Ecosystems of Bristol Bay, Alaska. Volume 2—Appendices A-D. EPA 910-R-12-004b, 457 p. http://www.hdgold.com/i/pdf/ndm/bbwa/BBWA_D1V2.pdf accessed on 25 September 2018.

EPA New Zealand. 2011. Reassessment of Antifouling Paints, Background Information, 194 p. https://www.epa.govt.nz/assets/FileAPI/hsno-ar/APP201051/APP201051-APP201051-Evaluation-and-Review-Final.pdf accessed 24 September 2018.

EPA. 2013a. Drinking Water Best Management Practices, 18 p. https://www.in.gov/idem/files/lead_epa_schools_pws.pdf accessed 23 September 2018.

Erftemeijer, P.L.A., B. Riegl, B.W. Hoeksema and P.A.Todd. 2012. Environmental impacts of dredging and other sediment disturbances on corals: A review. Mar. Poll. Bull. 64(9): 1737–1765. https://doi.org/10.1016/j.marpolbul.2012.05.008 accessed 23 October 2018.

Etzkorn, J. and B. Allan. 2007. Marine Antifoulants: Tributyltin. Case Study # 8. CHEM 301: Aqueous Environmental Chemistry, 8 p. https://web.viu.ca/krogh/chem301/CS8%20Marine%20Anti-Foulants.pdf accessed 09 August 2018.

Evans, L.K. and M.S. Edwards. 2011. Bioaccumulation of copper and zinc by the giant kelp Macrocystis pyrifera. Algae 26(3): 265–275. http://dx.doi.org/10.4490/algae.2011.26.3.265 accessed 18 September 2018.

Fair, L.F., S.D. Moffitt, M.J. Evenson and J. Erickson. 2008. Escapement goal review of Copper and Bering rivers, and Prince William Sound Pacific salmon stocks, 2008. Alaska Department of Fish and Game, Fishery Manuscript No. 08–02, Anchorage, 43 p. http://www.sf.adfg.state.ak.us/FedAidpdfs/fms08-02.pdf accessed 01 October 2018.

Fair, L.F., S.D. Moffitt, M.J. Evenson and J.W. Erickson. 2011. Escapement Goal Review of Copper and Bering Rivers, and Prince William Sound Pacific Salmon Stocks, 2011. Alaska Department of Fish and Game, Fishery Manuscript No. 11–07, Anchorage, 42 p. http://www.adfg.alaska.gov/FedAidpdfs/FMS11-07.pdf accessed 01 October 2018.

FAO. 2006. National Aquaculture Sector Overview. Visión general del sector acuícola nacional—Indonesia. National Aquaculture Sector Overview Fact Sheets. Texto de Sri Paryanti, T. *In*: Departamento de Pesca y Acuicultura de la FAO. http://www.fao.org/fishery/countrysector/naso_indonesia/fr accessed 19 August 2018.

FAO. 2008. Climate Change for Fisheries and Aquaculture. Technical Background Document from the Expert Consultation Held on 7 to 9 April 2008. 18p. http://www.fao.org/fileadmin/user_upload/foodclimate/HLCdocs/HLC08-bak-6-E.pdf accessed on 26 August 2018.

FAO. 2009. Integrated mariculture. A global review, 194 p. http://www.fao.org/docrep/012/i1092e/i1092e.pdf accessed 13 September 2018.

FAO. 2010. El estado mundial de la pesca y la acuicultura 2010, 242 p. http://www.zaragoza.es/contenidos/medioambiente/onu/093-spa-ed2010.pdf.

FAO. 2016. El estado mundial de la pesca y la acuicultura 2016. Contribución a la seguridad alimentaria y la nutrición para todos. Roma. 224 pp. http://www.fao.org/3/a-i5555s.pdf accessed 09 October 2018.

FAO. 2018. State of Fisheries and Aquaculture in the World 2018–FAO. http://www.fao.org/3/I9540EN/i9540en.pdf accessed 09 October 2018.

Fashola, M.O., V.M. Ngole-Jeme and O.O. Babalola. 2016. Heavy metal pollution from gold mines: Environmental effects and bacterial strategies for resistance. Int. J. Environ. Res. Public Health 13(11): 1047. https://dx.doi.org/10.3390%2Fijerph13111047.

FCSAP (Federal Contaminated Sites Action Plan). 2013. Supplemental Guidance for Ecological Risk Assessment. Module 4. Causality Assessment Module. Determining the Causes of Impairmment a Contaminated Sites: Are Observed effects Due to Exposure to Site-Related Chemicals or Due to Order Stressors? 78 p. https://www.canada.ca/content/dam/eccc/migration/fcs-scf/B15E990A-C0A8-4780-9124-07650F3A68EA/13-049-ERA_Module-204-ENG.pdf accessed 07 October 2018.

FEEDAP (EFSA Panel on Additives and Products or Substances used in Animal Feed). 2016. Revision of the currently authorised maximum copper content in complete feed. EFSA J. 14(8): 100 p. https://dx.doi.org/10.2903/j.efsa.2016.4563 accessed 17 October 2018.

Felipe, J.V.A. 2011. Evaluación de efectos de biocidas contenidos en recubrimientos "antifouling" (AF coatings) en ecosistemas marinos. Master Oficial en Investigación, Modelización y Análisis del Riesgo en Medio Ambiente. Universidad Politécnica de Madrid Escuela Técnica Superior de Ingenieros de Minas, 233 p. http://oa.upm.es/33952/1/TESIS_MASTER_JOSE_VICENTE_ALONSO_FELIPE.pdf accessed 16 October 2018.

Feroz, B. 2018. Saponins from marine macroalgae: A review. J. Marine Sci. Res. Dev. 8(4): 1–8. https://doi.org/10.4172/2155-9910.1000255.

Filipuci, I. 2011. The Effects of Environmental Stressors on Coastal Fish: *In Situ* and Experimental Approach. Thesis PhD. Géosciences, Ecologie, Paléontologie et Océanographie. ED 104 (SMRE) Ecole doctorale Sciences de la Matière, du Rayonnement et de l'Environnement. Universite du Littoral Cote D'Opale et Universite D'Ege. https://www.theses.fr/2011DUNK0399/abes accessed 17 October 2018.

Findlay, S., C. Wigand and W. Nieder. 2006. Submersed macrophyte distribution and function in the tidal freshwater hudson river. pp. 230–241. *In*: Jeffrey, S. Levinton and John R. Waldman (eds.). The Hudson River Estuary. Cambridge University Press, New York, NY. https://cfpub.epa.gov/si/si_public_record_report.cfm?dirEntryId=150750&keyword=University+AND+Park+AND+Press&actType=&TIMSType=+&TIMSSubTypeID=&DEID=&epaNumber=&ntisID=&archiveStatus=Both&ombCat=Any&dateBeginCreated=&dateEndCreated=&dateBeginPublishedPresented=&dateEndPublishedPresented=&dateBeginUpdated=&dateEndUpdated=&dateBeginCompleted=&dateEndCompleted=&personID=&role=Any&journalID=&publisherID=&sortBy=revisionDate&count=50 accessed 24 October 2018.

Fleeger, J.W., K.R. Carman and R.M. Nisbet. 2003. Indirect effects of contaminants in aquatic ecosystems. The Sci. Total Environ. 317: 207–233. https://doi.org/10.1016/S0048-9697(03)00141-4 accessed 17 October 2018.

Flores, L., S. Rodela, J .Abian, J. Claria and E. Esmatjes. 2004. F2Isoprostane is already increased at the onset of Type 1 Diabetes Mellitus: Effect of glycemic control. Metabolism 53(9): 1118–20. https://doi.org/10.1016/j.metabol.2004.04.005.

Flynn, E.E., B.E. Bjelde, N.A. Miller and A.E. Todgham. 2015. Ocean acidification exerts negative effects during warming conditions in a developing Antarctic fish. Conserv. Physiol. 3(1): cov033. https://doi.org/10.1093/conphys/cov033 accessed 06 October 2018.

Forstner, U. and W. Calmano. 1998. Characterization of dredged materials. Water Sci. Technol. 38(11): 149–157. https://doi.org/10.1016/S0273-1223(98)00650-7.

Fonseca, E.M.D., J.A.B. Neto, J. Mcalister, B. Smith, M.A. Fernandez and F.C. Balieiro. 2013. The role of the humic substances in the fractioning of heavy metals in Rodrigo de Freitas Lagoon, Rio de Janeiro–Brazil. An. Acad. Bras. Ciênc. 85(4) Rio de Janeiro. http://dx.doi.org/10.1590/0001-3765201371011.

Franklin, N.M., J.L. Stauber, S.J. Markich and R.P. Lim. 2000. pH-dependent toxicity of copper and uranium to a tropical freshwater alga (*Chlorella* sp.). Aquat. Toxicol. 48: 275–289. https://doi.org/10.1016/S0166-445X(99)00042-9.

Fraser, M.W. and G.A. Kendrick. 2017. Belowground stressors and long-term seagrass declines in a historically degraded seagrass ecosystem after improved water quality. Scientific Reports Vol. 7, Article number: 1446. https://doi.org/10.1038/s41598-017-14044-1 accessed on 26 August 2018.

Fresco, J.C. 2016. Estudio del sistema catecolaminérgico en el encéfalo anterior del pez cebra adulto (*Danio rerio*). Tesis de Maestría. Máster Biología Molecular, Celular y Genética. Universidade da Coruña.España.

Fu, Z., F. Wu, L. Chen, B. Xu, C. Feng, Y. Bai et al. 2016. Copper and zinc, but not other priority toxic metals, pose risks to native aquatic species in a large urban lake in Eastern China. Environ. Poll. 219: 1069–1076. http://dx.doi.org/10.1016/j.envpol.2016.09.007 accessed 10 October 2018.

Gadberry, B.A., J. Colt1, D. Maynard, D.C. Boratyn, K. Webb, R.B. Johnson et al. 2018. Intensive land-based production of red and green macroalgae for human consumption in the Pacific Northwest: an evaluation of seasonal growth, yield, nutritional composition, and contaminant levels. Algae 33(1): 109–125. https://doi.org/10.4490/algae.2018.33.2.21 accessed 18 September 2018.

Gaetke, L.M., H.S. Chow-Johnson and C.K. Chow. 2014. Copper: Toxicological relevance and mechanisms. Arch. Toxicol. 88(11): 1929–1938. https://dx.doi.org/10.1007%2Fs00204-014-1355-y accessed 24 September 2018.

Gamain, P., P. Gonzalez, J. Cachot, P. Pardon, N. Tapie, P.Y. Gourves et al. 2016. Combined effects of pollutants and salinity on embryo-larval development of the Pacific oyster, *Crassostrea gigas*. Mar. Environ. Res. 113: 31–38. http://dx.doi.org/10.1016/j.marenvres.2015.11.002.

Gamakaranage, C.S., C. Rodrigo, S. Weerasinghe, A. Gnanathasan, V. Puvanaraj and H. Fernando. 2011. Complications and management of acute copper sulphate poisoning; a case discussion. Journal of Occupational Medicine and Toxicology 6(1): 34. https://dx.doi.org/10.1186%2F1745-6673-6-34.

Gao, G., Y. Liu, X. Li, Z. Feng and J. Xu. 2016. An ocean acidification acclimatised green tide alga is robust to changes of seawater carbon chemistry but vulnerable to light stress. PLoS ONE 11(12): e0169040. https://doi.org/10.1371/journal.pone.0169040.

Gao, G., Y. Liua, X. Lia, Z. Fenga, Z. Xub, H. Wuc et al. 2017. Expected CO_2-induced ocean acidification modulates copper toxicity in the green tide alga Ulva prolifera. Environmental and Experimental Botany 135: 63–72. http://dx.doi.org/10.1016/j.envexpbot.2016.12.007.

Garcia-Ordiales, E., S. Covelli, J.M. Rico, N. Roqueñí, G. Fontolan, G. Flor-Blanco et al. 2018. Occurrence and speciation of arsenic and mercury in estuarine sediments affected by mining activities (Asturias, northern Spain). Chemos. 198: 281–289. https://doi.org/10.1016/j.chemosphere.2018.01.146 accessed 20 October 2018.

Gawande, M.B., A. Goswami, F.X. Felpin, T. Asefa, X. Huang, P. Silva et al. 2016. Cu and Cu-based nanoparticles: Synthesis and applications in catalysis. Chemical Reviews 116(6): 3722–3811. https://doi.org/10.1021/acs.chemrev.5b00482 accessed 07 October 2018.

Geer, T.D., C.M. Kinley, K.J. Iwinski, A.J. Calomeni and J.H. Jr Rodgers. 2016. Comparative toxicity of sodium carbonate peroxyhydrate to freshwater organisms. Ecotoxicol. Environ. Saf. 132: 202–11. https://doi.org/10.1016/j.ecoenv.2016.05.037.

Gende, S.M., T.R. Edwards Mary, M.F. Wilson and M.S. Wipfli. 2002. Pacific Salmon in aquatic and terrestrial ecosystems: Pacific salmon subsidize freshwater and terrestrial ecosystems through several pathways, which generates unique management and conservation issues but also provides valuable research opportunities. BioScience 52(10): 917–928. https://doi.org/10.1641/0006-3568(2002)052[0917:PSIAAT]2.0.CO;2 accessed 03 October 2018.

Gharedaashi, E., H. Nekoubin, M.R. Imanpoor and V. Taghizadeh. 2013. Effect of copper sulfate on the survival and growth performance of Caspian Sea kutum, *Rutilus frisii kutum*. Springerplus 2: 498. https://dx.doi.org/10.1186%2F2193-1801-2-498.

Gibbons, D., C. Morrissey and P. Mineau. 2015. A review of the direct and indirect effects of neonicotinoids and fipronil on vertebrate wildlife. Environ. Sci. Pollut. Res. Int. 22: 103–118. https://dx.doi.org/10.1007%2Fs11356-014-3180-5 accessed 17 October 2018.

Giordano, M., J. Beardall and J.A. Raven. 2005. CO_2 concentrating mechanisms in algae: mechanisms, environmental modulation, and evolution. Annu. Rev. Plant Biol. 56: 99–131. https://doi.org/10.1146/annurev.arplant.56.032604.144052.

Gordon, T.A.C., E.L. Wilding and D.C. Aldridge. 2016. Predation of freshwater gastropods (*Viviparus viviparus*) by brown rats (*Rattus norvegicus*). J. Mollus. Stud. 82(3): 1–7. https://doi.org/10.1093/mollus/eyw012 accessed 11 October 2018.

Grimm, C. and A. Gerhardt. 2018. Sensitivity towards copper: Comparison of stygal and surface water species' biomonitoring performance in water quality surveillance. Int. J. Sci. Res. Environ. Sci. Toxicol. 3(1): 1–15. https://pdfs.semanticscholar.org/b6bf/40397de779d87899d52ca58f09328f64aa03.pdf.

Guerrero, F.M., M. Lozano and J.M. Rueda-Cantuche. 2008. Spain's greatest and most recent mine disaster. Disasters 32(1): 19–40. https://doi.org/10.1111/j.1467-7717.2007.01025.x accessed 20 October 2018.

Guerreiro, M.J.R., J.A. Fraguela, G. González, E. Muñoz and L. Carral. 2009. Evaluación del impacto ambiental provocado por las pinturas antiincrustantes utilizadas en las embarcaciones de recreo en los puertos deportivos de Galicia (España), 13 p. https://dx.doi.org/10.13140/RG.2.1.2336.7123 acccessed 13 October 2018.

Gundersen, H., T. Bryan, W. Chen, F.E. Moy, A.N. Sandman, G. Sundblad et al. 2017. Ecosystem Services in the Coastal Zone of the Nordic Countries, 130 p. http://dx.doi.org/10.6027/TN2016-552 accessed on 27 August 2018.

Gutiérrez-Galindo, E.A., J.A. Villaescusa-Celaya and A. Arreola-Chimal. 1999. Bioaccumulation of metals in mussels from four sites of the coastal region of Baja California. Cienc. Mar. 25: 557–578. http://www.cienciasmarinas.com.mx/index.php/cmarinas/article/view/726/653 accessed 24 October 2018.

Gvozdenović, S., M. Mandić, V. Pešić, M. Nikolić, A. Pešić and Z. Ikica. 2017. Comparison between IMTA and monoculture farming of mussels (*Mytilus galloprovincialis* L.) in the Boka Kotorska Bay. Act. Adriat. 58(2): 271–284. https://hrcak.srce.hr/file/286164.

Hale, L., S. Wairepo, A. Bedford-Rolleston and C. Taiapa. 2013. An Annotated Bibliography: Health of Te Awanui Tauranga Harbour. Manaaki Taha Moana Research Report No. 8. Massey University, Palmerston North, 371 p. https://www.mtm.ac.nz/wp-content/uploads/2017/03/report-8.pdf.

Halpern, B.Ss. 2008. A global map of human impact on marine ecosystems. Science 319: 948–952. https://doi.org/10.1126 / science.1149345.

Hammarstrom, J.M., A.L. Meier, J.C. Jackson, R. Barden, P.J. Wormington, J.D. Wormington et al. 1999. Characterization of mine waste at the Elizabeth copper mine, orange county, vermont. Open-File Report, 99–564. P. 88. https://semspub.epa.gov/work/01/251669.pdf accessed 24 September 2018.

Hammer, A. and Lin Jones. 2012. China's dominance as a global consumer and producer of copper. *In*: Copper Market Forecast 2012–2013. 2p. USITC Executive Briefings on Trade. https://www.usitc.gov/publications/332/2012-08_ChinaCopper%28HammerLin%29.pdf.

Han, F.X., J.A. Hargreaves, W.L. Kingery, D.B. Huggett and D.K. Schlenk. 2001. Accumulation, distribution, and toxicity of copper in sediments of catfish ponds receiving periodic copper sulfate applications. J. Environ. Qual. 30(3): 912–919. http://rmag.soil.msu.ru/articles/385.pdf.

Han, T., S.H. Kang, J.S. Park, H.K. Lee and M.T. Brown. 2008. Physiological responses of *Ulva pertusa* and *U. armoricana* to copper exposure. Aquat. Toxicol. 86: 176–184. https://doi.org/10.1016/j.aquatox.2007.10.016.

Hand, J. 2004. Typical Water Quality Values for Florida's Lakes, Streams, and Estuaries, 101 p. http://pinellas.wateratlas.usf.edu/upload/documents/JoeHand_TypicalWQ_ValuesDraftFinalEdits10-26-04.pdf accessed 29 August 2018.

Harrold, M.N. and R.P. Guralnick. 2010. A Field Guide to the Freshwater Mollusks of Colorado. Colorado Division of Wildlife, 73 p. http://takeaim.org/wp-content/uploads/2016/11/Mollusk_GuidePocketGuideColorado.pdf acessed 11 October 2018.

Haskell, C.A., D.A. Beauchamp and S.M. Bollens. 2017. Linking functional response and bioenergetics to estimate juvenile salmon growth in a reservoir food web. PLoS ONE 12(10): e0185933. https://doi.org/10.1371/journal.pone.0185933 accessed 30 September 2018.

Haywood, D. 2015. Copper Flat Copper Mine Draft Environmental Impact Statement. 1: 529 p. http://www.emnrd.state.nm.us/MMD/MARP/documents/DEIS_CopperFlatCopperMine_Volume1.pdf accessed 15 August 2018.

He, Q., M.D. Bertness, J.F. Bruno, B. Li, G. Chen, T.C. Coverdale et al. 2014. Economic development and coastal ecosystem change in China. Scientific Reports 4(1). https://doi.org/10.1038/srep05995.

Hedberg, N. 2017. Sea cages, seaweeds and seascapes. Causes and consequences of spatial links between aquaculture and ecosystems. Academic dissertation for the Degree of Doctor of Philosophy in Marine Ecotoxicology at Stockholm University. http://urn.kb.se/resolve?urn=urn:nbn:se:su:diva-141009 accessed 30 September 2018.

Hernández, P.P., C. Undurraga, V.E. Gallardo, N. Mackenzie, M.L. Allende and A.E. Reyes. 2011. Sublethal concentrations of waterborne copper induce cellular stress and cell death in zebra fish embryos and larvae. Biol. Res. 44: 7–15. https://scielo.conicyt.cl/pdf/bres/v44n1/art02.pdf accessed 17 October 2018.

Herrero, A., J. Vila, E. Eljarrat, A. Ginebreda, S. Sabater, R.J. Batalla et al. 2018. Transport of sediment borne contaminants in a Mediterranean river during a high flow event. Sci. Total Environ. 633: 1392–1402. https://doi.org/10.1016/j.scitotenv.2018.03.205 accessed 16 October 2018.

Heuschele, J. and E. Selander. 2014. The chemical ecology of copepods. J. Plankton Res. 36(4): 895–913.

Hill, E.K. and L. Julang. 2017. Current and future prospects for nanotechnology in animal production. J. Anim. Sci. Biotechnol. 8: 26. https://dx.doi.org/10.1186%2Fs40104-017-0157-5.

Ho, Y.C., K.Y. Show, X.X. Guo, I. Norli, F.M. Alkarkhi Abbas and N. Morad. 2012. Industrial discharge and their effect to the environment, industrial waste, Prof. Kuan-Yeow Show (ed.). ISBN: 978-953-51-0253-3, InTech. Available from: http://www.intechopen.com/books/industrial-waste/industrial-emissions-and-theireffect-on-the-environment accessed 19 October 2018.

Ho, K.T., L. Portis, A.A. Chariton, M. Pelletier, M. Cantwell, D. Katz et al. 2018. Effects of micronized and nano-copper azole on marine benthic communities. Environ. Toxi. and Chem. 37: 362–375. https://doi.org/10.1002/etc.3954.

Hoang, T.C., E.C. Rogevich, G.M. Rand, P.R. Gardinali, R.A. Frakes and T.A. Bargar. 2008. Copper desorption in flooded agricultural soils and toxicity to the Florida apple snail (Pomacea paludosa): Implications in Everglades restoration. Environmental Pollution 154: 338e347. https://doi.org/10.1016/j.envpol.2007.09.024 accessed 29 August 2018.

Hogstrand, C. 2001. 3–Zinc. Homeostasis and Toxicology of Essential Metals. Fish Physiolo. pp. 135–200. *In*: Homeostasis and Toxicology of Essential Metals: Volume 31 A. FISH PHYSIOLOGYhttps://doi.org/10.1016/S1546-5098(11)31003-5 accessed 17 October 2018.

Hook, S.E., E.P. Gallagher and G.E. Batley. 2014. The role of biomarkers in the assessment of aquatic ecosystem health. Integr. Environ. Assess. Manag. 10(3): 327–341. https://dx.doi.org/10.1002%2Fieam.1530.

Hoppe, S., M. Sundbom, H. Borg and M. Breitholtz. 2015. Predictions of Cu toxicity in three aquatic species using bioavailability tools in four Swedish soft freshwaters. Environ. Sci. Eur. 27(1): 25. https://dx.doi.org/10.1186%2Fs12302-015-0058-1.

Hosokawa, S., S. Konuma and Y. Nakamura. 2016. Accumulation of trace metal elements (Cu, Zn, Cd, and Pb) in surface sediment via decomposed seagrass leaves:A mesocosm experiment using *Zostera marina* L. PLoS One 11(6): e0157983. https://dx.doi.org/10.1371%2Fjournal.pone.0157983.

Hostynek, J.J. and H.I. Maibach. 2003. Copper hypersensitivity: Dermatologic aspects: An overview. Rev. Environ. Health 18(3): 153–83. https://doi.org/10.1111 / j.1396-0296.2004.04035.x.

Hothem, R.L., J.T. May, J.K. Gibson and B.E. Brussee. 2015. Concentrations of metals and trace elements in aquatic biota associated with abandoned mine lands in the Whiskeytown National Recreation Area and nearby Clear Creek watershed, Shasta County, northwestern California, 2002–2003. Open-File Report 2015–1077. https://doi.org/10.3133/ofr20151077 accessed 24 September 2018.

Howarth, R., D. Anderson, J. Cloern, C. Elfring, C. Hopkinson, B. Lapointe et al. 2000. Nutrient pollution of coastal rivers, bays and seas. Issues Ecol. 7(1-14): 4. https://pubs.er.usgs.gov/publication/70185674.

Howley, C. 2001. An Evaluation of the Seagrass *Zostera Capricorni* as a Biomonitor of Metals contamination in Lake Illawarra. Unpublished PhD thesis. University of Wollongong, New South Wales, Australia, 102 p. https://ro.uow.edu.au/cgi/viewcontent.cgi?referer=https://www.google.co.ve/&httpsredir=1&article=3563&context=theses accessed 14 October 2018.

Hsu, L.C., C.Y. Huang, Y.H. Chuang, H.W. Chen, Y.T. Chan, H.Y. Teah et al. 2016. Accumulation of heavy metals and trace elements in fluvial sediments received effluents from traditional and semiconductor industries. Scien. Rep. Vol. 6, Article number: 34250. https://doi.org/10.1038/srep34250 accessed 20 October 2018.

Hua, L., X. Yang, Y. Liu, X. Tan and Y. Yang. 2018. Spatial distributions, pollution assessment, and qualified source apportionment of soil heavy metals in a typical mineral mining city in China. Sustainability 10: 3115. https://doi.org/10.3390/su10093115 accessed 24 September 2018.

Huang, X., C. Ke and W.X. Wang. 2010. Cadmium and copper accumulation and toxicity in the macroalga *Gracilaria tenuistipitata*. Aquat. Biol. 11: 17–26. https://doi.org/10.3354/ab00288 accessed 28 September 2018.

Huang, Z., J. Siwabessy, H. Cheng and S. Nichol. 2018. Using multibeam backscatter data to investigate sediment, acoustic relationships. J. Geophy. Res. https://doi.org/10.1029/2017JC013638.

Humann-Ziehank, E. 2016. Selenium, copper and iron in veterinary medicine-From clinical implications to scientific models. J. Trace Elem. Med. Biol. 37: 96–103. https://doi.org/10.1016/j.jtemb.2016.05.009 accessed 22 September 2018.

Icely, R.P.J., I. Galparsoro, K. Éva, A. Boyd, S. Bricker, A.M. Sequeira et al. 2018. AQUASPACE. Ecosystem Approach to making Space for Aquaculture EU Horizon 2020. 308 p. http://www.aquaspace-h2020.eu/wp-content/uploads/2018/09/AquaSpace-D4-2-revisedfinal-17Aug2018-1.pdf.

IMO (International Maritime Organization). 2001. International Conference on the Control of Harmful Anti-Fouling Systems for Ships, 31 p. https://www.state.gov/documents/organization/208110.pdf accessed 03 August 2018.

IMO (International Maritime Organization). 2002. Anti-fouling systems, 31 p. http://www.imo.org/en/OurWork/Environment/Anti-foulingSystems/Documents/FOULING2003.pdf accessed 13 October 2018.

IMO (International Maritime Organization). 2009. The Generation of Biocide Leaching Rate Estimates for Anti-Fouling Coatings and Their Use in the Development of Proposals to Amend Annex 1 of the AFS Convention, MEPC 60/13. http://www.imo.org/en/OurWork/Environment/Anti-foulingSystems/Pages/Default.aspx accessed 18 October 2018.

IMO (International Maritime Organization). 2011. Resolution MEPC.208(62). Guidelines for Inspection of Anti-Fouling Systems on Ships, 25 p. http://www.imo.org/en/KnowledgeCentre/IndexofIMOResolutions/Marine-Environment-Protection-Committee-(MEPC)/Documents/MEPC.208(62).pdf accessed 03 August 2018.

IMSP (Instituto Municipal de Salud Pública). 2017. La calidad sanitaria del agua de consumo humano en Zaragoza. 33 P. https://www.zaragoza.es/cont/paginas/noticias/aguas_consumo_2017.pdf accessed 28 September 2018.

IPCC. 2013. Climate change 2013: The physical science basis. *In*: Stocker, T.F., D.G-.K. Qin, M. Tignor, S.K. Allen, J. Boschung et al. (eds.). Working Group I Contribution to the Fifth Assessment Report of the Intergovernmental Panel on Climate Change. Cambridge Univ. Press, New York. https://www.ipcc.ch/pdf/assessment-report/ar5/wg1/WGIAR5_SPM_brochure_en.pdf accessed 26 September 2018.

Isani, G. and E. Carpenè. 2014. Metallothioneins, unconventional proteins from unconventional animals: A long journey from nematodes to mammals. Biomoléculas 4(2): 435–457. https://dx.doi.org/10.3390%2Fbiom4020435.

Islam, F., J. Wang, M.A. Farooq, M.S.S. Khan, L. Xu, J. Zhu et al. 2017. Potential impact of the herbicide 2,4-dichlorophenoxyacetic acid on human and ecosystems. Environ. Inter. 111: 332–351. http://dx.doi. org/10.1016/j.envint.2017.10.020 accessed 01 October 2018.

Ismail, A. and R. Ramli. 1997. Trace metals in sediments and molluscs from an estuary receiving pig farms effluent. Environmental Technology 18(5): 509–515. http://dx.doi.org/10.1080/09593331808616566 accessed 22 September 2018.

ISO 10890: 2010. Paints and varnishes—Modelling of biocide release rate from antifouling paints by mass-balance calculation, 7 p. https://www.iso.org/ru/standard/46281.html accessed 27 September 2018.

ISO. 2007. 15181-2 Paints and varnishes—Determination of release rate of biocides from antifouling paints—Part 2: Determination of copper-ion concentration in the extract and calculation of the release rate, 22 p. https://www. iso.org/standard/42868.html accessed 27 September 2018.

IUCN. 2012. An Environmental and Fisheries Profile of the Puttalam Lagoon System. Regional Fisheries Livelihoods Programme for South and Southeast Asia (GCP/RAS/237/SPA) Field Project Document 2011/LKA/CM/06. xvii+237 pp. http://www.fao.org/3-a-ar443e.pdf accessed 30 August 2018.

Jaishankar, M., Tenzin Tseten, Naresh Anbalagan, Blessy B. Mathew and Krishnamurthy N. Beeregowda. 2014. Toxicity, mechanism and health effects of some heavy metals. Interdiscip Toxicol. 7(2): 60–72. https://dx.doi. org/10.2478%2Fintox-2014-0009.

James M., N. Hartstein and H. Giles. 2018. 4. Assessment of ecological effects of expanding salmon farming in Big Glory Bay, Stewart Island—Part 2 Assessment of effects. Aqua. Environ. Sci. 45 p. https://www.es.govt. nz/services/consents-and-compliance/notified-consents/Documents/2018/Sanford%20Limited/Assessment%20 of%20Effects%20Volume%202%20-%20Aquatic%20Environmental%20Sciences(Mark%20James).pdf accessed29 September 2018.

Jan, A.T., M. Azam, K. Siddiqui, A. Ali, I. Choi and Q.M.R. Haq. 2015. Heavy metals and human health: Mechanistic insight into toxicity and counter defense system of antioxidants. Int. J. Mol. Sci. 16: 29592–29630. https://dx.doi.org/10.3390/ijms161226183 accessed 01 August 2018.

Jaramillo, M.F. and I. Restrepo. 2017. Wastewater reuse in agriculture: A review about its limitations and benefits. Sustainability 9: 1734. https://doi.org/10.3390/su9101734.

Jaysankar, D., K. Fukami, K. Iwasaki and K. Okamura. 2009. Occurrence of heavy metals in the sediments of Uranouchi Inlet, Kochi prefecture, Japan. Fish. Sci. 75(2). https://doi.org/10.1007/s12562-008-0054-0.

Jeyasanta, I., T.T. Lilly and J. Patterson. 2018. Macro and micro nutrients of seagrass species from Gulf of Mannar, India. MOJ Food Process Technol. 6(4): 391–398. https://doi.org/10.15406/mojfpt.2018.06.00193 accessed on 26 August 2018.

Ji, Y., Z. Xu, D. Zou and K. Gao. 2016. Ecophysiological responses of marine macroalgae to climate change factors. J. Appl. Phycol. 28(5): 2953–2967. http://dx.doi.org/10.1007/s10811-016-0840-5.

Jiang, D., Z. Hu, F. Liu, R. Zhang, B. Duo, J. Fu et al. 2014. Heavy metals levels in fish from aquaculture farms and risk assessment in Lhasa, Tibetan Autonomous Region of China. Ecotoxicology. 23(4): 577–83. https://doi. org/10.1007/s10646-014-1229-3. Epub 2014 Mar 27.

Jitar, O., C. Teodosiu, A. Oros, G. Plavan and M. Nicoara. 2015. Bioaccumulation of heavy metals in marine organisms from the Romanian sector of the Black Sea. New Biotech. 32(3): 369–378. https://doi.org/10.1016/j. nbt.2014.11.004.

Johnson, R.A., A.G. Gulick, A.B. Bolten and K.A. Bjorndal. 2017. Blue carbon stores in tropical seagrass meadows maintained under green turtle grazing. Scien. Rep. Vol. 7, Article number: 13545. http://dx.doi.org/10.1038/ s41598-017-13142-4 accessed on 26 August 2018.

Joshi, M., A. Mukherjee, S.C. Misra and U.S. Ramesh. 2013. Natural Biocides in Antifouling Paints. ICSOT: Technical Innovation in Shipbuilding, 12–13 December, Kharagpur, India. The Royal Institution of Naval Architects. https://pdfs.semanticscholar.org/eb61/f8f9a4b3996bc4a4fca48b06e2e9915181bb.pdf accessed 15 October 2018.

Juneja, A., R.M. Ceballos and G.S. Murthy. 2013. Effects of environmental factors and nutrient availability on the biochemical composition of algae for biofuels production: A review. Energies 6: 4607–4638. http://dx.doi. org/10.3390/en6094607 accessed 30 September 2018.

Karlsson, J., E. Ytreberg and B. Eklund. 2010. Toxicity of anti-fouling paints for use on ships and leisure boats to non-target organisms representing three trophic levels. Environ. Pollut. 158(3): 681–687. http://dx.doi. org/10.1016/j.envpol.2009.10.024.

Katagi, T. 2008. Surfactant effects on environmental behavior of pesticides. pp. 71–177. *In*: David M. Whitacre (ed.). Rev. Environ. Contamina. Toxicol. 194.

Katagi, T. and H. Tanaka. 2016. Metabolism, bioaccumulation, and toxicity of pesticides in aquatic insect larvae. J. Pestic. Sci. 41(2): 25–37. https://doi.org/10.1584/jpestics.D15-064.

Keen, C.L., M.S. Clegg, L.A. Hanna, L. Lanoue, J.M. Rogers, G.P. Daston et al. 2003. The plausibility of micronutrient deficiencies being a significant contributing factor to the occurrence of pregnancy complications. J. Nutr. 133(5 Suppl 2): 1597S–1605S. https://doi.org/10.1093/jn/133.5.1597S accessed 23 September 2018.

Kiaune, L. and N. Singhasemanon. 2011. Pesticidal Copper (I) Oxide: Environmental fate and aquatic toxicity. *In*: David M. Whitacre (ed.). Reviews of Environmental Contamination and Toxicology 213 Springer-Verlag. New York. https://doi.org/10.1007/978-1-4419-9860- 6_1.

Kibria, G. 2016. Traza metales/metales pesados y su impacto en el medio ambiente, la biodiversidad y la salud humanalth–Un corto revisión, 5 p. Available from: https://www.researchgate.net/publication/266618621_ Traceheavy_Metals_and_Its_Impact_on_Environment_Biodiversity_and_Human_Health-_A_Short_Review accessed 13 August 2018.

Kimball, B.E., R. Mathur, A.C. Dohnalkova, A.J. Wall, R.L. Runkel and S.L. Brantley. 2009. Geoch. Cosmochi. Acta. 73: 1247–1263. https://doi.org/10.1016/j.gca.2008.11.035 accessed 24 September 2018.

Kinney, H. 2017. Aquaculturists' Perceptions of Integrated MultiTrophic Aquaculture (IMTA). Master of Arts in Marine Affairs Thesis. University of Rhode Island, 85 p. https://digitalcommons.uri.edu/cgi/viewcontent. cgi?article=2022&context=theses accessed 17 Sept 2018.

Kiyani, V., M. Hosynzadeh and M. Ebrahimpour. 2013. Investigation acute toxicity some of heavy metals at different water. Int J. Adv. Biol. Biom. Res. 1(2): 134–142. http://www.ijabbr.com/article_6579_3c8dff184f17b 87eda63789595225adb.pdf accessed 24 October 2018.

Klein, C., R. Gebker, T. Kokocinski, S. Dreysse, B. Schnackenburg, E. Fleck et al. 2008. Combined magnetic resonance coronary artery imaging, myocardial perfusion and late gadolinium enhancement in patients with suspected coronary artery disease. JCMR 10: 45. https://doi.org/1186/1532-429X-10-45.

Knillmann, S. 2013. The influence of competition on effect and recovery from pesticides in freshwater zooplankton communities. PhD Thesis. Helmholtz Centre for Environmental Research—UFZ, 129 p. http://publications. rwth-aachen.de/record/229676/files/4762.pdf accessed 30 September 2018.

Komenda, J. 2000. Role of two forms of the D1 protein in the recovery from photoinhibition of photosystem II in the cyanobacterium Synechococcus PCC 7942. Biochim. et Biophys. Acta (BBA)—Bioenerg. 1457(3): 243–252. https://doi.org/10.1016/S0005-2728(00)00105-5.

Král'Ová, K., E. Masarovičová and K. Györyová. 2004. The physiological response of green algae (*Chlorella vulgaris*) to pH-dependent inhibitory activity of some zinc(II) compounds: Carboxylatoand halogenocarboxylatozinc(II) complexes. Chem. Pap. 58: 353–356. https://www.researchgate.net/publication/286722676_The_physiological_ response_of_green_algae_Chlorella_vulgaris_to_pH-dependent_inhibitory_activity_of_some_zincII_ compounds_Carboxylato-_and_halogenocarboxylatozincII_complexes accessed 12 October 2018.

Kułaga, E. 2014. Antimicrobial Coatings for Soft Materials. Thèse Docteur. Chimie-Matériaux. Université de Haute Alsace, 278 p. https://www.theses.fr/2014MULH5312.pdf accessed 15 October 2018.

Kumar, K.S., H.U. Dahms, J.S. Lee, H.C. Kim, W.C. Lee and K.H. Shin. 2014. Algal photosynthetic responses to toxic metals and herbicides assessed by chlorophyll a fluorescence. Ecotoxicol. Environ. Saf. 104: 51–71. https://doi.org/10.1016/j.ecoenv.2014.01.042.

Kumar, M., S.C. Subhash and M.K. Jha. 2016. Heavy metals concentration assessment in ground water and general public health aspects around granite mining sites of Laxman pura, U.P., Jhansi, India. International Research Journal of Environment Sciences 5(1): 1–6. http://www.isca.in/IJENS/Archive/v5/i1/1.ISCA-IRJEVS-2015-218. pdf accessed 30 July 2018.

Küpper, H., I. Šetliacutek., M. Spiller, F.C. Küpper and O. Prášil. 2002. Heavy metal induced inhibition of photosynthesis: targets of *in vivo* heavy metal chlorophyll formation. J. Phycol. 38: 429–441. https://doi. org/10.1046/j.1529-8817.2002.01148.x.

LaBarre, W.J. 2014. The Effectiveness of Bioretention Structures for Metal Retention and Toxicity Reduction of Copper Roof Runoff. A thesis presented to the faculty of Towson University in partial fulfillment of the requirements for the degree of MASTER OF SCIENCE. Department of Environmental Science. Towson University Towson, Maryland. https://mdsoar.org/bitstream/handle/11603/1886/TF2014LaBarre_redacted. pdf;sequence=1 accessed 30 July 2018.

Laberge, F. and T.J. Hara. 2003. Behavioural and electrophysiological responses to F-prostaglandins, putative spawning pheromones, in three salmonid fishes. J. Fish Biol. 62(1): 206–221. https://doi.org/10.1046/j.1095-8649.2003.00020.x.

Lage, O.M., H.M.V.M. Soares, M.T.S.D. Vasconcelos, A.M. Parente and R. Salema. 1996. Toxicity effects of copper (II) on the marine dinoflagellate *Amphidinium carterae*: Influence of metal speciation. European Journal of Phycology 31(4): 341–348. https://doi.org/10.1080/09670269600651571 670269600651571.

Lagerström, M., J.F. Lindgren, A. Holmqvist, M. Dahlström and E. Ytreberg. 2018. *In situ* release rates of Cu and Zn from commercial antifouling paints at different salinities. Marine Pollution Bulletin 127: 289–296. https:// doi.org/10.1016/j.marpolbul.2017.12.027.

Lal, R. 2015. Shifting cultivation versus sustainable intensification. Reference Module in Earth Systems and Environmental Sciences, 12 p. https://doi.org/10.1016/b978-0-12-409548-9.09295-2.

Landers, J.E. 2016. Aquatic food webs and heavy metal contamination in the upper blackfoot river, Montana. Thesis. Master of Science in Environmental Studies. University of Montana. Missoula, MT. USA, 65 p. https://scholarworks.umt.edu/etd/10719 accessed 19 October 2018.

Lane, K., K. Charles-Guzman, K. Wheeler, Z. Abid, N. Graber and T. Matte. 2013. Health effects of coastal storms and flooding in urban areas: A review and vulnerability assessment. J. Environ. Public. Health 2013: 913064. https://dx.doi.org/10.1155%2F2013%2F913064.

Lapointe, D., F. Pierron and P. Couture. 2011. Individual and combined effects of heat stress and squeous of diestary copper exposure in fathead minnows (*Pinephales promelas*). Aquatic Toxicology 1-2: 80–85. https://doi.org/10.1016/j.aquatox.2011.02.022.

Larkum, A.W.E. and R.J. West. 1982. Long-term changes of seagrass meadows in Botany Bay, Australia. Aquat. Bot. 37: 55–70.

Lavery, P.S., M.Á. Mateo, O. Serrano and M. Rozaimi. 2013. Variability in the carbon storage of seagrass habitats and its implications for global estimates of blue carbon ecosystem service. PLoS One 8(9): e73748. https://dx.doi.org/10.1371%2Fjournal.pone.0073748.

Lazzari, L., A.L.R. Wagener, C.O. Farias, A.P. Baêta, C.R. Mauad, A.M. Fernandes et al. 2016. Estuary adjacent to a megalopolis as potential disrupter of carbon and nutrient budgets in the coastal Ocean. J. Braz. Chem. Soc. Vol. 27 nº.10. http://dx.doi.org/10.5935/0103-5053.20160056 accesed 19 September 2018.

Leduc, A.O.H.C., P.L. Munday, G.E. Brown and M.C.O. Ferrari. 2013. Effects of acidification on olfactory-mediated behaviour in freshwater and marine ecosystems: a synthesis. Philos. Trans. R. Soc. Lond. B. Biol. Sci. 368(1627): 20120447. https://dx.doi.org/10.1098%2Frstb.2012.0447 accessed 03 October 2018.

Lee, M., E.A. Boyle, Y. Echegoyen-Sanz, J.N. Fitzsimmons, R. Zhang and R.A. Kayser. 2011. Analysis of trace metals (Cu, Cd, Pb, and Fe) in seawater using single batch nitrilotriacetate resin extraction and isotope dilution inductively coupled plasma mass spectrometry. Ana. Chimi. Acta 686: 93–101. http://dx.doi.org/10.1016/j.aca.2010.11.052 accessed 23 Sept 2018.

Lennox, R.J., J.M. Chapman, C.M. Souliere, C. Tudorache, M. Wikelski, J.D. Metcalfe et al. 2016. Conservation physiology of animal migration. Conserv. Physiol. 4(1): cov072. https://dx.doi.org/10.1093%2Fconphys%2Fcov072 acccessed 03 October 2018.

Levit, S.M. 2010. A literature review of effects of cadmium on fish. The Nature Conservancy, 16 p. https://www.conservationgateway.org/ConservationByGeography/NorthAmerica/UnitedStates/alaska/sw/cpa/Documents/L2010CadmiumLR122010.pdf accessed 01 October 2018.

Lewis, C., R.P. Ellis. E. Vernon, K. Elliot. S. Newbatt and R.W. Wilson. 2016. Ocean acidification increases copper toxicity differentially in two key marine invertebrates with distinct acid-base responses. Sci. Rep. 6: 21554. http://dx.doi.org/10.1038/srep21554.

Lewtas, K., M. Paterson, H.D. Venema and D. Roy. 2015. Manitoba Prairie Lakes: Eutrophication and In-Lake Remediation Treatments. International Institute for Sustainable Development (IISD), 121 p. https://www.iisd.org/sites/default/files/publications/manitoba-prairie-lakes-remediation-literature-review.pdf accessed 28 July 2018.

Li, L., X. Tian, X. Yu and S. Dong. 2016. Effects of acute and chronic heavy metal (Cu, Cd, and Zn) exposure on sea cucumbers (*Apostichopus japonicus*). BioMed. Res. Int. Vol. 2016, Article ID 4532697, 13 p. https://doi.org/10.1155/2016/4532697.

Lin, C. and L. Zhang. 2015. Chap. 18–Habitat Enhancement and Rehabilitation. *In*: Developments in Aquaculture and Fisheries Science 39: 333–351. https://doi.org/10.1016/B978-0-12-799953-1.00018-0.

Lin, H., T. Sun, S. Xue and X. Jiang. 2016. Heavy metal spatial variation, bioaccumulation, and risk assessment of Zostera japonica habitat in the Yellow River Estuary, China. Sci. Total Environ. 541: 435–443. http://dx.doi.org/10.1016/j.scitotenv.2015.09.050.

Lin, J.E., J.J. Hard, K.A. Naish, D. Peterson, R. Hilborn and L. Hauser. 2016. It's a bear market: evolutionary and ecological effects of predation on two wild sockeye salmon populations. Heredity (Edinb). 116(5): 447–457. https://dx.doi.org/10.1038%2Fhdy.2016.3 accessed 03 October 2018.

Lindgren, J.F., E. Ytreberg, A. Holmqvist, M. Dahlström, P. Dahl, M. Berglin et al. 2018. Copper release rate needed to inhibit fouling on the west coast of Sweden and control of copper release using zinc oxide. Biofouling 34(4): 453–463. https://doi.org/10.1080/08927014.2018.1463523 accesed 15 August 2018.

Liu, Y.M., Z.-M. Tang, X.-S. Li, Z.Z. Yang, D.-R. Yao, Z.-P. Liu et al. 2014. The responses of photosynthetic physiology in macroalga Ulva linza to temperature variation under ocean acidification. Chin. J. Ecol. 33(9): 2402–2407.

Liu, C., H. Lin, N. Mi, F. Liu, Y. Song, Z. Liu et al. 2017. Adsorption mechanism of rare earth elements in *Laminaria ochroleuca* and *Porphyra haitanensis*. J. Food Biochem. 2018: 1–6. http://dx.doi.org/10.1111/jfbc.12533.

Liu, S., Z. Jiang, Y. Deng, Y. Wu, J. Zhang, C. Zhao et al. 2018. Effects of nutrient loading on sediment bacterial and pathogen communities within seagrass meadows. MicrobiologyOpen, e00600. https://doi.org/10.1002/mbo3.600.

Llope, M., G.M. Daskalov, T.A. Rouyer, V. Mihneva, K.S. Chan and A.N. Grishin. 2011. Overfishing of top predators eroded the resilience of the Black Sea system regardless of the climate and anthropogenic conditions. Glob. Biol. 17(3): 1251–1265. https://dx.doi.org/10.1111%2Fj.1365-2486.2010.02331.x accessed on 26 August 2018.

López, M. 2018. Venenos en el cuerpo–Intoxicados por todas partes. El Blog de Mario, 13 p. https://lamatrixholografica.wordpress.com/2016/01/16/venenos-en-el-cuerpo-intoxicados-por-todas-partes/ accessed 15 July 2018.

Lorenzo, J.M., R. Agregán, P.E.S. Munekata, D. Franco, J. Carballo, S. Şahin et al. 2017. Proximate composition and nutritional value of three macroalgae: *Ascophyllum nodosum, Fucus vesiculosus* and *Bifurcaria bifurcate.* Mar. Drugs 15(11): 360. https://dx.doi.org/10.3390%2Fmd15110360.

Lu, Y., R. Wang, Y. Zhang, H. Su, P. Wang, A. Jenkins et al. 2015. Ecosystem health towards sustainability. Ecosystem Health and Sustainability 1(1): 2. http://dx.doi.org/10.1890/EHS14-0013.1 accessed 15 September 2018.

Luan, H. 2016. Impacts of Effluent and Stormwater Runoff Sources on Metal Lability and Bioavailability in Developed Streams. Doctoral Dissertations. University of Connecticut. 1306. 106 p. https://opencommons.uconn.edu/cgi/viewcontent.cgi?referer=https://www.google.co.ve/&httpsredir=1&article=7540&context=dissertations accessed 08 October 2018.

LWLI. 2011. Lake Worth Lagoon. Fixed Transect Seagrass Monitoring, 141 p. http://www.lwli.org/pdfs/2013ManagementPlan/2011FixedTransectSeagrassMonitoringReport.pdf.

LWLI. 2013. Lake Worth Lagoon Management Plan, 235 p. http://www.lwli.org/pdfs/2013ManagementPlan/2013LWLmanagementplanFINAL.pdf accessed 20 July 2018.

Lynch, N.R. 2014. Non-Additive Toxicity of Bi-Metal Mixtures to Fathead Minnows. Master's Theses. Paper 2241. Master of Science Program in Chemistry. Loyola University Chicago, 55 p. http://ecommons.luc.edu/luc_theses/2241 accessed 01 October 2018.

MacGill, M. 2018. Do copper bracelets help with arthritis? Medical News Today. https://www.medicalnewstoday.com/articles/305500.php accessed 23 September 2018.

Mandal, P. 2017. An insight of environmental contamination of arsenic on animal health. Emerging Contaminants 3(1): 17–22. http://dx.doi.org/10.1016/j.emcon.2017.01.004 accessed 19 October 2018.

Manikandan, S., S. Ganesapandian and K. Parthiban. 2011. Distribution and zonation of seagrasses in the Palk Bay, Southeastern India. J. Fish. Aqua. Sci. 6(2): 178–185. http://dx.doi.org/10.3923/jfas.2011.178.185 accessed on 26 August 2018.

Marín-Guirao, L., A. César, A. Marín and R. Vita. 2005. Assessment of sediment metal contamination in the Mar Menor coastal lagoon (SE Spain): Metal distribution, toxicity, bioaccumulation and benthic community structure. Cienc. Mar. 31: 413–428. http://www.scielo.org.mx/pdf/ciemar/v31n2/v31n2a9.pdf.

Márquez, A., O. Garcia, W. Senior, G. Martínez y and Á. González. 2012. Distribución de metales pesados en sedimentos superficiales del Orinoco Medio, Venezuela. Ciencias 20(1): 60–73. www.produccioncientifica.luz.edu.ve/index.php/ciencia/article/view/10038.

Márquez, A., G. Martínez, J. Figuera, W. Senior and Á. González. 2016. Aspectos geoquímicos y ambientales de los sedimentos del Río Cuchivero, Venezuela. Bol. Inst. Oceanogr. Venezuela 55(1): 41–53. www.ojs.udo.edu.ve/index.php/boletiniov/article/view/2273.

Marr, J.C.A., J. Lipton, D. Cacela, J.A. Hansen, H.L. Bergman, J.S. Meyer et al. 1998. Relationship between copper exposure duration, tissue copper concentration, and rainbow trout growth. Aqua. Tox. 36: 17–30. https://doi.org/10.1016/S0166-445X(96)00801-6.

Martinez-Porchas, M. and L.R. Martinez-Cordova. 2012. World aquaculture: Environmental impacts and troubleshooting alternatives. Scienti. World J. pp. 1–9. https://dx.doi.org/10.1100%2F2012%2F389623 accessed 11 August 2018.

Martorell, J.J. 2010. Biodisponibilidad de metales pesados en dos ecosistemas acuáticos de la costa Suratlántica andaluza afectados por Contaminación difusa. Tesis Doctoral Universidad de Cádiz. rodin.uca.es/xmlui/bitstream/handle/10498/15776/Tes_2010_06.pdf accessed 03 August 2018.

Massaut, L. 1999. Manejo de sabores/olores no deseados (off-flavor) en cultivo de camarón en el Ecuador. El Mundo Acuícola 5: 24–26.

Matheson, F.E., A.M. Dugdale, R.D.S. Wells, A. Taumoepeau and J.P. Smith. 2007: Efficacy of saltwater solutions to kill introduced freshwater species and sterilise freshwater fishing nets. DOC Research & Development Series 261. Department of Conservation, Wellington, 24 p. https://www.doc.govt.nz/documents/science-and-technical/drds261.pdf accessed 02 August 2018.

McConaghie, J. and A. Matzke. 2016. Technical Support Document: An Evaluation to Derive Statewide Copper Criteria Using the Biotic Ligand Model, 157 p. http://www.wca-environment.com/application/

files/4014/6425/1905/OR_Technical_Support_Document_An_evaluationto_derive_statewide_Cu_criteria_ using_the_BLM_final.pdf accessed 01 October 2018.

McCready, S., G. Spyrakis, C.R. Greely, G.F. Birch and E.L. Long. 2004. Toxicity of surficial sediments from Sydney Harbour and vicinity, Australia. Environ. Monit. Assess. 96: 53–83. https://doi.org/10.1023/ B:EMAS.0000031716.34645.71.

McCready, S., G.F. Birch, E.R. Long, G. Spyrakis and C.R. Greely. 2006a. An evaluation of Australian sediment quality guidelines. Arch. Environ. Contam. Toxicol. 50(3): 306–315. https://doi.org/10.1007/ s00244-004-0233-7.

McCready, S., G.F. Birch, E.L. Long, G. Spyrakis and C.R. Greely. 2006b. Predictive abilities of numerical sediment quality guidelines for Sydney Harbour, Australia and vicinity. Environ. Int. 32: 638–649. https://doi.org/10.1016/j.envint.2006.02.004.

McCready, S., G.F. Birch and E.R. Long. 2006c. Metallic and organic contaminants in sediments of Sydney Harbour and vicinity—a chemical dataset for evaluating sediment quality guidelines. Environ. Int. 32: 455–465. https:// doi.org/10.1016/j.envint.2005.10.006.

McGreer, J.C., K.V. Brix, J.M. Skeaff, D.K. DeForest and S.I. Brigham. 2003. Inverse relationship between bioconcentration factor and exposure concentration for metals: implications for hazard assessment of metals in the aquatic environment. Environ. Toxicol. Chem. 22(5): 1017–1037. https://doi.org/10.1002/etc.5620220509.

McMahon, K.M., R.D. Evans, K. van Dijk, U. Hernawan, G.A. Kendrick, P.S. Lavery et al. 2017. Disturbance is an important driver of clonal richness in tropical seagrasses. Front. Plant. Sci. 8: 2026. https://dx.doi. org/10.3389%2Ffpls.2017.02026.

Meador, J.P. 2014. Do chemically contaminated river estuaries in Puget Sound (Washington, USA) affect the survival rate of hatchery-reared Chinook salmon? Canad. J. Fish. Aqua. Sci. 71(1): 162–180. https://doi.org/10.1139/ cjfas-2013-0130 accessed 03 October 2018.

Medeiros, J., H. Medeiros, C. Mascarenhas, L.B. Davin, N.G.L. Medeiros and N.G. Lewis. 2002. Bioactive components of Hedera helix. Arquipélago, Life and Marine Sciences 19A: 27–32. http://www.horta.uac.pt/ intradop/images/stories/arquipelago/19a/3_Medeiros_et_al_19A.pdf accessed 23 September 2018.

Mehvar, S., T. Filatova, A. Dastgheib, E.R. van Steveninck and R. Ranasinghe. 2018. Quantifying economic value of coastal ecosystem services: A review. J. Mar. Sci. Eng. 6: 5. https://doi.org/10.3390/jmse6010005.

Méndez, N. and C. Green. 2005. Preliminary observations of cadmium and copper effects on juvenile of de polychaete Capitella sp. (Annelida: Polichaeta) from estero del Yugo, Mazatlán, México. Rev. Chil. Hist. Nat. 78: 701–710. http://dx.doi.org/10.4067/S0716-078X2005000400009.

MEP (Massachusetts Environmental Police). 2006. Your Guide to Boating Laws and Responsibilities, 84 p. https:// www.boat-ed.com/assets/pdf/handbook/ma_handbook_entire.pdf accessed 05 August 2018.

Mills, R. 2015. China Copper Con. http://www.kitco.com/ind/Mills/2015-03-06-China-Copper-Con.html.

Mitchell, A.J. and M.S. Hobbs. 2003. Effect of citric acid, copper sulfate concentration, and temperature on a pond shoreline treatment for control of the marsh rams-horn snail planorbella trivolvis and the potential toxicity of the treatment to channel catfish. North American Journal of Aquaculture 65: 306–313. https://eurekamag.com/ pdf/004/004119259.pdf accesed 29 August 2018.

Mkadmi, Y., O. Benabbi, M. Fekhaoui, R. Benakkam, W. Bjijou, M. Elazzouzi et al. 2018. Study of the impact of heavy metals and physico-chemical parameters on the quality of the wells and waters of the Holcim area (Oriental region of Morocco). J. Mater. Environ. Sci. 9(2): 672–679. https://doi.org/10.26872/jmes.2018.9.2.74.

Moe, S.J., K.D. Schamphelaere, W.H. Clements, M.T. Sorensen, P.J.V. Brink and M. Liess. 2013. Combined and interactive effects of global climate change and toxicants on populations and communities. Environ. Toxicol. Chem. 32(1): 49–61. https://dx.doi.org/10.1002%2Fetc.2045 accessed 17 October 2018.

Moenne, A., A. Gonzalez and C.A. Saez. 2016. Mechanisms of metal tolerance in marine macroalgae, with emphasis on copper tolerance in Chlorophyta and Rhodophyta. Aquat. Toxicol. 176: 30–37. https://doi.org/10.1016/j. aquatox.2016.04.015.

Mohamed, K.N., M.S.Y. May and N. Zainuddin. 2015. Water quality assessment of marine park Islands in Johor, Malaysia. BESM 3(2): 19–27. https://www.researchgate.net/profile/Khairul_Nizam_Mohamed/ publication/286927964_Water_Quality_Assessment_of_Marine_Park_Islands_in_Johor_Malaysia/ links/5670460c08aececfd5531641.pdf accessed 07 October 2018.

Mohammadbakir, S.M.H. 2016. Impacts of waterborne copper and silver on the early life stage (ELS) of zebrafish (Danio rerio): physiological, biochemical and molecular responses. Thesis. Doctor of Philosophy. University of Plymouth, U.K. School of Biological Sciences Faculty of Science and Engineering, 247 p. https://pdfs. semanticscholar.org/4e95/a58700d53fcd33ee6f3bc256019ff02a15b0.pdf accessed 04 October 2018.

Mojica, K. 2008. Marine Biofilms: Ecology and Impact, 11 p. https://pdfs.semanticscholar.org/0140/0023c7fba7fda7 e6bc696cfce1b674d53242.pdf accessed 15 October 2018.

Morrisey, D., J. Gadd, M. Page, O.C. Woods, J. Lewis, A. Bell et al. 2013. In-water cleaning of vessels: Biosecurity and chemical contamination risks. MPI Technical Paper No: 2013/11. 267p. http://www.mpi.govt.nz/news-resources/publications.aspx accessed 08 September 2018.

Moss, B. 2008. Water pollution by agricultura. Philos. Trans. R. Soc. Lond. B. Biol. Sci. 363(1491). https://dx.doi.org/10.1098%2Frstb.2007.2176.

Mostafa, T., G. ElKhouly and A. Hassan. 2012. Pheromones in sex and reproduction: Do they have a role in humans? J. Adv. Res. 3(1): 1–9. https://doi.org/10.1016/j.jare.2011.03.003 accessed 03 October 2018.

Munshi, M., K.N. Tumu, M.N. Hasan and M.Z. Amin. 2018. Biochemical effects of commercial feedstuffs on the fry of climbing perch (Anabas testudineus) and its impact on Swiss albino mice as an animal model. Toxicology Reports 5: 521–530. https://doi.org/10.1016/j.toxrep.2018.04.004.

NAS (National Academy of Sciences). 1977. Copper. Committee on Medical and Biologic Effects of Environmental Pollutants, National Research Council, National Academy of Sciences, Washington, D.C., 115 pp.

NAS (National Academy of Science). 1986. Health effects of excess copper. Chapter 5. In Copper in drinking water. National Acadamy Press. Washington DC.

Ndoro, T.O. and L.K. Witik. 2017. A review of the flotation of copper minerals. Int. J. Sci.: Basic Appl. Res. (IJSBAR) 34(2): 145–165. http://gssrr.org/index.php?journal=JournalOfBasicAndApplied.

Needhidasan, S., M. Samuel and R. Chidambaram. 2014. Electronic waste–an emerging threat to the environment of urban India. J. Environ. Health Sci. Eng. 12: 36. https://dx.doi.org/10.1186%2F2052-336X-12-36.

Negrato, C.A. and M.B. Gomes. 2013. Low birth weight: causes and consequences. Diabetol. Metab. Syndr. 5: 49. https://dx.doi.org/10.1186%2F1758-5996-5-49.

Nguyen, T., F.A. Roddick and L. Fan. 2012. Biofouling of water treatment membranes: A review of the underlying causes, monitoring techniques and control measures. Membranes (Basel). 2(4): 804–840. https://dx.doi.org/10.3390%2Fmembranes2040804 accessed 15 October 2018.

Nielsen, J.M. 2015. Species interactions and energy transfer in aquatic food webs. Doctoral Thesis. Stockholm: Department of Ecology, Environment and Plant Sciences, Stockholm University, p. 40. http://www.diva-portal.org/smash/get/diva2:875089/FULLTEXT01.pdf accessed 29 September 2018.

Nikitin, O.V., E.I. Nasyrova, V.R. Nuriakhmetova, N.Y. Stepanova, N.V. Danilova and V.Z. Latypova. 2018. Toxicity assessment of polluted sediments using swimming behavior alteration test with *Daphnia magna*. IOP Conf. Series: Earth Environ. Sci. 107: 012068. 7 p. https://doi.org/10.1088/1755-1315/107/1/012068.

Nilsson, J. and L. Gipperth. 2015. Antifouling for leisure boats in the Baltic Sea Mapping the legal situation. WP 3: National Study: Sweden, 33 p. https://law.handels.gu.se/digitalAssets/1648/1648953_national-study---sweden.pdf accessed 23 August 2018.

Nogués, P.A. 2017. Evolución en el empleo de las Pinturas Antiincrustantes en Acero de Construcción Naval. Trabajo Fin de Grado. Ingeniería Marítima. Escuela Técnica Superior de Náutica Universidad de Cantabria, 94 p. https://repositorio.unican.es/xmlui/bitstream/handle/10902/12462/Ruiz%20Nogu%C3%A9s%2C%20Pablo.pdf?sequence=1&isAllowed=y accessed 16 October 2018.

Nordlund, L.M., E.L. Jackson, M. Nakaoka, J. Samper-Villarreal, P. Beca-Carretero and J.C. Creed. 2017. Seagrass ecosystem services—What's next? Mar. Pollu. Bull. 1–7. https://doi.org/10.1016/j.marpolbul.2017.09.014.

Nriagu, J.O. 1979a. Global inventory of natural and anthropogenic emissions of trace metals to the atmosphere. Nature 279: 409–411. https://doi.org/10.1038 / 279409a0 26 September 2018.

Nriagu, J.O. 1979b. The global copper cycle. pp. 1–17. *In*: Nriagu, J.O. (ed.). Copper in the Environment. Part 1: Ecological Cycling. John Wiley, New York.

Nriagu, J.O. 1979c. Copper in the atmosphere and precipitation. pp. 45–75. *In*: Nriagu, J.O. (ed.). Copper in the Environment. Part 1: Ecological Cycling. John Wiley, NewYork.

Obreza, T.A. and K.T. Morgan. 2017. Nutrition of Florida Citrus Trees. 2nd Edition. U.S. Department of Agriculture, UF/IFAS Extension Service, University of Florida, IFAS, Florida A & M University Cooperative Extension Program, and Boards of County Commissioners Cooperating. Nick T. Place, dean for UF/IFAS Extension, 1000 p. https://edis.ifas.ufl.edu/pdffiles/SS/SS47800.pdf.

Odonkor, S.T. and J.K. Ampofo. 2013. Escherichia coli as an indicator of bacteriological quality of water: An overview. Microbiol. Res. 4: 5–11. https://doi.org/10.4081/mr.2013.e2 accessed 19 October 208.

O'farrel, I., R. Lombardo, P. De Tezanos and C. Loez. 2002. The assessment of water quality in the Lower Luján River (Buenos Aires, Argentina): phytoplankton and algal bioassays. Environ. Poll. 120: 207–218. https://doi.org/10.1016/S0269-7491(02)00136-7 accessed 04 August 2018.

Okocha, R.O. and O.B. Adedeji. 2012. Overview of copper toxicity to aquatic life. Rep. Opinion 4(8): 57–67. http://www.sciencepub.net/report/report0408/011_10314report0408_57_67.pdf.

Olaniran, A.O., A. Balgobind and B. Pillay. 2013. Bioavailability of heavy metals in soil: Impact on microbial biodegradation of organic compounds and possible improvement strategies. Int. J. Mol. Sci. 14(5): 10197–10228. https://dx.doi.org/10.3390%2Fijms140510197.

Oliveira-Filho, E.C., R. Matos Lopes and F.J.R. Paumgartten. 2004. Comparative study on the susceptibility of freshwater species to copper-based pesticides. Chemosphere 56: 369–374. https://doi.org/10.1016/j. chemosphere.2004.04.026.

Omae, I. 2003. General aspects of tin-free antifouling paints. Chem. Rev. 103: 3431−3448. https://pdfs.semanticscholar. org/06db/1b59d27dd84bb3651dc9028e5f32a63c55c3.pdf accessed 05 August 2018.

Osunde, I., S. Coyle, J. Tidwell and N. Russell. 2004. Acute toxicity of copper sulfate to juvenile freshwater prawn, *Macrobrachium rosenbergii.* J. Appl. Aquat. 14: 71–74. https://doi.org/10.1300/J028v14n03_06.

Otero, X.L., P. Vidal-Torrado, M.R. Calvo de Anta and F. Macías. 2005. Trace elements in biodeposits and sediments from mussel culture in the Ría de Arousa (Galicia, NW Spain). Environ. Pollut. 136: 119–134. https:// doi.org/10.1016/j.envpol.2004.11.026.

Paller, M.H. and A.S. Knox. 2013. Bioavailability of metals in contaminated sediments. E3S Web of Conference 1, 02001. https://doi.org/10.1051/e3sconf/20130102001.

Pandey, P.K., P.H. Kass, M.L. Soupir, S. Biswas and V.P. Singh. 2014. Contamination of water resources by pathogenic bacteria. AMB Express 4(51). https://dx.doi.org/10.1186/s13568-014-0051-x accessed 19 October 2018.

Pappas, S. 2014. Facts About Copper. Live Science. https://www.livescience.com/29377-copper.html accessed 23 Sept 2018.

Parent, S.M. and L.A. Burd. 2016. Notice of Violations of the Endangered Species Act Regarding Registration of Cuprous Iodide, 10 p. https://www.biologicaldiversity.org/campaigns/pesticides_reduction/atrazine/pdfs/ Notice_of_Intent_Cuprous_Iodide_2016_01_06.pdf accessed 17 October 2018.

Paul, D. 2017. Research on heavy metal pollution of river Ganga: A review. Ann. Agra. Sci. 15(2): 278–286. http:// dx.doi.org/10.1016/j.aasci.2017.04.001 accessed 24 September 2018.

Penttila, B. 2017. Report to the Legislature on Non-copper Antifouling Paints for Recreational Vessels in Washington, 27 p. https://fortress.wa.gov/ecy/publications/documents/1704039.pdf accessed 09 August 2018.

Pereira, A.M.M., A.M.V.M. Soares, F. Gonialves and R. Ribeiro. 2000. Water-column, sediment, and *in situ* chronic bioassays with cladocerans. Ecotoxi. Environ. Safety 47: 27–38. https://dx.doi.org/10.1006/eesa.2000.1926 accessed August 25 2018.

Pérez, L.D. and N.D. Pérez. 2012. Proyecto para el desarrollo del cultivo de especies marinas en instalaciones en mar abierto. Universidad Politécnica de Madrid. E.T.S.I. Navales. Proyecto n°105. 350p. http://oa.upm.es/13736/1/ PFC_Nuria_y_Laura_Dominguez_Perez.pdf accessed 12 August 2018.

Petänen, T. 2001. Assessment of Bioavailable Concentrations and Toxicity of Arsenite and Mercury in Contaminated Soils and Sediments by Bacterial Biosensors. Thesis PhD. Faculty of Science Department of Biosciences. Division of General Microbiology. University of Helsinki. Finland. http://ethesis.helsinki.fi/julkaisut/mat/bioti/ vk/petanen/assessme.pdf accessed 19 October 2018.

Piculell, B.J., J.D. Hoeksema and J.N. Thompson. 2008. Interactions of biotic and abiotic environmental factors in an ectomycorrhizal symbiosis, and the potential for selection mosaics. BMC Biol. 6: 23. https://dx.doi. org/10.1186%2F1741-7007-6-23 accessed 20 October 2018.

Playle, R.C., D.G. Dixon and K. Burnison. 1993. Copper and cadmium binding ot fish gills:modification by dissolved organic carbon and synthetic ligands. Canadian Journal of Fisheries and Aquatic Sciences 5050(12): 2667–2677. https://doi.org/10.1139/f93-290.

PlusMining. 2017. El impacto de la minería del cobre en Chile. Implicancias económicas y sociales para el país, 9 p. https://www.procobre.org/es/wp-content/uploads/sites/2/2018/04/ica-summary-document-el-impacto-de- la-mineria-del-cobre-en-chile-vf-04.04.2018.pdf accessed 03 July 2018.

Polidoro, B.A., K.E. Carpenter, L. Collins, N.C. Duke, A.M. Ellison, J.C. Ellison et al. 2010. The loss of species: Mangrove extinction risk and geographic areas of global concern. PLoS ONE 5(4): e10095. https://doi. org/10.1371/journal.pone.0010095 accessed on 26 August 2018.

Pope, K.L., S.E. Lochmann and M.K. Young. 2009. Methods for assessing fish populations. Chapter 11. pp. 325–351. https://www.fs.fed.us/rm/pubs_other/rmrs_2010_pope_k001.pdf accessed 13 October 2018.

Qie, Y., C. Chen, F. Guo, Y. Mu, F. Sun, H. Wang et al. 2017. Predicting criteria continuous concentrations of metals or metalloids for protecting marine life by use of quantitative ion characteristic–activity relationships– species sensitivity distributions (QICAR-SSD). Mar. Poll. Bull. 124(2): 639–644. https://doi.org/10.1016/j. marpolbul.2017.02.055.

Rajeshkuma, S. and X. Li. 2018. Bioaccumulation of heavy metals in fish species from the Meiliang Bay, Taihu Lake, China. Toxicol. Rep. 5: 288–295. https://dx.doi.org/10.1016%2Fj.toxrep.2018.01.007.

Ramesh, U.S., A. Mukherjee, S.C. Mirsa and M. Joshi. 2014. Failure analysis of antifouling paints on ships hull. Ind. J. Geo Mar. Sci. 43(11): 2060–2066. http://nopr.niscair.res.in/bitstream/123456789/34575/1/IJMS%20 43%2811%29%202060-2066.pdf accessed 09 August 2018.

Ramírez, R., S. Duran, J. Salazar and M. D'suze y T. Cabrera. 2002. Incidencia del sulfato de cobre (CuSO$_4$.5H$_2$O) en la sobrevivencia de postlarvas de *Litopenaeus vannamei*. VI Congreso Venezolano de Acuicultura, 22–25.

Rastogi, R.P., D. Madamwar and A. Incharoensakdi. 2015. Bloom Dynamics of cyanobacteria and their toxins: Environmental health impacts and mitigation strategies. Front Microbiol. 6: 1254. https://dx.doi.org/10.3389%2Ffmicb.2015.01254.

Raven, J.A., H. Elderfield, O. Hoegh-Guldberg, P. Liss, U. Riebesell, J. Shepherd et al. 2005. Ocean acidification due to increasing atmospheric carbon dioxide, 68 p. https://royalsociety.org/~/media/Royal_Society_Content/policy/publications/2005/9634.pdf.

Ray, B.R., W. Matthew, K.C. Johnson and D.L. Smee. 2014. Changes in seagrass species composition in northwestern gulf of mexico estuaries: Effects on associated seagrass Fauna. PLoS One 9(9): e107751. https://dx.doi.org/10.1371%2Fjournal.pone.0107751.

Redelstein, R., H. Zielke, D. Spira, U. Feiler, U. Erdinger, H. Zimmer et al. 2015. Bioacumulación y efectos moleculares de los metales unidos al sedimento en el pez cebra Tembryos. Environ Sci Pollut Res. https://doi.org/10.1007 / s11356-015-5328-3.

Reyes, Y.C., I. Vergara O.E. Torres and E.E. Díaz M. y González. 2016. Contaminación por Metales Pesados: implicaciones en salud, ambiente y seguridad alimentaria. Revista Ingeniería, Investigación y Desarrollo, Vol. 16 N° 2, Julio-Diciembre, pp. 66–77. ISSN Online 2422-4324.

Rhind, S.M. 2009. Anthropogenic pollutants: a threat to ecosystem sustainability? Philos. Trans. R. Soc. Lond. B. Biol. Sci. 364(1534): 3391–3401. https://dx.doi.org/10.1098%2Frstb.2009.0122.

Riba, I., J. Blasco, N. Jiménez-Tenorio and T.A. DelValls. 2005. Heavy metal bioavailability and effects. I. Bioaccumulation caused by mining activities in the Gulf of Cádiz (SW Spain). Chemosphere 58: 659–669. https://doi.org/10.1016/j.chemosphere.2004.02.015.

Riba, I., E. García-Luque, J. Kalman, J. Blasco and C. Vale. 2011. Effects of sediment acidification on the bioaccumulation of Zn in *R. Philippinarum*. *In*: Duarte, P. and J. Santana-Casiano (eds.). Oceans and the Atmospheric Carbon Content. Springer, Dordrecht. https://doi.org/10.1007/978-90-481-9821-4_6 accessed 20 October 2018.

Richards, R., M. Chaloupka, M. Sanò and R. Tomlinson. 2011. Modelling the effects of 'coastal' acidification on copper speciation. Ecol. Modell. 222: 3559–3567. Doi: 10.1016/j.ecolmodel.2011.08.017.

Rivas, C., N. Navarro, P. Huovinen and I. Gómez. 2016. Photosynthetic UV stress tolerance of the Antarctic snow alga *Chlorella* sp. modified by enhanced temperature? Rev. Chil. Hist. Nat. 89: 1–9. http://dx.doi.org/10.1186/S40693-016-0050-1 accessed 30 September 2018.

Roberts, D.A., S.N. Birchenough, C. Lewis, M.B. Sanders, T. Bolam and D. Sheahan.2013. Ocean acidification increases the toxicity of contaminated sediments. Glob. Chang. Biol. 19: 340–351. https://doi.org/10.1111/gcb.12048.

Robertson, D.M. and D.A. Saad. 2011. Nutrient inputs to the laurentian great lakes by source and watershed estimated using SPARROW watershed models. J. Am. Water Resour. Assoc. 47(5): 1011–1033. https://doi.org/10.1111/j.1752-1688.2011.00574.x.

Rodero, N.T. 2011. Contaminación ocasionada por las aguas de lastre en el Mediterráneo occidental. Trabajo final de carrera – Diplomatura Navegación Marítima, 150 p. https://upcommons.upc.edu/bitstream/handle/2099.1/12216/Neus%20T%C3%A9llez%20Rodero%20-%20TFC%20Diplomatura%20Navegaci%C3%B3n%20Mar%C3%ADtima.pdf accessed 16 October 2018.

Rodgers, J.H. 2008. Algal Toxins in Pond Aquaculture. SRAC Publication No. 4605. 8 pp. http://www2.ca.uky.edu/wkrec/AlgalToxinsPond.pdf.

Rodney, E., P. Herrera, J. Luxama, M. Boykin, A. Crawford, M.A. Carroll et al. 2007. Bioaccumulation and tissue distribution of arsenic, cadmium, copper and zinc in *Crassostrea virginica* grown at two different depths in Jamaica Bay, New York. *In Vivo* 29(1): 16–27. https://www.ncbi.nlm.nih.gov/pmc/articles/PMC3155416/pdf/nihms294615.pdf accessed 22 August 2018.

Röhr, M.E., C. Boström, P. Canal-Vergés and M. Holmer. 2016. Blue carbon stocks in Baltic Sea eelgrass (*Zostera marina*) meadows. Biogeosciences. 13: 6139–6153. https://doi.org/10.5194/bg-13-6139-2016 accessed 24 September 2018.

Romañach, S.S., D.L. DeAngelis, H.L.K.Y.L. Su Yean, T.R. Sulaiman, R. Barizan and L. Zhaig. 2018. Conservation and restoration of mangroves: Global status, perspectives, and prognosis. Ocean & Coast. Manage. 154: 72–82. https://doi.org/10.1016/j.ocecoaman.2018.01.009.

Ronan, J.M., D.B. Stengel, A. Raab, J. Feldmann, L. O'Hea, E. Bralatei et al. 2017. High proportions of inorganic arsenic in *Laminaria digitata* but not in Ascophyllum *nodosum* samples from Ireland, Chemosphere. https://doi.org/10.1016/j.chemosphere.2017.07.076 accessed 18 September 2018.

Rostern, N.T. 2017. The effects of some metals in acidified waters on aquatic organisms. Fish & Ocean Opj. 4(4): 555–645. https://doi.org/10.19080/OFOAJ.2017.04.555645.

RSA (Republic of South Africa). 2014. Essential Drugs Programme. Primary Healthcare Standard Treatment Guidelines and Essential Medicines List. 5th ed. Republic of South Africa: National Department of Health. http://apps.who.int/medicinedocs/documents/s23015en/s23015en.pdf accessed 23 September 2018.

Runkel, R.L., K.E. Bencala, B.A. Kimball, K. Walton-Day and P.L. Verplanck. 2009. A comparison of pre- and postremediation water quality, Mineral Creek, Colorado. Hydrological Processes 23: 3319–3333. https://doi.org/10.1002/hyp.7427.

Russell, S., C.A. Sullivan and A.J. Reichelt-Brushett. 2015. Aboriginal consumption of estuarine food resources and potential implications for health through trace metal exposure; a study in Gumbaynggirr Country, Australia. PLoS One 22.10(6): e0130689. https://doi.org/10.1371/journal.pone.0130689. eCollection 2015.

Ryder, J., K. Iddya and L. Ababouch. 2014. Assessment and management of seafood safety and quality Current practices and emerging issues. FAO Fisheries and Aquaculture Technical Paper 574. Rome, FAO, 432 pp. http://www.fao.org/3/a-i3215e.pdf accessed 29 August 2018.

Sandahl, J.F., D.H. Baldwin, J.J. Jenkins and N.L. Scholz. 2007. A sensory system at the interface between urban stormwater runoff and salmon survival. Environmental Science and Technology 14(8): 2998–3004. http://dx.doi.org/10.1021/es062287r.

Sang, H.M. 2010. Role of immunostimulants in the culture of decapod crustacean. Thesis presented for the Degree of Doctor of Philosophy of Curtin University. Aquaculture and Environment Centre Department of Environment and Agriculture, 182 p. https://espace.curtin.edu.au/bitstream/handle/20.500.11937/2426/162587_Huynh2010.pdf?sequence=2 accessed 02 September 2018.

Sansalone, J.J., S.G. Buchberger and ASCE Members. 1997. Partitioning and first flush of metals in urban road way storm water. Journal of Environmental Engineering 123: 134–143. https://doi.org/10.1061/(ASCE)0733-9372(1997)123:2(134).

Santos, R.O., D. Lirman and J.E. Serafy. 2011. Quantifying freshwater-induced fragmentation of submerged aquatic vegetation communities using a multi-scale landscape ecology approach. Mar. Ecol. Prog. Ser. 427: 233–246. https://www.int-res.com/articles/theme/m427p233.pdf.

Sappal, R., N. MacDonald, M. Fast, D. Stevens, F. Kibenge, A. Siah et al. 2014. Interactions of copper and thermal stress on mitochondrial bioenergetics in rainbow trout, *Oncorhynchus mykiss*. Aquat. Toxicol. 157: 10–20. https://doi.org/10.1016/j.aquatox.2014.09.007 accessed 04 October 2018.

Sarmiento, A.M., E. Bonnail, J.M. Nieto and Á. DelValls. 2016. Bioavailability and toxicity of metals from a contaminated sediment by acid mine drainage: linking exposure–response relationships of the freshwater bivalve *Corbicula fluminea* to contaminated sediment. Environ. Sci. Pollu. Res. 23(22): 22957–22967. https://doi.org/10.1007/s11356-016-7464-9 accessed 20 October 2018.

Scannell, P.W. 2009. Phyllis effects of copper on aquatic species: A review of the literature (Anchorage, Alaska: Alaska Department of Fish and Game Division of Habitat, 2009). https://www.adfg.alaska.gov/static/home/library/pdfs/habitat/09_04.pdf.

Schmitt, C.J., V.S. Blazer, G.M. Dethloff, D.E. Tillitt, T.S. Gross, W.L. Bryant Jr. et al. 1999. Biomonitoring of Environmental Status and Trends (BEST) Program: fields procedures for assessing the exposure of fish to environmental contaminants. U.S. Geological Survey, Biological Resources Division, Columbia (MO): Information and Technology USGS/BRD-1999-0007. iv + 35 pp. + appendices. https://www.cerc.usgs.gov/pubs/center/pdfDocs/91116.pdf accessed on 29 August 2018.

Schönberg, C.H.L., J.K.H. Fang, M. Carreiro-Silva, A. Tribollet and M. Wisshak. 2017. Bioerosion: the other ocean acidification problem. ICES J. Mar. Sci. 74(4): 895–925. https://doi.org/10.1093/icesjms/fsw254.

Scelzo, M. 1997. Toxicidad del cobre en larvas nauplii del camarón comercial *Artemesia longinaris* Bate (Crustacea, Decapoda, Penaeidae). Invest. Mar. Valparaíso 25: 177–185. http://dx.doi.org/10.4067/S0717-71781997002500013.

Selck, H., K. Drouillard, K. Eisenreich, A.A. Koelmans, A. Palmqvist, A. Ruus et al. 2011. Explaining differences between bioaccumulation measurements in laboratory and field data through use of a probabilistic modeling approach. Integr. Environ. Assess. Manage. 8(1): 42–63. https://doi.org/10.1002/ieam.217 accessed 21 October 2018.

Sen, I.S. and B. Peucker-Ehrenbrink. 2012. Anthropogenic disturbance of element cycles at the earth's surface. Environ. Sci. Technol. 46: 8601–8609. https://doi.org/10.1021/es301261x.

Sensarma, S., P. Chakraborty, R. Banerjee and S. Mukhopadhyay. 2016. Geochemical fractionation of Ni, Cu and Pb in the deep sea sediments from the Central Indian Ocean Basin: An insight into the mechanism of metal enrichment in sediment. Chemie Erde-Geochem. 76(1): 39–48. http://drs.nio.org/drs/bitstream/handle/2264/4971/Chemie_Erde-Geochem_76_39a.pdf?sequence=1.

Serrano, O., P. Lavery, P. Masque, K. Inostroza, J. Bongiovanni and C. Duarte. 2016. Seagrass sediments reveal the long-term deterioration of an estuarine ecosystem. Glob. Chang. Biol. 22(4): 1523–31. https://doi.or/10.1111/gcb.13195. Epub 2016 Jan 28.

Sevcikova, M., H. Modra, J. Blahova, R. Dobsikova, L. Plhalova, O. Zitka et al. 2016. Biochemical, haematological and oxidative stress responses of common carp (*Cyprinus carpio* L.) after sub-chronic exposure to copper. Vet. Med. 61(1): 35–50. http://dx.doi.org/10.17221/8681-VETMED accessed 27 August 2018.

Shah, T.K., I. Nazir, P. Arya and T. Pandey. 2017. Integrated multi trophic aquaculture (IMTA): An innovation technology for fish farming in India. I.J.F.B.S. 4(1): 12–14. http://www.faunajournal.com/archives/2017/vol4issue1/PartA/3-5-30-948.pdf.

Short, F., T. Carruthers, W. Dennison and M. Waycot. 2007. Global seagrass distribution and diversity: A bioregional model. J. Exp. Mar. Bio. Ecolo. 350: 3–20. http://dx.doi.org/10.1016/j.jembe.2007.06.012.

Show, P.L., S.Y. Malcolm, S.Y. Tang, D. Nagarajan,T.C. Ling, C.W. Ooi et al. 2017. A holistic approach to managing microalgae for biofuel applications. Inter. J. Mole. Sci. 18(1): 215. http://doi.org/10.3390/ijms18010215 acessed 01 de October 2018.

Shuhaimi-Othman, M., R. Nur-Amalina and Y. Nadzifah. 2012. Toxicity of metals to a freshwater snail, *Melanoides tuberculate.* The Scientific World Journal Volume 2012. 10 p. http://dx.doi.org/10.1100/2012/125785.

Siemering, G.S., D.J. Hayworth and B.K. Greenfield. 2008. Assessment of potential aquatic herbicide impacts to california aquatic ecosystems. Arch. Environ. Contam. Toxicol. http://dx.doi.org/10.1007/s00244-008-9137-2.

Silva, M.A., T.C.S. Motta, D.B. Tintor, T.A. Dourado, A.L. Alcântara, A.A. Menegário et al. 2017. Tilapia (*Oreochromis niloticus*) as a biondicator of copper and cadmium toxicity. A bioavailability approach. J. Braz. Chem. Soc. Vol. 28 No. 1 São Paulo Jan. 2017. http://dx.doi.org/10.5935/0103-5053.20160157.

Simons, A.M. 2015. A Fundamental Study of Copper and Cyanide Recovery from Gold Tailings by Sulfidisation. PhD. Thesis. Western Australian School of Mines. Department of Metallurgical and Minerals Engineering. Curtin University, 327 p. https://espace.curtin.edu.au/bitstream/handle/20.500.11937/1799/234250_Simons%20Andrew%202015.pdf?sequence=2 accessed 01 October 2018.

Sipaúba-Tavares, L.H., A.M.D.L. Segali, F.A. Berchielli-Morais and B. Scardoeli-Truzzi. 2017. Development of low-cost culture media for *Ankistrodesmus gracilis* based on inorganic fertilizer and macrophyte. Acta Limnol. Bras. 29: e5. http://dx.doi.org/10.1590/s2179-975x3916.

Smetacek, V. and A. Zingone. 2013. Green and golden seaweed tides on the rise. Nature 504: 84–88. https://doi.org/10.1038/nature12860.

Soissons, L.M., E.P. Haanstra, M.M. van Katwijk, R. Asmus, I. Auby, L. Barillé et al. 2018. Latitudinal patterns in european seagrass carbon reserves: influence of seasonal fluctuations versus short-term stress and disturbance events. Front. Plant. Sci. 9. https://doi.org/10.3389/fpls.2018.00088.

Solomon, F. 2008. Impacts of metals on aquatic ecosystems and human health. Environment & Communities, 14–19. https://pdfs.semanticscholar.org/6572/7277c6270165b2329e363ff645f3bec8c586.pdf.

Solomon, F. 2009. Impacts of copper on aquatic ecosystems and human health. Environment & Communities, 4 p. http://www.ushydrotech.com/files/6714/1409/9604/Impacts_of_Copper_on_Aquatic_Ecosystems_and_human_Health.pdf.

Sorensen, W. and N.E. Stacey. 2004. Brief review of fish pheromones and discussion of their possible uses in the control of non-indigenous teleost fishes. New Zealand Journal of Marine and Freshwater Research 38(3): 399–417. https://doi.org/10.1080/00288330.2004.9517248 accessed 03 October 2018.

Spokas, E.G., B.W. Spur, H. Smith, F.W. Kemp and J.D. Bogden. 2008. Tissue lead concentration during chronic exposure of *Pimephales promelas* (Fathead Minnow) to lead nitrate in aquarium water. Environ. Sci. Technol. 1. 40(21): 6852–6858. https://www.ncbi.nlm.nih.gov/pmc/articles/PMC2527373/pdf/nihms-63309.pdf.

Squadrone, S., P. Brizio, C. Stella, M. Prearo, P. Pastorino, L. Serracca et al. 2016. Presence of trace metals in aquaculture marine ecosystems of the northwestern Mediterranean Sea (Italy). Environmental Pollution 215: 77–83. https://doi.org/10.1016/j.envpol.2016.04.096.

Stern, B.R., M. Solioz, D. Krewski, P. Aggett, T.C. Aw, S. Baker et al. 2007. Copper and human health: Biochemistry, genetics, and strategies for modeling dose-response relationships. J. Toxicol. Environ. Health B Crit. Rev. 10(3): 157–222. https://doi.org/10.1080 / 10937400600755911.

Su, C., L.Q. Jiang and W.J. Zhang. 2014. A review on heavy metal contamination in the soil worldwide: Situation, impact and remediation techniques. Environ. Skeptics and Critics 3(2): 24–38. http://www.iaees.org/publications/journals/environsc/articles/2014-3(2)/a-review-on-heavy-metal-contamination-in-the-soil-worldwide.pdf.

Su, C., Y. Lu, A.C. Johnson, Y. Shi, M. Zhang, Y. Zhang et al. 2017. Which metal represents the greatest risk to freshwater ecosystem in Bohai Region of China? Ecosystem Health and Sustainability 3(2): e01260. https://doi.org/10.1002/ehs2.1260 accessed 10 October 2018.

Subirats, H.L. 2015. Estudio a escala piloto del efecto de diferentes condiciones de operación sobre la eliminación de nutrientes en un cultivo de microalgas. Máster en Ingeniería Ambiental. UPV Universitat Politècnica de

València. Escuela Técnica Superior de Ingenieros de Caminos, Canales y Puertos, 124 p. https://riunet.upv. es/bitstream/handle/10251/57909/TFM-Hector_Lores_Subirats.pdf?sequence=1 accessed 25 September 2018.

Sundström, L.F., M. Lõhmus and R.H. Devlin. 2010. Migration and growth potential of coho salmon smolts: implications for ecological impacts from growth-enhanced fish. Ecological Applications 20(5): 1372–1383. https://doi.org/10.1890/09-0631.1 accessed 03 October 2018.

Sylvander, P. 2007. Increased sensitivity to antifouling paints in Fucus vesiculosus growing under salinity stress in the Baltic Sea. Degree project in biology. Biology Education Centre and Department of Plant Ecology. UPPSALA Universitet, 21 p. http://www.ibg.uu.se/digitalAssets/177/c_177012-l_3-k_sylvander-peter-arbete. pdf accessed 15 October 2018.

Szymański, P., T. Frączek, M. Markowicz and E. Mikiciuk-Olasik. 2012. Development of copper based drugs, radiopharmaceuticals and medical materials. Biometals 25(6): 1089–1112. https://dx.doi.org/10.1007%2Fs10534-012-9578-y accessed 22 September 2018.

Tacon, A.G.J. 1987. The Nutrition and Feeding of Farmed Fish and Shrimp—A Training Manual 1. The Essential Nutrients. Food and Agriculture Organization of The United Nations. FAO Brasilia, Brazil June 1987. http://www.fao.org/docrep/field/003/ab470e/AB470E00.htm#TOC.

Tait, T.N. 2013. Determination of Copper Speciation, Bioavailability and Toxicity in Saltwater Environments. Thesis Submitted to the Department of Chemistry in partial fulfillment of the requirements for the degree of Master of Science. Wilfrid Laurier University Waterloo, Ontario, Canada, 207 p. https://scholars.wlu.ca/cgi/viewcontent. cgi?referer=https://www.google.co.ve/&httpsredir=1&article=2699&context=etd accessed 02 September 2018.

Telegdi, J., L. Trif and L. Románszki. 2016. Smart anti-biofouling composite coatings for naval applications. Smart Composite Coatings and Membranes, 123–155. http://dx.doi.org/10.1016/b978-1-78242-283-9.00005-1 acceseed 29 August 2018.

Telesca, L., A. Belluscio, A. Criscoli, G. Ardizzone, E.T. Apostolaki, S. Fraschetti et al. 2015. Seagrass meadows (*Posidonia oceanica*) distribution and trajectories of change. Scient. Rep. Vol. 5, Article number: 12505. https://doi.org/10.1038/srep12505.

Taylor, L.N., J.C. McGeer, C.M. Wood and D.G. McDonald. 2000. Physiological effects of chronic copper exposure to rainbow trout (Oncorhynchus mykiss) in hard and soft water: Evaluation of chronic indicators. Environmental Toxicology and Chemistry 19(9): 2298–2308. https://doi.org/10.1002/etc.5620190920.

Tchounwou, P.B., C.G. Yedjou, A.K. Patlolla and D.J. Sutton. 2012. Heavy metals toxicity and the environment EXS. pp. 133–164. *In*: Luch, A. (ed.). Molecular, Clinical and Environmental Toxicology, Experientia Supplementum 101. https://doi.org/10.1007/978-3-7643-8340-4_6.

Tett, P., S. Benjamins, K. Black, M. Coulson, K. Davidson, F.T. Fernandes et al. 2018. Review of the environmental impacts of salmon farming in Scotland. Executive Summary and Main Report. Issue 01. The Scottish Parliament, 197 p. https://pure.uhi.ac.uk/portal/files/3214103/20180125_SAMS_Review_of_Environmental_Impact_of_Salmon_Farming_Report.pdf accessed 18 September 2018.

Thangam, Y., S. Jayaprakash and M. Perumayee. 2014. Effect of copper toxicity on hematological parameters to fresh water fish *Cyprinus Carpio* (Common Carp). IOSR J. Environ. Sci. Toxi. Food Tech. 8(9) Ver. I: 50–60. http://www.iosrjournals.org/iosr-jestft/papers/vol8-issue9/Version-1/H08915060.pdf.

Thomas, G. 2014. Effects of nanomolar copper on water plants–comparison of biochemical and biophysical mechanisms of deficiency and sublethal toxicity under environmentally relevant conditions. Dissertation submitted for the degree of Doctor of Natural Sciences. Univesitat de Konstant. https://d-nb.info/1113109394/34.

Thomas, O.R.B., N.C. Barbee, K.L. Hassell and S.E. Swearer. 2016. Smell no evil: Copper disrupts the alarm chemical response in a diadromous fish, *Galaxias maculatus*. Environ. Toxi. Chem. 35(9): 2209–2214. https://doi.org/10.1002/etc.3371 accessed 03 October 2018.

Tiam, S.K., V. Fauvelle, S. Morin and N. Mazzella. 2016. Improving toxicity assessment of pesticide mixtures: The use of polar passive sampling devices extracts in microalgae toxicity tests. Front Microbiol. 7: 1388. https://dx.doi.org/10.3389%2Ffmicb.2016.01388 accessed on 25 September 2018.

Tierney, K.B., D.H. Baldwin, T.J. Hara, P.S. Ross, N.L. Scholz and C.J. Kennedy. 2010. Aqua. Toxi. 96: 2–26. https://doi.org/10.1016/j.aquatox.2009.09.019 accessed 17 October 2018.

Tornero, V., A.M. Arias and J. Blasco. 2014. Trace element contamination in the Guadalquivir River Estuary ten years after the Aznalcóllar mine spill. Mar. Pollut. Bull. 15 86(1-2): 349–360. https://doi.org/10.1016/j. marpolbul.2014.06.044 accessed 20 October 2018.

Traganos, D. and Peter Reinartz. 2018. Interannual change detection of mediterranean seagrasses using rapid eye image time series. Front. Plant. Sci. 9: 96. https://dx.doi.org/10.3389%2Ffpls.2018.00096.

Tull, M., S.J. Metcalf and H. Gray. 2016. The economic and social impacts of environmental change on fishing towns and coastal communities: A historical case study of Geraldton, Western Australia. ICES J. Mar. Sci. 73(5): 1437–1446. https://doi.org/10.1093/icesjms/fsv196 accessed 10 October 2018.

Turner, A. and G. Millward. 2002. Suspended particles: Their role in estuarine biogeochemical cycles. Estuarine, Coast. and Shelf Sci. 55(6): 857–883. https://doi.org/10.1006/ecss.2002.1033.

Urien, N., E. Uher, E. Billoir, O. Geffard, L.C. Fechner and J.D. Lebrun. 2015. A biodynamic model predicting waterborne lead bioaccumulation in Gammarus pulex: Influence of water chemistry and *in situ* validation. Environmental Pollution, Elsevier 203: 22–30. https://hal.archives-ouvertes.fr/hal-01153660 accessed 19 October 2018.

USEPA (US Environmental Protection Agency). 1980. Ambient water quality criteria for copper. USEPA Report 440/5-80-036. 162 pp. https://nepis.epa.gov/Exe/ZyPURL.cgi?Dockey=P1007V43.TXT.

USEPA (US Environmental Protection Agency). 1999. 25 Years of the Safe Drinking Water Act: History and Trends. EPA-816-R-99-007. Washington, DC. 57 p. https://nepis.epa.gov/Exe/ZyPURL.cgi?Dockey=200027R1.TXT.

USEPA. 2004. Comments on Baseline Human Health Risk Assessment Report, Wells G&H Superfund Site, Aberjona River Study, Operable Unit 3, Woburn, MA, USEPA Region 1, March, 2003. 227 p. https://www3.epa.gov/region1/superfund/sites/industriplex/65233.pdf.

USEPA. 2007. Aquatic life ambient freshwater quality criteria: Copper, 2007 revision. EPA 822-R-07-001. https://nepis.epa.gov/Exe/ZyPURL.cgi?Dockey=P1007V43.TXT.

Van der Werf, H.M.G., J. Petit and J. Sanders. 2005. The environmental impacts of the production of concentrated feed: the case of pig feed in Bretagne. Agricultural Systems 83: 153–177. https://doi.org/10.1016/j.agsy.2004.03.005.

Van Zwieten, M., G. Stovold and L. Van Zwieten. 2007. Alternatives to Copper for Disease Control in the Australian Organic Industry. A report for the Rural Industries Research and Development Corporation, 24 p. https://rirdc.infoservices.com.au/downloads/07-110.pdf accessed 29 August 2018.

Veum, T.L., M.S. Carlson, C.W. Wu, D.W. Bollinger and M.R. Ellersieck. 2004. Copper proteinate in weanling pig diets for enhancing growth performance and reducing fecal copper excretion compared with copper sulfate. J. Anim. Sci. 82(4): 1062–1070. https://doi.org/10.2527/2004.8241062x accessed 22 September 2018.

Vilches, A., G.D. Pérez, J.C. Toscano and O. Macías. 2014. Lucha contra la contaminación. http://www.oei.es/decada/accion.php?accion=8 accessed 03 October 2018.

Vineeta, S., M. Dhankhar and J. Prakash. 2007. Copper removal from aqueous solution by marine green alga *Ulva reticulata* bioaccumulation of Zn Cu and Cd in *Channa punctatus*. J. Environ. Biol. 28: 395–397.

Viriyatum, R. 2013. Effectiveness of Coated, Controlled-Release Copper Sulfate as an Algicide for Phytoplankton Control in Ponds. A dissertation submitted to the Graduate Faculty of Auburn University in partial fulfillment of the requirements for the Degree of Doctor of Philosophy, 91 p. https://pdfs.semanticscholar.org/f04b/bd31e594da5a7a05d410530c5775e5dc28b6.pdf.

Wagner, J.L., A.K. Townsend, A.E. Velzis and E.A. Paul. 2017. Temperature and toxicity of the copper herbicide (NautiqueTM) to freshwater fish in field and laboratory trials. Cogent Environmental Science 3(1). https://doi.org/10.1080/23311843.2017.1339386.

Waite, R., M. Beveridge, R. Brummett, S. Castine, N. Chaiyawannakarn, S. Kaushik et al. 2014. Improving productivity and environmental performance of aquaculture. Working Paper, 60 p .http://www.aquacultuurvlaanderen.be/sites/aquacultuurvlaanderen.be/files/public/ImprovingAquaculture2014.pdf.

Walker, D.I. and A.J. McComb. 1992. Seagrass degradation in Australian Coastal Waters. Mar. Pollut. Bull. 25: 191–195. https://doi.org/10.1016/0025-326X(92)90224-T.

Wang, W.X. 2002. Interactions of trace metals and different marine food chains. Mar. Ecol. Prog. Ser. 243: 295–309. https://www.int-res.com/articles/meps2002/243/m243p295.pdf.

Wang, L., H.M. Espinoza and E.P. Gallagher. 2013. Brief exposure to copper induces apoptosis and alters mediators of olfactory signal transduction in coho salmon. Chemosphere 93(10): 2639–2643.

Wang, H., R. Sathasivam and J.S. Ki. 2017. Physiological effects of copper on the freshwater alga *Closterium ehrenbergii* Meneghini (Conjugatophyceae) and its potential use in toxicity assessments. Algae 32(2): 131–137. https://doi.org/10.4490/algae.2017.32.5.24 accessed August 25 2018.

Wartenberg, R., L. Feng, J.J. Wu, Y.L. Mak, L.L. Chan, T.C. Telfer et al. 2017. The impacts of suspended mariculture on coastal zones in China and the scope for integrated multi-trophic aquaculture. Ecosystem Health and Sustainability 3:6, 1340268. http://dx.doi.org/10.1080/20964129.2017.1340268.

Watson, C. and R.P.E. Yanong. 2011. Use of Copper in Freshwater Aquaculture and Farm Ponds. Fact Sheet FA-13. Department of Fisheries and Aquatic Sciences, Florida Cooperative Extension Service, Institute of Food and Agricultural Sciences, University of Florida. http://fisheries.tamu.edu/files/2013/09/Use-of-Copper-in-Freshwater-Aquaculture-and-Farm-Ponds.pdf.

Wells, M.L. and V.L. Trainer. 2016. International Scientific Symposium on Harmful algal blooms and climate change. L. PICES Press; Sidney Tomo 24, N.º 1. pp. 16–17. https://search.proquest.com/openview/784184b4d32d603190f6c7aab1709cac/1?pq-origsite=gscholar&cbl=666306.

Weltje, L., P. Simpson, M. Gross, M. Crane and J.R. Wheeler. 2013. Comparative acute and chronic sensitivity of fish and amphibians: A critical review of data. Environ. Toxi. Chem. 32(5): 984–994. https://doi.org/10.1002/etc.2149.

West, G. and R.J. Williams. 2008. A preliminary assessment of the historical, current and future cover of seagrass in the estuary of the Parramatta River. *In*: NSW Department of Primary Industries–Fisheries Final Report Series No. 98, 61 p.

WHO (World Health Organization). 1998. Copper. Environmental Health Criteria 200. IPCS-International Programme on Chemical Safety, WHO, Geneva.

WHO (World Health Organization). 2004. Copper in Drinking-water. Background document for development of WHO. Guidelines for Drinking-water Quality. WHO/SDE/WSH/03.04/88. 31 p. http://www.who.int/water_sanitation_health/dwq/chemicals/copper.pdf accessed September 2018.

WHOI. 2012. The Chemistry of Ocean Acidification, 4 p. http://www.whoi.edu/OCB-OA/page.do?pid=112136 accessed 05 October 2018.

Wichard, T., B. Charrier, F. Mineur, J.H. Bothwell, O. De Clerck and J.C. Coates. 2015. The green seaweed *Ulva*: A model system to study morphogenesis. Front. Plant. Sci. 6: 2. https://dx.doi.org/10.3389%2Ffpls.2015.00072.

Williams, A. 2018. Comments on the Proposed Pebble Mine in the Bristol Bay Region of Southwest Alaska, 65 p. https://static1.squarespace.com/static/55319e94e4b02842e0615731/t/5b898f7c575d1fcb79b0283f/1535741828299/TU+Scoping+Comments+to+Corps+-+20180628+Final+with+Attachments.pdf accessed 10 October 2018.

Willsher, J. 2007. The Effect of Biocide Free Foul Release Systems on Vessel Performance, 6 p. https://www.ship-efficiency.org/onTEAM/pdf/WILLSHER.pdf.

Winemiller Kirk, O. 1990. Spatial and temporal variation in tropical fish trophic networks. Ecol. Monographs 60(3): 31–367. https://doi.org/ 10.2307/1943061.

Witeska, M., P. Sarnowski, K. Ługowska and E. Kowal. 2014. The effects of cadmium and copper on embryonic and larval development of ide *Leuciscus idus* L. Fish Physiol. Biochem. 40(1): 151–163. https://dx.doi.org/10.1007%2Fs10695-013-9832-4.

Wong, B.B.M. and U. Candolin. 2015. Behavioral responses to changing environments. Behavioral Ecology 26(3): 665–673. https://doi.org/10.1093/beheco/aru183 accessed 03 October 2018.

Wood, C.M., H.A. Al-Reasi and D.S. Smith. 2011. The two faces of DOC. Aquatic Toxicology Vol.105 Special Issue: SI Supplement: 3–4 Pages: 3–8 https://doi.org/10.1016/j.aquatox.2011.03.007.

Woody, C.A. and S.L. O'Neal. 2012. Effects of Copper on Fish and Aquatic Resources. The Nature Conservancy, 27 p. http://www.pebblescience.org/pdfs/2012-December/16%20June%202012_FINAL_%20Effects%20of%20Copper%20on%20Fish.pdf 01 Juin 2018.

Wu, W.H., G. Wei, X. Tan, L. Li and M. Li. 2017. Species-dependent variation in sensitivity of *Microcystis* species to copper sulfate: implication in algal toxicity of copper and controls of blooms. Sci. Rep. 7(1): 40393. https://dx.doi.org/10.1038/srep40393 accessed August 25 2018.

Xiao, L. 2014. Influence of Surface Topography on Marine Biofouling. Thesis Doctor of Natural Sciences. Combined Faculty of Natural Sciences and Mathematics at the Ruprecht-Karls University of Heidelberg, Germany, 128 p. https://core.ac.uk/download/pdf/32584427.pdf accessed 15 October 2018.

Xu, X., S. Pin, M. Gathinji, R. Fuchs and Z.L. Harris. 2004. Aceruloplasminemia: An inherited neurodegenerative disease with impairment of iron homeostasis. Ann. N Y Acad. Sci. 1012(1): 299–305. https://doi.org/10.1196/annals.1306.024.

Yan, X., M. Liu, J. Zhong, J. Guo and W. Wu. 2018. How human activities affect heavy metal contamination of soil and sediment in a long-term reclaimed area of the Liaohe River Delta, North China. Sustainability 10: 338, 19 p. https://doi.org/10.3390/su10020338.

Yanong, R.P.E. 2013. Use of Copper in Marine Aquaculture and Aquarium Systems. FA165, 5 p. http://edis.ifas.ufl.edu/pdffiles/FA/FA16500.pdf accessed 01 September 2018.

Yiğit, M., M. Osienski, J. DeCew, B. Çelikkol, O.S. Kesbiç, M. Karga et al. 2018. Construction, assembly and system deployment of a fish cage with copper alloy mesh pen: challenging work load and estimation of man-power. Aqua. Res. 1(1): 38–49. https://doi.org/10.3153/AR18005.

Ying, F., Z. Xiao-dong, S. De-shuai and L. Xin. 2018. Release rate of cuprous oxide in seawater. China Environ. Sci. 38(05). Accessed on 25 September 2018.

Younger, P.L., S.A. Banwart and R.S. Hedin. 2002. Mine Water: Hydrology, Pollution, Remediation. Kluwer Academic Publishers, Dordrecht. (ISBN 1-4020-0137-1). 464 pp. https://doi.org/10.1007/978-94-010-0610-1.

Ytreberg, E., M.A. Bighiu, L. Lundgren and B. Eklund. 2016. XRF measurements of tin, copper and zinc in antifouling paints coated on leisure boats. Environ. Pollu. 213: 594–599. https://doi.org/10.1016/j.envpol.2016.03.029 accessed 15 October 2018.

Zaharudin, N. 2017. Seaweed Bioactivity. Effects on Glucose Liberation. Ph.D. Thesis. University of Copenhagen. Faculty of Science, 155 p.

Zhang, Y., W. Zang, L. Qin, L. Zheng, Y. Cao, Z. Yan et al. 2017. Water quality criteria for copper based on the BLM approach in the freshwater in China. PLoS One 12(2): e0170105. https://dx.doi.org/10.1371%2Fjournal.pone.0170105 accessed on 25 Sept 2018.

Zhao, Q., A.N. Chen, S.X. Hu, Q. Liu, M. Chen, L. Liu et al. 2018. Microalgal microscale model for microalgal growth inhibition evaluation of marine natural products. Scient. Rep. 8(1). https://doi.org/10.1038/s41598-018-28980-z acccessed 01 October 2018.

Zou, D. and K. Gao. 2010. Physiological responses of seaweeds to elevated atmospheric CO_2 concentrations. pp. 115–126. *In*: Seckbach, J., R. Einav and A. Israel (eds.). Seaweeds and their Role in Globally Changing Environments. Springer, Dordrecht. https://doi.org/10.1007/978-90-481-8569-6_7.

6

Behavioral Responses of Marine Animals to Metals, Acidification, Hypoxia and Noise Pollution

Judith S. Weis

1. Introduction

Human activities have been changing ocean chemistry, both in coastal waters and in the open ocean. Many decades of pollution, accompanied by overfishing, bottom trawling and other kinds of habitat destruction, are having devastating effects on the marine environment. Increasing demand for seafood around the world has depleted commercial fish populations and damaged the economy of many coastal communities. Furthermore, climate change is altering the oceans in major ways that we are just beginning to understand.

Land-based sources discharge nutrients, sediments, pathogens, and thousands of toxic chemicals, including metals, pesticides, industrial products, and pharmaceuticals, into estuaries and coastal waters. Materials are released into the environment from chemical industries, sewage treatment plants, and agriculture, eventually reaching marine ecosystems. Highly visible disastrous events such as the *Exxon Valdez* and the Gulf of Mexico "gusher" raised public awareness of some types of marine pollution, but chronic pollution that does not get press coverage is less well known. There is growing scientific evidence demonstrating serious, and occasionally disastrous, effects of pollution in the marine environment. The kinds of pollutants that are of greatest concern are those that are widespread and persistent in the environment, accumulate in biota, and produce toxic effects at low concentrations. Toxic chemicals are varied and often difficult to detect. In recent years, attention is being devoted to newly recognized threats to the environment—such as ocean acidification, hypoxia, and noise.

Most pollutants of the marine environment come from the land. While the source of many pollutants is industrial or residential areas, others are produced in agricultural areas. Factories and

Dept. of Biological Sciences, Rutgers University, Newark, NJ 07102.
Email: jweis@newark.rutgers.edu

sewage treatment plants discharge wastes into receiving waters via a pipe—a "point source"—that can be readily monitored and regulated by governmental agencies. Recently, concern has moved from end-of-pipe discharges to more diffuse pathways, such as runoff and atmospheric deposition. The sources of contaminants that wash into the water during rainfall are diffuse and enter the water in many places, as do pollutants from the atmosphere that come down during rainfall. This diffuse "non-point source" pollution is not easily regulated. Non-point sources, such as farms, roadways, and urban or suburban land, remain largely uncontrolled and are significant sources of continuing inputs of pollution.

In addition to toxic chemicals, oceans are subject to ocean acidification from carbon dioxide dissolving from the atmosphere, to hypoxia, and to noise pollution from human activities near or in the ocean.

Behavior is a very sensitive response to stresses, including pollution. Noticeable changes in behavior can be found at low concentrations of chemicals, often lower than concentrations affecting other biomarkers, such as biochemistry. Since behavior links physiological and other individual-level responses to ecological processes, it is a very important type of response. In addition to high sensitivity, behavior responses are likely to occur in nature (Scott and Sloman 2004) and can cause ecological effects at the population and community level. Much of the early behavioral studies focused on avoidance, tremors, or coughs, but more recently complex behaviors such as predator/prey interactions and burrowing, reproductive, and social behaviors have been studied. These behaviors are more relevant to ecological impacts. Most animals respond to most contaminants with a reduction in feeding. Decreased food intake places energetic demands on the organism, which may be responsible for decreases in other physiological functions (e.g., respiration). Often, when food intake is reduced, animals reduce their activity in order to conserve energy; this in turn may make it harder to find and get food—which intensifies the problem.

2. Metals

Metals from industrial activities and mining are among the major contaminants in coastal environments. They accumulate in sediments and in organisms. Among the metals of greatest concern are mercury, cadmium, copper, zinc, and silver. Since mercury is present in coal and is released into the atmosphere when coal is burned, it can be transported in the atmosphere for long distances before being deposited, in many cases, far from its source. While some metals (e.g., copper and zinc) are essential in low concentrations for life, most metals play no normal biological role. While most metal contaminants come from sources on land, some metals are used in anti-fouling paints for ships. Since fouling organisms can accumulate on ship bottoms (and thus reduce streamlining and increase fuel consumption), antifoulant coatings have been developed for treating ship hulls. Paints containing copper have been used for many years. Starting in the 1940s, organotin-based paints were developed and are among the most effective and long-lasting. Tributyltin, which is one of the most effective, is also extremely toxic to non-target organisms.

The chemical form of the metal plays an important role in its toxicity. In aquatic environments, copper exists in particulate, colloidal and soluble states, predominantly as metallic (Cu^0) and cupric copper (Cu^{2+}). It forms complexes with both inorganic and organic ligands. The toxicity of copper is directly associated with the free ion, as is the toxicity of Cd, so measurements of total Cu or total Cd in the water overestimate the amount that is bioavailable (Sunda et al. 1978, Sunda and Lewis 1978).

Mercury is a highly toxic element. Although its potential for toxicity in highly contaminated areas such as Minamata Bay, Japan, in the 1950s and 1960s, is well documented, mercury can also be a threat to the health of people and marine life in environments that are not so obviously polluted. Its risk depends on the form of mercury and the geochemical and biological factors that influence how it moves and changes in the environment. Mercury can exist in three oxidation states in the water:

Hg^0, Hg^{1+} and Hg^{2+}. The distribution of these different forms of mercury depends on the pH, redox potential, and availability of anions to form complexes with it. Furthermore, inorganic mercury can be transformed into organic mercury compounds by bacteria in the environment and in organisms. Methylmercury (meHg) is the most toxic form, and inorganic Hg can be converted to meHg by bacteria in marine sediments. Bacteria capable of methylating Hg^{2+} have been found in sediments, water, and fish tissue. However, little is known about the physiology and the mechanisms that control methylation. MeHg is not only far more toxic than inorganic mercury, it also is biomagnified up the food chain, so tissue concentrations increase as it passes up to higher trophic levels. Humans are exposed to meHg primarily by eating fish that are fairly high on aquatic food chains.

Another organometal of concern is tributyltin (TBT), used in antifouling paints for ships. Unlike Hg, TBT breaks down in the environment, losing its butyl groups over time, and becoming less toxic as it becomes dibutyltin, monobutyltin, and eventually inorganic tin, which is not toxic. However, the breakdown is not as rapid as was originally thought, so the toxic effects of butyltins can persist for some years.

Metals bind to bottom sediments, from where they are to some degree available to marine organisms, particularly benthic ones, which accumulate the metals, and from which the metals can be moved up the food chain. Bioavailability of metals bound to sediments is a critical issue for their effects on organisms. Acid volatile sulfide (AVS) can be used to predict the toxicity of divalent metals in sediments, including copper (Cu), cadmium (Cd), nickel (Ni), lead (Pb) and zinc (Zn) (Ankley et al. 1996, Berry et al. 1996). The rationale behind this is that the AVS in sediment reacts with the simultaneously extracted metal (the reactive metal fraction that is measured in the cold acid extract). This reaction forms an insoluble metal sulfide that is considered non-available for uptake by benthic animals. Estuarine sediments tend to have high levels of sulfide, and therefore there is relatively low bioavailability of sediment-bound metals. Ironically, a higher oxygen level in overlying water, which is otherwise good for organisms, increases the redox potential of the sediment and decreases AVS, thus increasing the availability of metals in the pore water of the sediment. Thus, increased oxygen from water quality improvements can increase metal mobility and may cause sediment-bound metals to leach into the overlying water. In contrast, prolonged hypoxia promotes the release of iron and manganese from contaminated estuarine sediments (Banks et al. 2012).

2.1 Crustaceans

Activity

Studies of effects of low concentrations of metals on behavior of crustacean larvae were started in the early 1970s by the Vernberg group. Fiddler crab (*Uca pugilator*) zoeae exposed to 18 μg L^{-1} $HgCl_2$ exhibited reduced activity levels and erratic spiral swimming, swimming on their sides or darting up from the bottom, then settling slowly back down (DeCoursey and Vernberg 1972). Their metabolic rate (O_2 consumption) was also reduced. Mud crab (*Eurypanopeus depressus*) zoeae were exposed to 10 μg Cd L^{-1} or 1.8 μg Hg^{-1}. Cd exposure increased the swimming rates of the older stages, while Hg depressed swimming rates of the early stages (Mirkes et al. 1978). Harayashiki et al. (2016) investigated effects of dietary exposure of mercury (0.56 μg g^{-1} and 1.18 μg g^{-1}) on post-larvae of the shrimp *Penaeus monodon* over 96 h to evaluate changes in behavior (swimming activity and risk taken) and biochemical biomarkers. Results showed a decrease in swimming activity with increased mercury exposure, but no changes were observed in risk taken, as measured by the time spent in the inner part of the chamber (Fig. 1).

Nauplius larvae of barnacles *Balanus improvisus* initially increased their swimming speeds at 20–80 μg Cu L^{-1}, but after 72 h of exposure, their swimming speed was depressed in all concentrations (Lang et al. 1981). This may be an example of hormesis (cases where low levels stimulate activity, but higher levels reduce activity). Phototactic behavior (swimming toward the light) was also altered

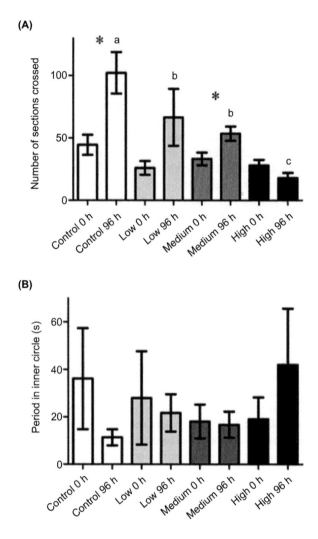

Fig. 1. (a) Number of sections crossed at beginning of experiment (0 h), and after 96 h in shrimp exposed to low, medium, and high levels of dietary mercury. (b) Time spent by shrimp in inner part of circular chamber (risk taking). From Harayashiki et al. (2016). Courtesy of Elsevier Inc.

at higher concentrations of copper. Likewise, exposure to Cd at 50–100 μg L^{-1} initially increased swimming speed but subsequently depressed it at 100 μg L^{-1}. Sullivan et al. (1983) determined that the swimming speed of nauplii of the copepod *Eurytemora affinis* was affected by Cu at concentrations greater than 10 μg L^{-1}; their escape responses were depressed, and thus, they were more rapidly captured by larval striped bass. Adult crustaceans have also been the subject of many behavioral investigations. Roast et al. (2001) studied effects of 7 d of exposure to 0.5 and 1.0 μg L^{-1} Cd^{2+} on swimming speed of the mysid *Neomysis integer* in a flume. The lower concentration caused fewer mysids to move into the current (normal behavior) at low current speeds. When in the current, some animals maintained their position while others were swept away. At the higher concentration of Cd more mysids spent more time in the water column. Being in the water column instead of on the bottom could displace them from their optimum habitat in an estuary. Depressed locomotion was seen in grass shrimp (*Palaemonetes pugio*) exposed to 0.56 but not to 0.3 mg L^{-1} Cd (Hutcheson et al. 1985). Wallace and Estephan (2004) found that while control amphipods (*Gammarus lawrencianus*) were active 61% of the time, the time spent in horizontal swimming decreased to only 0.3% of the time in 125 and 500 µg Cd L^{-1}. Their vertical swimming was also reduced by cadmium, with significant decreases seen at 12 µg L^{-1} and 62 µg L^{-1}. Vertical swimming was more sensitive, presumably because it takes more energy to swim upward rather than sideways.

Burrowing

Burrowing into the substrate is protective behavior for many animals. An animal on the surface is more visible and vulnerable to predation than one burrowed under the surface of the substrate. Reduced burrowing in contaminated sediments may be due to avoiding contact with contamination, which is protective, or could reflect abnormal behavior. Not all studies distinguish between these different mechanisms. A good study ought to provide animals with both clean and contaminated sediments to burrow in. Burrowing by the isopod *Saduria entomon* decreased in metal-contaminated sediments 10 µg g^{-1} dry sediment Cu, 35 µg g^{-1} Cd, or 299 µg g^{-1} Fe). This was an example of avoidance, as animals burrowed normally in clean sediments. Pre-exposure at levels similar to those of contaminated parts of the Baltic Sea impaired the animals' ability to avoid contaminated sediments and also reduced their feeding rate (Pynnönen 1996). The amphipod *Rhepoxinius* avoided burrowing in sediments with levels of sewage, Zn, or Cd typical of waste-water discharge sites (Oakden et al. 1984).

Predator/prey behaviors

Reduced food consumption is a very common response of crustaceans to many toxicants (Maltby and Crane 1994, Wallace et al. 2000). Chronic exposure to Cu (> 85 µg L^{-1}) and Zn (> 106 µg L^{-1}) reduced growth of shrimp larvae (*Farfantepenaeus paulensis*) as a result of decreased feeding (Santos et al. 2000). Both metals reduced the number of brine shrimp, *Artemia,* consumed by the larval shrimp in 30 min. Oxygen consumption also was reduced by about 30% in all metal concentrations. Similarly, the gut fullness of larvae of the shrimp *Metapenaeus ensis* feeding on the diatom *Gracilis* was diminished by a 2 h exposure to Cu at 0.25 mg L^{-1} (Wong et al. 1993). In contrast, gut fullness of *M. ensis* larvae was not altered after up to 24 h of exposure to Cr or Ni at concentrations close to the 48 h LC$_{50}$ (5.41 and 1.28 mg L^{-1}, respectively). However, postlarval shrimp exposed for 24 h to those concentrations of Cr, Cu, or Ni consumed fewer *Artemia* nauplii. While younger stages are generally more sensitive to contaminants than older stages, this obviously depends on the species and pollutant.

Blue crabs, *Callinectes sapidus*, were fed grass shrimp *Palaemonetes pugio* contaminated with 1.8 µg TBT, 0.09 µg dibutyltin and 0.03 µg monobutyltin g^{-1} wet weight tissue. Feeding rates for exposed and control crabs were equivalent during the 16 d test (Rice et al. 1989). Growth, molting success and feeding rates were also not affected. Catabolism reduced tissue concentrations of TBT, thereby increasing the tolerance of the blue crabs to TBT. Wallace et al. (2000) investigated prey capture by grass shrimp (*P. pugio*) after being fed Cd-contaminated prey (field-exposed oligochaetes or laboratory-exposed *Artemia salina*). Shrimp that had been fed contaminated *A. salina* (with Cd body burdens of 4, 16, and 40 mg g^{-1} wet weight, respectively, which would correspond to shrimp ingesting about 0.08, 0.32, or 0.80 mg Cd d^{-1}, if they consumed all of the food) demonstrated impaired ability to capture prey; after feeding on contaminated oligochaetes effects were not as severe, perhaps because the Cd in the oligochaetes was less available, stored in insoluble granules. Shrimp exposed to Cd produced a low molecular weight Cd-binding metallothionein protein; their prey capture ability decreased along with increased amounts of Cd bound to high molecular weight proteins, i.e., enzymes.

Treatment with inorganic Hg or meHg (0.01 mg L^{-1}) affected predator avoidance ability of *Palaemonetes pugio* (Kraus and Kraus 1986). While exposure caused shrimp from a reference site to be more vulnerable to predation by *F. heteroclitus*, shrimp from a contaminated site (Piles Creek in New Jersey, USA) were unaffected by HgCl$_2$ treatment and were less affected by meHg treatment than the reference population, suggesting that the shrimp in the polluted environment had acquired some tolerance to Hg. Brief exposure of the copepods *Notodiaptomus conifer* and *Argyrodiaptomus falcifer* to Cu and Cr (15 min) affected their escape behavior in an unexpected way. Exposed copepods had higher escape ability than controls (Gutierrez et al. 2012). This may have been an example of hormesis, which might have turned into reduced escape ability after a longer exposure time, but this study did not include longer exposures.

Reproductive behavior

Effects of Cu (0, 0.1 or 0.5 mg Cu(II) L^{-1} for 5 d) on mating behavior of male shore (green) crabs, *Carcinus maenas,* were investigated by Krang and Ekerholm (2006). Cu exposure altered the response to a pheromone (pre-molt female urine) presented alone or presented together with a "dummy" female (a sponge injected with pre-molt female urine) or with a real female *C. maenas.* Male crabs that had been exposed to the highest Cu concentration took more than twice as long to initiate search activity following the pheromone introduction and their searching behavior was less directed. When presented with a "dummy" female, they had reduced responses in both low and high Cu treatments. Stroking behavior was reduced, and those exposed to higher Cu often pinched the dummy female or the real female, and took longer to establish cradle-carrying behavior, which normally precedes mating. Therefore, the Cu exposure affected the males' ability to detect female pheromones, perform specific mating behaviors, and form pairs.

2.2 Mollusks

Activity level

Valve closing speed of juvenile Catarina scallops (*Argopecten ventricosus*) exposed to Cd (0.02, 0.1, 0.2 mg L^{-1}), Cr (0.1, 0.5, 1.0 mg L^{-1}), or Pb (0.2, 0.4 mg L^{-1}) and metal mixtures was studied by Sobrino-Figueroa and Cáceres-Martínez (2009). The average closing speed was under 1 sec in controls, but 2–3.6 sec in Cd, 1.4–3.4 sec in Cr, 3–12 sec in Pb, and 12–15 sec in the mixtures. Therefore, all three metals retarded closing at all the concentrations. In combination, effects of the metals appeared to be additive.

Octopuses were exposed to mercuric chloride to investigate effects on locomotor response and expansion of chromatophores (Adams et al. 1988). A dose-sensitive relationship was seen for the loss and subsequent recovery of locomotor response and chromatophore expansion in *Octopus joubini, O. maya* and *O. bimaculoides* following exposure to mercuric chloride. For each species of octopus, the LC_{50} for chronic exposure was 1,000 mg L^{-1} for 3 h of exposure.

Burrowing

Burial by the bivalve *Scrobicularia plana* in clean sediments was reduced after a 4 d exposure to 25 to 150 µg Cu L^{-1} (Bonnard et al. 2009). At the end of exposure, the burrowing kinetics in clean sediment were determined after 1 and 2 d. Even the lowest tested concentration reduced burying. Other reports document reduced burying in 5.8 µg g^{-1} Cu-contaminated sediments (avoidance) by the clams *Protothaca staminea* (Phelps et al. 1985) and *Macomona liliana* (Roper and Hickey 1994), as well as by the gastropod *Polinices sordidus* (Hughes et al. 1987). Cd exposure also reduced burying in the short-neck clam *Ruditapes philippinarum.* Clams took the longest time to burrow in sediments from sites with the highest concentrations of Cd, Cr, Cu, Ni and Zn (Shin et al. 2002) There was reduced burrowing when sediment Cd concentrations were 1 mg kg^{-1} (dry weight, dw) or higher. Juveniles of the clam *Macomona liliana* were exposed to Cu- and Zn-dosed sediments and sediments that had been collected from contaminated sites. The number of clams burying by 10 min was reduced at 25 mg Cu kg^{-1} (dw). After 96 h exposure, clams crawled away from sediment containing 10 mg Cu kg^{-1} (dw), and when a weak current was provided they left this sediment by drifting

(Roper et al. 1995). Zn-dosed sediment slowed clam burial at 80 mg Zn kg^{-1} (dw), and stimulated crawling and drifting at 40 mg Zn kg^{-1} (dw). Some field-collected contaminated sediments slowed the burial rate and stimulated drifting away, which could affect the distribution of these clams in natural habitats. Buffet et al. (2015) examined effects of CdS quantum dots (nanoparticles) vs soluble Cd on the marine clam *Scrobicularia plana* exposed for 14 d to these nanomaterials (10 µg Cd L^{-1}). Measurement of Cd released from CdS quantum dots indicated that 52% of CdS quantum dots remained in the nanoparticulate form. While clams accumulated comparable levels of Cd from both forms, their behavior in sediment showed impairments of foot movements in those exposed to CdS quantum dots but not soluble Cd.

Predator/prey behavior

Feeding behavior of snails and bivalves has been studied in relation to metal contaminants. Feeding of the snail *Potamopyrgus antipodarum* was studied by Alonso and Valle-Torres (2018). They found percentage of actively feeding animals was reduced by the four cadmium treatments (0.009, 0.026, 0.091 and 0.230 mg Cd/L) with stronger effects in the highest concentrations.

Filtration rates of the mussel *Mytilus edulis* and clam *Mya arenaria* were reduced by exposure to chromium (1 mg L^{-1}), to sediments from a Cr-contaminated site, or to artificial sediments enriched in Cr (Capuzzo and Sasner 1977). Both dissolved and particulate Cr (CrCl$_3$) reduced the rate of filtration; *Mya* was less affected than *Mytilus* by particulate Cr. *M. edulis* were exposed for 5 d to Cu (18–56 µg L^{-1}). While molecular biomarkers were affected at concentrations of 56 µg L^{-1} Cu, the filtration rate showed a significant decrease at the much lower concentration of 18 µg L^{-1} (Al-Subiai et al. 2011), supporting observations that feeding is a more sensitive response than biochemical biomarkers.

Mollusks are particularly sensitive to copper, which is used as a molluscicide. Effects of 20 µg Cu^{2+} L^{-1} and lowered salinity (20 psu) were studied on the grazing snail *Trochus maculatus* provided with the macroalgae *Gracilaria tenuistipitata* and *Enteromorpha intestinalis* (Elfwing and Tedengren 2002). The two factors were applied both separately and in combination to evaluate interactions. Results indicated that moderate salinity reduction and ecologically relevant amounts of Cu reduced the grazing rate of the snails but not the productivity of the algae, and therefore could promote algal growth and potential dominance on coral reefs.

2.3 Fishes

Activity level

MeHg (10 µg L^{-1}) reduced activity level and swimming performance in mummichogs, *Fundulus heteroclitus* (Weis and Khan 1990, Zhou and Weis 1998), gobies *Pomatoschistus microps* (Viera et al. 2009), and Atlantic croaker. Alvarez et al. (2006) fed adult croakers (*M. undulatus*) meHg-contaminated food for a month, induced spawning, and analyzed swimming speed and startle response of the larvae. Maternally transferred meHg impaired these behaviors, which are considered survival skills, in the next generation.

Cu exposure can either increase or decrease activity, depending on the concentration and species. Swimming velocity of silversides (*Menidia menidia*) was increased after a short exposure to 100 µg L^{-1} (Koltes 1985), but 50 µg L^{-1} reduced swimming speed in the goby *Pomatoschistus microps* (Viera et al. 2009). Scarfe et al. (1982) found that 72 h exposure to 0.1 mg ml^{-1} Cu decreased activity level in Atlantic croaker (*M. undulatus*) and pinfish (*Lagodon rhomboids*) but increased the activity of sheepshead (*Archosargus probatocephalus*) and sea catfish (*Ariopsis felus*). Exposure of larval sand gobies (*Pomatoschistus minutus*) to Cu at 1, 10, or 100 µg L^{-1} resulted in hyperactivity (Asnicar et al. 2018).

Newly hatched *F. heteroclitus* larvae were exposed to 0.1, 0.3, or 1.0 mg L^{-1} Pb, which reduced spontaneous activity and swimming stamina after 1 wk, but when larvae were subsequently returned to clean sea water for 4 wk, behaviors were no longer statistically different from controls, showing that effects were reversible (Weis and Weis 1998).

Behavioral development occurs in association with the development of the nervous system and early life stages are generally more sensitive to contaminants than adults. Embryonic exposures to chemicals can affect subsequent predator/prey and other behaviors later in life. Early life stages may be exposed to contaminants passed on from females via the egg as well as directly from the water and food. Understanding the physiological mechanisms that underlie effects on behavior early in life has not received much attention, possibly because physiological measurements are difficult to perform on small specimens. The physiology underlying behavioral disruption in early life stages is similar to that in juveniles and adults: sensory impairment, altered neurogenesis and altered neurotransmitters (Sloman and McNeil 2012).

Predator/prey behavior

Feeding of mummichog larvae was examined after embryonic exposure to 5 or 10 μg L^{-1}. *Fundulus heteroclitus* larvae were maintained in clean water. Feeding by early larvae was reduced by the embryonic mercury exposure, but after about 1 wk after hatching the feeding rate was comparable to controls, showing that this effect was temporary (Weis and Weis 1995a). The exposure may have caused retardation of development of the nervous system that was later compensated for. After mercury exposure of both embryos and larvae, deleterious effects on feeding of *F. heteroclitus* were more severe than after exposure of only embryos, i.e., lower concentrations could cause reduced feeding (Zhou et al. 2001).

Detection of predators and of prey by fish is often visual or olfactory. Some species produce an alarm substance that warns conspecifics of danger. Injured fish release a chemical that alerts other fish. Individuals with impaired ability to detect an alarm substance would be at greater risk of predation (Sandahl et al. 2004). Copper and some other metals can suppress the olfactory responses and reduce the response of salmon to water-borne alarm substances, thus making them more vulnerable to predation (McIntyre et al. 2012). Juvenile coho salmon generally freeze when they detect alarm substances, making it harder for predators to detect them. However, salmon in water with 5 μg L^{-1} Cu failed to detect the alarm substance and kept swimming; they were readily attacked by predators (McIntyre et al. 2012). Copper and some other metals can suppress the olfactory response to amino acids that are detected as food odors by coho salmon (Sandahl et al. 2004).

Weis and Khan (1990) found that exposure of adult mummichogs (*F. heteroclitus*) to 10 μg L^{-1} of either HgCl$_2$ or meHg for 1 wk reduced their prey capture ability. Sticklebacks (*Gasterosteus aculeatus*) exposed to 3 μg L^{-1} TBT swam in more exposed regions of the water column and had longer latency times before performing antipredator behavior in response to a simulated predator (a fake heron bill) (Wibe et al. 2001), thereby increasing their predation risk. TBT is also able to alter predatory behavior. Yu et al. (2013) examined effects of 10, 100 and 1000 ng L^{-1} TBT on prey capture ability of *Sebastiscus marmoratus*. TBT depressed predatory activity after 50 d of exposure. Accompanying the behavioral changes, dopamine levels in the fish brains increased in a dose-dependent manner and 5-hydroxytryptamine and norepinephrine levels decreased in the exposed group relative to the control.

Fundulus heteroclitus exposed to 5 and 10 μg L^{-1} meHg as embryos and subsequently reared in clean water had impaired prey capture (see above) and predator avoidance. Larvae that had been exposed as embryos were more susceptible to predation by adult *Palaemonetes pugio* or by adult mummichogs (Weis and Weis 1995b) and had increased activity levels, making them more susceptible to predation (Zhou and Weis 1998). After both embryonic and larval exposure, deleterious effects were greater than after embryonic exposure alone (Zhou et al. 2001).

Newly hatched *F. heteroclitus* larvae were exposed to 0, 0.1, 0.3, or 1.0 mg L^{-1} Pb and tested for prey capture (*Artemia*) and predator avoidance. Prey capture was significantly decreased after 4 wk and susceptibility to predation by grass shrimp increased. However, when larvae were returned to clean sea water for another 4 wk these behaviors were no longer statistically different from controls (Weis and Weis 1998).

Reproductive behavior

Successful reproduction in fishes requires the performance of behaviors that may include selection of spawning site, nest building, courtship and spawning, and later behaviors such as nest guarding and fanning, depending on the species. Contaminants can disrupt any of these behaviors and decrease reproductive success. However, despite considerable research on pollution and fish behavior, there are relatively few articles that focus on reproductive behavior (Jones and Reynolds 1997), and of these, most are on freshwater species (guppies, fathead minnows, mosquitofish, cichlids) that tend to have more complex behaviors than most marine fishes. Nest-building and courtship in male sticklebacks, which live in both fresh and salt water, have been the most studied, along with gobies.

Matta et al. (2001) found that dietary methylmercury (0.2–11 μg g^{-1}) altered male behavior in killifish (*F. heteroclitus*), increasing aggression in some individuals and lethargy in others. Furthermore, their offspring were less able to reproduce successfully and had an altered sex ratio. This was one of the first papers that discussed "personality" difference among different individual fish.

Social behavior

Schooling behavior is a social interaction, in which fish of the same size and same species swim close together and at the same speed, in a coordinated fashion. Schooling, which decreases susceptibility to predation, can be altered by exposure to a variety of contaminants. Copper (100 μg L^{-1}) exposure of Atlantic silversides *Menidia menidia* caused them to become hyperactive and increased the school cohesion (Koltes 1985). Social interactions were impaired in larval *F. heteroclitus* after they had been exposed to 5 and 10 μg meHg L^{-1} as embryos (Ososkov and Weis 1996). Treated larvae collided into one another more frequently. Young-of-the-year bluefish *Pomatomus saltatrix* that were fed contaminated diets (killifish and juvenile menhaden, *Brevoortia tyrannus*, collected from contaminated Hackensack Meadowlands of New Jersey, an estuary with multiple contaminants including Hg and PCBs) disrupted their normal schooling behavior more often than fish fed the same prey species collected from a cleaner estuary (Candelmo et al. 2010).

Homing behavior

Olfaction is vital for the migration of adult salmonids from the ocean to their natal river to spawn. If the olfactory sense is impaired by toxicants, this homing behavior can be affected. Smolting is the juvenile stage specialized for downstream migration, seawater entry, and marine residence; smolting is controlled by hormones and includes many physiological and behavioral changes that take place in fresh water and that prepare smolts for their migration into marine waters (McCormick et al. 1998). As mentioned previously, copper is particularly damaging to the olfactory system, critical for migration and homing in salmon. Saucier et al. (1991) found that after exposure to 22 ug L^{-1} copper, salmonids could no longer discriminate among different types of water. When controls were given a choice between their own rearing water or either well water or heterospecific water, they clearly preferred their own rearing water, but exposed fish showed no preference. These behavioral responses of copper-exposed fish indicate impairment of their olfactory discrimination ability.

2.4 Other taxa

Polychaete worms reduce their burrowing in contaminated sediments and, after exposure to contaminants, reduce their burrowing in clean sediments. Exposure to 25 µg Cu L^{-1} reduced burrowing in *Nereis diversicolor* (Bonnard et al. 2009). Behavioral impairments were not related to acetylcholinesterase inhibition but may have been due to metabolic or physiological disturbances. In a field study, Diaz-Jaramillo et al. (2013) translocated polychaetes *Perinereis gualpensis* from a reference site (Raqui estuary, Chile) to an estuarine site with significant sediment Hg concentrations (Lenga estuary: 1.78–9.89 mg/kg). Individual worms were exposed in polluted and non-polluted sediments for 21 d and sampled every 7 d with cages deployed at three different depths. Differences in burrowing behavior were observed; control polychaetes exhibited more homogenous vertical distributions, whereas in polluted Lenga, worms reduced their burrowing and tended to remain in upper layers.

Colvin et al. (2016) investigated effects of water flow velocity (6 or 15 cm/s) and Cu (0 or 85 µg L^{-1}) on feeding, growth, and Cu accumulation in juvenile polychaetes *Polydora cornuta*. Animals had best growth in the faster flow and in the absence of Cu. In the slow flow, the time spent feeding decreased as Cu increased, but Cu did not significantly affect feeding time in the faster flow. In all treatments, there was a direct relationship between time spent feeding and growth rate. Worms used suspension feeding more often in the fast flow and used deposit feeding more often in the slow flow, but Cu did not affect the proportion of time spent in either feeding mode.

Feeding of coral polyps that normally feed on zooplankton can be suppressed by metals. Peng et al. (2004) found that Cu-exposed corals *Subergorgia suberosa* were unable to catch or consume brine shrimp effectively. While the rate of successful feeding of control polyps was 85%, it was only 57% in those exposed to 0.2 µg Cu L^{-1} and only 24% in those exposed to 0.5 µg Cu L^{-1}. This is one of the most sensitive responses to Cu. Other metals (Zn, Cd, Pb) did not produce this sublethal effect in corals.

DeForest et al. (2018) reviewed effects of Cu on neurobiology and behavior of many marine species to see if the sublethal criteria used in the USA were protective. EC$_{20}$ values (20% effect concentrations) for the embryo-larval life stage of the blue mussel (*Mytilus edulis*) have historically been used to derive saltwater Cu criteria. Only eight of the 35 tests studied had sufficient toxicity and chemistry data to support definitive conclusions (that is, a Cu EC$_{20}$ or a no-observed-effect concentration could be derived). In the remaining 27 tests, the analysis was limited by several factors. The Cu toxicity threshold was a "less than" value in 19 tests, because only a lowest-observed-effect concentration, and not a no-observed-effect concentration could be calculated. In other cases, Cu and/or DOC (dissolved organic carbon) concentrations were not measured. In two of those tests, the criteria would not have been protective if based only on a conservatively high upper-bound DOC estimate. Thus, US regulatory criteria for copper appear to be not generally protective against chemosensory and behavioral impairment in aquatic organisms.

3. Ocean Acidification

Ocean acidification (OA) occurs when carbon dioxide, emitted to the atmosphere by burning fossil fuels, dissolves in the ocean. Much of this CO_2 is absorbed by the ocean, where it is converted to carbonic acid, which releases hydrogen ions into the water, making it more acidic. Since the beginning of the industrial age, the pH of the oceans has declined by 0.1 pH unit, which, because the pH scale is logarithmic, represents a 30% increase in acidity. According to projections of the Intergovernmental Panel on Climate Change, pH values will decrease another 0.2–0.3 units by 2100, doubling the current acidity. While the most obvious deleterious effects of acidification have been impaired shell formation in mollusks and other organisms that make calcareous shells, effects on behavior have also been reported.

The extent to which OA has increased varies with season, year and region (Feeley et al. 2009). Direct observations go back only 30 years. Friedrich and colleagues (2012), combining computer modeling with observations, concluded that CO_2 emissions over the last 100 to 200 years have raised ocean acidity far beyond the range of natural variation. The excess hydrogen ions reduce concentrations of carbonate ions in sea water. They studied changes in the saturation level of aragonite (a form of calcium carbonate that is used by many shell-forming animals), which is often used to measure OA. As seawater becomes more acid, the saturation level of aragonite decreases. Their models captured variations of aragonite in coral reefs, where levels of aragonite saturation are already five times lower than in the pre-industrial range. The saturation state (denoted by the Greek letter Ω) indicates the degree to which seawater is saturated with carbonate and is inversely proportional to the mineral's solubility. The saturation state is determined by the concentration of calcium and carbonate ions in relation to the solubility coefficient for the calcium carbonate mineral (aragonite or calcite). Aragonite saturation is particularly sensitive to acidity because it is more soluble. Since aragonite is also the form of carbonate generally used by mollusks in shell formation, it can affect the growth of these animals. When the saturation state equals 1, there is an equal chance of dissolution or formation of calcium carbonate; when it is less than 1, dissolution is greater than formation, and when it is greater than 1, formation of calcium carbonate is greater than dissolution. If the aragonite saturation state falls below 1, shells that have been formed will begin to dissolve. If it falls below 1.5, some animals cannot build new shells. The saturation state is highest in shallow warm tropical waters and lowest in deep and cold high latitude waters (Feely et al. 2009), which suggests that effects of OA will be more severe in cold high latitudes.

Acidification interacts with eutrophication in estuaries and coastal waters. When nutrient-rich water enters coastal waters, phytoplankton (algae) bloom. When the algae die, they sink to the bottom and are decomposed by bacteria. The decomposition process is a respiratory one in which carbon dioxide is released and oxygen is depleted. The CO_2 generated from bacterial respiration reduces the pH of the water. The combination of these two sources of CO_2 increases coastal and estuarine acidity beyond what would be expected from the individual processes (Cai et al. 2011).

The co-occurrence of higher CO_2 and lower oxygen (hypoxia) in coastal waters and estuaries is one reason to discuss hypoxia as well as acidification in this paper. Hypoxia is not a pollutant, but rather a deficiency of a required substance. It is most often caused by pollution from excessive nutrients that stimulate algal blooms that die and use up the oxygen.

3.1 Crustaceans

Briffa et al. (2012) reviewed studies of elevated CO_2 and the behavior of tropical reef fishes and hermit crabs. Three main routes were identified by which behavior could be altered. These were (1) elevated metabolic load, (2) "info-disruption" (transfer of chemical information between organisms) and (3) avoidance of polluted sites. They stated that there is ample evidence that exposure to high CO_2 disrupts the ability of animals to find settlement sites and shelters as well as the ability to detect predators and food, and that these behavioral changes appear to occur primarily via info-disruption.

Reduced pH affected chemosensation associated with feeding in hermit crabs *Pagurus bernhardus* (de la Haye et al. 2012). Crabs maintained in 6.8 pH water along with a food odor had less antennular flicking ("sniffing" response), were less successful in locating the source of the odor, and had reduced activity compared to hermit crabs in normal pH. De la Haye et al. (2011) tested the chemosensory responses of *P. bernhardus* to a food odor under reduced pH (6.8). Acidifying the odor did not affect its attractiveness. De la Haye et al. (2012) investigated effects of reduced pH on shell assessment and selection by *P. bernhardus*. At pH 6.8, crabs were less likely to change from a suboptimal to an optimal shell than crabs in normal pH, and those that did change shells took longer to do so. Thus, a reduction in pH disrupts resource assessment and decision-making of these crabs, reducing their ability to acquire a vital resource.

Appelhaus et al. (2012) investigated the impact of increased pCO_2 levels (650, 1250 and 3500 µatm) on the shore crab *Carcinus maenas* and tested whether the amount or the size of prey consumed was affected by the acidity. They exposed both the predators and their prey, the blue mussel *Mytilus edulis*, over a period of 10 wk, then performed feeding experiments. Intermediate acidification levels did not affect growth or consumption, but the highest acidification level reduced feeding by 41% over the experimental period. In contrast, Hildebrandt et al. (2016) incubated the copepod *Calanus glacialis* at 390, 1120, and 3000 µatm for 16 d with the diatom *Thalassiosira weissflogii* as a food source. Elevated pCO_2 did not directly affect grazing activities or body mass, suggesting that the copepods did not have additional energy demands for coping with acidification, neither during long-term exposure nor after immediate changes in pCO_2.

Lord et al. (2017) studied interactions of crabs, whelks, juvenile abalone, and mussels to determine how feeding, growth, and interactions between species could be shifted by changing ocean conditions. The 10 wk experiment found complex results, which indicate the unpredictability of community-level responses. Contrary to predictions, the largest impact of elevated CO_2 was reduced crab feeding and survival, with a pH drop of 0.3 units. Surprisingly, whelks showed no response to higher temperatures or CO_2 levels, while abalone shells grew 40% less under high CO_2 conditions.

3.2 Corals

Coral larvae (*Acropora millepora*) normally settle on the crustose calcareous alga *Titanoderma* when they are ready for metamorphosis. Doropoulos et al. (2012) found that as the pCO_2 increased to 800 and 1300 µatm, the larvae began to avoid this alga and started to settle elsewhere. *Titanoderma* also became less abundant in the environment. They concluded that acidification had three major impacts: it reduced the number of larvae settling, disrupted their normal preference for settlement substrate, and reduced the availability of the most desirable algal substrate for their future survival.

3.3 Echinoderms

Appelhaus et al. (2012) investigated the impact of increased pCO_2 levels (650, 1250 and 3500 µatm) on the sea star *Asterias rubens* and tested whether the quantity or size of prey consumed was affected. They exposed both the predators and their prey, the blue mussel *Mytilus edulis*, over 10 wk and subsequently performed feeding experiments. Intermediate acidification levels had no significant effect on growth or consumption of mussels, but the highest level reduced feeding and growth rates in sea stars by 56%.

3.4 Mollusks

Vargas et al. (2015) found significant negative effects of elevated pCO_2 on the filter feeding, i.e., clearance and ingestion, of both larvae of the gastropod *Concholepas concholepas* and juveniles of the mussel *Perumytilus purpuratus*. In the latter, the filtering rate dropped 15–70% under high pCO_2 conditions (700 and 1,000 ppm). The study also noted large variations in the sensitivities of *C. concholepas* larvae from different local populations. They suggested that the influence of both acidic upwelling waters and of freshwater discharges in the Las Cruces region might explain the relatively minor effects of high pCO_2 on larvae from that area, which would suggest that they have adapted to be more tolerant to OA.

Amaral et al. (2012) investigated the susceptibility of oysters from acidified areas (receiving runoff from acid sulfate soils) and reference areas to predation by the gastropod *Morula marginalba*. Oyster shells were weaker at acidified sites and were more vulnerable because *M. marginalba* could drill through them more quickly. Many other predators consume prey at rates inversely proportional to their shell strength. While it is not a behavioral response, this effect of OA alters predator/prey relationships. Greater susceptibility to increased predation due to thinner or weaker shells may be particularly severe in sessile prey that depend on their shells for defense (Amaral et al. 2012).

Kroeker et al. (2014) discussed probable effects of acidification on predator/prey interactions of coastal mollusks. Since acidification alters both predator detection and predator avoidance behaviors of prey species, it could lead to significant population changes in these mollusks, but the outcome depends on to what extent OA alters the prey detection and prey capture ability of specific predators. In experimental work, predation by the gastropod *Thais clavigera* on two size groups of the mussel *Brachidontes variabilis* was studied by Yu et al. (2017) under three pCO_2 levels: 380, 950, and 1250 µatm. At 950 µatm pCO_2, the prey handling time decreased significantly, and the predator preferred large mussels. The consumption rate was independent of pCO_2 levels, however, although the searching time increased at elevated pCO_2. These findings indicated that the predator/prey interaction between *T. clavigera* and *B. variabilis* was altered under OA, which could have long-term impacts on the populations of the two species. Glaspie et al. (2017) investigated predator/prey interactions of blue crabs (*C. sapidus*) and soft-shell clams (*M. arenaria*). Acidified clams had lighter shells, indicating that shell dissolution occurred. They had reduced responsiveness to a mechanical disturbance that simulated an approaching predator. However, crabs took longer to find the first clam, and often ate only a portion of the prey available. As a result, there was no net change in predation-related clam mortality in acidified trials.

Watson et al. (2013) found that the predator avoidance behavior ("jumping") of the snail *Gibberulus gibberulus gibbosus*, including whether or not to jump away from the predator (cone snail), the speed of jumping, and the distance jumped, were all significantly reduced by exposure to elevated CO_2. When exposed to cues from a predator, the percentage of snails that jumped was almost halved in high-CO_2 conditions, and it took them almost twice as long to jump (Fig. 2).

Pteropods, tiny pelagic gastropods, are known to be highly sensitive to effects of OA on shell formation. They are also highly sensitive in terms of effects on behavior. The combined effects of lower salinity and lower pH impaired the ability of the pteropod *Limacina retroversa* to swim upwards (Manno et al. 2012). The results suggested that the energy costs of maintaining ion balance

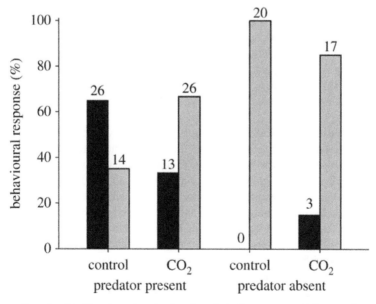

Fig. 2. Percentage of jumping (black) and non-jumping (grey) snails in the presence and absence of a cone snail predator during 5 min trials. The number of snails that jumped away from the predator was reduced in snails in elevated CO_2. Numbers of replicates given above the bars. From Watson et al. 2013.

and avoiding sinking (in low salinity), combined with the extra energy costs needed to counteract shell dissolution (in high pCO_2), exceeded the available energy budget of the animals, causing them to change their swimming behavior.

3.5 Fish

Maneja et al. (2015) studied the kinematics of swimming behavior of Atlantic herring (*Clupea harengus*) larvae 34 d post-hatch cultured under three pCO_2 conditions (control 370, medium 1800, and high 4200 μatm) from swim path recordings. The swim paths were analyzed for move duration, speed and length, stop duration, and horizontal and vertical turn angles. The swimming kinematics and postures in larvae that had survived to 34 d post-hatch were not significantly altered by highly elevated levels of pCO_2, indicating that at least some larvae in the population are resilient to OA.

Effects of elevated pCO_2 were tested on prey and predator fishes by letting one predatory reef fish (dottyback *Pseudochromis fuscus*) interact for 24 h with eight small or large juvenile damselfishes (*Pomacentrus amboinensis)* (Ferrari et al. 2011). Both prey and predator were exposed to control (440 μatm) or elevated (700 μatm) CO_2. Small juveniles of all species had higher mortality from predation at high pCO_2 due to reduced anti-predator behavior, while larger juveniles were unaffected. For larger prey, the pattern of prey selection by predators was reversed under elevated CO_2. The results demonstrate effects of acidification on the behavior of juvenile damselfish, which was probably caused by altered neurological function. Juvenile damselfish *P. amboinensis* exposed to high pCO_2 responded differently to a threat, the sight of a predator, the chromis *Acanthochromis polyacanthus* placed inside a watertight bag (Ferrari et al. 2012a). Juvenile damselfish in 440 (control), 550 or 700 μatm pCO_2 did not differ in their responses to the chromis, but those in 850 μatm showed reduced antipredator responses; they failed to reduce their foraging activity and use of the area. They moved closer to the chromis, suggesting that their response to visual cues from the predator was impaired. Ferrari et al. (2012a) compared the behavior of pre-settlement damselfish *P. amboinensis* that were exposed to 440 μatm pCO_2 (current levels) or 850 μatm pCO_2, a level predicted to occur before the end of this century and found that fish exposed to elevated pCO_2 did not learn to respond appropriately to a predator, the dottyback *P. fuscus*. To determine if the lack of response was due to a failure in learning or a short-term shift in trade-offs preventing the fish from displaying antipredator behavior, they conditioned 440 or 700 μatm-CO_2 fish to learn to recognize a dottyback as a predator using injured conspecific cues, as in the first experiment. When tested either 1 d or 5 d after conditioning, high pCO_2 fish again failed to respond to the predator odor, even though both control and treated fish responded to a general risk cue (injured conspecific cues). Thus, exposure to OA appeared to impair the cognitive abilities of juvenile fish and rendered their learning ineffective (Ferrari et al. 2012b).

Nowicki et al. (2012) found CO_2 level did not have a significant effect on foraging behavior in juvenile anemonefish *Amphiprion melanopus*, but there was an interaction with temperature. At high temperature (31.5°C) at either control or moderate (530 μatm) pCO_2, food consumption and foraging activity were reduced, while at high temperature and high pCO_2 (960 μatm) these behaviors were increased. Maintaining or increasing foraging activity in high temperature and high CO_2 may reduce energy efficiency if the thermal optimum for food assimilation and for growth have been exceeded. OA affected olfactory preferences, activity, and feeding behavior of the brown dottyback (*Pseudochromis fuscus*) (Cripps et al. 2011). Fish were exposed to present-day and elevated pCO_2 levels (~ 600 μatm or ~ 950 μatm). Exposed fish switched from preference to avoidance of the smell of injured prey, spending 20% less time than controls in a stream containing prey odor. General activity level was higher in the high pCO_2 treatment, and feeding was lower in the medium pCO_2 treatment. Elevated activity in the high pCO_2 treatment may be an attempt to compensate for reduced olfaction, as visual detection of food might improve.

Exposure to OA can affect homing and migration. Larvae of the clownfish (*Amphiprion percula*) reared in acidified conditions (pH 7.8) were unable to locate safe habitats (Munday et al. 2009). They had damaged sensory abilities, were unable to recognize the odor and sound of predators and thus were less likely to detect and swim away from threats, which lowered their fitness and chances for survival. Adults can also be impaired, as well as juveniles. The ability of adult cardinalfish *Cheilodipterus quinquelineatus* to home to their diurnal resting sites after nocturnal feeding was impaired by elevated pCO_2 (550, 700 or 950 mg L^{-1}) (Devine et al. 2012). Exposed fish had reduced ability to distinguish between home vs foreign-site odors. Exposed fish showed reduced homing success when released 200 m away from home sites. Their behavior at the home sites was also affected, with exposed fish exhibiting higher activity and venturing further from shelter. Disruption of chemosensory mechanisms was responsible for impairing these critical behaviors.

Elasmobranch fishes are also affected. Dixson et al. (2014) found that projected future CO_2 levels impaired odor tracking behavior of the smooth dogfish (*Mustelus canis*). Adults were held for 5 d in a control (405 µatm), medium (741 µatm), or high pCO_2 (1064 µatm) treatment consistent with the projections for the year 2100 on a "business as usual" scenario. Control and medium pCO_2-treated individuals showed normal odor tracking behavior, while the high pCO_2 group avoided the odor cues that indicated food. While control sharks spent over 60% of their time in the water containing the food stimulus, time in water with the food stimulus was below 15% for the high pCO_2-treated sharks. In addition, sharks maintained in medium and high pCO_2 conditions reduced their attack behavior compared with controls.

Larval Indian Ocean medaka (*Oryzias melastigma*) were investigated for the impact of OA on development and escape behavior in newly hatched larvae (Wang et al. 2017). No significant effects on growth were observed in the larvae when exposed to pCO_2 levels expected in the next 100 years. Larvae reared under control conditions readily produced typical C-start escape behavior in response to mechanical stimuli; however, in the moderate and high pCO_2 treatment groups, the probability of C-start was lower than in the control group. Therefore, the sensory integration needed for the C-start escape behavior appears to be vulnerable to OA.

Decades before any awareness of OA, reduced pH in freshwater systems was a concern because of acid rain, and many researchers studied the effects of this environmental problem. Acidity and accompanying aluminum in freshwater can impair the development of gill Na^+, K^+-ATP-ase and thus salinity tolerance, which is essential for salmonid fishes to develop before they migrate into more saline waters. Exposures as short as 12 h to relatively mild acidity (pH 5.2, 31 mg L^{-1} aluminum) could compromise salinity tolerance (Staurnes et al. 1996a). Atlantic salmon smolts (*S. salar*) released into an acidic river in Norway often had no adult returns and had only one-tenth the number of returns of fish released at the mouth of the river or in a nearby limed river with normal pH (Staurnes et al. 1996b). These return rates were strongly correlated with the effect of acidity on gill Na^+, K^+-ATPase activity and osmotic balance of fish held in cages at the different sites. Therefore, the smolt stage of salmonids appears to be particularly sensitive to acidity and other pollutants. By affecting the development of salinity tolerance, migration to the sea can be impaired.

4. Hypoxia

While not a pollutant itself, since it is a deficiency rather than an excess of something, hypoxia can result from pollution by excess nutrients, particularly nitrogen, in coastal waters and estuaries, a process called eutrophication. The nutrients come from sewage, from fertilizer runoff from farm fields, lawns, golf courses and other land parcels, and from atmospheric deposition. The initial response is stimulation of an algal bloom, which subsequently dies and sinks to the bottom to be decomposed by bacteria. The bacterial respiration uses up oxygen in the water and releases carbon dioxide. Thus, in coastal waters, as mentioned in the previous section, hypoxia is accompanied by intensified OA in which the acidification due to absorption of CO_2 from the atmosphere is greatly magnified by the CO_2 produced locally by bacterial respiration. Negative effects of OA can be intensified by negative effects of low oxygen. While normal levels of dissolved oxygen are ~ 8 mg L^{-1}, most organisms can tolerate somewhat lower levels. A general response to hypoxia is reduced respiration and activity.

4.1 Mollusks

Liu et al. (2011) found that respiration rates of larvae of the gastropods *Nassarius siquijorensis* and *N. conoidalis* were depressed at 4.5 mg O_2 L^{-1} and swimming speed was reduced in 10-day-old larvae exposed to < 2.0 mg O_2 L^{-1} for *N. siquijorensis* and < 1.0 mg O_2 L^{-1} for *N. conoidalis*, suggesting that the latter species is more tolerant of low oxygen.

Hypoxia can cause bivalves in sediments to move up to the sediment surface to find oxygen, which makes them more susceptible to predation. Long and Seitz (2008) performed experiments varying predator access to marked *Macoma balthica* clams at deep and shallow sites in the York River, Virginia, before and during episodes of hypoxia. During hypoxia, predation rates at hypoxic sites were more than double those in normoxic sites. Ambient clam densities were lower at the deep sites, and lower in August than in June. The authors concluded that hypoxia (which had increased over the summer as the water warmed) increased the susceptibility of benthic clams to predation. Chronic hypoxia also impaired anti-predator responses in the green-lipped mussel, *Perna viridis* (Wang et al. 2012), by reducing its normal responses of shell thickening, byssus thread production, and adductor muscle growth. Hypoxia affected predation by the non-native rapa whelk (*Rapana venosa*) on bivalve prey: the non-native ark shell (*Scapharca inaequivalvis*) and Manila clam (*Tapes philippinarum*), and the native cockle *Cerastoderma glaucum* (Munari and Mistri 2011). Under normal oxygen conditions, *R. venosa* preferred to eat *S. inaequivalvis*. However, short-term hypoxia increased the vulnerability of *T. philippinarum*, and the whelks switched their preference to this species. Altered prey preferences can affect community structure.

4.2 Fish

A common behavioral response to hypoxia in more mobile animals is avoidance. Craig (2012) investigated behavior of fishes in the Gulf of Mexico in relation to the large hypoxic region (Dead Zone) in the Gulf. On average, DO avoidance thresholds were low (1 to 3 mg L^{-1}), suggesting fish avoid only the lowest, lethal DO levels. They stayed close to the edges of the hypoxic zone, indicating that effects of hypoxia are most intense inside a relatively narrow region. Avoidance thresholds were similar in brown shrimp and several species of finfish.

When juvenile turbot (*Scophthalmus maximus)* and European sea bass (*Dicentrarchus labrax*) were fed to satiation, feeding rate and therefore growth were reduced in hypoxic conditions (3.2 and 4.5 mg O_2 L^{-1}) (Pichavant et al. 2001). Growth was similar in fish in hypoxia that were fed to satiation and fish in normoxia that were given restricted rations. Decreased food intake is a likely mechanism by which hypoxia reduces growth, and can be considered adaptive, as it can be a way to reduce energy demand and thus oxygen demand under conditions of low oxygen. The northern pipefish (*Syngnathus fuscus*) and dusky pipefish (*S. floridae*) were kept in normal (> 5 mg L^{-1} O_2) and hypoxic (2 and 1 mg L^{-1} O_2) conditions (Ripley and Foran 2007). Both pipefish normally produce short, high-frequency clicks when they are feeding. In hypoxic conditions, reduced feeding correlated with decreased sound production. Declines in both behaviors occurred after one day in hypoxic conditions and continued as long as hypoxia was maintained.

Growth rates of winter flounder (*Pseudopleuronectes americanus*) and summer flounder (*Paralichthys dentatus*) declined as the DO decreased and as temperature increased. Summer flounder were the more tolerant species (Steirhoff et al. 2006). A significant correlation between feeding rate and growth rate indicated that reduced feeding behavior was the major cause of reduced growth under hypoxia. European sea bass (*Dicentrarchus labrax* L.) were exposed to 40% air saturation, with oscillations between 40% and 86% with a period of 770 min and 86% saturation (control) for one month (Thetmeyer et al. 2001). Fish in hypoxia consumed less food, had reduced growth, and had a lower condition factor. Groups maintained in oscillating conditions were intermediate. Growth was correlated with food intake, again suggesting that reduced growth is primarily due to reduced appetite.

In contrast, bottom-feeding spot (*Leiostomus xanthurus*) and hogchoker (*Trinectes maculatus*) in Chesapeake Bay benefited indirectly from hypoxia, because their benthic prey became more available (Pihl et al. 1992). During or immediately after hypoxic events, their gut contents contained larger and deeper-burrowing prey than during periods of normoxia. Spot consumed more biomass (45–73%) of polychaetes than other prey. Thus, in areas where hypoxia is intermittent, its effects on behavior of benthic animals (moving up towards the sediment surface) can be advantageous to more oxygen-tolerant bottom-feeding fishes. However, in the Neuse River estuary, intermittent hypoxia had negative effects on feeding by croakers *M. undulatus* by restricting the fish to shallower oxygenated areas where their prey were less abundant and, in addition, by killing benthic prey in deeper waters, thus greatly reducing their numbers (Eby et al. 2005).

Female common gobies (*Pomatoschistus microps*) usually prefer to spawn with males that have already been chosen by and mated with other females, and whose nests already contain some eggs. However, this preference was reversed in low DO conditions (Reynolds and Jones 1999). In 35% saturation, males exhibited a threefold increase in egg ventilation and spent less time near a female. The authors thought that, under low DO, females tended to avoid those males that would be less likely to be able to meet the demands of rearing a second clutch. In contrast, naked gobies (*Gobiosoma bosc*) subjected to hypoxia in Chesapeake Bay and in the laboratory were quite resistant to low DO in terms of their reproductive behavior and continued to guard eggs until DO levels approached lethal levels (< 1 mg L^{-1}) (Breitburg 1992).

4.3 Crustaceans

Behavioral reactions of the shrimp *Crangon crangon* were studied in individuals exposed to varying degrees of hypoxia at different temperatures. At 20°C, the normally buried shrimp emerged from the sand at 40–50% saturation; at 9°C this emersion response occurred at 20% saturation (Hagerman and Szaniawska 1986). Thus, the shrimp withstood lower DO at a lower temperature. Feeding and predation by the mud crab *Neopanope sayi* and juvenile blue crab *C. sapidus* decreased during hypoxia (1.0 and 0.5 mg O$_2$ L^{-1}), suggesting that short hypoxic episodes may create predation refuges for their prey (Sagasti et al. 2001). (However, as described above, benthic prey often come closer to the sediment surface and become more susceptible to predation.) Bell et al. (2003) used biotelemetry with measurements of DO to monitor feeding and movement of free-ranging *C. sapidus* in the Neuse River estuary, North Carolina, USA, during hypoxic upwelling and subsequent relaxation events. The amount of feeding declined in mild (2–4 mg L^{-1}) and severe (< 2 mg L^{-1}) hypoxia. Blue crabs in hypoxia reduced the proportion of time they spent feeding, but during relaxation events when DO increased, their feeding time did not increase, and crabs did not reinvade deep water, as had been hypothesized.

Predator-prey dynamics between the blue crab *Callinectes sapidus* and the infaunal clam *Mya arenaria* were examined by Taylor and Eggleston (2000) to assess effects of hypoxia on foraging rates and prey mortality. The relationship between predator consumption rate and prey density were evaluated in normoxia, moderate hypoxia (3.0 to 4.0 mg L^{-1}) after acclimation to high DO, and moderate hypoxia after acclimation to low DO. *Mya arenaria* burial depth decreased and siphon extension increased in severe hypoxia. When moderate hypoxia occurred after normoxia, blue crab foraging was changed. Low DO affected the interaction between *C. sapidus* and *M. arenaria* by either (1) hindering blue crab foraging or (2) increasing clam vulnerability by altering their siphon extension and depth in the sediment. Predator preference for certain prey can be modified by low DO

(Munari and Mistri 2012). The crab *Carcinus aestuarii* normally prefers to eat the mussel *Musculista senhousia* as prey, but after hypoxia, its preference was influenced by the presence of another prey, *Tapes philippinarum.*

Hermit crabs frequently fight over possession of the gastropod shells that they inhabit. *Pagurus bernhardus* attackers rap their shell against that of the shell defender in bouts while defenders remain tightly withdrawn into their shells. At the end of a fight the attacker may either evict the defender from its shell or give up; the defender's decision is to either maintain a grip on its shell or allow itself to be evicted. Briffa and Elwood (2000) found that the vigor of rapping and the likelihood of eviction were reduced if the attacker was subjected to low DO, but that low DO did not affect shell defenders in terms of eviction rates. *Pagurus bernhardus* in hypoxia spent less time investigating new shells before entering them and selected lighter shells than did crabs in normal DO levels. This shift in shell preference may reduce energy expenditure but was at the expense of internal spaciousness of the shell, since lighter shells were smaller. These smaller shells may make the hermit crabs more vulnerable to predation (Côté et al. 1998).

5. Noise Pollution

Many organisms occupying the oceans actively use and produce sound. Marine mammals use sound as their primary method for communicating over large distances, fish do the same over shorter spatial scales. Marine invertebrates also produce sound both as a behavioral display and as a result of feeding or movement. Anthropogenic sound in parts of the ocean has increased considerably in recent decades due to human activities such as resource acquisition, global shipping, construction, sonar, and recreational boating, and there is increasing concern over deleterious effects this noise pollution may have on marine biota. Kunc et al. (2016) reviewed such effects and categorized behavioral changes as (1) overlapping with the hearing range of species, (2) overlapping with the bandwidth of acoustic information, i.e., masking acoustic information, (3) distracting individuals even if acoustic information is not energetically masked, and (4) affecting behavior across sensory modalities: cuttlefish, for example, changed their visual signals when exposed to anthropogenic noise (Kunc et al. 2014), and aquatic mammals may alter their vocalization (Nowacek et al. 2007).

5.1 Mollusks

Charifi et al. (2017) examined responses of oysters *Crassostrea gigas* to different levels of noise. At high enough acoustic energy, oysters closed their shells in response to frequencies in the range of 10 to < 1000 Hz, with maximum sensitivity from 10 to 200 Hz. This response could be maladaptive since oysters need their valves to be open in order to feed and respire, so their health can depend on how much time they spend with their valves open. Roberts et al. (2015) investigated sound detection of blue mussels *Mytilus edulis,* also using valve closure as an indicator of sound perception. They studied the impact of loud substrate-borne vibrations—similar to pile driving and blasting—and established the threshold sensitivity of mussels on a sand and gravel bottom. They found that responses were relatively constant from 5 to 90 Hz but showed a sharp decrease in sensitivity at 210 Hz. Day et al. (2017) studied physiological and behavioral effects of air gun noise on the scallop *Pecten fumatus.* Following exposure to air gun noise in the field, scallops had significantly increased mortality, disrupted behaviors and reflex responses, both during and after exposure to the noise, as well as changed hemolymph biochemistry, physiology, and osmoregulation.

Cuttlefish changed their visual signals when exposed to anthropogenic noise (Kunc et al. 2014). *Sepia officinalis,* which uses highly complex visual signals, adjusted their visual displays by changing their color more frequently during a playback of anthropogenic noise, compared with before and after the playback.

5.2 Crustaceans

Wale et al. (2013) investigated behavioral responses of shore crabs to ship noise, the most common source of underwater noise. Ship noise affected foraging and antipredator behavior in *Carcinus maenas*. Ship noise playback was more likely than ambient noise playback to disrupt feeding, although crabs in the two sound treatments had equal speed at finding food. While crabs exposed to ship noise were just as likely as ambient noise controls to detect and respond to a simulated predatory attack, they were slower to retreat to shelter. These results demonstrated that anthropogenic noise has the potential to increase the risks of both starvation and predation in crabs. Being distracted by noise also made hermit crabs more vulnerable to predation. When exposed to boat motor playback, *Coenobita clypeatus* allowed a simulated predator to approach closer than usual before initiating the normal hiding response (Chan et al. 2010).

Exposure of the lobster *Palinurus elephas* to boat noise caused significant changes in locomotor behaviors (Filiciotto et al. 2014). Exposed lobsters significantly increased their locomotor activities. They also exhibited altered hemolymph, with elevated indicators of stressful conditions, such as glucose and total proteins. Zhou et al. (2018) conducted experiments with a hydrophone, transducer, and video recording system, using juvenile mud crabs *Scylla paramamosain*. Three acoustic stimulations with ascending levels were tested separately. Results showed that sound power greater than 110 dB re 1 $\mu Pa^2/Hz$ in total bandwidth (100–1000 Hz), and 155 dB re 1 $\mu Pa^2/Hz$ within 600–800 Hz increased both locomotor activities and HSP70 gene expression significantly.

5.2 Various invertebrates

Solan et al. (2016) exposed three sediment-dwelling species—the langoustine (*Nephrops norvegicus*), the Manila clam (*Ruditapes philippinarum*) and the brittlestar (*Amphiura filiformis*)—to either continuous broadband noise (CBN) similar to shipping traffic or intermittent broadband noise (IBN) similar to marine construction activity. They found that exposure to underwater broadband sound fields that resemble offshore shipping and construction activity altered sediment processing behavior. The langoustine, which disturbs sediment to create burrows, reduced the depth of sediment redistribution when exposed to either IBN or CBN. Under CBN and IBN there was evidence that bioirrigation increased. The Manila clam, which lives in the sediment and connects to the overlying water through a retractable siphon, reduced its surface activity under CBN. Bioirrigation was greatly reduced by CBN and slightly reduced by IBN. In contrast, the exposure had little impact on the brittlestar's behavior.

5.3 Fishes

Some fish communicate using sounds. Vasconcelos et al. (2007) investigated effects of ship noise on the detection of conspecific vocalizations by the Lusitanian toadfish, *Halobatrachus didactylus*. Ambient and ferry boat noises were recorded, as well as toadfish sounds. Hearing was measured under quiet conditions in the laboratory and in the presence of these noises at levels found in the field. In the presence of ship noise, auditory thresholds increased considerably because the boat noise was within the most sensitive hearing range of this species. The ship noise decreased the fish's ability to detect conspecific sounds, which are important in agonistic encounters and mate attraction.

There is concern that the noise from offshore wind farms may also decrease the effective range for sound communication of fish, but little is known about to what extent it may occur. Windmill noise is not destructive to hearing, even within short distances. It is estimated that fish are frightened away from windmills at distances less than 4 m, and only at high wind speeds. Thus, the acoustic impacts of windmills may involve masking communication signals rather than causing physiological damage or avoidance (Wahlberg and Westerberg 2005). However, data are very limited and further studies of fish behavior around offshore windmills are needed.

Stanley et al. (2017) estimated effective communication spaces at spawning locations for populations of the Atlantic cod (*Gadus morhua*) and haddock (*Melanogrammus aeglefinus*). Ambient sound came primarily from shipping activity since increases in sound pressure level were positively correlated with the number of vessels tracked by the US Coast Guard's Automatic Identification System at the sites. The almost constant high levels of low-frequency sound from shipping caused a reduction in the communication space observed during times of high vocalization activity (spawning). There is concern that communication may be compromised during these critical periods.

Sebastianutto et al. (2011) evaluated the effects of boat noise on the goby *Gobius cruentatus,* which normally emit sounds during agonistic interactions. Investigators played a field-recorded diesel engine boat noise during aggressive encounters between an intruder and a resident fish in the laboratory. Agonistic behavior of the resident fish was reduced by the noise: they were more submissive and won fewer encounters. The authors suggested that sound production is used for territorial defense, and since it was impaired by the boat noise, the ability of the resident to maintain its territory was reduced.

Motorboat noise elevated the metabolic rate in Ambon damselfish (*Pomacentrus amboinensis*), which when stressed responded less often and less rapidly to simulated predatory strikes (Simpson et al. 2016). In the laboratory, they were captured more readily by their predator, the dusky dottyback (*Pseudochromis fuscus*), during exposure to motorboat noise compared with ambient conditions. Furthermore, more than twice as many were consumed by the predator in field experiments when motorboats were passing (Fig. 3). This study suggests that boat noise in the marine environment could have major impacts on fish populations and predator prey interactions.

5.4 Cetaceans

Nowacek et al. (2007) reviewed responses of cetaceans to noise and found three types of responses: behavioral, acoustic and physiological. Behavioral responses involve changes in surfacing, diving, and swimming patterns. Acoustic responses include changes in the timing or type of vocalizations.

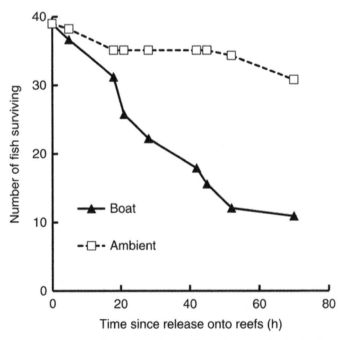

Fig. 3. Survival of *Pomacentrus amboinensis* on reefs with and without playback of boat noise. From Simpson et al. 2016.

Physiological responses include shifts in auditory thresholds. In this study they documented responses of cetaceans to various noise sources but were concerned about the relative absence of knowledge about effects of noise sources such as commercial sonars, depth finders and acoustics gear used by fisheries. Romano et al. (2004) measured blood parameters of the beluga whale (*Delphinapterus leucas*) and bottlenose dolphin (*Tursiops truncates*) exposed to noise. Norepinephrine, epinephrine, and dopamine levels, neurotransmitters related to stress, increased with increasing sound levels, and were significantly higher after high-level sound exposures compared with low-level sound exposures or controls.

Noise from ship traffic and commercial, research and military activities has increased greatly over the past century and has caused changes in the vocalizations and behaviors of many marine mammals, including beluga whales (*Delphinapterus leucas*) (Lesage et al. 1999), manatees (*Trichechus manatus*) (Miksis-Olds and Tyack 2009) and right whales (*Eubalaena glacialis, E. australis*) (Parks et al. 2007). The calls of killer whales are prolonged in the presence of noise from boats, probably to compensate for the acoustic pollution (Foote et al. 2004), while humpback whales (*Megaptera novaeangliae*) increase the repetition of phrases in their songs when exposed to low-frequency sonar (Miller et al. 2000). Several species of dolphins change their behavior and vocalizations when there is boat noise (Buckstaff 2004). Parks et al. (2011) documented changes in calling behavior by individual North Atlantic right whales (*Eubalaena glacialis*) in the presence of increased background noise. Right whales respond to periods of increased noise by increasing the amplitude of their calls, which may help to maintain their communication range with conspecifics during periods of increased noise. This may be interpreted as an adaptive response. However, periods of high noise are increasing and have reduced the ability of right whales to communicate with each other by about two-thirds. *Eubalaena glacialis* were studied by Hatch et al. (2012) in the 10,000 km² Stellwagen Bank marine sanctuary for one month using vessel-tracking data from the US Coast Guard's Automatic Identification System to quantify acoustic signatures of large commercial vessels and evaluate the noise from vessels inside and outside the sanctuary. By comparing current noise levels from commercial ships with the lower noise conditions a half-century ago, the authors concluded that right whales have lost about 63–67% of their communication space in the sanctuary and surrounding waters. Humpback whales in this sanctuary completely stopped their singing during an Ocean Acoustic Waveguide Remote Sensing (OAWRS) experiment that was located approximately 200 km away (Risch et al. 2012). Blackwell et al. (2013) assessed effects of air gun sounds on bowhead calling behavior during the autumn migration. At the onset of air gun use, calls dropped significantly at sites near the air guns, but remained unchanged at sites distant from the air guns. This drop could result from a cessation of calling, deflection of whales around seismic activities, or both. Castellote et al. (2012) documented acoustic and behavioral changes by fin whales in response to shipping and air gun noise. They compared acoustic features in areas with different shipping noise levels and different traffic intensities, and during a seismic air gun array survey. In high noise conditions of 20 Hz, note duration shortened, bandwidth decreased, and frequency decreased. Similar results were obtained during a 10 d seismic survey. During the survey, whales moved away from the detection area, and this displacement persisted well beyond the 10 d duration of seismic air gun activity. In contrast, Di Iorio and Clark (2009) investigated blue whale (*Balaenoptera musculus*) vocal behavior during a seismic survey and found that whales consistently called more on seismic exploration days and during periods when the air gun was operating. This increase was observed for the calls that are emitted during social encounters and feeding. The authors felt this response represents a compensatory behavior for the elevated noise from seismic survey operations.

A common behavioral response of fishes and cetaceans to noise is to leave the area. Olesiuk et al. (2002) assessed impacts of the sound generated by an acoustic harassment device (AHD) on the relative abundance and distribution of harbor porpoises (*Phocoena phocoena*). Abundance declined quickly when the device was activated. The mean number of re-sightings of porpoises while tracking their movements also declined, suggesting that the few porpoises that went into the study area spent less time within it when the AHD was activated. The effect of the AHD diminished with distance, but no porpoises were observed within 200 m of the device when it was activated.

Conducting experiments with cetaceans is very challenging and opportunities are limited, so studies should include rigorous measurements and modeling of exposure (Nowacek et al. 2015).

6. Neurotoxicology Underlying Behavior

Pollutant effects on behavior are due to underlying changes in the nervous and endocrine systems. Interference with these systems can involve neurological development and levels of neurotransmitters or effects on sensory receptors. Chemoreception is important for responses to the environment; olfaction, vision, and hearing are important senses that are often impaired by contaminants. In addition, many behaviors can be affected by alterations in the endocrine system, which is very sensitive to environmental factors, including contaminants. Some of these responses have been discussed previously. These underlying mechanisms are reviewed here briefly.

6.1 Crustaceans

Various metals and organic pollutants can affect the nervous system of crustaceans. Increased acetylcholinesterase in thoracic ganglia of the crab *Barytelphusa guerini* was noted (Reddy and Venugopal 1991) after 4 d of exposure to Cd (0.6 mg L^{-1}), but after 15 d the enzyme was inhibited. The eyestalks of decapod crustaceans are the central neuroendocrine coordinators, controlling almost all aspects of their lives including the molt cycle, color change, maturation, blood sugar, and nerve function. Cd (10 mg L^{-1}) was found to damage neurosecretory cells in the brain and eyestalk ganglia of fiddler crabs *Uca pugilator* (Reddy and Fingerman 1995). Impacts on color change are easily measured responses to disruption of these hormones. For example, crabs exposed to Cd were less able to disperse pigment in their melanophores because the Cd depleted the neurosecretory material (Reddy and Fingerman 1995).

Crustaceans depend on chemical senses for feeding and social interactions, and their chemoreceptors are on the surface of the body and are therefore exposed directly to the environment, including contaminants that may be present. Disrupted chemoreception can be responsible for changes in settlement of larvae, shell acquisition of hermit crabs, and reproductive interactions (Krang and Eckerholm 2006).

6.2 Mollusks

Low-frequency sounds were found to cause acoustic trauma in the ears of Cephalopods (Andre et al. 2011).

6.3 Fishes

Neurotransmitters

Brain neurotransmitter levels are related to behaviors, so it is likely that altered neurotransmitters induced by toxicants will result in altered behaviors. One of the most common indicators of changes in neural function is altered acetylcholinesterase (AchE) in the brain. AchE breaks down the transmitter acetylcholine after it diffuses across synapses of cholinergic neurons. Fish with altered behavior have been shown to have altered brain neurotransmitters. Mummichogs (*F. heteroclitus*) after exposure to mercury in the laboratory or collected from polluted sites exhibit reduced activity and prey capture; they also have reduced levels of serotonin in their brains (Smith et al. 1995, Zhou et al. 1999a). Yu et al. (2013) found that 10, 100 and 1000 ng L^{-1} of TBT reduced the prey capture ability of

Sebastiscus marmoratus and altered their neurotransmitters. In this case, dopamine levels in the fish brains increased in a dose-dependent manner and 5-hydroxytryptamine and norepinephrine levels decreased in TBT-exposed fish compared to controls. It is interesting that different organometals can cause alteration of different neurotransmitters in the two fish species, resulting in the same type of behavioral change.

Exposure to $PbNO_3$ (1.6 mg L^{-1}) reduced the feeding periods of the cleaner fish *Thalassoma pavo* after 24 h exposure, while the fish exhibited hyperactive swimming episodes (Zizza et al. 2013). The abnormal behaviors were correlated to up-regulated orexin receptor (ORXR) mRNA expression in the lateral thalamic nucleus and optic tectum of the brain. These transcriptional effects were reduced when the exposed fish received either 100 ng g^{-1} of ORX-A or 0.1 µg g^{-1} of γ-aminobutyric $acid_A$ receptor ($GABA_AR$) agonist muscimol (MUS). The neurodegenerative processes noted after Pb exposure were not observed after treatment with MUS, but addition of the $GABA_AR$ antagonist bicuculline (BIC 1 µg g^{-1}) enhanced the behavioral and neurodegenerative effects of Pb. Thus, there are various neurotransmitters that can be affected by various contaminants that are associated with altered behaviors.

The underlying mechanism linking OA (high CO_2) to behavioral responses has also been shown to be neurotransmitters (Nilsson et al. 2012). Abnormal olfactory preferences of fish exposed to high pCO_2 can be reversed by treatment with an antagonist of the receptor for $GABA_A$, a major neurotransmitter receptor in the brain. This demonstrates that elevated CO_2 interferes with neurotransmitter function, which underlies many behaviors. Since these receptors are widespread in animals, rising pCO_2 levels could cause sensory and behavioral impairment in a wide range of marine species. Hamilton et al. (2013) used a camera-based tracking software system to examine whether OA-dependent changes in $GABA_A$ receptors affected anxiety in juvenile Californian rockfish (*Sebastes diploproa*). They evaluated anxiety by using tests that measure light/dark preference (scototaxis) and proximity to an object. After 1 wk in OA conditions projected for the next century along the California shore (1125 µatm CO_2, pH 7.75), anxiety was significantly increased relative to controls (483 µatm CO_2, pH 8.1). The $GABA_AR$ agonist muscimol, but not the antagonist gabazine, also caused a significant increase in anxiety. Experimental fish remained more anxious, even after 7 d back in control seawater, but resumed normal behavior by day 12.

Olfaction

The olfactory system of fishes is open to the environment and is particularly sensitive to metals including Hg (Baatrup et al. 1990), although inorganic Hg and meHg localize in different parts of the olfactory system. Many metals directly enter the olfactory system, where they can disrupt normal function. By accumulating in and damaging cells in the olfactory system, toxicants can disrupt transmission of information from olfactory lobes to higher levels of the brain. Olfactory neurons can be a transport route of contaminants to the olfactory bulbs and the brain, with resulting effects on the functioning of the nervous system. Cd appears to move along olfactory neurons by axonal transport (Scott and Sloman 2004). Some studies have shown a connection between altered behavior and altered olfactory system. Rehnberg and Schreck (1986) showed reduced avoidance of the amino acid L-serine (a potent odor to fish) by coho salmon (*Oncorhynchus kisutch*) after exposure to Cu and Hg. Hg but not Cu inhibited serine binding to the olfactory epithelium.

Copper produced anatomical lesions in the olfactory, taste, and lateral line receptors (Brown et al. 1982). Cu exposure of juvenile coho salmon (*Oncorhynchus kisutch*) (30 min exposure to 20 µg L^{-1} Cu) reduced the olfactory response to a natural odorant (10^{-5} M L-serine) by 82% (McIntyre et al. 2008). Kennedy et al. (2012) found that Cu exposure inhibited the ability of chinook salmon to detect and avoid the odorant L-histidine in a concentration-dependent manner, and Cu toxicity (olfactory inhibition) decreased with increasing dissolved organic carbon (DOC) concentration. These findings suggest that DOC concentration is an important factor affecting impacts of Cu on fish olfaction. However, as discussed earlier, a meta-analysis by DeForest et al. (2018) indicated that the sublethal criteria used by the US Environmental Protection Agency for copper is frequently not protective against behavioral and neurotoxicological responses.

Toxicants or acidified water can easily disrupt transmission of information from the olfactory lobes to higher levels of the brain, which may cause larval fish to ignore or be attracted to odors they normally avoid, including those from predators and unfavorable habitats. Impaired olfactory function may cause larval fish in OA to be attracted to odors they normally avoid, including those from predators and unfavorable habitats. The underlying mechanism linking high CO_2 (ocean acidification) to these responses has been shown to be neurotransmitters (Nilsson et al. 2012).

Hearing and lateral line

Many fishes rely on hearing for orientation, habitat selection, predator avoidance and/or communication. Simpson et al. (2011) studied the influence of enriched CO_2 on directional responses of juvenile clownfish (*A. percula*) to daytime reef noise. Juveniles in ambient CO_2 conditions avoided reef noise, as expected, but this avoidance behavior was not exhibited by fish living in high CO_2 (700 and 900 µatm pCO_2), showing that OA affects auditory responses, with potentially detrimental impacts on early survival. The lateral line can also be a target for toxicants, such as Cd, the effects of which can impair behavior (Faucher et al. 2006).

Anthropogenic noise can lead to physiological damage to the auditory systems of marine animals. McCauley et al. (2003) found that the noise created by an operating air gun (used in seismic testing) severely damaged the ears of the pink snapper (*Pagrus auratus*), causing ablated hair cells in the sensory epithelia. The damaged cells of the cochlea were not repaired or replaced until 58 d after exposure to the air gun.

Endocrine system

A large number of studies have shown linkages between hormones and behavior and many pollutants act as agonists or antagonists to hormones. Thyroid hormones influence many processes in fishes, including neural development, metabolism, maturation (smoltification in salmonids and metamorphosis in flatfish), and behavior. Many chemicals exert effects on the fish thyroid gland, which has implications for behavior. About 40 fish species have been shown to have thyroid responses to contaminants (reviewed by Brown et al. 2004). Mummichogs (*F. heteroclitus*) from polluted sites in New Jersey (Piles Creek, contaminated with metals, PCBs and more) tend to be sluggish and have poor prey capture and reduced predator avoidance. They have abnormal thyroid glands, with extremely enlarged follicles and follicular cell heights, and contain elevated plasma thyroxine (T4), but not plasma or tissue T3 (Zhou et al. 1999b). Fish from relatively unpolluted reference sites held in conditions simulating Piles Creek also developed elevated T4.

7. Conclusions

Behavioral ecotoxicology can link disturbances at the biochemical level (e.g., altered neurological function or thyroid hormones) to effects at the population level. Types of behaviors that have been studied and reviewed include individual behaviors such as swimming activity, burrowing, and migration, and interactions among individuals, including prey capture, predator avoidance, reproductive behaviors, aggression, and social behaviors. Effects on behavior may be direct, such as impairment of reproductive behavior, habitat evaluation, prey capture, and avoidance of predators. Indirect effects include reduction of activity or reproductive success due to reduced feeding and, therefore, less energy. The sensitivity of behavioral responses can be useful in ecological risk assessments, since behavior can be affected at levels lower than those that affect physiology.

Behavior can be altered in larvae or older stages after exposure to contaminants at the embryo stage. These delayed effects should be considered in risk assessments, although they are not generally taken into consideration. Behavior evolves in response to natural selection, maximizing an organism's fitness. Since few species show behavioral changes that are beneficial, most behavioral alterations in response to contaminants are deleterious to an organism's fitness.

Acknowledgement

Thank you, Dr. Peddrick Weis, for reviewing this paper.

References

Adams, P.A., R. Hanlon and J.W. Forsythe. 1988. Toxic exposure to ethylene dibromide and mercuric chloride: Effects on laboratory-reared octopuses. Neurotox. Teratol. 10: 519–523.

Alonso, S. and G. Valle-Torres. 2018. Feeding behavior of an aquatic snail as a simple endpoint to assess the exposure to cadmium. Bull. Environ. Contamin. Toxicol. 100: 82–88.

Al-Subiai, N., A.J. Moody, S.A. Mustafa and A.N. Jha. 2011. A multiple biomarker approach to investigate the effects of copper on the marine bivalve mollusc, *Mytilus edulis*. Ecotoxicol. Environ. Safety 74: 1913–1920.

Alvarez, M.C., C.A. Murphy, K.A. Rose, I.D. McCarthy and L.A. Fuiman 2006. Maternal body burdens of methylmercury impair survival skills of offspring in Atlantic croaker (*Micropogonias undulatus*). Aquat. Toxicol. 80: 329–337.

Amaral, V., H.N. Cabral and M.J. Bishop. 2012. Effects of estuarine acidification on predator–prey interactions. Mar. Ecol. Prog. Ser. 445: 117–127.

André, M., M. Solé, M. Lenoir, M. Durfort, C. Quero, A. Mas et al. 2011. Low-frequency sounds induce acoustic trauma in cephalopods. Front. Ecol. Environ. 9: 489–493.

Ankley, G.T., D.M. Di Toro, D.J. Hansen and W.J. Berry. 1996. Technical basis and proposal for deriving sediment quality criteria for metals. Environ. Toxicol. Chem. 15: 2056–2066.

Appelhaus, Y., J. Thomsen, C. Pansch, F. Maelzner and M. Wahl. 2012. Sour times: seawater acidification effects on growth, feeding behaviour and acid–base status of *Asterias rubens* and *Carcinus maenas*. Mar. Ecol. Prog. Ser. 459: 85–98.

Asnicar, D., G. Asmonaite, L. Birgersson, C. Kvarnemo, O. Svensson and J. Sturve. 2018. Sand Goby—An ecologically relevant species for behavioural exotoxicology. Fishes DOI: 10.3390/fishes3010013.

Baatrup, E., K.B. Doving and S. Winberg. 1990. Differential effects of mercurial compounds on the electro-olfactogram (EOG) of salmon (*Salmo salar* L.). Ecotox. Environ. Safety 20: 269–276.

Banks, J., D.J. Ross and M.J. Keough. 2012. Short-term (24 h) effects of mild and severe hypoxia (20% and 5% dissolved oxygen) on metal partitioning in highly contaminated estuarine sediments. Estuar. Coast. Shelf Sci. 99: 121–131.

Bell, G.W., D.B. Eggleston and T.G. Wolcott. 2003. Behavioral responses of free-ranging blue crabs to episodic hypoxia. II. Feeding. Mar. Ecol. Prog. Ser. 259: 227–235.

Berry, W., D. Hansen, W. Boothman, J. Mahoney, D. Robson, D. DiToro et al. 1996. Predicting the toxicity of metals-spiked laboratory sediments using acid-volatile sulfide and interstitial water normalization. Environ. Toxicol. Chem. 15: 2067–2079.

Blackwell, S.D., C. Nations, T. McDonald, C. Greene, A. Thode, M. Guerra et al. 2013. Effects of airgun sounds on bowhead whale calling rates in the Alaskan Beaufort Sea. Mar. Mamm. Sci. 29: E342–E365.

Bonnard, M., M. Romeo and C. Amiard-Triquet. 2009. Effects of copper on the burrowing behavior of estuarine and coastal invertebrates, the polychaete *Nereis diversicolor* and the bivalve *Scrobicularia plana*. Hum. Ecol. Risk Assess 15: 11–26.

Breitburg, D.L. 1992. Episodic hypoxia in Chesapeake Bay: interacting effects of recruitment, behavior, and physical disturbance. Ecol. Monogr. 62: 525–546.

Briffa, M. and R.W. Elwood. 2000. Cumulative or sequential assessment during hermit crab shell fights: effects of oxygen on decision rules. Proc. Royal Soc. 267: 2445–2452.

Briffa, M., K. de la Haye and P.L. Munday. 2012. High CO_2 and marine animal behaviour: Potential mechanisms and ecological consequences. Mar. Poll. Bull. 64: 1519–1528.

Brown, S.B. 1982. Chemoreception and aquatic pollutants. pp. 363–393. *In*: Chemoreception in Fishes, Elsevier Pub.

Brown, S., B.A. Adams, S. Cyr and J.G. Eales. 2004. Contaminant effects on the teleost fish thyroid. Envir. Toxicol. Chem. 23: 1680–1701.

Buckstaff, K.C. 2004. Effects of watercraft noise on the acoustic behavior of bottlenose dolphins *Tursiops truncatus*, in Sarasota Bay, Florida. Marine Mamm. Sci. 20: 709–725.

Buffet, P.E., A. Zalouk-Vergnoux, L. Poirer, C. Lopes, J.C. Amiard, P. Gaudin et al. 2015. Cadmium sulfide quantum dots induce oxidative stress and behavioral impairments in the marine clam *Scrobicularia plana*. Environ. Toxic. Chem. 34: 1659–1664.

Cai, W.-J., X. Hu, W.J. Huang, M. Murrell, J.C. Lehrter, S. Lohrenz et al. 2011. Acidification of subsurface coastal waters enhanced by eutrophication. Nature Geosci. 4: 766–770.

Candelmo, A., A. Deshpande, B. Dockum, P. Weis and J.S. Weis 2010. The effect of contaminated prey on feeding, activity, and growth of young-of-the-year bluefish, *Pomatomus saltatrix*, in the laboratory. Estuaries and Coasts 33: 1025–1038.

Capuzzo, J. and J.J. Sasner. 1977. The effect of chromium on filtration rates and metabolic activity of *Mytilus edulis* L. and *Mya arenaria* L. pp. 226–237. *In*: Vernberg, J., A. Calabrese, F. Thurberg and W. Vernberg (eds.). Physiological Reponses of Marine Biota to Pollution. Acad. Press, NY.

Castellote, M., C.W. Clark and M.O. Lammers. 2012. Acoustic and behavioural changes by fin whales (*Balaenoptera physalus*) in response to shipping and airgun noise. Biol. Conserv. 147: 115–122.

Chan, A.A., P. Giraldo-Perez, S. Smith and D.T. Blumstein. 2010. Anthropogenic noise affects risk assessment and attention: The distracted prey hypothesis. Biol. Lett. 6: 458–461.

Charifi, M., M. Sow, P. Ciret, S. Benomar and J.-C. Massabuau. 2017. The sense of hearing in the Pacific oyster, *Magallana gigas*. PLoS ONE 12(10): e0185353. https://doi.org/10.1371/journal.pone.0185353.

Colvin, M.A., B.T. Hentschel and D.D. Deheyn. 2016. Combined effects of water flow and copper concentration on the feeding behavior, growth rate, and accumulation of copper in tissue of the infaunal polychaete *Polydora cornuta*. Ecotoxicology 25: 1720–1729.

Côté, I.M., B. Reverday and P.K. Cooke. 1998. Less choosy or different preference? Impact of hypoxia on hermit crab shell assessment and selection. Anim. Behav. 56: 867–873.

Craig, J.K. 2012. Aggregation on the edge: effects of hypoxia avoidance on the spatial distribution of brown shrimp and demersal fishes in the Northern Gulf of Mexico. Mar. Ecol. Prog. Ser. 445: 75–95.

Cripps, I.L., P.L. Munday and M.I. McCormick. 2011. Ocean acidification affects prey detection by a predatory reef fish. PLoS ONE 6(7): art. e22736.

Day, R.D., R.D. McCauley, Q.P. Fitzgibbon, K. Hartmann and J.M. Semmens. 2017. Exposure to seismic air gun signals causes physiological harm and alters behavior in the scallop *Pecten fumatus*. Proc. Nat. Acad. Sci. 114(40): E8537–E8546. Doi: 10.1073/pnas.1700564114.

DeCoursey, P. and W.B. Vernberg. 1972. Effect of mercury on survival, metabolism and behaviour of larval *Uca pugilator*. Oikos 23: 241–247.

DeForest, D.K., R.W. Gensemer, J.W. Gorsuch, J.S. Meyer, R.C. Santore, B.K. Shephard et al. 2018. Effects of copper on olfactory, behavioral, and other sublethal responses of saltwater organisms: Are estimated chronic limits using the biotic ligand model protective? Environ. Toxicol. Chem. In press. Doi: 10.1002/etc.4112.

De la Haye, K., J.I. Spicer, S. Widdicombe and M. Briffa. 2011. Reduced seawater pH disrupts resource assessment and decision-making in the hermit crab *Pagurus bernhardus*. Anim. Behav. 82: 495–501.

De la Haye, K.L., J.I. Spicer, S. Widdicombe and M. Briffa. 2012. Reduced pH sea water disrupts chemo-responsive behaviour in an intertidal crustacean. J. Exper. Mar. Biol. Ecol. 412: 134–140.

Devine, B.M., P.L. Munday and G.P. Jones. 2012. Homing ability of adult cardinalfish is affected by elevated carbon dioxide. Oecologia 168: 269–276.

Diaz-Jaramillo, M., A.M. Rocha, G. Chiang, D. Buchwalter, J.M. Monserrat and R. Barra. 2013. Biochemical and behavioral responses in the estuarine polychaete *Perinereis gualpensis* (Nereididae) after *in situ* exposure to polluted sediments. Ecotoxic. Environ. Safety 89: 182–188.

Di Iorio, L. and C.W. Clark. 2009. Exposure to seismic survey alters blue whale acoustic communication. Biol. Lett. Doi: 10.1098/rsbl.2009.0651.

Dixson, D., A.R. Jennings, J. Atema and P. Munday. 2014. Odor tracking in sharks is reduced under future ocean acidification conditions. Global Change Biology 21: 1454–1462.

Doropoulos, C., S. Ward, G. Diaz-Pulido, O. Hoegh-Guldberg and P.J. Mumby. 2012. Ocean acidification reduces coral recruitment by disrupting intimate larval-algal settlement interactions. Ecol. Lett. 15: 338–346.

Eby, L.A., L.B. Crowder, C.M. McClellan, C.H. Peterson and M.J. Powers. 2005. Habitat degradation from intermittent hypoxia: impacts on demersal fishes. Mar. Ecol. Prog. Ser. 291: 249–261.

Elfwing, T. and M. Tedengren. 2002. Effects of copper and reduced salinity on grazing activity and macroalgae production: a short-term study on a mollusc grazer, *Trochus maculatus*, and two species of macroalgae in the inner Gulf of Thailand. Mar. Biol. 140: 913–919.

Faucher, K., D. Fichet, P. Miramand and J.P. Lagardère. 2006. Impact of acute cadmium exposure on the trunk lateral line neuromasts and consequences on the "C-start" response behaviour of the sea bass (*Dicentrarchus labrax* L.; Teleostei, Moronidae). Aquat. Toxicol. 76: 278–294.

Feeley, R.A., S.C. Doney and S.R. Cooley. 2009. Ocean acidification: Present conditions and future changes in a high CO2 world. Oceanography 22: 36–47.

Ferrari, M.C., M.I. McCormick, P.L. Munday, M.G. Meekan, D.L. Dixson, Ö. Lonnstedt et al. 2011. Putting prey and predator into the CO_2 equation—qualitative and quantitative effects of ocean acidification on predator–prey interactions. Ecol. Lett. 14: 1143–1148.

Ferrari, M.C.O., M.I. McCormick, P.L. Munday, M.G. Meekan, D.L. Dixson, O. Lönnstedt et al. 2012a. Effects of ocean acidification on visual risk assessment in coral reef fishes. Funct. Ecol. 26: 553–558.

Ferrari, M., R. Manassa, D. Dixson, P.L. Munday, M. McCormick, M. Meekan et al. 2012b. Effects of ocean acidification on learning in coral reef fishes. PLOS One 7(2): e31478. Doi: 10.1371/journal.pone.0031478.

Filiciotto, F., M. Vazzana, M. Celi, V. Maccarrone, M. Ceraulo, G. Buffa et al. 2014. Behavioural and biochemical stress responses of *Palinurus elephas* after exposure to boat noise pollution in tank. Mar. Poll. Bull. 84: 104–11.

Foote, A., R.W. Osborne and A.R. Hoelzel. 2004. Environment: Whale call response to masking boat noise. Nature 428: 10. Doi: 10.1038/428910a.

Friedrich, T., A. Timmermann, A. Abe-Ouchi, N.R. Bates, M.O. Chikamoto, M.J. Church et al. 2012. Detecting regional anthropogenic trends in ocean acidification against natural variability. Nature Climate Change 2: 167–171.

Glaspie, C.N., K. Longmire and R.D. Seitz. 2017. Acidification alters predator-prey interactions of blue crab *Callinectes sapidus* and soft-shell clam, *Mya arenaria*. J. Exper. Mar. Biol. Ecol. 489: 58–65.

Gutierrez, M.F., J.C. Paggi and A.M. Gagneten. 2012. Microcrustaceans escape behavior as an early bioindicator of copper, chromium and endosulfan toxicity. Ecotoxicology 21: 428–438.

Hagerman, L. and A. Szaniawska. 1986. Behaviour, tolerance and anaerobic metabolism under hypoxia in the brackish water shrimp Crangon crangon. Mar. Ecol. Prog. Ser. 34: 125–132.

Hatch, L.T., C.W. Clark, S.M. Van Parijs, A.S. Frankel and D.W. Ponirakis 2012. Quantifying loss of acoustic communication space for right whales in and around a U.S. National Marine Sanctuary. Conserv. Biol. DOI: 10.1111/j.1523-1739.2012.01908.x.

Hamilton, T.J., A. Holcombe and M. Tresguerres. 2013. CO_2-induced ocean acidification increases anxiety in Rockfish via alteration of $GABA_A$ receptor functioning. Proc. Roy. Soc. B DOI: 10.1098/rspb.2013.2509.

Harashiyaki, C.A., A. Reichelt-Brushett, L. Liu and P. Butcher. 2016. Behavioural and biochemical alterations in *Penaeus monodon* post-larvae diet-exposed to inorganic mercury. Chemosphere 164: 241–247.

Hildebrandt, N., F. Sartoris, K. Schulz, U. Riebesell and B. Niehoff. 2016. Ocean acidification does not alter grazing in the calanoid copepods *Calanus finmarchicus* and *Calanus glacialis*. ICES Journal of Marine Science 73: 927–936.

Hughes, J.M., H.F. Chapman and R.L. Kitching. 1987. Effects of sublethal concentrations of copper and freshwater on behaviour in an estuarine gastropod *Polinices sordidus*. Mar. Poll. Bull. 18: 127–131.

Hutcheson, M., D.C. Miller and A.Q. White. 1985. Respiratory and behavioral responses of the grass shrimp *Palaemonetes pugio* to cadmium and reduced dissolved oxygen. Mar. Biol. 88: 59–66.

Jones, J.C. and J.D. Reynolds. 1997. Effects of pollutants on reproductive behaviour in fishes. Rev. Fish Biol. Fisheries 7: 463–491.

Kennedy, C.J., P. Stecko, B. Truelson and D. Petkovich. 2012. Dissolved organic carbon modulates the effects of copper on olfactory-mediated behaviors of chinook salmon. Environ. Toxicol. Chem. 31: 2281–2288.

Koltes, K.H. 1985. Effects of sublethal copper concentrations on the structure and activity of Atlantic silverside schools. Trans. Amer. Fish Soc. 114: 413–422.

Krang, A.-S. and M. Ekerholm. 2006. Copper reduced mating behaviour in male shore crabs (*Carcinus maenas* L.) Aquat. Toxicol. 80: 60–69.

Kraus, M. and D.B. Kraus. 1986. Differences in the effects of mercury on predator avoidance in two populations of the grass shrimp *Palaemonetes pugio*. Mar. Environ. Res. 18: 277–289.

Kroeker, K., E. Sanford, B. Jellison and B. Gaylord. 2014. Predicting the effects of ocean acidification on predator/prey interactions: A conceptual framework based on coastal mollusks. Biol. Bull. 226: 211–222.

Kunc, H.P., G.N. Lyons, J.D. Sigwart, K.E. McLaughlin and J.D.R. Houghton. 2014. Anthropogenic noise affects behavior across sensory modalities. Am. Nat. 184: E93–E100 (Doi: 10.1086/677545).

Kunc, H.P., K.E. McLaughlin and R. Schmidt. 2016. Aquatic noise pollution: implications for individuals, populations, and ecosystems. Proc. Roy. Soc. 283(1836): 20160839. Doi: 10.1098/rspb.2016.0839.

Lang, W.H., D.C. Miller, P.J. Ritacco and M. Marcy. 1981. The effects of copper and cadmium on the behavior and development of barnacle larvae. pp. 165–203. *In*: Vernberg, F.J., A. Calabrese, F.P. Thurberg and W.B. Vernberg (eds.). Biological Monitoring of Marine Pollutants. Acad. Press NY.

Lesage, V., C. Barrette, M.C.S. Kingsley and B. Sjare. 1999. The effect of vessel noise on the vocal behaviour of Belugas in the St. Lawrence River estuary, Canada. Mar. Mamm. Sci. 15: 65–84.

Liu, C.C., J.M.Y. Chiu, L. Li, P.K.S. Shin and S.G. Cheung. 2011. Respiration rate and swimming activity of larvae of two sub-tidal nassariid gastropods under reduced oxygen levels: Implications for their distributions in Hong Kong waters. Mar. Poll. Bull. 63: 230–236.

Long, W.C. and R. Seitz. 2008. Trophic interactions under stress: hypoxia enhances foraging in an estuarine food web. Mar. Ecol. Prog. Ser. 362: 59–68.

Lord, J.P., J. Barry and D. Graves. 2017. Impact of climate change on direct and indirect species interactions. Mar. Ecol. Prog. Ser. 571: 1–11.

Maltby, L. and M. Crane. 1994. Responses of *Gammarus pulex* (Amphipoda, Crustacea) to metalliferous effluents: identification of toxic components and the importance of interpopulation variation. Environ. Poll. 84: 45–52.

Maneja, R., A. Frommel, H. Browman, A. Geffen, A. Folkward, U. Piatkowski et al. 2015. The swimming kinematics and foraging behavior of larval Atlantic herring (*Clupea harengus* L.) are unaffected by elevated pCO$_2$. J. Exper. Mar. Biol. Ecol. 466: 42–48.

Manno, C., N. Morata and R. Primicerio. 2012. *Limacine retroversa*'s response to combined effects of ocean acidification and sea water freshening. Estuar. Coast. Shelf Sci. 113: 163–171.

Matta, M.B., L. Linse, C. Cairncross, L. Francendese and R.M. Kocan. 2001. Reproductive and transgenerational effects of methylmercury or aroclor 1268 on *Fundulus heteroclitus*. Environ. Toxicol. Chem. 20: 327–335.

McCauley, R.D., J. Fewtrell and A.N. Popper. 2003. High intensity anthropogenic sound damages fish ears. J. Acoust. Soc. Am. 113: 638–642.

McCormick, S.D., L.P. Hansen, T.P. Quinn and R.L. Saunders. 1998. Movement, migration, and smolting of Atlantic salmon (*Salmo salar*). Ca. J. Fish. Aquat. Sci. 55(supp. 1): 77–92.

McIntyre, J., D.H. Baldwin, J.P. Meador and N.L. Scholz. 2008. Chemosensory deprivation in juvenile coho salmon exposed to dissolved copper under varying water chemistry conditions. Envir. Sci. Tech. 42: 1352–1358.

McIntyre, J., D.H. Baldwin, D.A. Beauchamp and N.L. Scholz. 2012. Low-level copper exposures increase visibility and vulnerability of juvenile coho salmon to cutthroat trout predators. Ecol. Applic. 22: 1460–1471.

Miksis-Olds, J. and P. Tyack. 2009. Manatee (*Trichechus manatus*) vocalization usage in relation to environmental noise levels. J. Acoustic Soc. Amer. 125. Doi: 10.1121/1.3068455.

Miller, P.J.O., N. Biassoni, A. Samuels and P.L. Tyack. 2000. Whale songs lengthen in response to sonar. Nature 405: 903.

Mirkes, D.Z., W.B. Vernberg and P.J. DeCoursey. 1978. Effects of cadmium and mercury on the behavioral responses and development of *Eurypanopeus depressus* larvae. Mar. Biol. 47: 143–147.

Munari, C. and M. Mistri. 2011. Short-term hypoxia modulates *Rapana venosa* (Muricidae) prey preference in Adriatic lagoons. J. Exper. Mar. Biol. Ecol. 407: 166–170.

Munari, C. and M. Mistri. 2012. Short-term sublethal hypoxia affects a predator-prey system in northern Adriatic transitional waters. Estuar. Coast. Shelf Sci. 97: 136–140.

Munday, P.L., D.L. Dixson, J.M. Donelson, G.P. Jones, M.S. Pratchett, G.V. Devitsina et al. 2009. Ocean acidification impairs olfactory discrimination and homing ability of a marine fish. Proc. Nat. Acad. Sci. 106: 1848–1852.

Nilsson, G.E., D.L. Dixson, P. Domenici, M.I. McCormick, C. Sørensen, S.-A. Watson et al. 2012. Near-future carbon dioxide levels alter fish behaviour by interfering with neurotransmitter function. Nature Climate Change 2: 201–204.

Nowacek, D.P., L.H. Thorne, D.W. Johnston and P.L. Tyack. 2007. Responses of cetaceans to anthropogenic noise. Mammal. Rev. 37: 81–115.

Nowacek, D.P., C.W. Clark, D. Mann, P.J. Miller, H.C. Rosenbaum, J.S. Golden et al. 2015. Marine seismic surveys and ocean noise: time for coordinated and prudent planning. Front. Ecol. Environ. 13: 378–386.

Nowicki, J.P., G.M. Miller and P.L. Munday. 2011. Interactive effects of elevated temperature and CO$_2$ on foraging behavior of juvenile coral reef fish. J. Exper. Mar. Biol. Ecol. 412: 46–51.

Oakden, J.M., J.S. Oliver and A.R. Flegal. 1984. Behavioral responses of a phoxocephalid amphipod to organic enrichment and trace metals in sediment. Mar. Ecol. Prog. Ser. 14: 253–257.

Olesiuk, P.F., L.M. Nichol, M.J. Sowden and J.K. Ford. 2002. Effect of the sound generated by an acoustic harassment device on the relative abundance and distribution of harbor porpoises (*Phocoena phocoena*) in Retreat Passage, British Columbia. Mar. Mamm. Sci. 18: 843–862.

Ososkov, I. and J.S. Weis. 1996. Development of social behavior in larval mummichogs after embryonic exposure to methylmercury. Trans. Amer. Fish. Soc. 125: 983–987.

Parks, S.E., C.W. Clark and P.L. Tyack. 2007. Short- and long-term changes in right whale calling behaviour: The potential effects of noise on acoustic communication. J. Acoust. Soc. Amer. 122: 3725–3731.

Parks, S.E., M. Johnson, D. Nowacek and P.L. Tyack. 2011. Individual right whales call louder in increased environmental noise. Biol. Lett. 7: 33–35.

Peng, S., J. Huang, T. Xiong and M. Huang. 2004. Effects of four heavy metals on toxicity and feeding behavior of soft coral (*Subergorgia suberosa*). J. Oceanog. Taiwan Strait. 23: 293–301.

Phelps, H., W.H. Pearson and J.T. Hardy. 1985. Clam burrowing behaviour and mortality related to sediment copper. Mar. Poll. Bull. 16: 309–313.

Pichavant, K., J. Person-Le-Ruyet, N. Le Bayon, A. Severe, A. Le Roux and G. Boeuf. 2001. Comparative effects of long-term hypoxia on growth, feeding and oxygen consumption in juvenile turbot and European sea bass. J. Fish Biol. 59: 875–883.

Pihl, L., S.P. Baden, R.J. Diaz and L.C. Schaffner. 1992. Hypoxia-induced structural changes in the diet of bottom-feeding fish and Crustacea. Mar. Biol. 112: 349–361.

Pynnönen, K. 1996. Heavy metal-induced changes in the feeding and burrowing behaviour of a Baltic isopod, *Saduria (Mesidotea) entomon* L. Mar. Environ. Res. 41: 145–156.

Reddy, S. and N. Venugopal. 1991. *In vivo* effects of cadmium chloride on certain aspects of protein metabolism in tissues of a freshwater field crab *Barytelphusa guerini*. Bull. Environ. Contamin. Toxicol. 46: 583–590.

Reddy, P.S. and M. Fingerman. 1995. Effect of cadmium chloride on physiological color changes of the fiddler crab *Uca pugilator*. Ecotox. Environ. Safety 31: 69–75.

Rehnberg, B.C. and C.B. Schreck. 1986. Acute metal toxicology of olfaction in coho salmon: behavior, receptors, and odor-metal complexation. Bull. Environ. Contam. Toxicol. 36: 579–586.

Reynolds, J.D. and J.C. Jones. 1999. Female preference for preferred males is reversed under low oxygen conditions in the common goby (*Pomatoschistus microps*). Behav. Ecol. 10: 149–154.

Rice, S.D., J.W. Short and W.B. Stickle. 1989. Uptake and catabolism of tributyltin by blue crabs fed TBT contaminated prey. Mar. Envir. Res. 27: 137–145.

Ripley, J. and C. Foran. 2007. Influence of estuarine hypoxia on feeding and sound production by two sympatric pipefish species (Syngnathidae). Mar. Environ. Res. 63: 350–367.

Risch, D., P.J. Corkeron, W.T. Ellison and S.M. Van Parijs. 2012. Changes in humpback whale song occurrence in response to an acoustic source 200 km away. PLoS ONE7: art. e29741.

Roast, S.D., J. Widdows and M.B. Jones. 2001. Impairment of mysid (*Neomysis integer*) swimming ability: An environmentally realistic assessment of the impact of cadmium exposure. Aquat. Tox. 52: 217–227.

Roberts, L., S. Cheesman, T. Breithaupt and M. Elliott. 2015. Sensitivity of the mussel *Mytilus edulis* to substrate-borne vibration in relation to anthropogenically generated noise. Mar. Ecol. Prog. Ser. 538: 185–195.

Romano, T.A., M.J. Keogh, C. Kelly, P. Feng, L. Berk, C. Schlundt et al. 2004. Anthropogenic sound and marine mammal health: measures of the nervous and immune systems before and after intense sound exposure. Can. J. Fish. Aquat. Sci. 61: 1124–1134.

Roper, D.S. and C.W. Hickey. 1994. Behavioural responses of the marine bivalve *Macomona liliana* exposed to copper- and chlordane-dosed sediments. Mar. Biol. 118: 673–680.

Roper, D.S., M.G. Nipper, C.W. Hickey, M.L. Martin and M.A. Weatherhead. 1995. Burial, crawling and drifting behaviour of the bivalve *Macomona liliana* in response to common sediment contaminants. Mar. Poll. Bull. 31: 471–478.

Sagasti, A., L.C. Schaffner and J.E. Duffy. 2001. Effects of periodic hypoxia on mortality, feeding and predation in an estuarine epifaunal community. J. Exper. Mar. Biol. Ecol. 258: 257–283.

Sandahl, J., D. Baldwin, J. Jenkins and N. Scholz. 2004. Odor-evoked field potentials as indicators of sublethal neurotoxicity in juvenile coho salmon (*Oncorhynchus kisutch*) exposed to copper, chlorpyrifos, or esfenvalerate. Can. J. Fish. Aquat. Sci 61: 404–413.

Santos, M.H.S., N.T. da Cunha and A. Bianchini. 2000. Effects of copper and zinc on growth, feeding and oxygen consumption of *Farfantepenaeus paulensis* postlarvae (Decapoda: Penaeidae). J. Exper. Mar. Biol. Ecol. 247: 233–242.

Saucier, D., L. Astic and P. Riouz. 1991. The effects of early chronic exposure to sublethal copper on the olfactory discrimination of rainbow trout, *Oncorhynchus mykiss*. Environ. Biol. Fish 30: 345–351.

Scarfe, A.D., K.A. Jones, C.E. Steele, H. Kleerekoper and M. Corbett. 1982. Locomotor behavior of four marine teleosts in response to sublethal copper exposure. Aquat. Toxicol. 2: 335–353.

Scott, G.R. and K.A. Sloman. 2004. The effects of environmental pollutants on complex fish behaviour: integrating behavioural and physiological indicators of toxicity. Aquat. Toxicol. 68: 369–392.

Sebastianutto, L., M. Picciulin, M. Costantini and E.A. Ferrero. 2011. How boat noise affects an ecologically crucial behaviour: the case of territoriality in *Gobius cruentatus* (Gobiidae). Environ. Biol. Fishes 92: 207–215.

Shin, P., A.W.M. Ng and R.Y.H. Cheung. 2002. Burrowing responses of the short-neck clam *Ruditapes philippinarum* to sediment contaminants. Mar. Poll. Bull. 45: 133–139.

Simpson, S.D., P.L. Munday, M.L. Wittenrich, R. Manassa, D.L. Dixson, M. Gagliano et al. 2011. Ocean acidification erodes crucial auditory behavior in marine fish. Biol. Lett. 10: 1098.

Simpson, S.D., A.N. Radford, S.L. Nedelec, M.C.O. Ferrari, D.P. Chivers, M.I. McCormick et al. 2016. Anthropogenic noise increases fish mortality by predation. Nat. Commun. 7: 10544.

Sloman, K.A. and P.L. McNeil. 2012. Using physiology and behavior to understand the responses of fish early life stages to toxicants. J. Fish Biol. 81: 2175–2198.

Smith, G.M., A.T. Khan, J.S. Weis and P. Weis. 1995. Behavior and brain correlates in mummichogs (*Fundulus heteroclitus*) from polluted and unpolluted environments. Mar. Environ. Res. 39: 329–333.

Sobrino-Figueroa, A. and C. Carceres-Martinez. 2012. Alterations of valve-closing behavior in juvenile Caterina scallops (*Argopecten ventricosus* Sowerby 1842) exposed to toxic metals. Ecotoxicology 18: 983–987.

Solan, M., C. Hauton, J.A. Godbold, C.L. Wood, T.G. Leighton and P. White. 2016. Anthropogenic sources of underwater sound can modify how sediment-dwelling invertebrates mediate ecosystem properties. Sci. Rep. 6: 20540. DOI: 10.1038/srep20540.

Stanley, J.A., S.M. Van Parijs and L.T. Hatch. 2017. Underwater sound from vessel traffic reduces the effective communication range in Atlantic cod and haddock. Sci. Reports 7: 14633. DOI: 10.1038/s41598-017-14743-9.

Staurnes, M., F. Kroglund and B.O. Rosseland. 1996a. Water quality requirement of Atlantic salmon (*Salmo salar*) in water undergoing acidification or liming in Norway. Water, Air, Soil Poll. 85: 347–352.

Staurnes, M., L.P. Hansen, K. Fugelli and O. Haraldstad. 1996b. Short-term exposure to acid water impairs osmoregulation, seawater tolerance, and subsequent marine survival of smolt of Atlantic salmon (*Salmo salar* L.). Can. J. Fish. Aquat. Sci. 53: 1695–1704.

Steirhoff, K.L., T. Targett and K. Miller. 2006. Ecophysiological responses of juvenile summer and winter flounder to hypoxia: experimental and modeling analyses of effects on estuarine nursery quality. Mar. Ecol. Prog. Ser. 325: 255–266.

Sullivan, B.K., E. Buskey, D.C. Miller and P.J. Ritacco. 1983. Effects of copper and cadmium on growth, swimming and predator avoidance in *Eurytemora affinis* (Copepoda). Mar. Biol. 77: 299–306.

Sunda, W.G., D.W. Engel and R.M. Thuotte. 1978. Effect of chemical speciation on toxicity of cadmium to the grass shrimp, *Palaemonetes pugio*: Importance of free cadmium ion. Envir. Sci. Tech. 12: 409–413.

Sunda, W.G. and J.A. Lewis. 1978. Effect of complexation by natural organic ligands on the toxicity of copper to the unicellular alga *Monochrisis lutheri*. Limnol. Oceanog. 23: 870–876.

Taylor, D.L. and D.B. Eggleston. 2000. Effects of hypoxia on an estuarine predator-prey interaction: foraging behavior and mutual interference in the blue crab *Callinectes sapidus* and the infaunal clam prey *Mya arenaria*. Mar. Ecol. Prog. Ser. 196: 221–237.

Thetmeyer, H., U. Waller, K.D. Black, S. Inselmann and H. Rosenthal. 2001. Growth of European sea bass (*Dicentrarchus labrax* L.) under hypoxic and oscillating oxygen conditions. Aquaculture 174: 355–367.

Vargas, C., V. Aguilera, V. San Martín, P.H. Manríquez, J. Navarro, C. Duarte et al. 2015. CO_2-Driven ocean acidification disrupts the filter feeding behavior in Chilean gastropod and bivalve species from different geographic localities. Estuar. Coast. 38: 1163–1177.

Vasconcelos, R.O., M.C. Amorim and F. Ladich. 2007. Effects of ship noise on the detectability of communication signals in the Lusitanian toadfish. J. Exper. Biol. 210: 2104–2112.

Viera, L.R., C. Gravato, A.M. Soares, F. Morgado and L. Guilhermina. 2009. Acute effects of copper and mercury on the estuarine fish *Pomatoschistus microps*: Linking biomarkers to behavior. Chemosphere 76: 1416–1427.

Wahlberg, M. and H. Westerberg. 2005. Hearing in fish and their reactions to sounds from offshore wind farms. Mar. Ecol. Prog. Ser. 288: 295–309.

Wale, M.A., S.D. Simpson and A.N. Radford. 2013. Noise negatively affects foraging and antipredator behaviour in shore crabs. Anim. Behav. 86: 111–118.

Wallace, W.G., T.M.H. Brouwer, M. Brouwer and G. Lopez. 2000. Alterations in prey capture and induction of metallothioneins in grass shrimp fed cadmium-contaminated prey. Environ. Toxicol. Chem. 19: 962–971.

Wallace, W.G. and A. Estephan. 2004. Differential susceptibility of horizontal and vertical swimming activity to cadmium exposure in a gammaridean amphipod (*Gammarus lawrencianus*). Aquat. Toxicol. 69: 289–297.

Wallace, W.G., T.M. Hoexum Brouwer, M. Brouwer and G.R. Lopez. 2009. Alterations in prey capture and induction of metallothioneins in grass shrimp fed cadmium-contaminated prey. Envir. Toxicol. Chem. 19: 962–971.

Wang, L., I. Song, Y. Chen, H. Ran and J. Song. 2017. Impact of ocean acidification on the early development and escape behavior of marine medaka (*Oryzias melastigma*) Mar. Environ. Res. 131: 10–18.

Wang, Y., M. Hu, S.G. Cheung, P.K. Shin, W. Lu and J. Li. 2012. Chronic hypoxia and low salinity impair anti-predatory responses of the green-lipped mussel *Perna viridis*. Mar. Environ. Res. 77: 84–89.

Watson, S.-A., S. Lefevre, M.I. McCormick, P. Domenici, G.E. Nilsson and P.L. Munday. 2013. Marine mollusc predator-escape behaviour altered by near-future carbon dioxide levels. Proc. R. Soc. Lond. B Biol. Sci. 281: 20132377.

Weis, J.S. and A.A. Khan. 1990. Effects of mercury on the feeding behavior of the mummichog, *Fundulus heteroclitus* from a polluted habitat. Mar. Environ. Res. 30: 243–249.

Weis, J.S. and P. Weis. 1995a. Effects of embryonic exposure to methylmercury on larval prey capture ability in the mummichog, *Fundulus heteroclitus*. Environ. Toxicol. Chem 14: 153–156.

Weis, J.S. and P. Weis. 1995b. Effects of embryonic and larval exposure to methylmercury on larval swimming performance and predator avoidance in the mummichog, *Fundulus heteroclitus*. Can. J. Fish. Aquat. Sci. 52: 2168–2173.

Weis, J.S. and P. Weis. 1998. Effects of lead on behaviors of larval mummichogs, *Fundulus heteroclitus*. J. Exper. Mar. Biol. Ecol. 222: 1–10.

Wibe, A.E., T. Nordtug and B.M. Jenssen. 2001. Effects of bis(tributyltin)oxide on antipredator behavior in threespine stickleback *Gasterosteus aculeatus* L. Chemosphere 44: 475–481.

Wong, C., K.H. Chu, K.W. Tang and T.W. Tam. 1993. Effects of chromium, copper and nickel on survival and feeding behaviour of *Metapenaeus ensis* larvae and postlarvae (Decapoda: Penaeidae). Mar. Envir. Res. 36: 63–78.

Yu, A., X. Wang, Z. Zuo, J. Cai and C. Wang. 2013. Tributyltin exposure influences predatory behavior, neurotransmitter content and receptor expression in *Sebastiscus marmoratus*. Aquat. Toxicol. 128-129: 158–162.

Yu, X.Y., K.R. Yip, P.K.S. Shin and S.G. Cheung. 2017. Predator–prey interaction between muricid gastropods and mussels under ocean acidification. Mar. Poll. Bull. 124: 911–916.

Zhou, T. and J.S. Weis. 1998. Swimming behavior and predator avoidance in three populations of *Fundulus heteroclitus* larvae after embryonic and/or larval exposure to methylmercury. Aquat. Toxicol. 43: 131–148.

Zhou, T., H.B. John-Alder, P. Weis and J.S. Weis. 1999. Thyroidal status of mummichogs (*Fundulus heteroclitus*) from a polluted versus a reference habitat. Environ. Toxicol. Chem. 18: 2817–2823.

Zhou, T., D. Rademacher, R.E. Steinpreis and J.S. Weis. 1999. Neurotransmitter levels in two populations of larval *Fundulus heteroclitus* after methylmercury exposure. Comp. Biochem. Physiol. Part C 124: 287–294.

Zhou, T., R. Scali and J.S. Weis. 2001. Effects of methylmercury on ontogeny of prey capture ability and growth in three populations of larval *Fundulus heteroclitus*. Arch. Environ. Contam. Toxicol. 41: 47–54.

Zhou, W., X. Huang and X. Xu. 2018. Changes of movement behavior and HSP70 gene expression in the hemocytes of the mud crab (*Scylla paramamosain*) in response to acoustic stimulation. Mar. Freshwat. Behav. Physiol. DOI: 10.1080/10236244.2018.1439337.

Zizza, M., G. Giusi, M. Crudo, M. Canonaco and R.M. Facciolo. 2013. Lead-induced neurodegenerative events and abnormal behaviors occur *via* ORXRergic/GABA$_A$Rergic mechanisms in a marine teleost. Aquat. Toxicol. 126: 231–241.

7

Eco-friendly Strategies of Remediation in the Marine System

Bioremediation and Phytoremediation

Vanesa L. Negrin,[1,2,]* *Lautaro Gironés*[1] and *Analía V. Serra*[1]

1. Introduction

Several anthropogenic activities are responsible for the release of large amounts of inorganic and organic pollutants into the marine ecosystem both directly (by effluent discharges to the sea) and indirectly (Jacob et al. 2018, Marques 2016). Indirect pollution (i.e., coming from the terrestrial ecosystem) is responsible for most of the environmental issues in the marine ecosystem. Land-derived pollution is transferred to the ocean by the continent-ocean interface, through the atmosphere and watersheds, which make coastal areas particularly vulnerable. This interface encompasses approximately 100 km of shoreline, which corresponds to less than 25% of the planet's surface. However, nearly 50% of the human population and 75% of megacities are located there, with more than 10 million inhabitants. In addition, coastal areas are the sites of industrial and tourism development (Marcovecchio and de Lacerda 2017, Barletta et al. 2019).

Marine systems are polluted by metals, oil, pesticides, radionuclides, gasoline additives and other substances that pose several threats to the environment. Fortunately, over the last century, public concern about pollution has greatly increased, but remediation of sediments and waters is a complex issue, even in rich countries. There is a wide range of remediation strategies including chemical, physico-chemical, thermal and, more recently, biological techniques (e.g., Lim et al. 2016, Jacob et al. 2018, Marques 2016). Traditional techniques are still widely used, and they are the only option in some cases, but biological strategies, namely bioremediation and phytoremediation, have gained importance since they are eco-friendly and usually low-cost and, hence, acceptable to the public (Jacob et al. 2018, Marques 2016). Therefore, here we will discuss the main biological techniques that are

---

[1] Instituto Argentino de Oceanografía (IADO, CONICET, CCT-Bahía Blanca), Camino La Carrindanga km 7,5, Edificio E-1 CC 804 (8000) Bahía Blanca, Buenos Aires, Argentina.

[2] Departamento de Biología, Bioquímica y Farmacia, UNS (8000) Bahía Blanca, Buenos Aires, Argentina.
  Emails: lgirones@iado-conicet.gob.ar; avserra@iado-conicet.gob.ar

* Corresponding author: vlnegrin@criba.edu.ar

or can be applied to marine ecosystems to remediate both inorganic pollutants (namely metals and metalloids, hereafter metals) and organic pollutants (e.g., oil, pesticides). Our goal is to provide an integrative approach of biological remediation of most important pollutants in marine ecosystems, since most studies focus on agricultural soils.

2. Biological Remediation

These techniques involve the use of living organisms for the treatment of contaminated soil and water and reduce pollution to a safe or acceptable level. They are usually preferred over conventional remediation but it should be considered that they are only applicable for low or moderate levels of pollutants and because of that, sometimes, they need to be used in conjunction with different traditional methods to reach those levels (Liu et al. 2018, Yadav et al. 2018). Biological techniques include bioremediation and phytoremediation.

2.1 Bioremediation

The US Environmental Protection Agency (US EPA 2005) defined bioremediation as "the use of living organisms to clean up oil spills or remove other pollutants from soil, water, or wastewater; use of organisms such as non-harmful insects to remove agricultural pests or counteract diseases of trees, plants, and garden soil." In other words, bioremediation is the use of microorganisms or their products to remediate polluted sites. Bioremediation was developed in the 1940s, but only gained popularity in the 1980s because of the famed Exxon Valdez oil spill (Lim et al. 2016).

In general, bioremediation proceeds through different microbial processes such as bioaccumulation, biosorption, biotransformation or biomineralization (Khalid et al. 2017, Jacob et al. 2018). Bioaccumulation refers to the retention and concentration of the pollutant within the microorganism, whereas in biosorption the substances are associated with the cell wall, without entering the cell. On the other hand, biotransformation implies the enzymatic transformation of pollutants to a non-toxic or less toxic form and biomineralization refers to the use of microbially generated ligands to precipitate pollutants (Tabak et al. 2005).

2.1.1 Microorganisms involved in bioremediation

Bacteria. Among microorganisms, bacteria play a key role in the mitigation of environmental pollution caused by inorganic and organic pollutants, given their wide distribution worldwide and their ability to grow easily and quickly (dos Santos and Maranho 2018, Yin et al. 2019). Bacteria-based bioremediation requires that competent bacterial strains function in the presence of the target pollutants, as well as other pollutants (Abatenh et al. 2017).

Fungi (mycoremediation). Filamentous fungi are ubiquitous microorganisms that are present in a wide range of ecological niches since they have the ability to adapt to varying carbon and nitrogen sources (Jacob et al. 2018). Fungi are important organisms in the remediation of oil, due to their enzymatic activities, and for the treatment of some metals (Congeevaram et al. 2007, dos Santos and Maranho 2018). One advantage of the use of fungal biomass is that they can be easily and inexpensively cultured on a large scale using simple fermentation techniques (Congeevaram et al. 2007).

Algae (phycoremediation). Phycoremediation is the use of micro- or macroalgae to remediate different pollutants such as nutrients, metals, or organic compounds in water or to remove CO_2 from the air (Deniz and Karabulut 2017). Algae are photosynthetic and highly adaptive microorganisms, having fast growth and wide niches (i.e., freshwater and marine environments as well as moist soil) (Yin et al. 2019).

Microbial mats. Microorganisms are not usually isolated but mixed in a consortium with coordinated and synergistic metabolisms. Microbial mats are the aggregation of microorganisms (bacteria, fungi and mainly algae) arranged in multilayer cohesive structures. These assemblages of cells, sediments and biological extracellular polymeric substances (EPS) are remarkably cosmopolitan, encountered in practically every country of the world (e.g., Tabak et al. 2005, Serra et al. 2017). Several studies have revealed that these microbial mats might be useful for the bioremediation of organic and inorganic pollutants (e.g., Cohen 2002, Serra et al. 2017). The success of this consortium relies on the association of mixed populations with different enzymatic abilities to degrade complex mixtures of pollutants (dos Santos and Maranhao 2018).

2.1.2 Factors affecting bioremediation

Microbial action depends not only on the type of microorganism but also on a wide range of abiotic factors.

Temperature greatly impacts the efficiency of bioremediation, first by its direct effect on physicochemical state of the pollutant and second through its effect on the metabolic activity of microorganisms. The rate of microbial activities increases with temperature and reaches its maximum at an optimum temperature (Adams et al. 2015, Jacob et al. 2018).

pH is another important factor in bioremediation due to its ability to control microbial growth, enzyme activity, metal complexation chemistry and behavior of functional groups in the cell surface of microorganisms (Abatenh et al. 2017). Although biodegradation can occur in a wide range of pH, generally a pH between 6.5 and 8.5 is optimal for biodegradation in most aquatic and terrestrial systems (Adams et al. 2015, Venosa and Zhu 2003). Moreover, the pH of the pollutants has an impact on microbial metabolism, increasing or decreasing the removal process (Abatenh et al. 2017).

Oxygen is another key factor in bioremediation since most microbial processes involve oxidation (Jacob et al. 2018). The availability of oxygen in soils depends on soil characteristics and rates of microbial oxygen consumption (Tabak et al. 2005, dos Santos and Maranho 2018). Moreover, redox status influences the process related to metal binding (sorption and desorption), precipitation, complex formation and speciation of pollutants (Jacob et al. 2018).

The content of organic matter and clay in soils increases the absorption of organic pollutants and metals and decreases their bioavailability to be taken up by microorganisms (Kabata-Pendias 2010, Perelo 2010, Jacob et al. 2018). Soil moisture should also be considered since microorganisms require adequate water to accomplish their growth and because the rate of metabolization of pollutants is influenced by the kind and amount of soluble materials that are available as well as the osmotic pressure of terrestrial and aquatic systems (Venosa and Zhu 2003, Abatenh et al. 2017).

The availability of essential nutrients such as C, N, P, S, Ca and Mg is directly involved in microbial survival and growth, increasing the metabolic activity of microorganisms and thus the biodegradation rate (Abatenh et al. 2017). Nutrient balancing, especially the supply of N and P, can improve the biodegradation efficiency by optimizing the C:N:P ratio.

The degree and mechanisms of toxicity vary with the pollutant, its concentration, and the exposed microorganisms, but removal efficiency generally increases as the initial pollutant concentration increases (Abatenh et al. 2017). However, at elevated levels, pollutants seriously affect the growth and metabolic activities of microorganisms and can slow down remediation (Adams et al. 2015). That is why bioremediation is recommended for systems with low or moderate pollution.

2.1.3 Main bioremediation technologies

There is a wide variety of bioremediation techniques (Cristaldi et al. 2017) but we will focus here on the ones that are or could be used in marine environments (Fig. 1).

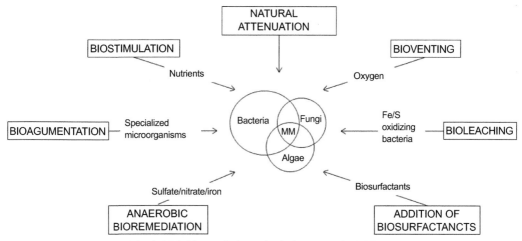

Fig. 1. Main bioremediation technologies. MM: microbial mat.

Natural attenuation involves leaving contaminated sediments in place and allowing ongoing natural processes to degrade or immobilize the contaminant *in situ*. Although no action is required, natural attenuation should be the result of a thoughtful decision following detailed site characterization and monitoring. This technique is not expensive and is usually the most effective for sites with low-level or diffuse contamination (Perelo 2010). It is the appropriate strategy where the contaminants can be remediated in an acceptable time and where the ecological impacts of other techniques would have greater negative impacts.

Bioaugmentation is defined as the introduction of allochthonous species to enhance remediation of pollutants. The added microorganism could be a particular mixture of species (consortium) or genetically engineered species (with specific catabolic activities) (Perelo 2010). However, it should be considered that non-indigenous cultures are not good competitors with an indigenous species to develop and sustain useful population levels. Bioaugmentation is an effective bioremediation technology for removal of persistent organic pollutants and metals (Lim et al. 2016, Marques 2016).

Biostimulation is based on the stimulation of indigenous populations of microorganisms by removing the factors limiting their growth (Adams et al. 2015). In other words, there is an improvement of environmental requirements (i.e., pH, temperature, oxygen) and/or the injection of limiting nutrients such as N, P and C (Margesin and Schinner 2001, Abatenh et al. 2017). This strategy has an advantage over bioaugmentation: native microorganisms are well adapted and distributed in the environment. The challenge lies in the delivery of additives in a form that is available to microorganisms (Adams et al. 2015).

Bioventilation or bioventing is an *in situ* technology that involves the addition of oxygen to the soil to stimulate the growth of indigenous or introduced microorganisms (Abatenh et al. 2017). The presence of oxygen could enhance the metabolism of organic matter and generate more energy for the microorganisms (Lim et al. 2016). Bioventing uses low air flow rates to provide only enough oxygen to sustain microbial activity; oxygen is usually supplied through direct air injection into the soil by means of wells (Abatenh et al. 2017).

Anaerobic bioremediation involves the remediation of pollutants through the stimulation of indigenous anaerobic microorganisms using alternative acceptor donors (other than oxygen). Nitrate, sulfate or iron can be added to the contaminated site to stimulate anaerobic bioremediation (Balagurusamy 2005). This method is considered feasible especially in contaminated sites that are already anaerobic in nature, where it is a cost-effective method as compared to aerobic bioremediation process, as the delivery of oxygen to the contaminated site could be expensive (Balagurusamy 2005, Lim et al. 2016).

Biosurfactants are microbe-derived compounds that lower the surface tension between two liquids or between a liquid and a solid and thus are more likely to interact with other compounds, which improve the solubility and bioavailability of hydrophobic contaminants (Dell'Anno et al. 2018). Compared to synthetic surfactants, biosurfactants have lower toxicity, biodegradability, selectivity and specific activity at extreme temperatures, pH and salinity (Yang et al. 2016). Addition of biosurfactants, which can be produced by different microbial taxa, can improve bioremediation performance, especially in sites contaminated by organic pollutants (Dell'Anno et al. 2018).

Bioleaching is the leaching of metals mediated by microorganisms. In other words, it is the application of acidophilic Fe/S-oxidizing microorganisms to promote solubilization of metals from solid matrices (Fonti et al. 2016). During this process, the microorganisms use iron- or sulfur-reduced compounds as substrate for their growth, producing soluble metal sulfates (White et al. 1998). Sediments with high contents of metal sulfides and other reduced forms of metals are useful for this technique (Akcil et al. 2015).

2.2 Phytoremediation

Phytoremediation is defined as the use of plants to remove pollutants from the environment. It is a technique driven by solar energy, which is particularly useful in the treatment of large areas of soil (Mahar et al. 2016, Yadav et al. 2018). Phytoremediation is used mainly in metal-contaminated soils but can also be used for radionuclides, pesticides, and other pollutants (Yadav et al. 2018). It is worth noting that plants generally handle the pollutants without affecting soil, and they can even improve soil fertility through the input of organic matter. Moreover, phytoremediation has low installation and maintenance costs compared to other remediation options and has a secondary aesthetic role (Ali et al. 2013).

Phytoremediation is a relatively recent technology. Studies have been conducted mostly since 1990. The concept of phytoremediation (as phytoextraction) was suggested by Chaney (1983). The discovery of hyperaccumulators—plants capable of absorbing metals at a level 50 to 500 times that of normal plants—has led to the revolutionary progress of phytoextraction technology (Ali et al. 2013, Mahar et al. 2016). Hyperaccumulators have the potential to accumulate metals to levels that would be lethal for non-accumulating plants (Reeves et al. 2018).

Despite the numerous advantages, this technology has some limitations. For instance, it requires the knowledge of some biological aspects of the species to be used (e.g., phytotoxicity limits, accumulation capacity, mechanisms of tolerance). The ideal plant species should be fast-growing and easy to cultivate and handle, produce large root biomass, be resistant to pathogens, and be adapted to the climatic conditions of the region (Yadav et al. 2018). Therefore, the choice of the appropriate plant species is a key issue in phytoremediation.

2.2.1 Factors affecting phytoremediation

As with bioremediation, several abiotic factors influence the success of phytoremediation. Redox and pH of sediments behave similarly to that in bioremediation, modifying the form of pollutants, mainly metals, and thus affecting their bioavailability for plants. High organic matter and clay content reduce pollutant bioavailability for plant uptake, reducing the potential for phytoextraction, for example (Ghosh and Singh 2005). Root exudates as well as nutrient and moisture content of the soil also have implications for the properties and bioavailability of pollutants, as well as for the proper development of the plant. The time of the year, which rules temperature and solar irradiance, is a key factor in the photosynthetic rate and thus in the production of plant biomass (Yadav et al. 2018).

2.2.2 Phytoremediation strategies

Depending on the strategy of action of the plant to remove the pollutant, phytoremediation can take various forms (Fig. 2).

Phytoextraction is the uptake of pollutants by plants through the roots and their translocation to above-ground tissues, such as leaves and shoots, where they are accumulated (Weis and Weis 2004). The above-ground tissues are then harvested for safe disposal (Yadav et al. 2018). This technique can be considered for long-term clean-up of metal-contaminated soils, and its efficiency can be enhanced by increasing metal bioavailability, which in turn can be achieved by modifying soil properties (Ghosh and Singh 2005, Yadav et al. 2018). Phytoextraction can also be used to remove radionuclides, perchlorate and other organic compounds (Herath and Vithanage 2015). The plants used for phytoextraction should be hyperaccumulators of pollutants in above-ground tissues, easy to harvest and unpalatable for herbivores to avoid transference within the food web (Mahar et al. 2016, Yadav et al. 2018).

Phytostabilization (or phytoimmobilization) is the immobilization of pollutants by plants and their storage in underground tissues and/or soil (Weis and Weis 2004). Phytostabilization does not reduce the concentration of pollutants present in the contaminated soil but prevents their off-site movement.

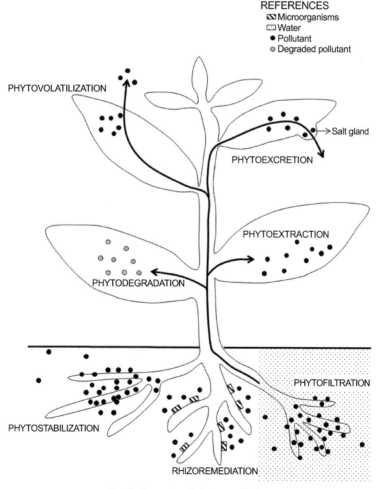

Fig. 2. Strategies of phytoremediation.

This is particularly useful for metals, promoting immobilization through metal sorption by roots, precipitation, complexation, and other means (Ghosh and Singh 2005, Ali et al. 2013). It has also been reported to be efficient in removal of phenols and chlorinated solvents (Herath and Vithanage 2015). Phytostabilization is highly effective in sites having fine-textured soil and with high organic matter (Yadav et al. 2018).

Phytovolatilization is the uptake of pollutants from soil by plants, their conversion to volatile forms and their subsequent release into the atmosphere through transpiration (Ali et al. 2013). It has been demonstrated to be useful for chlorinate solvents and for elements such as As, Hg and Se (Mahar et al. 2016). Metals are converted into gaseous/volatilized forms inside the plants via specific mechanisms, which are usually governed by some specific enzymes or genes, but there are a very low number of naturally occurring plants capable of converting metals into volatilized form. Therefore, genetic engineering is generally used to enhance the ability of plants to volatilize metals (Khalid et al. 2017).

Phytofiltration (or rhizofiltration) is the use of plants for removing pollutants, primarily metals, from water (Rezania et al. 2016, Yadav et al. 2018). In phytofiltration, the pollutants are absorbed or adsorbed to plant roots and thus their movement is minimized; hence, they are prevented from mixing with water streams or groundwater (Ali et al. 2013, Limmer and Burken 2016). Wetland plants are suitable material for phytofiltration and constructed wetlands are the best environments for removing metals from wastewater and nutrient-recycling systems (Rezania et al. 2016). Phytofiltration remediates metals such as Pb, Cd, Ni, Cu, Cr, V and radionuclides (U, Cs and Sr) (Yadav et al. 2018).

Phytodegradation (or phytotransformation) is the degradation of organic pollutants by plant enzymes such as dehalogenase and oxygenase; it is not dependent upon rhizospheric microorganisms (Ali et al. 2013). Phytodegradation is limited to removal of organic impurities since metals are non-biodegradable (Mahar et al. 2016). Recent research has focused on phytodegradation of various organic compounds such as synthetic herbicides, insecticides, munitions, DDT, TNT, and chlorinated solvents. Genetically modified plant species are currently being tested for their phytodegradation capabilities (Ali et al. 2013, Yadav et al. 2018).

Rhizoremediation (or rhizosphere bioremediation) is an enhanced remediation of pollutants by root-associated bacteria and fungi (Ghosh and Singh 2005). In the rhizosphere there is a symbiotic interaction where plants supply the carbon sources—through the production of root exudates containing sugar, amino acids, flavonoids—that stimulate the activity of microorganisms, which in turn benefit the plant by supplying vitamins, amino acids, and other nutrients (Cristaldi et al. 2017, Yadav et al. 2018). Remediation is done by microorganisms within the rhizosphere. It is suitable mainly for organic pollutants such as petroleum hydrocarbons; in this case it is usually called rhizodegradation (dos Santos and Maranho 2018). Rhizoremediation could also be applied to metals, promoting the conversion from toxic to non-toxic or less toxic forms; for instance, reduction of Fe(III) and Mn(IV) to Fe(II) and Mn(III), respectively (Jacob et al. 2018). Symbiosis between plants and microorganisms may also modify soil conditions, allowing increased metal bioavailability or promoting aerobic degradation of some organic pollutants. Besides the removal of pollutants, microbe-assisted phytoremediation also supports plant growth (Yadav et al. 2018).

Phytoexcretion is a novel phytoremediation technology that is only applicable with some halophytes, which are plants that tolerate high concentrations of salt in the soil (Manousaki and Kalogerakis 2011). A halophyte has different mechanisms to deal with high salinity and some species excrete the excess of salt through glands in the leaves. These glands are not always specific to Na^+ and Cl^-, and other toxic elements such as some metals (Cd, Zn, Pb, Cu) are accumulated and excreted by salt glands. The term "phytoexcretion" has been recently introduced based on this property of some halophytes (Kadukova et al. 2008, Manousaki and Kalogerakis 2011).

3. Biological Remediation in Marine Environments

As already mentioned, the use of bioremediation and phytoremediation is highly dependent on a number of factors, such as the type of pollutant (mainly organic versus inorganic), the species of microorganism or plant available, and physicochemical variables. Despite the general physicochemical factors previously mentioned (i.e., pH, Eh, nutrient content, temperature), in the marine ecosystem some of these factors are more relevant and still others emerge as ruling ones. Salinity and redox conditions are especially important in intertidal sediments that are subjected to regular tidal flooding.

Intertidal sediments, main deposits of pollutants of marine ecosystems, are usually subjected to high salinity and reductive conditions due to flooding with tidal water. Salinity in sediments and waters is a key factor for pollutant distribution, especially in estuaries. Salinity may affect biodegradation through alteration of the microbial population and the physicochemical state of the pollutants (Venosa and Zhu 2003). In addition, salinity enhances the sorption of organic pollutants in sediments, whereas it affects metal speciation, flocculation and sedimentation of fine particles (Barletta et al. 2019). Although oxygen is not limiting in the water column in marine environments, intertidal sediments show oxic conditions only in a thin layer at the surface of the sediment. The depletion of oxygen in depth due to tidal inundation may give rise to anoxic conditions that are responsible for the immobilization of pollutants such as metals (Weis and Weis 2004, Kabata-Pendias 2010).

The type of pollutant is also very important in the selection of the type of remediation technique. Phytoremediation and bioremediation technologies have been applied to both organic and inorganic pollutants present in soil and water. Metals are not degraded but can be transformed to non-toxic or less toxic forms, or they can be immobilized in plant tissues or surrounding sediments. Organic pollutants are usually degraded to a variable degree with time. Hence, here different pollutants will be explained separately, although we are aware that most of the time more than one type of pollutant is present in a specific ecosystem, which makes the selection of the remediation action more difficult.

3.1 Metals

The marine environment has been increasingly subjected to metal pollution from natural processes (i.e., geochemical) and anthropogenic activities (de Souza Machado et al. 2016, Marques 2016). Sediments constitute the major and final phase of pollution by metals in the aquatic systems. Sediments are the compartment in which particulate and dissolved substances in the water column (including metals) tend to accumulate (Akcil et al. 2015) and, hence, remediation actions are focused there. Changes in sediment Eh, pH, salinity, and other variables can cause changes in metal speciation and solubility, which can result in a flux from sediments to porewater and then to overlying water.

Conventional remediation strategies of metal-polluted sediments include *in situ* and relocation actions. *In situ* techniques involve capping, where the sediment is isolated by the environment by inert or reactive barriers, and natural attenuation. Relocation actions involve mainly dredging followed by landfill or dumping the sediment at sea. However, effective treatment of dredged material prior to dumping is needed. Traditional techniques that are used for aquatic dredged sediments are: (1) sediment washing, where metals are solubilized by aqueous solutions of extracting chemical or chelating agents; (2) electro-chemical treatments, where metal cations are separated by applying an electromagnetic field; and (3) thermal treatments, where heat is applied to dewatered sediment (US EPA 2005). All these techniques raise concerns about their sustainability and environmental impact; thus, there is an active search for alternatives for biological treatment of metal-polluted sediments. Bioremediation, and mainly phytoremediation, for metals is under exhaustive study, although most works refer to laboratory or pilot-scale studies.

3.1.1 Bioremediation

Heavy metals cannot be degraded but can be remediated. Microorganisms can remediate metals by several mechanisms: (1) changing their oxidation state (e.g., Cr(VI) to the less toxic Cr(III)); (2) biosorption to cell surface; (3) production of siderophores (Fe-complexing molecules), which facilitates extraction of metals from the soil; (4) extracellular chemical precipitation (e.g., conversion of sulfate to hydrogen sulfate by sulfate-reducing bacteria, which subsequently react with heavy metals such as Cd and Zn and precipitate them); and (5) volatilization, through the microbial reduction toxic forms of Hg and Se to volatile less toxic forms (e.g., Chibuike and Obiora 2014, Liu et al. 2018).

As stated previously, bioremediation can be performed by different microorganisms. The most commonly used microorganism for metal remediation in marine environments is bacteria, although fungi and algae have also proved to be useful. Living or dead biomass of marine fungi species such as *Aspergillus*, for example, were successful in the biosorption of Cu, Pb and especially Cr (Marques 2016 and references therein). Several species of macroalgae from marine environments have also been pointed out as useful biosorbents due to the large quantities of biopolymers that bind metals that are usually present in their cell walls (Jacob et al. 2018). Although brown algae are more suitable for bioremediation of polluted waters due to the presence of alginates in their cell walls (Mazur et al. 2018, Yin et al. 2019), some species of green and red algae have also been reported as good biosorbents (Deniz and Karabulut 2017, Yin et al. 2018). Regarding consortia, microbial mats are well documented for removal of metals in marine sediments (e.g., Burgos et al. 2013, Serra et al. 2017).

Several techniques of bioremediation can be applied to the treatment of metals in marine sediments.

Biosurfactants. Although biosurfactants are mainly used for the remediation of organic pollutants, they could also represent an alternative for the removal of metals from sediments (Dell'Anno et al. 2018). The biosurfactants most commonly used in metal bioremediation include molecules with an electric charge, i.e., cationic and anionic biosurfactants that bind metals having opposite charge (Lawniczak et al. 2013). There are several biosurfactants that can be produced by marine bacteria, rhamnolipids being the most studied, which can desorb metals such as Cd, Cu, Pb, Zn and Ni from contaminated sediments (Mulligan 2005, Dell'Anno et al. 2018).

Bioleaching. The inoculation of bioleaching bacteria (i.e., Fe/S acidophilic oxidizing bacteria) can be used for the treatment of dredged sediments (e.g., White et al. 1998, Akcil et al. 2015). In marine sediments, the studies are poor but promising. Beolchini et al. (2009) reported a solubilization of 90–100% for Cd, Cu, Zn and Hg and 40% for Pb and Ni by bioleaching. In addition, Fonti et al. (2013), after a pre-treatment (washing for removing salt), found solubilization efficiencies always over 50% for Zn and variable results for Cd and Ni, but usually higher than 20%.

Biostimulation. Although biostimulation is mostly employed for the biodegradation of organic pollutants it can equally be used for remediation of metals, especially in soils, through alteration of soil pH (Chibuike and Obiora 2014). In marine sediments, the only example of biostimulation application to remediate metals comes from microcosm experiments with addition of sodium acetate, lactose and/or inorganic nutrients that were performed by Fonti et al. (2015). Those authors found that the carried out treatments enhanced the mobility of Zn, Pb, Cd and As by an increase of metal concentrations in the exchangeable/carbonatic fraction of the sediment.

Bioaugmentation. The use of microbial consortia is the most common and classic method of bioaugmentation in marine systems (Marques 2016). For instance, Beolchini et al. (2009), in a microcosm experiment with sediments dredged from a harbor, evaluated the bioremediation efficiency based on different marine consortia. They concluded that the mixture of two consortia consisting of autotrophic and heterotrophic Fe/S-oxidizing bacteria achieved metal extraction yields of 90% for Cu, Cd, Hg and Zn.

The addition of genetically engineered microorganisms has gained importance in recent decades. Producing resistant bacteria through the introduction of metallothionein (MT) genes is an upcoming technology. MTs are thiol-containing proteins synthesized by some species of all types of organisms (from bacteria to animals) and induced by metals. They serve to transport and sequester metals to reduce their toxicity for the organism (Cobbett and Goldsbrough 2002). However, there are still only a few studies on the genetic engineering of marine bacteria. As an example, Sode et al. (1998) successfully introduced an MT-encoding gene from a freshwater *Synechococcus* into a marine *Synechococcus* allowing, in the latter, a significantly higher tolerance to Cd.

3.1.2 Phytoremediation

Phytoremediation is the preferred green technology for treatment of metal pollution. Several of the phytoremediation strategies mentioned (e.g., phytostabilization, phytovolatilization) are suitable for metals. However, phytoremediation in marine environments is still in an early phase, identifying the appropriate species through field work and performing the first laboratory or mesocosm experiments.

Phytoremediation is well developed in wetlands. The Ramsar Convention defines wetlands as areas of marsh, fen, peatland, or water, whether natural or artificial, permanent or temporary, with water that is static or flowing, fresh, brackish, or salty, including areas of marine water, the depth of which at low tide does not exceed 6 m (Mitsh and Gosselink 2007). Wetlands play several ecological functions within the environment, among other things acting as natural filters of nutrients and pollutants and improving the quality of water (Constanza et al. 1997, Mitsh and Gosselink 2007). Therefore, natural wetlands as well as constructed ones have been used for improvement of water quality. The construction of freshwater wetlands for water improvement relies on the principle of phytoremediation and has been used since the late 1960s for remediation of wastewater from various sources. Constructed wetlands proved to be efficient in the removal of excess of nutrients, organic matter and metals from wastewater from industries and sewage worldwide (e.g., Stefanakis 2018 and references therein, Maine et al. 2019).

Salt marshes and the importance of halophytes in phytoremediation

Although most examples of constructed wetlands (and hence use of phytoremediation) come from freshwater environments, salt marshes as a type of wetland are also important sites for natural phytoremediation. Salt marshes are defined as halophytic grasslands on alluvial sediments bordering saline water bodies where water level fluctuates either tidally or non-tidally (Mitsch and Gosselink 2007). Those environments are usually located in estuaries from medium and high latitudes and offer many ecosystem services to coastal populations, such as coastal protection, erosion control, water purification, carbon sequestration, and areas for tourism and recreation (Costanza et al. 1997, Barbier et al. 2011).

Salt marsh plants are halophytes, able to survive and reproduce in environments where the salt concentration is around 200 mM NaCl or more and tolerate salt concentrations that kill 99% of other species (Flowers and Colmer 2008; Fig. 3). Halophytic plants are of special interest in phytoremediation because it has been proved that they can tolerate not only salinity stress but also stress imposed by metals, since tolerance to salt and to metals relies partly on common physiological mechanisms (Anjum et al. 2014). Many halophytes are hyperaccumulators of metals, although they are usually not considered in hyperaccumulator databases (Reeves et al. 2018). Halophytes are naturally better adapted to cope with high levels of heavy metals than crop plants commonly chosen for phytoremediation purposes (Anjum et al. 2014 and references therein).

Fig. 3. Some prevalent salt marsh halophytes mentioned in this chapter: (a) *Halimione portucaloides*, (b) *Sarcocornia perennis*, (c) *Spartina maritima*, (d) *Spartina alterniflora*, (e) *Spartina densiflora*.

Understanding the tolerance mechanisms of salt marsh plants to cope with metal stress might be useful for choosing the best phytoremediation strategy and even to develop combined and new technologies. There are different modes of tolerance, which are associated to different organization levels. The most common adaptation is element partitioning between above-ground and underground organs, with higher levels of metals in non-photosynthetic tissues and restricted movement towards shoots and leaves. Most salt marsh plants accumulate metals in roots and other underground structures rather than shoots and leaves, and hence they have potential for phytostabilization (Table 1). Another tolerance mechanism is to have higher metal concentration in the cell wall rather than in vacuoles or cytoplasm. Immobilization of trace elements in cell walls was reported in salt marsh species *Halimione portulacoides* and *Sarcocornia perennis* (Castro et al. 2009, Válega et al. 2009). Another mechanism of metal tolerance in some salt marsh plants is excretion through salt glands, and this is the basis of phytoexcretion (Manousaki and Kalogerakis 2011).

Another level of response to metal stress is the biogeochemical. Under environmental stress (e.g., metal, salinity), there is an increased production of reactive oxygen species (ROS) in plant cells, but halophytes have an efficient antioxidant defense system for the removal of ROS and thus maintain cellular homeostasis. This is the ascorbate-glutathione (ASA-GSH) pathway, which includes both enzymatic and non-enzymatic components; thoughtful reviews of the ASA-GSH pathway in salt marsh plants were made by Anjum and coworkers (2014, 2016). The synthesis of all the components of the ASA-GSH pathway is stimulated by different environmental stresses, but GSH has also a specific role in metal tolerance. GSH is a monothiol (molecule with a cysteine group), which is the precursor of phytochelatins (PCs), specific binding peptides of some metals (Pb, Cu, Zn, Hg, Ag, and particularly Cd and As) found in some plants. Plants exposed to high concentrations of these metals may respond rapidly by the induction of the synthesis of PCs, which are polithiols (2 to 6 cysteine groups), from GSH. Metals can be bound to cysteine groups either in GSH or PCs and the complex thiol-metal is accumulated in vacuoles and later can be transformed into more stable

Table 1. Phytoremediation strategies used for salt marsh plants to remediate metals.

Phytoremediation mechanism	Plant species	Metals	Reference
Phytostabilization	*Sarcocornia fruticosa*	Hg, Zn, Cu, Cd, Co	Canario et al. 2010, Duarte et al. 2010
	Halimione portulacoides	Hg, Zn, Cu, Cd, Co, Cr, Ni, Pb, As	Canario et al. 2010, Duarte et al. 2010, Negrin et al. 2017
	Spartina maritima	Hg, Zn, Cu, Cd, Co, Cr, Ni, Pb, As, Fe, Al	Canario et al. 2010, Duarte et al. 2010, Cuoto et al. 2013, Curado et al. 2014a, Negrin et al. 2017
	Sarcocornia perennis	Zn, Cu, Cd, Co, Cr, Ni, Pb, As, Fe, Al	Duarte et al. 2010, Curado et al. 2014b, Negrin et al. 2016, 2017, Idaszkin et al. 2017
	Spartina alterniflora	Pb, Cu, Zn, Cr, Ni	Negrin et al. 2016, 2019 (and references therein)
	Spartina densiflora	Cu, Ni, Cr, Pb, Fe, Zn	Idaszkin et al. 2014, 2015
Phytoextraction	*Spartina argentinensis*	Cr	Redondo-Gómez et al. 2011
Phytovolatilization	*Salicornia bigelovii*	Se	Shrestha et al. 2006
Phytoexcretion	*Spartina alterniflora*	Hg	Windham et al. 2001
Rhizoremediation	*Halimione portulacoides*	Cu	Mucha et al. 2011
	Sarcocornia fruticosa	Hg	Figueiredo et al. 2014
	Spartina maritima	Hg	Figueiredo et al. 2014

compounds (Cobbett and Goldsbrough, 2002 and references therein). Despite their importance in metal tolerance, PC production by salt marsh plants has been poorly studied. Padinha et al. (2000) found high values of total thiol-containing compounds in *Spartina maritima*, Válega et al. (2009) found PC production in tissues of *H. portulacoides*, and Negrin et al. (2017) reported the synthesis of different types of PCs and monothiols in *S. maritima, H. portulacoides* and *S. perennis*, probably induced by As, Zn and Pb. Metallothioneins could also be found in plants for metal tolerance but, to the best of our knowledge, there are no studies of salt marsh plants.

As already mentioned, most of the research on the use of phytoremediation with salt marsh plants is based on the identification of potential species. Most salt marsh species are suitable for phytostabilization of metals, but others have been suggested for phytoextraction, phytovolatilization, phytoexcretion and rhizoremediation (Table 1). Rhizoremediation deserves a special analysis due to the importance of the processes that take place in the rhizosphere of salt marsh plants. Some salt marsh plants have the ability to oxygenate the rhizosphere by the release of oxygen by roots, promoting the existence of oxygenated microzones within mainly reduced sediment (Colmer 2003). This enhances the coexistence of aerobic and anaerobic microorganisms within millimeters, allowing the remediation of a large number of contaminants. The importance of microbial communities associated with *S. maritima* and *H. portulacoides* has been studied (Reboreda and Caçador 2008, Machado et al. 2012); however, there are few works that deal with the effects of these communities on metal mobility, and those pertain only to Hg and Cu (Table 1).

Mangroves are also important in phytoremediation

Mangroves are coastal wetlands located in low latitudes. They are defined as subtropical and tropical coastal ecosystems dominated by halophytic trees and shrubs growing in brackish to saline tidal waters (Mitsch and Gosselink 2007). As intertidal ecosystems, mangroves have many features in common with salt marshes, such as sediment characteristics of high salinity and usually low redox potential. Moreover, being halophytes, many mangrove plants hyperaccumulate metals in their tissues. Mangrove sediments can also immobilize metals in unavailable forms through complexation with organic matter and/or precipitation with sulfides (Zhou et al. 2011). Therefore, several mangrove species were proposed as phytoremediation agents.

Avicennia marina is one of the most widely studied mangrove species with respect to metal accumulation worldwide. Although some authors (e.g., Ghasemi et al. 2018) found low accumulation of Zn, Cu, Pb, and Cd in tissues of *A. marina*, others reported high levels of some metals in this species, mainly in above-ground tissues. For instance, Cu and Pb concentrations were higher in root and leaf tissues of *A. marina* than in sediments (MacFarlane et al. 2003) and high uptake and translocation to leaves of Cu, Zn, Ni, Cd and Pb was also reported (Parvaresh et al. 2011, Kaewtubtim et al. 2016). Other mangrove species that proved to accumulate metals are *Avicennia officinalis, Sonneratia apetala* and *Excoecaria agallocha*. *Avicennia officinalis* and *S. apetala* exhibited high concentration of Co and Mn in above-ground tissues (Chowdhury et al. 2017), whereas *E. agallocha* accumulated high levels of Cd in leaves (Chowdhury et al. 2015).

The above-mentioned studies highlight the potential of some mangrove species in phytoremediation, either through phytostabilization or phytoextraction. Moreover, metal excretion through salt glands has been observed in *Armeria maritima* and *Avicennia germinans* (Manousaki and Kalogerakis 2011), showing a potential of these species for phytoexcretion. However, in either case, phytoremediation should proceed naturally due to the inconvenience of planting trees. The size of trees and the time needed for their development make the construction of mangroves unviable. However, under natural conditions they can be useful for phytoremediation due to their high biomass above and belowground.

Enhanced strategies of phytoremediation

In some cases, a single method of phytoremediation may not be effective in removing metals. Some improvements and combination of strategies are being developed, with potential for use in marine environments.

Chelate-assisted phytoremediation. Addition of synthetic chelates stimulates the release of metals into soil solution and therefore enhances the potential for plant uptake, which can enhance phytoextraction or other phytoremediation strategies (Liang et al. 2016, Yadav et al. 2018). Chelates such as EDTA (ethylenediamine tetraacetic acid) have been proved to enhance phytoextraction of several metals but present several disadvantages, such as leaching risk, low efficiency and high cost (Ghosh and Singh 2005, Khalid et al. 2017, Yadav et al. 2018). Thus, the use of natural chelates has been proposed (for example, low molecular weight organic acids, produced by salt marsh plants) to enhance phytoremediation (Duarte el al. 2011).

Bioaugmentation-assisted phytoremediation. As already stated, rhizoremediation is based on the symbiotic interaction between plants and microorganisms. If the microorganisms are deliberately introduced in the sediments, the technique could be called bioaugmentation-assisted phytoremediation. However, studies about the use of this technique are scarce and mostly experimental. In salt marshes the role of addition of both bacteria and fungi has been explored. For instance, Said et al. (2018) found that the phytoremediation potential of *Sarcocornia fruticosa* increased in the presence of the yeast *Saccharomyces cerevisiae*. Almeida et al. (2017) studied the role of a consortium of bacteria in the phytoremediation of Cu in a microcosm and concluded that bioaugmentation improved the process of phytoremediation.

Genetic engineering-assisted phytoremediation. Hyperaccumulating plants do not always have a good growth rate. The improvement of these plants could be achieved by conventional breeding or genetic engineering. The former involves hybridization between small, slow-growing hyperaccumulators and fast-growing, high-biomass non-accumulator plants (Bhargava et al. 2012, Yadav et al. 2018). For instance, Brewer et al. (1999) have raised somatic hybrids between the Zn hyperaccumulator

Typha caerulescens and *Brassica napus* and found that the hybrids accumulated high levels of Zn that would otherwise have been toxic.

Genetic engineering is gaining importance and there are several examples of transgenic plants used in phytoremediation. One of the main approaches is the introduction of the genes involved in glutathione, phytochelatin and metallothionein synthesis to increase metal tolerance in plants with good production of biomass (Ghosh and Singh 2005, Bhargava et al. 2012, Yadav et al. 2018). For example, Zhu et al. (1999) introduced the *Escherichia coli* gene for glutathione to *Brassica juncea* and found that the transgenic seedlings showed increased tolerance to Cd and had higher concentrations of phytochelatins and glutathione.

Another application of genetic engineering in phytoremediation is through the production of transgenic plants encoding genes from Hg-detoxifying bacteria to enhance phytovolatilization (Khalid et al. 2017). The introduction of key genes for Hg remediation was evaluated in salt marsh species. Czako et al. (2006) co-inoculated embryogenic callus of *S. alterniflora* with a pair of *Agrobacterium* strains with the organomercurial lyase (merB) and mercuric reductase (merA) genes. The inoculation proved to be successful and promoted higher Hg resistance, promising results for the future of genetic engineering in phytoremediation in marine ecosystems.

3.2 Organic compounds

3.2.1 Oil spills

Oil spills are generated by accidents in the development of productive activities such as structural failures, operational errors, negligence, war conflicts or natural catastrophes. The most recognized worldwide events were (1) the intentional release of more than 72 million tons of oil into the Gulf of Kuwait during the Gulf War in 1991, (2) spills from the explosion of the Deepwater Horizon oil rigs in 2010, dumping about 779,000 tons of crude oil in the Gulf of Mexico near the mouth of the Mississippi River, and (3) the release of approximately 476,000 tons of oil in 1979 from the Ixtoc I well in the Bay of Campeche, Gulf of Mexico. Fortunately, these catastrophic incidents are rare. However, small to moderate spills occur daily in the world in all environmental compartments, the final destination of which normally is the sea or the marine coasts (Fingas 2010).

The protocols to apply immediately after a spill to contain, eliminate and/or recover oil are regulated internationally by agreements such as the International Convention for the Prevention of Pollution from Ships (MARPOL 73/78) and Ship Oil Pollution Emergency Plan (SOPEP) of the International Marine Organization (IMO, a specialized agency of the UN). Conventional methods include physical removal with barriers, skimmers and absorbent materials together with chemical removal or dispersion with surfactants and solvents that disperse the oil into droplets. However, these technologies have disadvantages: they do not eliminate all the oil from the environment and chemical methods can also be toxic (Nikolopoulou and Kalogerakis 2011, Zahed et al. 2011). Therefore, new technologies and stricter controls have been developed to reduce the occurrence of spills or chronic releases. Nevertheless, the sources of contamination continue to be active at present, so the remediation treatments are vital to mitigate the environmental impacts.

The components of oil tend to have different behavior when spilled in seawater. However, in general, most of the components of the mixture, mainly light hydrocarbons, float. If they are not removed immediately they can dissolve, spread, precipitate, evaporate, photooxidate, biodegrade, enter the trophic network or persist (NRC 2003). When spills occur near the coast, the degradation or dispersion processes cannot be completed; thus, the oil usually reaches the shoreline. Once on the coast, the oil comes into contact with sediments and a complex phenomenon of adsorption begins. Degradation times are extended by the decrease in the degree of interaction between water and the compounds.

Oil pollution on the coasts and in the sea generates devastating effects on natural ecosystems. Consequently, one of the objectives of companies, governments and scientists is to find and use techniques to avoid or remedy this problem. Therefore, several technologies have been developed that use different principles of elimination, degradation or recovery of oil to be used in specific situations.

Biorremediation

Many microorganisms possess the enzymatic capability to degrade oil. The initial steps in biodegradation by bacteria and fungi involve the oxidation of the substrate by oxygenases and continue according to the type of hydrocarbon (Atlas 1995). The remediation of very large coastal areas, often difficult to access, is a challenge for modern engineering, and bioremediation is a promising technique for those areas (Fingas 2010).

Bioremediation in sediments

Several techniques of bioremediation can be applied to the treatment of oil in marine sediments.

Natural attenuation. As already stated, there are cases of pollution in which actively intervening can cause more damage than letting the environment recover naturally; for oil spills in mangroves, salt marshes and corals, natural attenuation may have better outcomes (Fingas 2010).

Biostimulation. The incorporation of nutrients is the most commonly used bioremediation technique for oil-polluted sediments, since nutrient concentrations are usually low in marine sediments and oil is mainly composed of C (Floodgate 1984, Venosa and Zhu 2003). Therefore, by adding nutrients it is possible to maintain an adequate ratio of C:N:P for the optimal development of microorganisms. However, the fate of the added nutrients must be considered to avoid eutrophication of adjacent water bodies, especially in sensitive environments such as salt marshes.

The incorporation of nutrients presents several challenges. First, it should be taken into account that nutrients can be added in different forms and each one can act in a different way and have different elimination efficiencies. For example, NH_4-N shows a better elimination efficiency than NO_3-N (Emami et al. 2014). Second, it must be borne in mind that increasing the doses of fertilizers does not necessarily increase the biodegradation of oil proportionately. For instance, Chaineau et al. (2005) observed that higher fertilizer doses produce greater bacterial stimulation, but this does not coincide with an increase in the degradation of hydrocarbons; on the contrary, the best results were reported under low doses of fertilizers, especially in saturated compounds. In line with this, Venosa et al. (1996) indicated that if the natural concentration of N is high enough to support high rates of biodegradation of hydrocarbons, the addition of exogenous nutrients may not be appropriate. Moreover, nutrient ceases to be a limiting factor also when, after degradation of most compounds, only more resistant ones continue, reaching 60–70% of the total mass of the hydrocarbons because the available C is low and the C:N:P ratio is adequate for optimal microbiological development (Bragg et al. 1994).

Moreover, whenever the frequency of application is increased, the degradation efficiency will be better (Bragg et al. 1994); hence, another goal is to find simple and economical methods that allow determining the optimal frequency of application or increasing the residence of fertilizers in the area. Some techniques are based on the use of slow-release fertilizers in the contaminated area (Xu et al. 2005) and others on the more efficient application of traditional inorganic fertilizers. The latter was studied by Li et al. (2007) through mathematical modeling for sandy marine beaches. The authors concluded that the optimal application of fertilizers should be made after high tide and during the middle of the tidal cycle by means of a perforated pipeline in the high tide line.

In general, nutrients are added in the form of inorganic fertilizers, but they can cause eutrophication. An alternative is the use of organic matter derived from plants or animals, but it must be taken into account that the efficiency of degradation of oil varies substantially according to the type of amendment used. Some studies have observed higher removal rates with organic or natural fertilizer treatments than with inorganic fertilizer treatments. For example, Horel et al. (2015)

reported that the organic matter derived from fish tissues of the region increased degradation rates by 104% compared to a 57% increase for the addition of inorganic nutrients, both in comparison with the control treatments. Another way to incorporate organic matter is through the use of composted vegetable waste. Some works have observed an 88% removal of petroleum hydrocarbons with the use of compost in contaminated soils (Bastida et al. 2016). In addition, the application of imported organic matter from rich soil cannot be ignored (Wu et al. 2015).

Bioaugmentation. The addition of bacteria to soils contaminated with oil is a technique that has been closely studied and used worldwide (Venosa et al. 2003, Lim et al. 2016). In general the use of bacterial consortia produces an improvement in removal efficiency over the use of bacteria alone (Szulc et al. 2014). This is because there is no single species of bacteria that has the capacity to efficiently degrade all types of petroleum hydrocarbons (Varjani et al. 2015, Rahman et al. 2002). In addition, the presence of organisms adapted to different types of habitats allows them to act successfully in heterogeneous and varying natural environments (Cohen 2002). Although bacterial consortia give the best removal rates, there are some isolated microorganisms that have very high removal efficiencies alone (Rahman et al. 2002).

Bioaugmentation with autochthonous bacteria usually corresponds with very high degradation rates. This is a technique based on the isolation, multiplication and subsequent sowing of the pollutant-degrading bacteria of the affected site (Ueno et al. 2007). However, it should be noted that exotic bacteria can usually be purchased commercially and are improved in the laboratory to increase their ability to adapt to different environments, which means a saving of time and resources compared to the use of native ones, because the latter must be isolated and selected and their removal efficiency must be quantified to determine if their application in contaminated soil is useful.

As with biostimulation, when oil concentrations are high, degradation efficiency of bioaugmentation tends to decrease. For example, Rahman et al. (2002) found that the degradation rate varied from 78% to 52% for 1% and 10% of the concentration of oil respectively. Therefore, it is emphasized that the best strategy is to carry out a gross removal of the oil with some mechanical, chemical or thermal technique and then apply bioremediation techniques such as bioaugmentation (Fingas 2010).

Biostimulation with bioaugmentation. It has been shown that the combination of biostimulation and bioaugmentation produces a higher rate of oil degradation than the use of individual technologies (Suja et al. 2014), especially in poor soils with adverse environmental conditions (Gómez and Sartaj 2013). Some authors observed degradation rates of petroleum hydrocarbons of 55%, 52% and 82% after 92 d for a soil with microbial consortia, soil with compost, and soil with both treatments, respectively (Gómez and Sartaj 2013). However, it must be taken into account that this work was carried out *ex situ* and more studies are needed to estimate the efficiency of the application of this technology *in situ*.

Addition of biosurfactants. The removal efficiency of biosurfactants varies greatly with environmental conditions, the pollutant and the type and concentration of the biosurfactant. As successful studies, we can mention the findings of Bezza et al. (2015), who observed a degradation of 70% of oil with the use of biosurfactant produced by *Ochrobactrum intermedium* CN_3, and those of Molnar et al. (2005), who reported an oil removal efficiency of up to 57% with the use of cyclodextrin. However, some studies showed a very low increase in removal efficiency in relation to natural degradation. For instance, Gruiz et al. (1996) reported an increase of only < 5% with the addition of RAMEB in relation to the control, but they pointed out that the use of this biosurfactant decreases the toxicity of the soil and allows the growth of plant roots.

Bioventilation. Many studies have shown that the lack of oxygen in marine sediments reduces the activity and degradation efficiency of aerobic microorganisms, which represent the main route of oil degradation (Atlas 1995, Bossert and Bartha 1984). Anaerobic conditions usually occur in the subsurface sedimentary layers of the shoreline and in the fine-grained coastal wetlands (Venosa et al. 2003). To mitigate this problem, air is usually injected into contaminated sediments by mixing or moving sediments in order to improve the removal efficiency of aerobic microorganisms and at the same time to allow the elimination of volatile soil compounds (Lee and Swindoll 1993). This is a very simple, economical and effective technology for oil spill control, although it has some disadvantages: (1) the need to periodically mix to maintain the aerobic microbial population (Zouboulis and Moussas 2011); (2) the low versatility, since it can only be applied on coasts with soft sediments and in the upper layers of the soil (Genou et al. 1994); (3) the slowness with which they degrade hydrocarbons, and (4) the potential negative environmental impact related to the escape of toxic volatile compounds to the atmosphere.

Anaerobic bioremediation. The anaerobic degradation of oil in marine sediments has scarcely been documented because of its low rate of biotransformation per unit of substrate in relation to aerobic degradation (Yang et al. 2009). In addition, in marine sediments, anaerobic bioremediation is not adequate because of its high cost, since the places where it could be applied are generally inaccessible environments for biota, and because of the uncertain effects that the use of bioremediation technologies can have on sensitive environments such as coastal wetlands (Lim et al. 2016). For example, in naturally anaerobic environments, such as salt marshes, natural attenuation uses anaerobic degradation as the main remediation strategy (Atlas 1995, Overton et al. 2016). In addition, Venosa et al. (2010), studying the 1989 Exxon Valdez spill, pointed out that this technology can be used in places where aerobic bioremediation was performed to finish degrading the hydrocarbons that are adsorbed in the anoxic sediments.

Bioremediation of seawater

Despite the importance of bioremediation, in the treatment of seawater polluted with oil the use of artificial dispersants is also worth mentioning. In particular, Corexit 9500 is not toxic to marine biota. Dispersants aim to reduce the interfacial tension between the water and the oil, breaking the spots into fine droplets and accelerating the dispersion in the water column and the dissolution of the hydrophobic components of the oil (NRC 2005), which produces an increase in the surface that may come into contact with the oil-degrading microorganisms. Most of the dispersants studied and produced so far are toxic to biota (Fingas 2010), which would counteract the beneficial effect just described. However, in 2010, for the Deepwater Horizon spill, Corexit 9500 was developed, which was harmless to marine biota, including oil-degrading bacteria. It was also found that various components of the dispersant can be easily degraded by the marine microfauna (Chakraborty et al. 2012).

Biostimulation. Many authors have proposed biostimulation as a bioremediation technique, but in open waters the incorporation of hydrophilic inorganic fertilizers would not be effective because of the great effects of dilution. Swannell et al. (1996) proposed the use of oleophilic fertilizers supported in paraffin to keep them in the oil-water interface, where the biodegradation activity takes place. However, later research affirms that neither oleophilic nor hydrophilic fertilizers are effective in open systems and that their applications are very expensive; hence, the use of hydrophobic N and P fertilizers such as uric acid was proposed (Ron and Rosenberg 2014). Moreover, some authors pointed out that the combination of fertilizers of different types such as organic or inorganic in sites contaminated with diesel in seawater can produce higher removal rates than if they were used alone (Piehler and Paerl 1996).

Biostimulation with bioaugmentation. When biostimulation and bioaugmentation are combined, the results are usually successful. In 1976, Atlas and Budosh, through experiments in mesocosms, achieved a 65% removal of crude oil with oleophilic fertilizers and a degrading *Pseudomonas* against a 51% removal in the presence of oleophilic fertilizers alone and 25% without fertilizer. More recently, some researchers studied the use of microbial consortia linked to a biostimulated carrier material. This method is based on the use of the carrier material as a means for the formation of biofilms with microbial communities of several species working together, giving protection to the microbial consortia and giving buoyancy to the whole, which allows the microorganisms to interact with the oil through the water column (Sheppard et al. 2014). Several studies selected shell grit as carrier material due to its buoyancy, absorbance capacity, simplicity of application and generation of protective niches of microbial consortia (Simons et al. 2012, Sheppard et al. 2014).

Phytoremediation

Phytoremediation is a viable alternative to the treatment of oil-polluted sites (dos Santos and Maranho 2018). Few studies have been carried out on marine coasts but some general conclusions can be drawn from studies in terrestrial ecosystems that could be applied to these environments. From these, it could be inferred that the most suitable plants to carry out a successful phytoremediation are trees and grasses, which have large coverage of roots and biomass (Cook and Hesterberg 2013) and demonstrated high removal efficiency (Olson et al. 2007). Among these, grasses would be the most suitable for marine coasts because of their tolerance to droughts, acid soils and a wide temperature range, typical of coastal wetlands (Parrish et al. 2004, Muratova et al. 2008). In addition, grasses have higher rates of oil degradation (Euliss et al. 2008) and a more uniform soil cover, which reduces runoff and surface erosion (Collins 2007). However, the efficiency of removal of phytoremediation with trees of *Rhizophora* mangle in mesocosms and in the field has been studied and the results were very encouraging, with oil removal rates of up to 87% in 90 d, mainly due to an increase in microbial concentration (Moreira et al. 2011).

One of the main problems that must be overcome by the phytoremediation of hydrocarbons in marine environments is the damage produced by high concentration of oil spill in plants. This can lead to the death of the plants or negative impacts on their growth, mainly due to physical action, since the oil covers the plant and prevents the exchange of gases, increases the temperature and thus results in stress, reduces the photosynthetic rate, and decreases the diffusion of oxygen to the roots, a fundamental factor for the growth and survival of wetland plants (Mendelssohn and McKee 1988, Pezeshki et al. 2000, Lin and Mendelssohn 2009, Redondo-Gómez et al. 2014). Halophyte species such as *S. argentinensis* can continue growing in soils contaminated with diesel up to 3 v/w% after 250 d of exposure (Redondo-Gómez et al. 2014).

The death of plants generates another very important problem: water erosion (Pezeshki et al. 2000). This in turn prevents the vegetation from returning to normal cover and, consequently, there is also an increase in the tidal prism and erosion in healthy adjacent areas (Krebs and Tanner 1981). In addition, oil spills generally cover the soil surface, resulting in a lower rate of oxygen diffusion to the deeper strata, which makes sediments anaerobic (Ranwell 1968).

Strategies of phytoremediation. Many plants have the ability to immobilize oil in the rhizosphere, promoting phytostabilization of those pollutants (Germida et al. 2002). For instance, it has been observed that the concentrations of hydrocarbons in the rhizosphere of *Salix viminalis* was twice the initial concentration applied 10 h before, so it was concluded that the mechanism of exclusion of the roots concentrated the hydrocarbons in the surrounding sediments (Byström and Hirtz 2002). Since most hydrocarbons are hydrophobic, the amount that the plant can absorb is low. The compounds that are absorbed have different destinations according to their own physicochemical characteristics. Non-volatile hydrocarbons are retained and accumulate in vacuoles or cell walls; if they are accumulated in above-ground biomass, pollutants can be removed along with biomass promoting phytoextraction (Gerhardt et al. 2009, Haritash and Kaushik 2009). Low molecular weight and volatile hydrocarbons such as benzene or toluene, once they enter the plant, can be remediated via phytovolatilization (Limmer and Burken 2016, Yavari et al. 2015).

Since hydrocarbons can be degraded, phytodegradation and rhizodegradation are the main strategies of phytoremediation for oil-polluted sites. Some of the non-volatile compounds that enter the plant can be degraded or rendered non-toxic via enzymatic modification (phytodegradation) or be degraded in the rhizosphere through the actions of microorganisms (rhizodegradation) (Gerhardt et al.

2009). The main plant enzymes involved in degradation of petroleum hydrocarbons are peroxidase, nitrilase, nitroreductase and laccase (Germida et al. 2002, Yavari et al. 2015). Rhizodegradation is very important in marine sediments since the incorporation of oxygen by plants to the rhizosphere in coastal wetlands is essential to promote the action of aerobic microorganisms and achieve high rates of degradation of oil. Some studies indicate that the presence of submerged macrophytes such as *Zostera marina* (Huesemann et al. 2009), marsh grasses such as *Juncus roemerianus* (Lin and Mendelssohn 2009) or mangrove trees such as *Rhizophora* (Moreira et al. 2011) in contaminated marine sediments improve biodegradation of organic compounds through multiplying the density of bacteria in the rhizosphere.

Enhanced strategies of phytoremediation. As for metals, phytoremediation alone may not always be effective, and in recent years many scientists around the world have studied strategies that improve the performance of phytoremediation. The combination of several technologies such as biostimulation, bioaugmentation and phytoremediation substantially improves the efficiency of oil degradation.

Biostimulation-assisted rhizoremediation basically consists in the addition of fertilizer as an additional source of nutrients for microorganisms and plants. On sea coasts it has not been well studied yet, but in a mesocosm study it was reported that this "biostimulated phytoremediation" with marsh grasses such as *S. alterniflora* and *Spartina patens* accelerates the rate of degradation of oil (Mendelssohn et al. 1995, Wright et al. 1997). Commonly, inorganic fertilizers are used but the application of organic wastes such as plant residues or compost to the soil has also proved to be efficient. For example, a removal between 91% and 99% of diesel was observed in phytoremediation with trees (Dadrasnia and Agamuthu 2013) and removal of 46% of pirene with grasses (Wang et al. 2012), both amended with organic wastes. Many studies have been carried out to determine the ideal quantities and proportions of fertilizers to apply, but there is not a single answer to this question mainly because they tend to vary between different plant species (Dadrasnia and Agamuthu 2013, Jagtap et al. 2014, Merkl et al. 2005). However, most authors conclude that the incorporation of nutrients, regardless of their concentration, significantly increases the rates of oil degradation (Jagtap et al. 2014), contributes to soil improvement and could promote long-term ecological recovery (Leewis et al. 2013).

In bioaugmentation-assisted rhizoremediation, rhizobacteria are used to promote the survival and growth of plants or to improve the conditions for the development of the oil-degrading bacteria. Hou et al. (2015), applying this technique in mesocosms, observed removal rates of up to 67.9% after 4 mon in *Festuca arundinacea*. Another option is the incorporation of mycorrhizal fungi that promote the growth of plants and improve the characteristics of the rhizosphere. Alarcón et al. (2008), in a greenhouse study, found higher degradation rates when plants (*Lolium multiflorum*), mycorrhizal fungi (*Glomus intraradices*), and oil-degrading microorganisms were combined in comparison with the same treatment without mycorrhizal fungi. Kuo et al. (2013) also found an increase in oil degradation when inoculating mycorrhizal fungi in some plant species such as *Biden pilosa*.

3.2.2 Persistent organic pollutants

Persistent organic pollutants (POPs) are a diverse group of compounds with high stability and persistence in the environment. They are hydrophobic and lipophilic and can be bioaccumulated in the trophic web. Because of these characteristics, they are ubiquitous in nature, found in places far away from their sources such as deep ocean sediments or polar regions (Passatore et al. 2014). As a consequence of their extended distribution and proven harmful effects on human health (Kallenborn et al. 2012, Venkatesan and Halden 2014), global agreements were made to prohibit or restrict the use and production of POPs, such as the Stockholm Convention in 2001 (UN) and the Aarhus Protocol in 1998 (EU, US and Canada). However, POPs continue to enter the environment because of lack of industrial control, waste disposal, illegal use, spills or accidental leaks or lack of regulation throughout the world, especially in developing countries (Pariatamby and Kee 2016). The oceans and seas are the main sinks of POPs, due to the contribution of rivers and atmospheric deposition.

POPs can be classified according to their origin as intentional or unintentional. Among the former we find: (1) organochlorine pesticides (OCPs) such as DDT, Mirex, Endrin, Dieldrin, Heptachlor, Chlordane, Endosulfan, HCH and HCB, which were widely used during most of the 20th century for phytosanitary and zoosanitary control throughout the world; (2) polychlorinated biphenyls (PCBs), widely used as heat exchangers and dielectric fluids in electrical systems; and (3) polybrominated diphenyl ethers (PBDEs), used as flame retardants in plastics of electronic devices and building materials. The unintended POPs, which are byproducts of the combustion of organic compounds, are polycyclic aromatic hydrocarbons (PAHs), dioxins and furans. PAHs are caused by the fusion of two or more benzene rings during incomplete combustion at high temperatures or by diagenesis of the organic material. Dioxins (polychlorinated dibenzo-p-dioxins) and furans (polychlorinated dibenzo-furans) are caused by the incomplete combustion of chlorinated organic compounds (El-Shahawi et al. 2010). Another, more practical, classification is between PAHs and organochlorine contaminants (OCs). The latter include PCBs, OCPs, dioxins and furans (Mateju 2003).

Currently, the intervention techniques most used worldwide in marine sediments contaminated with POPs are dredging followed by incineration or filling in landfills for highly contaminated sites (Zeller and Cushing 2005, Passatore et al. 2014). The dredging or removal of sediments presents some disadvantages such as its high cost and the magnification of the problem of pollution, since dredged sediments are not usually treated. In recent years attention has been given to biological remediation as a viable alternative for the remediation of sediments contaminated by POPs, mainly because of its low cost and milder impact on the environment and because it can treat large areas of sediments with low concentrations of POPs, a characteristic situation of marine environments that cannot be treated with other methods.

Bioremediation

Besides natural attenuation (see below), there is no cost-effective *in situ* bioremediation technology for marine sediments contaminated with POPs (Huesemann et al. 2009, Passatore et al. 2014), mainly because of their low bioavailability due to the strong adsorption in the sediment and because PAHs with more than four benzene rings and OCs with many chlorines are very difficult to degrade (Bamforth and Singleton 2005, Harmsen et al. 1997). This combination of characteristics means that it can take a lot of time and resources to reduce them successfully. However, scientists continue to look for effective bioremediation techniques that allow *in situ* treatment in a cost-effective way. Some bioremediation techniques that could be applied to POPs are explained below.

Natural attenuation. Currently, natural attenuation is the only bioremediation technology that is applied in the management of sediments contaminated by POPs (Magar 2001, Perelo 2010), since it is considered very effective for low-risk sites and where the treatment times are not limiting, such as most of the marine coasts (Perelo 2010). The microorganisms capable of degrading POPs are found in most of the marine coasts (Mateju 2003). However, there are more microorganisms capable of degrading PAHs of low molecular weight than PAHs of high molecular weight and OCs; hence, the activation of the biodegradation process of the latter will have a delay period ranging from one to several months (Vasilyeva and Strijakova 2007).

Biostimulation. Carbon levels in sediment contaminated with PAHs are usually high by the very nature of PAHs, but available nutrients can quickly be depleted during microbial metabolism (Breedveld and Sparrevik 2000). The optimal C:N ratio for the degradation of PAHs in sediments is different from that necessary for general microbial development—25:1 and 5:1 to 12:1, respectively (Atagana et al. 2003). Thus, the incorporation of nutrients is an option for biostimulation. However, excessively high nutrient loads can inhibit microbial metabolism in some cases or produce eutrophication and unnecessarily increase the cost of remediation (Bamforth and Singleton 2005).

Regarding OCs, most of the research was conducted on agricultural soils and combining the application of fertilizers with single aerobic or aerobic/anaerobic sequential treatment. No *in situ* studies have been found in marine environments, but some conclusions related to the use of this technology can be drawn from experiences in mesocosms and agricultural environments. Among them, it can be highlighted that good degradation rates have been obtained by combining biostimulation with bioventing in the treatment of soils contaminated with HCH (Rubinos et al. 2007, Phillips et al. 2006) and DDT (McGhee and Burns 1995). In the case of DDT, since there are no known microorganisms that can degrade it without another source of C and energy, it is proposed to add usable organic C as fresh soil (Varca and Magallona 1994).

Bioaugmentation. Microorganisms capable of degrading PAHs of low molecular weight are ubiquitous in nature, but few microorganisms can degrade high molecular weight PAHs and OCs and they are not always found in contaminated soil or are found in very low concentrations (Adrian et al. 1998). Therefore, the incorporation of specialized bacteria was suggested as a strategy to accelerate the degradation of these compounds in soils (Heitkamp and Cerniglia 1988, Mueller et al. 1989). In general, bioaugmentation showed successful results for all POPs (Passatore et al. 2014, Mateju 2003); however, most of the studies were performed in microcosms, bioreactors or industrial soils; hence, the feasibility of *in situ* application on marine coasts is uncertain.

Bioaugmentation of marine sediments polluted with PAHs has been studied more than that of sediments polluted with OCs. It is known that the success of biaugmentation for remediation of OCs depends on pre-existing microbiological characteristics. It is successful in sediments with low or no intrinsic degradation potential (Juhasz and Naidu 2000, Major et al. 2002) and without improvement in soils with high intrinsic degradation potential such as mangroves (Tam and Wong 2008). As for other pollutants, it has also been proved that it is more effective to add native microbial consortia than exogenous ones (Wu et al. 2013) or individual microorganisms (Gupta et al. 2016).

The main challenges presented by bioaugmentation for sediments with POPS are: (1) the impossibility of large-scale application of microorganisms degrading OCs and PAHs of high molecular weight, since they are hydrophobic and water is not a feasible vehicle; (2) the low survival of exogenous strains in contaminated sediments, mainly due to nutrient limitations and competition with the indigenous microflora (Rhee et al. 1989, Saavedra et al. 2010); and (3) the inability of most systems to sustain communities of OC-specialized microorganisms because of the low level of compounds usually found. To allow large-scale application, it was proposed to inoculate cells adhered to a solid substrate (Payne et al. 2011) or to combine bioaugmentation with phytoremediation, because the plant contributes to improve the penetration of the inoculum in the sediment (Singer et al. 2003). To sustain the microbial communities specialized in OC, the use of inductors that mimic the behavior of those pollutants was proposed, as well as the structural analogues of the OCs contained in the root exudates (Chaudhry et al. 2005).

In addition to the use of bacteria, the use of white-rot fungi has been proposed to remedy sediments contaminated with PAHs and OCs (Bhatt et al. 2002, Cajthaml et al. 2008). These organisms can degrade POPs in very low concentrations and can access less bioavailable pollutants, since the induction of enzymes is independent of the presence of the pollutant (Canet et al. 1999). The disadvantages of adding fungi are the high cost of the large amount of fungus inoculum needed to achieve high remediation rates, the heterogeneous distribution of mycelium in the sediment, the slow and inefficient colonization of the soil, and the need for a woody substrate (Mateju 2003).

Biosurfactants. One of the greatest difficulties in the bioremediation of sediments with POPs is their low bioavailability. For this, one of the most widespread techniques for mitigation is the use of surfactants (either synthetic or biosurfactants) that allow the water solubility of POPs (Ahrens et al. 2001, Nakajima et al. 2005). By increasing bioavailability, in general, surfactants produce an increase in the biodegradation of pollutants, including compounds difficult to degrade such as DDT or PCB (Parfitt et al. 1995, Walters and Aitken, 2001, Luo and Hu 2013). However, some factors can reduce their effectiveness, such as competition with POPs for the possibility of being used as C and energy sources by microorganisms (Luo and Hu 2013, Berselli et al. 2004), their potential toxicity (Arostein et al. 1991), and the excessive increase of the solubility of the contaminant in the water (Zhang and Miller 1995). As for metals, rhamnolipid is a biosurfactant that can be effective in bioremediation of PAHs (Chen et al. 2013, Dell'Anno et al. 2018).

Aerobic/anaerobic bioremediation. Biodegradation rates of POPs are similar in aerobic and anaerobic conditions (McNally et al. 1998, Bamforth and Singleton 2005, Mateju 2003). Most marine sediments lack oxygen; thus, it is convenient to enhance anaerobic bioremediation (Perelo 2010). However, if the aerobic bioremediation should be stimulated, oxygen can be supplied to the sediment with simple technology such as tillage or mixing the soil with specialized machinery (Bamforth and Singleton 2005). For high molecular weight PAHs and most OCs, especially the most chlorinated ones, a combination of the two bioremediation techniques (aerobic and anaerobic) is usually the most effective strategy (Mateju 2003). In the case of OCs, such as DDT, dioxins, furans and PCBs, reductive dechlorination that eliminates the chlorine atoms of more complex molecules under anaerobic conditions can be stimulated (Carmona et al. 2009), reducing their toxicity and making them more easily degradable by aerobic bacteria; then bioventilation treatment can be applied (e.g., Nojiri et al. 2001, Passatore et al. 2014).

Phytoremediation

Phytoremediation can completely mineralize or transform POPs into less toxic or non-toxic compounds by different routes of action. Main strategies are phytodegradation and rhizoremediation (Nzengung et al. 1999, Gregory et al. 2005). Rhizoremediation is promising for the treatment of dredged sediments in landfills or in shallow waters, but the results are variable. In marine sediments dredged from a port, 20 species of plants were tested: some reached up to 90% degradation of PAHs while others showed low rates or did not survive the experiment (Paquin et al. 2002). In a mesocosm experience the ability to degrade PAHs and PCBs in marine sediments vegetated by *Zostera marina* for 14 mon was analyzed, and the results showed a remarkable increase in the degradation rates of both groups of compounds, with a reduction of 73% in PAHs and 60% of PCBs in the vegetated sediments (Huesemann et al. 2009). In this study the importance of both phytodegradation and rhizodegradation was evaluated, and the results showed that the translocation of PCB and PAHs to above-ground tissues was insignificant; it was concluded that the main remediation was mediated by the microbial stimulation in the rhizosphere by the release of root exudates, plant enzymes and oxygen (Huesemann et al. 2009).

As the reductive dechlorination necessary for the degradation of OCs occurs in anaerobic environments, the incorporation of oxygen into the sediment can decrease the degradation rates of these compounds. For this reason, the use of low transpiration plants such as *Carex aquatalis* and *Spartina pectinata* is more efficient (Smith et al. 2007).

4. Conclusions and Future Perspectives

Despite the relatively scarce examples of real application of biological techniques in marine environments the information available until now is enough to affirm that bioremediation and phytoremediation are promising techniques for remediation of both organic and inorganic pollutants in those ecosystems. Regardless of the special physicochemical characteristics of those environments, such as high salinity and reduced sediments, the mechanism of bioremediation and phytoremediation proved to be efficient on several occasions and with a large number of microorganisms and plant species. Here we approached bioremediation and phytoremediation together, and we even analyzed some cases of interaction between the two techniques to achieve better results, ecologically and economically. In this chapter we also considered both organic and inorganic pollutants, showing the usefulness of some techniques for both kinds of pollutants, which allows a better choice for real situations of mixed pollutants. The actual possibilities for eco-friendly remediation actions in coastal systems encourage further research in the area.

More research is also desirable in the field of new biological technologies such as vermiremediation. Vermiremediation is the use of earthworms to clean up pollutants from the soil and might be a promising technique in intertidal sediments polluted by both metals and oil. Moreover, the study of new pollutants, such as pharmaceutical compounds, is rising and the biological remediation of such pollutants should not be neglected. More research is also needed to understand the mechanisms that microorganisms and especially plants have to overcome pollution and serve as efficient agents of remediation. The more we understand about the underlying mechanisms the more efficient the application of bioremediation or phytoremediation will be. The challenge to come is to find techniques easily applicable to marine environments, systems subject to important and increasing pollution of diverse origin.

References

Abatenh, E., B. Gizaw, Z. Tsegay and M. Wassie. 2017. The role of microorganisms in bioremediation—A Review. Open J. Environ. Biol. 2(1): 038–046.

Adams, G.O., P.T. Fufeyin, S.E. Okoro and I. Ehinomen. 2015. Bioremediation, biostimulation and bioaugmention: A review. Int. J. Environ. Bioremediation & Biodegradation 3(1): 28–39.

Adrian, L., W. Manz, U. Szewczyk and P. Görisch. 1998. Physiological characterization of a bacterial consortium reductively dechlorinating 1,2,3- and 1,2,4-trichlorobenzene. Appl. Environ. Microbiol. 64(2): 496–503.

Ahrens, M.J., J. Hertz, E.M. Lamoureux, G.R. Lopez, A.E. McElroy and B.J. Brownawell. 2001. The role of digestive surfactants in determining bioavailability of sediment-bound hydrophobic organic contaminants to 2 deposit-feeding polychaetes. Mar. Ecol. Prog. Ser. 212: 145–157.

Akcil, A., C. Erust, S. Ozdemiroglu, V. Fonti and F. Beolchini. 2015. A review of approaches and techniques used in aquatic contaminated sediments: metal removal and stabilization by chemical and biotechnological processes. J. Clean. Prod. 86: 24–36.

Alarcón, A., F.T. Davies, R.L. Autenrieth and D.A. Zuberer. 2008. Arbuscular mycorrhiza and petroleum degrading microorganisms enhance phytoremediation of petroleum-contaminated soil. Int. J. Phytoremediation 10(4): 251–263.

Ali, H., E. Khan and M.A. Sajad. 2013. Phytoremediation of heavy metals-concepts and applications. Chemosphere 91(7): 869–881.

Almeida, C.M.R., T. Oliveira, I. Reis, C.R. Gomes and A.P. Mucha. 2017. Bacterial community dynamic associated with autochthonous bioaugmentation for enhanced Cu phytoremediation of salt-marsh sediments. Mar. Environ. Res. 132: 68–78.

Anjum, N.A., I. Ahmad, M. Válega, I. Mohmood, S.S. Gill, N. Tuteja et al. 2014. Salt marsh halophyte services to metal–metalloid remediation: assessment of the processes and underlying mechanisms. Crit. Rev. Environ. Sci. Technol. 44(18): 2038–2106.

Anjum, N.A., B. Duarte, I. Caçador, N. Sleimi, A.C. Duarte and E. Pereira. 2016. Biophysical and biochemical markers of metal/metalloid-impacts in salt marsh halophytes and their implications. Front. Environ. Sci. 4: 1–13.

Aronstein, B.N., Y.M. Calvillo and M. Alexander. 1991. Effect of surfactant at low concentrations on the desorption and biodegradation of sorbed aromatic compounds in soil. Environ. Sci. Technol. 25: 1728–1731.

Atagana, H.I., R.J. Haynes and F.M. Wallis. 2003. Optimization of soil physical and chemical conditions for the bioremediation of creosote-contaminated soil. Biodegradation 14: 297–307.

Atlas, R.M. and M. Budosh. 1976. Microbial degradation of petroleum in the Arctic. pp. 79–85. *In*: Sharpley, J.M. and A.M. Kaplan (eds.). Proceedings of the 3rd International Biodegradation Symposium. Applied Science Publishers, London.

Atlas, R.M. 1995. Petroleum biodegradation and oil spill bioremediation. Mar. Pollut. Bull. 31: 178–182.

Balagurusamy, N. 2005. Anaerobic bioremediation—An untapped potential. Revista Mexicana de Ingeniería Química. 4: 273–287.

Bamforth, S.M. and I. Singleton. 2005. Review Bioremediation of polycyclic aromatic hydrocarbons: current knowledge and future directions. J. Chem. Technol. Biotechnol. 80: 723–736.

Barbier, E.B., S.D. Hacker, C. Kennedy, E.W. Koch, A.C. Stier and B.R. Silliman. 2011. The value of estuarine and coastal ecosystem services. Ecol. Monogr. 81(2): 169–193.

Barletta, M., A.R. Lima and M.F. Costa. 2019. Distribution, sources and consequences of nutrients, persistent organic pollutants, metals and microplastics in South American estuaries. Sci. Total Environ. 651: 1199–1218.

Bastida, F., N. Jehmlich, K. Lima, B. Morris, H. Richnow, T. Hernández et al. 2016. The ecological and physiological responses of the microbial community from a semiarid soil to hydrocarbon contamination and its bioremediation using compost. J. Proteomics 135: 162–169.

Beolchini, F., A. Dell'Anno, L. De Propris, S. Ubaldini, F. Cerrone and R. Danovaro. 2009. Auto- and heterotrophic acidophilic bacteria enhance the bioremediation efficiency of sediments contaminated by heavy metals. Chemosphere 74: 1321–1326.

Berselli, S., G. Milone, P. Canepa, D. Di Gioia and F. Fava. 2004. Effects of cyclodextrins, humic substances, and rhamnolipids on the washing of a historically contaminated soil and on the aerobic bioremediation of the resulting effluents. Biotechnol. Bioen. 88: 111–120.

Bezza, F.A., M. Beukes and E.M.N. Chirwa. 2015. Application of biosurfactant produced by Ochrobactrum intermedium CN3 for enhancing petroleum sludge bioremediation. Process Biochem. 50(11): 1911–1922.

Bhargava, A., F.F. Carmona, M. Bhargava and S. Srivastava. 2012. Approaches for enhanced phytoextraction of heavy metals. J. Environ. Manage. 105: 103–120.

Bhatt, M., T. Cajthaml and V. Sasek. 2002. Mycoremediation of PAH-contaminated soil. Folia Microbiol. 47(3): 255–258.

Bossert, I. and R. Bartha. 1984. The fate of petroleum in soil ecosystems. pp. 435–476. *In*: Atlas R.M. (ed.). Petroleum Microbiology. Macmillan, New York, USA.

Bragg, J.R., R.C. Prince, E.J. Harner and R.M. Atlas. 1994. Effectiveness of bioremediation for the Exxon Valdez oil spill. Nature 368: 413–418.

Breedveld, G.D. and M. Sparrevik. 2000. Nutrient limited biodegradation of PAHs in various soil strata at a creosote contaminated site. Biodegradation 11: 391–399.

Brewer, E.P., J.A. Saunders, J.S. Angle, R.L. Chaney and M.S. McIntosh. 1999. Somatic hybridization between the Zn accumulator *Thlaspi caerulescens* and *Brassica napus* Theor. Appl. Genet. 99: 761–771.

Burgos, A., J. Maldonado, A. de los Rios, A. Solé and I. Esteve. 2013. Effect of copper and lead on two consortia of phototrophic microorganisms and their capacity to sequester metals. Aquat. Toxicol. 140: 324–336.

Byström, A. and M. Hirtz. 2002. Phytoremediation of Petroleum Hydrocarbon-Contaminated. Soils with Salix Viminalis. Thesis. Chalmers University of Technology. Goteborg.

Cajthaml, T., P. Erbanova, A. Kollmann, C. Novotny, V. Sasek and C. Mougin. 2008. Degradation of PAHs by ligninolytic enzymes of Irpex lacteus. Folia Microbiol. 53(4): 289–294.

Canário, J., C. Vale, L. Poissant, M. Nogueira, M. Pilote and V. Branco. 2010. Mercury in sediments and vegetation in a moderately contaminated salt marsh (Tagus Estuary, Portugal). J. Environ. Sci. 22(8): 1151–1157.

Canet, R., J.M. Lopez-Real and A.J. Beck. 1999. Overview of polycyclic aromatic hydrocarbon bioremediation by white-rot fungi. Land Contam. Reclam. 7(3): 191–197.

Carmona, M., M.T. Zamarro, B. Blazquez, G. Durante-Rodriguez, J.F. Juarez, J.A. Valderrama et al. 2009. Anaerobic catabolism of aromatic compounds: a genetic and genomic view. Microbiol. Mol. Biol. Rev. 73: 71–133.

Castro, R., S. Pereira, A. Lima, S. Corticeiro, M. Válega, E. Pereira et al. 2009. Accumulation, distribution and cellular partitioning of mercury in several halophytes of a contaminated salt marsh. Chemosphere 76(10): 1348–1355.

Chaineau, C., G. Rougeux, C. Yepremian and J. Oudot. 2005. Effects of nutrient concentration on the biodegradation of crude oil and associated microbial populations in the soil. Soil Biol. Biochem. 37: 1490–1497.

Chakraborty, R., C.H. Wu and T.C. Hazen. 2012. Systems biology approach to bioremediation. Curr. Opin. Biotechnol. 23: 483–490.

Chaney, R.L. 1983. Plant uptake of inorganic waste constituents. pp. 50–76. *In*: Parr, J.F.E.A. (ed.). Land Treatment of Hazardous Wastes. Noyes Data Corp., Park Ridge, NJ, USA.

Chaudhry, Q., M. Blom-Zandstra, S.K. Gupta and E. Joner. 2005. Utilising the synergy between plants and rhizosphere microorganisms to enhance breakdown of organic pollutants in the environment. Environ. Sci. Pollut. Res. 12: 34–48.

Chen, Q., M. Bao, X. Fan, S. Liang and P. Sun. 2013. Rhamnolipids enhance marine oil spill bioremediation in laboratory system. Mar. Pollut. Bull. 71: 269–275.

Chibuike, G.U. and S.C. Obiora. 2014. Heavy metal polluted soils: effect on plants and bioremediation methods. Appl. Environ. Soil Sci. 2014.

Chowdhury, R., P.J. Favas, J. Pratas, M.P. Jonathan, P.S. Ganesh and S.K. Sarkar. 2015. Accumulation of trace metals by mangrove plants in Indian Sundarban Wetland: Prospects for Phytoremediation. Int. J. Phytoremediation. 17(9): 885–894.

Chowdhury, R., P.J. Favas, M.P. Jonathan, P. Venkatachalam, P. Raja and S.K. Sarkar. 2017. Bioremoval of trace metals from rhizosediment by mangrove plants in Indian Sundarban Wetland. Mar. Pollut. Bull. 124(2): 1078–1088.

Cobbett, C. and P. Goldsbrough. 2002. Phytochelatins and metallothioneins: roles in heavy metal detoxification and homeostasis. Annu. Rev. Plant Biol. 53(1): 159–182.

Cohen, Y. 2002. Bioremediation of oil by marine microbial mats. Int. Microbiol. 5: 189–193.

Collins, C.D. 2007. Implementing phytoremediation of petroleum hydrocarbons. pp. 99–108. *In*: Willey, N. (ed.). Phytoremediation: Methods and Reviews. Humana Press, NY, USA.

Colmer, T.D. 2003. Long-distance transport of gases in plants: a perspective on internal aeration and radial oxygen loss from roots. Plant Cell Environ. 26: 17–36.

Congeevaram, S., S. Dhanarani, J. Park, M. Dexilin and K. Thamaraiselvi. 2007. Biosorption of chromium and nickel by heavy metal resistant fungal and bacterial isolates. J. Hazard. Mater. 146(1-2): 270–277.

Cook, R.L. and D. Hesterberg. 2013. Comparison of trees and grasses for rhizoremediation of petroleum hydrocarbons. Int. J. Phytoremediation 15(9): 844–860.

Costanza, R., R. d'Arge, R. de Groot, S. Farber, M. Grasso, B. Hannon et al. 1997. The value of the world's ecosystem services and natural capital. Nature 387: 353–360.

Couto, T., B. Duarte, D. Barroso, I. Caçador and J.C. Marques. 2013. Halophytes as sources of metals in estuarine systems with low levels of contamination. Funct. Plant Biol. 40(9): 931–939.

Cristaldi, A., G.O. Conti, E.H. Jho, P. Zuccarello, A. Grasso, C. Copat et al. 2017. Phytoremediation of contaminated soils by heavy metals and PAHs. A brief review. Environ. Biotechnol. Innov. 8: 309–326.

Curado, G., B.J. Grewell, E. Figueroa and J.M. Castillo. 2014a. Effectiveness of the aquatic halophyte *Sarcocornia perennis* spp. perennis as a biotool for ecological restoration of salt marshes. Water Air Soil Pollut. 225(9): 2108.

Curado, G., A.E. Rubio-Casal, E. Figueroa and J.M. Castillo. 2014b. Potential of *Spartina maritima* in restored salt marshes for phytoremediation of metals in a highly polluted estuary. Int. J. Phytoremediation 16: 1209–1220.

Czako, M., X. Feng, Y. He, D. Liang and L. Marton. 2006. Transgenic *Spartina alterniflora* for phytoremediation. Environ. Geochem. Health 28(1-2): 103–110.

Dadrasnia, A. and P. Agamuthu. 2013. Organic wastes to enhance phyto-treatment of diesel contaminated soil. Waste Manag. Res. 31: 1133–1139.

de Souza Machado, A.A., K. Spencer, W. Kloas, M. Toffolon and C. Zarfl. 2016. Metal fate and effects in estuaries: a review and conceptual model for better understanding of toxicity. Sci. Total Environ. 541: 268–281.

Dell'Anno, F., C. Sansone, A. Ianora and A. Dell'Anno. 2018. Biosurfactant-induced remediation of contaminated marine sediments: Current knowledge and future perspectives. Mar. Environ. Res. 137: 196–205.

Deniz, F. and A. Karabulut. 2017. Biosorption of heavy metal ions by chemically modified biomass of coastal seaweed community: studies on phycoremediation system modeling and design. Ecol. Eng. 106: 101–108.

dos Santos, J.J. and L.T. Maranho. 2018. Rhizospheric microorganisms as a solution for the recovery of soils contaminated by petroleum: A review. J. Environ. Manag. 210: 104–113.

Duarte, B., M. Caetano, P.R. Almeida, C. Vale and I. Caçador. 2010. Accumulation and biological cycling of heavy metal in four salt marsh species, from Tagus estuary (Portugal). Environ. Pollut. 158(5): 1661–166.

Duarte, B., J. Freitas and I. Caçador. 2011. The role of organic acids in assisted phytoremediation processes of salt marsh sediments. Hydrobiologia 674(1): 169–177.

El-Shahawi, M., A. Hamza, A. Bashammakh and W. Al-Saggaf. 2010. An overview on the accumulation, distribution, transformations, toxicity and analytical methods for the monitoring of persistent organic pollutants. Talanta 80(5): 1587–1597.

Emami, S., A.A. Pourbabaei and H.A. Alikhani. 2014. Interactive effect of nitrogen fertilizer and hydrocarbon pollution on soil biological indicators. Environ. Earth Sci. 72(9): 3513–3519.

Euliss, K., C. Ho, A. Schwab, S. Rock and M.K. Banks. 2008. Greenhouse and field assessment of phytoremediation for petroleum contaminants in a riparian zone. Bioresour. Technol. 99(6): 1961–1971.

Figueiredo, N.L., A. Areias, R. Mendes, J. Canário, A. Duarte and C. Carvalho. 2014. Mercury-resistant bacteria from salt marsh of Tagus Estuary: the influence of plants presence and mercury contamination levels. J. Toxicol. Environ. Health A 77(14-16): 959–971.

Fingas, M. 2010. Oil Spill Science and Technology: Prevention, Response, and Clean Up. 1st ed. Elsevier Inc. USA.

Floodgate, G.D. 1984. The fate of petroleum in marine ecosystems. pp. 355–397. *In*: Arias R.M. (ed.). Petroleum Microbiology. Macmillan, New York, USA.

Flowers, T.J. and T.D. Colmer. 2008. Salinity tolerance in halophytes. New Phytol. 179(4): 945–963.

Fonti, V., A. Dell'Anno and F. Beolchini. 2016. Does bioleaching represent a biotechnological strategy for remediation of contaminated sediments? Sci. Total Environ. 563: 302–319.

Fonti, V., A. Dell'Anno and F. Beolchini. 2013. Influence of biogeochemical interactions on metal bioleaching performance in contaminated marine sediment. Water Res. 47(14): 5139–5152.

Fonti, V., F. Beolchini, L. Rocchetti and A. Dell'Anno. 2015. Bioremediation of contaminated marine sediments can enhance metal mobility due to changes of bacterial diversity. Water Res. 68: 637–650.

Genou, G., F. de Naeyer, P. van Meenen, H. van der Wert, W. de Nijis and W. Verstraete. 1994. Degradation of oil sludge by landfarming—a case study at Ghent Harbour. Biodegradation 5: 37–46.

Gerhardt, K.E., X.D. Huang, B.R. Glick and B.M. Greenberg. 2009. Phytoremediation and rhizoremediation of organic soil contaminants: potential and challenges. Plant Sci. 176: 20–30.

Germida, J., C. Frick and R. Farrell. 2002. Phytoremediation of oil-contaminated soils. Dev. Soil Sci. 28: 169–186.

Ghasemi, S., S.S. Moghaddam, A. Rahimi, C.A. Damalas and A. Naji. 2018. Phytomanagement of trace metals in mangrove sediments of Hormozgan, Iran, using gray mangrove (*Avicennia marina*). Environ.Sci. Pollut. Res. 25(28): 28195–28205.

Ghosh, M. and S.P. Singh. 2005. A review on phytoremediation of heavy metals and utilization of it's by products. Asian J. Energy Environ. 6(4): 18.

Gomez, F. and M. Sartaj. 2013. Field scale ex-situ bioremediation of petroleum contaminated soil under cold climate conditions. Int. Biodeterioration & Biodegradation 85: 375–382.

Gregory, S.T., D. Shea and E. Guthrie-Nichols. 2005. Impact of vegetation on sedimentary organic matter composition and PAH attenuation. Environ. Sci. Technol. 39: 5285–5292.

Gruiz, K., E. Fenyvesi, E. Kriston, M. Molnar and B. Horvath. 1996. Potential use of cyclodextrins in soil bioremediation. J. Incl. Phenom. Mol. Recognit. Chem. 25: 233–236.

Gupta, G., V. Kumar and A.K. Pal. 2016. Biodegradation of polycyclic aromatic hydrocarbons by microbial consortium: a distinctive approach for decontamination of soil. Soil Sedim. Contam. Int. J. 25(6): 597–623.

Harmsen, J.A., H.J.J. Wieggers, J.H.H. Van de Akker, O.M. Van Dijk-Hooyer, A. Van den Toorn and A.J. Zweers. 1997. Intensive and extensive treatment of dredged sediments on landfarms. pp. 153–158. *In*: Allemana, B.C. and A. Leeson (eds.). *In-situ* and On-Site. Bioremediation, Vol. 2. Battelle Press, Columbus, OH.

Haritash, A.K. and C.P. Kaushik. 2009. Biodegradation aspects of polycyclic aromatic hydrocarbons (PAHs): A review. J. Hazard. Mater. 169: 1–15.

Heitkamp, M.A. and C.E. Cerniglia. 1988. Mineralization of polycyclic aromatic hydrocarbons by a bacterium isolated from sediment below an oil field. Appl. Environ. Microbiol. 54: 1612–14.

Herath, I. and M. Vithanage. 2015. Phytoremediation in constructed wetlands. pp. 243–263. *In*: Ansari, A.A., S.S. Gill, R. Gill, G.R. Lanza and L. Newman (eds.). Phytoremediation: Management of Environmental Contaminants (Vol. 2). Cham: Springer International Publishing, NY, USA.

Horel, A., B. Mortazavi and P.A. Sobecky. 2015. Input of organicmatter enhances degradation of weathered diesel fuel in sub-tropical sediments. Sci. Total Environ. 533: 82–90.

Hou, J., W. Liu, B. Wang, Q. Wang, Y. Luo and A.E. Franks. 2015. PGPR enhanced phytoremediation of petroleum contaminated soil and rhizosphere microbial community response. Chemosphere 138: 592–598.

Huesemann, M.H., T.S. Hausmann, T.J. Fortman, R.M. Thom and V. Cullinan. 2009. *In situ* phytoremediation of PAH-and PCB-contaminated marine sediments with eelgrass (*Zostera marina*). Ecol. Eng. 35(10): 1395–1404.

Idaszkin, Y.L., P.J. Bouza, C.H. Marinho and M.N. Gil. 2014. Trace metal concentrations in *Spartina densiflora* and associated soil from a Patagonian salt marsh. Mar. Pollut. Bull. 89(1-2): 444–450.

Idaszkin, Y.L., J.L. Lancelotti, P.J. Bouza and J.E. Marcovecchio. 2015. Accumulation and distribution of trace metals within soils and the austral cordgrass *Spartina densiflora* in a Patagonian salt marsh. Mar. Pollut. Bull. 101(1): 457–465.

Idaszkin, Y.L., J.L. Lancelotti, M.P. Pollicelli, J.E. Marcovecchio and P.J. Bouza. 2017. Comparison of phytoremediation potential capacity of *Spartina densiflora* and *Sarcocornia perennis* for metal polluted soils. Mar. Pollut. Bull. 118(1-2): 297–306.

Jacob, J.M., C. Karthik, R.G. Saratale, S.S. Kumar, D. Prabakar, K. Kadirvelu et al. 2018. Biological approaches to tackle heavy metal pollution: A survey of literature. J. Environ. Manag. 217: 56–70.

Jagtap, S.S., S.M. Woo, T.S. Kim, S.S. Dhiman, D. Kim and J.K. Lee. 2014. Phytoremediation of diesel-contaminated soil and saccharification of the resulting biomass. Fuel 116: 292–298.

Juhasz, A.L. and R. Naidu. 2000. Bioremediation of high molecular weight polycyclic aromatic hydrocarbons: a review of the microbial degradation of benzo[a]pyrene. Int. Biodeterior. Biodegr. 45: 57–88.

Kabata-Pendias, A. 2010. Trace Elements in Soils and Plants, 2nd ed. CRC Press, Boca Ratón, Florida, USA.

Kadukova, J., E. Manousaki and N. Kalogerakis. 2008. Pb and Cd accumulation and phyto-excretion by salt cedar (Tamarix smyrnensis Bunge). Int. J. Phytoremediation 10: 31–46.

Kaewtubtim, P., W. Meeinkuirt, S. Seepom and J. Pichtel. 2016. Heavy metal phytoremediation potential of plant species in a mangrove ecosystem in Pattani Bay, Thailand. Appl. Ecol. Environ. Res. 14: 367–382.

Kallenborn, R., C. Halsall, M. Dellong and P. Carlsson. 2012. The influence of climate change on the global distribution and fate processes of anthropogenic persistent organic pollutants. J. Environ. Monit. 14(11): 2854–2869.

Khalid, S., M. Shahid, N.K. Niazi, B. Murtaza, I. Bibi and C. Dumat. 2017. A comparison of technologies for remediation of heavy metal contaminated soils. J. Geochem. Explor. 182: 247–268.

Krebs, C.T. and C.E. Tanner. 1981. Restoration of oiled marshes through sediment stripping and *Spartina* propagation. *In*: Proceedings of the 1981 Oil Spill Conference. American Petroleum Institute, Washington, D.C.

Kuo, H.C., D.F. Juang, L. Yang, W.C. Kuo and Y.M. Wu. 2013. Phytoremediation of soil contaminated by heavy oil with plants colonized by mycorrhizal fungi. Int. J. Environ. Sci. Technol. 11(6): 1661–1668.

Ławniczak, Ł., R. Marecik and Ł. Chrzanowski. 2013. Contributions of biosurfactants to natural or induced bioremediation. Appl. Microbiol. Biotechnol. 97: 2327–2339.

Lee, M.D. and C.M. Swindoll. 1993. Bioventing for *in situ* remediation. Hydrol. Sci. J. 38(4): 273–282.

Leewis, M.C., C.M. Reynolds and M.B. Leigh. 2013. Long-term effects of nutrient addition and phytoremediation on diesel and crude oil contaminated soils in subarctic Alaska. Cold Reg. Sci. Technol. 96: 129–137.

Li, H., Q. Zhao, M.C. Boufadel and A.D. Venosa. 2007. A universal nutrient application strategy for the bioremediation of oil-polluted beaches. Mar. Pollut. Bull. 54: 1146–1161.

Liang, L., W. Liu, Y. Sun, X. Huo, S. Li and Q. Zhou. 2016. Phytoremediation of heavy metal contaminated saline soils using halophytes: current progress and future perspectives. Environ. Rev. 25(3): 269–281.

Lim, M.W., E.V. Lau and P.E. Poh. 2016. A comprehensive guide of remediation technologies for oil contaminated soil—Present works and future directions. Mar. Pollut. Bull. 109: 14–45.

Limmer, M. and J. Burken. 2016. Phytovolatilisation of organic contaminants. Environ. Sci. Technol. 50: 6632–6643.

Lin, Q. and I.A. Mendelssohn. 2009. Potential of restoration and phytoremediation with Juncus roemerianus for diesel-contaminated coastal wetlands. Ecol. Eng. 35: 85–91.

Liu, L., W. Li, W. Song and M. Guo. 2018. Remediation techniques for heavy metal-contaminated soils: Principles and applicability. Sci. Total Environ. 633: 206–219.

Luo, W. and C. Hu. 2013. Interaction of plant secondary metabolites and organic carbon substrates affected on biodegradation of polychlorinated biphenyl. J. Environ. Biol. 34(2): 337.

MacFarlane, G., A. Pulkownik and M. Burchett. 2003. Accumulation and distribution of heavy metals in the grey mangrove, *Avicennia marina* (Forsk.) Vierh: Biological indication potential. Environ. Pollut. 123: 139–151.

Machado, A., C. Magalhães, A.P. Mucha, C.M.R. Almeida and A.A. Bordalo. 2012. Microbial communities within saltmarsh sediments: Composition, abundance and pollution constrains. Estuar. Coast. Shelf Sci. 99: 145–152.

Magar, V.S. 2001. Natural recovery of contaminated sediments. J. Environ. Eng. 127: 473–474.

Mahar, A., P. Wang, A. Ali, M.K. Awasthi, A.H. Lahori, Q. Wang et al. 2016. Challenges and opportunities in the phytoremediation of heavy metals contaminated soils: a review. Ecotoxicol. Environ. Saf. 126: 111–121.

Maine, M.A., G.G. Sanchez, H.R. Hadad, S.E. Caffaratti, M.C. Pedro, M.M. Mufarrege et at. 2019. Hybrid constructed wetlands for the treatment of wastewater from a fertilizer manufacturing plant: Microcosms and field scale experiments. Sci. Total Environ. 650: 297–302.

Major, D.W., M.L. McMaster, E.E. Cox, E.A. Edwards, S.M. Dworatzek, E.R. Hendrickson et al. 2002. Field demonstration of successful bioaugmentation to achieve dechlorination of tetrachloroethene to ethene. Environ. Sci. Technol. 36: 5106–5116.

Manousaki, E. and N. Kalogerakis. 2011. Halophytes—An emerging trend in phytoremediation. Int. J. Phytoremediation 13(10): 959–969.

Marcovecchio, J.E. and L.D. de Lacerda. 2017. Continent derived metal pollution through time challenges of the global ocean. pp. 99–117. *In*: Arias, A.H. and J.E. Marcovecchio (eds.). Marine Pollution and Climate Change, CRC Press, Taylor and Francis Group. Boca Ratón, Florida, USA.

Margesin, R. and F. Schinner. 2001. Bioremediation (natural attenuation and biostimulation) of diesel-oil-contaminated soil in an alpine glacier skiing area. Appl. Environ. Microbiol. 67: 3127–3133.

Marques, C.R. 2016. Bio-rescue of marine environments: on the track of microbially-based metal/metalloid remediation. Sci. Total Environ. 565: 165–180.

Mateju, V. 2003. Bioremediation of persistent organic pollutants—A review. pp. 30–74. *In*: Holoubek, I. (ed.). Inventory Check of POPs in the Czech Republic. Cz. Press, Czech Republic.

Mazur, L.P., M.A. Cechinel, S.M.G.U de Souza, R.A. Boaventura and V.J. Vilar. 2018. Brown marine macroalgae as natural cation exchangers for toxic metal removal from industrial wastewaters: A review. J. Environ. Manage. 223: 215–253.

McGhee, I. and R.G. Burns. 1995. Biodegradation of 2,4-dichlorophenoxyacetic acid (2,4-D) and 2-methyl-4-chlorophenoxyacetic acid (MCPA) in contaminated soil. Appl. Soil Ecol. 2: 143–154.

McNally, D.L., J.R. Mihelcic and D.R. Lueking. 1998. Biodegradation of three- and four-ring polycyclic aromatic hydrocarbons under aerobic and denitrifying conditions. Environ. Sci. Technol. 32: 2633–2639.

Mendelssohn, I.A. and K.L. McKee. 1988. *Spartina alterniflora* dieback in Louisiana: time-course investigation of soil waterlogging effects. J. Ecology 76: 509–521.

Mendelssohn, I.A., M.W. Hester and J.W. Pahl. 1995. Environmental Effects and Effectiveness of *In-situ* Burning in Wetlands: Considerations for Oil Spill Cleanup. Louisiana Oil Spill, applied Oil Spill Research and Development Program, Baton Rouge, LA, USA.

Merkl, N., R. Schultze-Kraft and M. Arias. 2005. Influence of fertilizer levels on phytoremediation of crude oil-contaminated soils with the tropical pasture grass Brachiaria brizantha (Hochst. ex A. Rich.) Stapf. Int. J. Phytoremediation 7: 217–230.

Mitsch, W.J. and J.G. Gosselink. 2007. Wetlands, 4º ed. John Wiley & Sons, NJ, USA.

Molnar, M., L. Leitgib, K. Gruiz, E. Fenyvesi, N. Szaniszlo, J. Szejtli et al. 2005. Enhanced biodegradation of transformer oil in soils with cyclodextrin -from the laboratory to the field. Biodegradation 16: 159–168.

Moreira, I.T., O.M. Oliveira, J.A. Triguis, A.M. dos Santos, A.F. Queiroz, C.M. Martins et al. 2011. Phytoremediation using Rhizophora mangle L. in mangrove sediments contaminated by persistent total petroleum hydrocarbons (TPH's). Microchem. J. 99(2): 376–382.

Mucha, A.P., C.M.R. Almeida, C.M. Magalhães, M.T.S. Vasconcelos and A.A. Bordalo. 2011. Salt marsh plant–microorganism interaction in the presence of mixed contamination. Int. Biodeterior. Biodegrad. 65(2): 326–333.

Mueller, J.G., P.J. Chapman and P.H. Pritchard. 1989. Action of a fluoranthene-utilizing bacterial community on polycyclic aromatic hydrocarbon components of creosote. Appl. Environ. Microbiol. 55: 3085–3090.

Mulligan, C.N. 2005. Environmental applications for biosurfactants. Environ. Pol. 133: 183–198.

Muratova, A.Y., T. Dmitrieva, L. Panchenko and O. Turkovskaya. 2008. Phytoremediation of oil-sludge contaminated soil. Int. J. Phytoremediation 10(6): 486–502.

Nakajima, F., A. Baun, A. Ledin and P.S. Mikkelsen. 2005. A novel method for evaluating bioavailability of polycyclic aromatic hydrocarbons in sediments of an urban stream. Water Sci. Technol. 51: 275–281.

National Research Council Committee on Oil in the Sea. 2003. Oil in the Sea III: Inputs, Fates, and Effects, National Research Council Ocean Studies Board and Marine Board Divisions of Earth and Life Studies and Transportation Research Board. Washington, DC: National Academy Press.

National Research Council. 2005. Oil Spill Dispersants: Efficacy and Effects. The National Academies Press, Washington, DC.

Negrin, V.L., S.E. Botté, P.D. Pratolongo, G. González Trilla and J.E. Marcovecchio. 2016. Ecological processes and biogeochemical cycling in salt marshes: synthesis of studies in the Bahía Blanca estuary (Argentina). Hydrobiologia 774(1): 217–235.

Negrin, V.L., B. Teixeira, R.M. Godinho, R. Mendes and C. Vale. 2017. Phytochelatins and monothiols in salt marsh plants and their relation with metal tolerance Mar. Pollut. Bull. 121(1-2): 78–84.

Negrin, V.L., S.E. Botté, N.S. La Colla and J.E. Marcovecchio. 2019. Uptake and accumulation of metals in *Spartina alterniflora* salt marshes from a South American estuary. Sci. Total Environ. 649: 808–820.

Nikolopoulou, M. and N. Kalogerakis. 2011. Petroleum spill control with biological means. pp. 263–274. *In*: Moo-Young, M. (ed.). Comprehensive Biotechnology. 2nd Edition. Elsevier Inc. USA.

Nojiri, H., H. Habe and T. Omori. 2001. Bacterial degradation of aromatic compounds via angular dioxygenation. J. Gen. Appl. Microbiol. 47: 279–305.

Nzengung, V.A., L.N. Wolfe, D.E. Rennels, S.C. McCutcheon and C. Wang. 1999. Use of aquatic plants and algae for decontamination of waters polluted with chlorinated alkanes. Int. J. Phytoremediation 1: 203–226.

Olson, P.E., A. Castro, M. Joern, N.M. DuTeau, E.A.H. Pilon-Smits and K.F. Reardon. 2007. Comparison of plant families in a greenhouse phytoremediation study on an aged polycyclic aromatic hydrocarbon contaminated soil. J. Environ. Qual. 36(5): 1461–1469.

Overton, E.B., T.L. Wade, J.R. Radović, B.M. Meyer, M.S. Miles and S.R. Larter. 2016. Chemical composition of Macondo and other crude oils and compositional alterations during oil spills. Oceanography 29(3): 50–63.

Padinha, C., R. Santos and M.T. Brown. 2000. Evaluating environmental contamination in Ria Formosa (Portugal) using stress indexes of *Spartina maritima*. Mar. Environ. Res. 49(1): 67–78.

Paquin, D., R. Ogoshi, S. Campbell and Q.X. Li. 2002. Bench-scale phytoremediation of polycyclic aromatic hydrocarbon-contaminated marine sediment with tropical plants. Int. J. Phytoremediation 4: 297–313.

Parfitt, R.L., J.S. Whitton and S. Susarla. 1995. Removal of DDT residues from soil by leaching with surfactants. Comm. Soil Sci. Plant Anal. 26: 2231–2241.

Pariatamby, A. and Y.L. Kee. 2016. Persistent organic pollutants management and remediation. The Tenth International Conference on Waste Management and Technology (ICWMT). Procedia Environ. Sci. 31: 842–848.

Parrish, Z.D., M.K. Banks and A.P. Schwab. 2004. Effectiveness of phytoremediation as a secondary treatment for polycyclic aromatic hydrocarbons (PAHs) in composted soil. Int. J. Phytoremediation 6(2): 119–137.

Parvaresh, H., Z. Abedi, P. Farshchi, M. Karami, N. Khorasani and A. Karbassi. 2011. Bioavailability and concentration of heavy metals in the sediments and leaves of grey mangrove, Avicennia marina (Forsk.) Vierh, in SirikAzini Creek, Iran. Biol. Trace Elem. Res. 143(2): 1121–1130.

Passatore, L., S. Rossetti, A.A. Juwarkar and A. Massacci. 2014. Phytoremediation and bioremediation of polychlorinated biphenyls (PCBs): state of knowledge and research perspectives. J. Hazard. Mater. 278: 189–202.

Payne, R.B., H.D. May and K.R. Sowers. 2011. Enhanced reductive dechlorination of polychlorinated biphenyl impacted sediment by bioaugmentation with a dehalorespiring bacterium. Environ. Sci. Technol. 45: 8772–8779.

Perelo, L.W. 2010. Review: *In situ* and bioremediation of organic pollutants in aquatic sediments. J. Hazard. Mater. 177: 81–89.

Pezeshki, S.R., M.W. Hester, Q. Lin and J.A. Nyman. 2000. The effects of oil spill and clean-up on dominant US Gulf coast marsh macrophytes: a review. Environ. Pollut. 108: 129–139.

Phillips, T.M., H. Lee, J.T. Trevors and A.G. Seech. 2006. Full-scale *in situ* bioremediation of hexachlorocyclohexane-contaminated soil. J. Chem. Technol. Biotechnol. 81: 289–298.

Piehler, M.E. and H.W. Paerl. 1996. Enhanced biodegradation of diesel fuel through the addition of particulate organic carbon and inorganic nutrients in coastal marine waters. Biodegradation 7: 239–247.

Rahman, K., J. Thahira-Rahman, P. Lakshmanaperumalsamy and I. Banat. 2002. Towards efficient crude oil degradation by a mixed bacterial consortium. Bioresour. Technol. 85: 257–261.

Ranwell, D.S. 1968. Extent of damage to coastal habitats due to the Torrey Canyon incident. pp. 39–47. *In*: The Biological Effects of Oil Pollution in Littoral Communities. Field Studies Council, London.

Reboreda, R. and I. Caçador. 2008. Enzymatic activity in the rhizosphere of *Spartina maritima*: potential contribution for phytoremediation of metals. Mar. Environ. Res. 65(1): 77–84.

Redondo-Gómez, S., E. Mateos-Naranjo, I. Vecino-Bueno and S.R. Feldman. 2011. Accumulation and tolerance characteristics of chromium in a cordgrass Cr-hyperaccumulator, *Spartina argentinensis*. J. Hazard. Mater. 185(2-3): 862–869.

Redondo-Gómez, S., M.C. Petenello and S.R. Feldman. 2014. Growth, nutrient status, and photosynthetic response to diesel-contaminated soil of a cordgrass, *Spartina argentinensis*. Mar. Pollut. Bull. 79: 34–38.

Reeves, R.D., A.J. Baker, T. Jaffré, P.D. Erskine, G. Echevarria and A. van der Ent. 2018. A global database for plants that hyperaccumulate metal and metalloid trace elements. New Phytologist 218(2): 407–411.

Rezania, S., S.M. Taib, M.F.M. Din, F.A. Dahalan and H. Kamyab. 2016. Comprehensive review on phytotechnology: heavy metals removal by diverse aquatic plants species from wastewater. J. Hazard. Mater. 318: 587–599.

Rhee, G., B. Bush, M. Brown, M. Kane and L. Shane. 1989. Anaerobic biodegradation of polychlorinated biphenyls in Hudson River sediments and dredged sediments in clay encapsulation. Water Res. 23: 957–964.

Ron, E.Z. and E. Rosenberg. 2014. Enhanced bioremediation of oil spills in the sea. Curr. Opin. Biotechnol. 27: 191–194.

Rubinos, D.A., R. Villasuso, S. Muniategui, M.T. Barral and F. Diaz Fierros. 2007. Using the landfarming technique to remediate soils contaminated with hexachlorocyclohexane isomers. Water Air Soil Pollut. 181: 385–399.

Saavedra, J.M., F. Acevedo, M. González and M. Seeger. 2010. Mineralization of PCBs by the genetically modified strain Cupriavidus necator JMS34 and its application for bioremediation of PCBs in soil. Appl. Microbiol. Biotechnol. 87: 1543–1554.

Said, O.B., M.M. da Silva, F. Hannier, H. Beyrem and L. Chícharo. 2018. Using *Sarcocornia fruticosa* and *Saccharomyces cerevisiae* to remediate metal contaminated sediments of the Ria Formosa lagoon (SE Portugal). Ecohydrology & Hydrobiology. Doi: 10.1016/j.ecohyd.2018.10.002

Serra, A.V., S.E. Botté, D.G. Cuadrado, N.S. La Colla and V.L. Negrin. 2017. Metals in tidal flats colonized by microbial mats within a South-American estuary (Argentina). Environ. Earth Sci. 76(6): 254.

Sheppard, P.J., K.L. Simons, E.M. Adetutu, K.K. Kadali, A.L. Juhasz, M. Manefield et al. 2014. The application of a carrier-based bioremediation strategy for marine oil spills. Mar. Pollut. Bull. 84(1-2): 339–346.

Shrestha, B., S. Lipe, K.A. Johnson, T.Q. Zhang, W. Retzlaff and Z.Q. Lin. 2006. Soil hydraulic manipulation and organic amendment for the enhancement of selenium volatilization in a soil–pickleweed system. Plant Soil 288(1-2): 189–196.

Simons, K., A. Ansar, K. Kadali, A. Bueti, E. Adetutu and A. Ball. 2012. Investigating the effectiveness of economically sustainable carrier material complexes for marine oil remediation. Bioresour. Technol. 126: 202–207.

Singer, A.C., D. Smith, W.A. Jury, K. Hathuc and D.E. Crowley. 2003. Impact of the plant rhizosphere and augmentation on remediation of polychlorinated biphenyl contaminated soil. Environ. Toxicol. Chem. 22: 1998–2004.

Smith, K.E., A.P. Schwab and M.K. Banks. 2007. Phytoremediation of polychlorinated biphenyl (PCB)-contaminated sediment: a greenhouse feasibility study. J. Environ. Qual. 36: 239–244.

Sode, K., Y. Yamamot and N. Hatano. 1998. Construction of a marine cyanobacterial strain with increased heavy metal ion tolerance by introducing exogenic metallothionein gene. J. Mar. Biotechnol. 6: 174–177.

Stefanakis, A.I. 2018. Constructed Wetlands for Industrial Wastewater Treatment. John Wiley & Sons, NJ, USA.

Suja, F., F. Rahim, M.R. Taha, N. Hambali, M. Rizal Razali, A. Khalid et al. 2014. Effects of local microbial bioaugmentation and biostimulation on the bioremediation of total petroleum hydrocarbons (TPH) in crude oil contaminated soil based on laboratory and field observations. Int. Biodeterior. Biodegrad. 90: 115–122.

Swannell, R.P.J., K. Lee and M. McDonagh. 1996. Field evaluations of marine oil spill bioremediation. Microbiol. Rev. 60: 342–365.

Szulc, A., D. Ambrożewicz, M. Sydow, L. Ławniczak, A. Piotrowska-Cyplik, R. Marecik et al. 2014. The influence of bioaugmentation and biosurfactant addition on bioremediation efficiency of diesel-oil contaminated soil: feasibility during field studies. J. Environ. Manag. 132: 121–128.

Tabak, H.H., P. Lens, E.D. van Hullebusch and W. Dejonghe. 2005. Developments in bioremediation of soils and sediments polluted with metals and radionuclides–1. Microbial processes and mechanisms affecting bioremediation of metal contamination and influencing metal toxicity and transport. Rev. Env. Sci. Biotech. 4(3): 115–156.

Tam, N.F.Y. and Y.S. Wong. 2008. Effectiveness of bacterial inoculum and mangrove plants on remediation of sediment contaminated with polycyclic aromatic hydrocarbons. Mar. Pollut. Bull. 57: 716–726.

Ueno, A., Y. Ito, I. Yumoto and H. Okuyama. 2007. Isolation and characterization of bacteria fromsoil contaminated with diesel oil and the possible use of these in autochthonous bioaugmentation. World J. Microbiol. Biotechnol. 23: 1739–1745.

US EPA (United States Environmental Protection Agency). 2005. Contaminated Sediment Remediation Guidance for Hazardous Waste Sites. Office of Solid Waste and Emergency Response OSWER.

Válega, M., A.I.G. Lima, E.M.A.P. Figueira, E. Pereira, M.A. Pardal and A.C. Duarte. 2009. Mercury intracellular partitioning and chelation in a salt marsh plant, *Halimione portulacoides* (L.) Aellen: strategies underlying tolerance in environmental exposure. Chemosphere 74(4): 530–536.

Varca, L.M. and E.D. Magallona. 1994. Dissipation and degradation of DDT and DDE in Philippine soil under field conditions. J. Environ. Sci. Health Part B 29: 25–35.

Varjani, S.J., D.P. Rana, A.K. Jain, S. Bateja and V.N. Upasani. 2015. Synergistic *ex-situ* biodegradation of crude oil by halotolerant bacterial consortium of indigenous strains isolated from on shore sites of Gujarat, India. Int. Biodeterior. Biodegrad. 103: 116–124.

Vasilyeva, G. and E. Strijakova. 2007. Bioremediation of soils and sediments contaminated by polychlorinated biphenyls. Microbiology 76: 639–653.

Venkatesan, A.K. and R.U. Halden. 2014. Wastewater treatment plants as chemical observatories to forecast ecological and human health risks of manmade chemicals. Sci. Rep. 4: 3731.

Venosa, A.D., M.T. Suidan, B.A. Wrenn, K.L. Strohmeier, J.R. Haines, B.L. Eberhart et al. 1996. Bioremediation of an experimental oil spill on the shoreline of Delaware Bay. Environ. Sci. Technol. 30: 1764–1775.

Venosa, A.D., D.J. Feldhake, E.L. Holder and K.M. Koran. 2003, April. Biodegradability of orimulsion in saltwater and freshwater environments. pp. 663–668. *In*: International Oil Spill Conference, American Petroleum Institute.

Venosa, A.D. and X. Zhu. 2003. Biodegradation of crude oil contaminating marine shorelines and freshwater wetlands. Spill Sci. Technol. Bull. 8: 163–178.

Venosa, A.D., P. Campo and M.T. Suidan. 2010. Biodegradability of lingering crude oil 19 years after the Exxon Valdez oil spill. Environ. Sci. Technol. 44: 7613–7621.

Walters, G.W. and M.D. Aitken. 2001. Surfactant-enhanced solubilization and anaerobic biodegradation of 1,1,1-trichloro-2,2-bis(p-chlorophenyl)-ethane (DDT) in contaminated soil. Water Environ. Res. 73(1): 15–23.

Wang, M.C., Y.T. Chen, S.H. Chen, S.C. Chien and S.V. Sunkara. 2012. Phytoremediation of pyrene contaminated soils amended with compost and planted with ryegrass and alfalfa. Chemosphere 87(3): 217–225.

Weis, J.S. and P. Weis. 2004. Metal uptake, transport and release by wetland plants: implications for phytoremediation and restoration. Environ. Int. 30(5): 685–700.

White, C., A.K. Shaman and G.M. Gadd. 1998. An integrated microbial process for the bioremediation of soil contaminated with toxic metals. Nat. Biotechnol. 16(6): 572.

Windham, L., J.S. Weis and P. Weis. 2001. Patterns and processes of mercury release from leaves of two dominant salt marsh macrophytes, *Phragmites australis* and *Spartina alterniflora*. Estuaries 24: 787–795.

Wright, A.L., R.W. Weaver and J.W. Webb. 1997. Oil bioremediation in salt march mesocosms as influenced by N and P fertilization, flooding, and season. Water Air Soil Pollut. 95(1-4): 179–191.

Wu, G., X. Zhu, H. Ji and D. Chen. 2015. Molecular modeling of interactions between heavy crude oil and the soil organic matter coated quartz surface. Chemosphere 119: 242–249.

Wu, M., L. Chen, Y. Tian, Y. Ding and W. Dick. 2013. Degradation of polycyclic aromatic hydrocarbons by microbial consortia enriched from three soils using two different culture media. Environ. Pollut. 178: 152–158.

Xu, R., L.C. Yong, Y.G. Lim and J.P. Obbard. 2005. Use of slow-release fertilizer and biopolymers for stimulating hydrocarbon biodegradation in oil-contaminated beach sediments. Mar. Pollut. Bull. 51: 1101–1110.

Yadav, K.K., N. Gupta, A. Kumar, L.M. Reece, N. Singh, S. Rezania et al. 2018. Mechanistic understanding and holistic approach of phytoremediation: A review on application and future prospects. Ecol. Eng. 120: 274–298.

Yang, S.Z., H.J. Jin, Z. Wei, R.X. He, Y.J. Ji, X.M. Li et al. 2009. Bioremediation of oil spills in cold environments: a review. Pedosphere 19(3): 371–381.

Yang, Z., Z. Zhang, L. Chai, Y. Wang, Y. Liu and R. Xiao. 2016. Bioleaching remediation of heavy metal-contaminated soils using *Burkholderia* sp. Z-90. 2016. J. Hazard. Mater. 30: 145–152.

Yavari, S., A. Malakahmad and N.B. Sapari. 2015. A review on phytoremediation of crude oil spills. Water Air Soil Pollut. 226: 279.

Yin, K., Q. Wang, M. Lv and L. Chen. 2019. Microorganism remediation strategies towards heavy metals. Chem. Eng. J. 360: 1553–1563.

Zahed, M., H. Aziz, M. Isa, L. Mohajeri, S. Mohajeri and S. Kutty. 2011. Kinetic modelling and half life study on bioremediation of crude oil dispersed by Corexit 9500. J. Hazard. Mater. 185: 1027–1031.

Zeller, C. and B. Cushing. 2005. Panel discussion: remedy effectiveness: what works, what doesn't. Integr. Environ. Assess. Manag. 2: 75–79.

Zhang, Y. and R.M. Miller. 1995. Effect of rhamnolipid (biosurfactant) structure on solubilization and biodegradation of n-alkanes. Environ. Microbiol. 61: 2247–2251.

Zhou, Y.W., Y.S. Peng, X.L. Li and G.Z. Chen. 2011. Accumulation and partitioning of heavy metals in mangrove rhizosphere sediments. Environ. Earth Sci. 64(3): 799–807.

Zhu, Y.L., E.A.H. Pilon-Smits, A.S. Tarun, S.U. Weber, L. Jouanin and N. Terry. 1999. Cadmium tolerance and accumulation in Indian mustard is enhanced by overexpressing glutamylcysteine synthetase. Plant Physiol. 121: 1169–1177.

Zouboulis, A.I. and P.A. Moussas. 2011. Groundwater and soil pollution: Bioremediation. pp. 1037–1044. *In*: Nriagu J.O. (ed.). Encyclopaedia of Environmental Health. Elsevier Science. Amsterdam, London.

8

Lesser-known Metals with Potential Impacts in the Marine Environment

Pedro L. Borralho Aboim de Brito,[1,*] *Rute Cesário*[1,2]
and *Carlos Eduardo S.S. Monteiro*[1,2]

1. General Introduction

Metals and the mining of metals have been an important pillar in the development of civilization. Technological advances led to the search and exploitation of new mineral resources, with materials expected to be more resistant, durable and light, as well as the development of new forms of "green" energy. This development has not always been accompanied by adequate studies of environmental impacts, which studies have generally been confined to mining and ore processing activities. The exploitation and use of rare earth elements (REE) and platinum group elements (PGE) has increased significantly in the last decades, while information on their biogeochemical cycles and impacts is still scarce. According to the International Union of Pure and Applied Chemistry (IUPAC 2005), the REE consist of 17 elements comprising the lanthanide series, from lanthanum to lutetium (atomic numbers 57–71), yttrium (atomic number 39) and scandium (atomic number 21). Platinum group elements are rare and noble transitional elements that include ruthenium, rhodium and palladium (Ru, Rh and Pd; atomic numbers 44–46, respectively), and osmium, iridium and platinum (Os, Ir and Pt; atomic numbers 76–78, respectively).

This chapter begins with a general introduction of these two groups of lesser-known metals and includes a brief historical context, a characterization and description of the main properties of those elements, their natural and anthropogenic sources, and applications. In the second part, a brief review of the analytical methodologies developed for the study of those elements is presented. Finally, the various pathways from the sources to the transitional (estuaries) and coastal areas and into the deep ocean are described.

[1] Instituto Português do Mar e da Atmosfera, Av. Alfredo Magalhães Ramalho, 6, 1495-165 Lisboa.
[2] Chemistry for the Environment, Centro de Química Estrutural, Instituto Superior Técnico – Universidade de Lisboa, Av. Rovisco Pais, 1; 1049-001 Lisboa, Portugal.
  Emails: rcesario@ipma.pt; carlos.monteiro@ipma.pt
* Corresponding author: pbrito@ipma.pt

2. Historical Context

The REE were discovered in the late 18th century, when Axel Fredrik Cronstedt, a Swedish mineralogist and chemist, first described an extremely heavy, reddish mineral (later called cerite) in an abandoned mine in Bastnäs, Sweden (Greinacher 1981).

The major use of the first REE discovered (lanthanum and cerium) was for the manufacture of gas blankets, invented by Auer von Welsbach in 1885, for the production of light in factories and later used in street lighting (Greinacher 1981). That was the beginning of the exploitation of these mineral resources, which has been growing significantly since then.

With the Industrial Revolution, and the constant search for new materials and technologies to cope with the development and growth of our civilization, there has been an increasing demand for these natural resources. The first mineral resources exploited for Welsbach blanket production were located in northern Europe (mainly Scandinavia) and the United States (North and South Carolina), and later in India, Brazil, and China (Inner Mongolia) (Klinger 2015).

The first known use of PGE goes back thousands of years to Ancient Egypt, in jewellery. Despite being a metal of antiquity, Pt was discovered in the 16th century, followed by Pd, Rh, Os and Ir in the 18th century, approximately when Pt was brought to Europe (Hartley 1991). Ruthenium was the last element found and isolated, in the 19th century (Hartley 1991, Griffith 2008). They have had increasing and varied applications since the Industrial Revolution, especially in recent decades.

3. Characterization and General Properties

Lanthanides are elements of Group IIIA of the periodic table (Fig. 1), with similar physical and chemical properties, due to the nature of their electronic configurations. These electron configurations give a stable 3^+ oxidation state and a small but stable decrease in ionic radius with increase in atomic number. Yttrium is an element of the same group, with properties similar to REE and usually included with them. The lightest element of this group, scandium, has a distinct chemistry because of the relatively small radius of its 3^+ ion.

Rare earth elements are usually divided into two distinct subgroups: those from lanthanum to samarium, with low atomic numbers and masses, known as light-REE (LREE); and those from gadolinium to lutetium, with higher atomic numbers and masses, referred to as heavy-REE (HREE). Some scientists do not agree with this separation; another group from samarium to terbium sometimes appears, usually called the middle-REE (MREE).

The REE are the first example of the Oddo-Harkins Rule (Piper and Bau 2013), where even-numbered elements are more abundant than odd-numbered elements. The saw-shaped curve of the measured REE concentrations, when plotted against the atomic number, is converted into a smooth curve by normalizing each element with one of several rock patterns. This procedure makes it easier to visually compare the relative concentrations of REE in different samples, also allowing the identification of possible anomalies of one or a group of elements. Among the various patterns of rocks that are used to normalize, the most common are the World Shale Average, calculated by Piper (1974a) from analyses published by Haskin and Haskin (1966) and Hans Wedepohl (1995); the North American Shale Composite, analysed by Gromet et al. (1984); the Upper Continental Crust; the Australian Post-Archean Shale (PAAS), proposed by McLennan (2001); and, finally, a mean of chondrites (Schmitt et al. 1963).

The PGE are located in groups 8 to 10 in the periodic table (Fig. 1), with the "triad" Ru-Rh-Pd in the V period and Os-Ir-Pt in the VI period (IUPAC 2005, Griffith 2008). The PGE are highly siderophile, with a strong affinity to Fe, and also chalcophile, with high affinity to S, opposing the low affinity to oxygen. These similar characteristics among PGE set them apart from other elements.

Group IA	IIA	III A	IV A	V A	VI A	VII A	Group VIII			I B	II B	III B	IV B	V B	VI B	VII B	VIII B
1 H																	2 He
3 Li	4 Be											5 B	6 C	7 N	8 O	9 F	10 Ne
11 Na	12 Mg											13 Al	14 Si	15 P	16 S	17 Cl	18 Ar
19 K	20 Ca	21 **Sc**	22 Ti	23 V	24 Cr	25 Mn	26 Fe	27 Co	28 Ni	29 Cu	30 Zn	31 Ga	32 Ge	33 As	34 Se	35 Br	36 Kr
37 Rb	38 Sr	39 **Y**	40 Zr	41 Nb	42 Mo	43 Tc	44 **Ru**	45 **Rh**	46 **Pd**	47 Ag	48 Cd	49 In	50 Sn	51 Sb	52 Te	53 I	54 Xe
55 Cs	56 Ba	57 **La**	72 Hf	73 Ta	74 W	75 Re	76 **Os**	77 **Ir**	78 **Pt**	79 Au	80 Hg	81 Tl	82 Pb	83 Bi	84 Po	85 At	86 Rn
87 Fr	88 Ra	89 Ac	104 Rf	105 Db	106 Sg	107 Bh	108 Hs	109 Mt	110 Ds	111 Rg	112 Cn	113 Nh	114 Fl	115 Mc	116 Lv	117 Ts	118 Og

58 **Ce**	59 **Pr**	60 **Nd**	61 **Pm**	62 **Sm**	63 **Eu**	64 **Gd**	65 **Tb**	66 **Dy**	67 **Ho**	68 **Er**	69 **Tm**	70 **Yb**	71 **Lu**
90 Th	91 Pa	92 U	93 Np	94 Pu	95 Am	96 Cm	97 Bk	98 Cf	99 Es	100 Fm	101 Md	102 No	103 Lr

Fig. 1. The rare earth elements and the platinum group elements (in bold) in the IUPAC periodic table.

Owing to their particular physical and chemical properties, the most remarkable features of PGE are their excellent catalytic properties and high resistance to corrosion. Despite also having a strong tendency to form complexes, their kinetics is usually very slow, with the exception of Pd. The chemical similarities of the above-mentioned triads were initially observed by Johann Dobereiner but in 1853 John Hall Gladstone identified the relation between the two triads, in which the atomic weights of Pt-Ir-Os were roughly twice those of Rh-Ru-Pd (Griffith 2008). The PGE are also included in the densest elements of the periodic table: Pt, Os and Ir have $\approx$ 22 g cm^{-3}, being denser than Au (19 g cm^{-3}), while Pd, Rh and Ru are lighter, $\approx$ 12.5 g cm^{-3}, in the same order of Hg and Pb. Yet, Pt and Pd are soft and ductile, thus easier to work with and often used with the addition of other metals or in metal alloys. Rh and Ir are difficult to work with but valuable either alone or in alloys. The hardest and most brittle metals are Ru and Os, almost unworkable in the metallic state and with poor oxidation state. However, they are valuable as additions to other metals (Matthey 2016).

Both groups, along with some other elements, are named technology-critical elements because of their current importance in energy and other technology-based industries, for which the global demand continues to increase (Cobelo-García et al. 2015). As such, and because of their increasing concentrations in different environmental compartments, many of these elements are often considered anthropogenic emerging contaminants in the environment, particularly in transitional environments such as estuaries, which are a strategic location for such industries, with easy access to worldwide transportation. Significant changes in the concentrations and their natural cycle in the earth's surface are due to extensive use and are likely to cause impacts on their biogeochemical cycles, thus posing potential hazards to biota and humans in the future.

4. Natural Sources

Despite their name, REE are abundant in the earth's crust, although the economically exploitable concentrations are lower than for most other ores (Fig. 2). The average content of REE in the earth's crust ranges from 150 to 200 mg kg^{-1}, exceeding the concentrations of many other metals mined for industrial use (e.g., copper at 55 mg kg^{-1} and zinc at 70 mg kg^{-1}) (Long et al. 2010). Unlike the REE and other metals, PGE occur at very low concentrations in the earth's upper continental crust, with global average concentrations of 0.4 ng g^{-1} for Pt and Pd; 0.1 ng g^{-1} for Ru; 0.06 ng g^{-1} for Rh; and

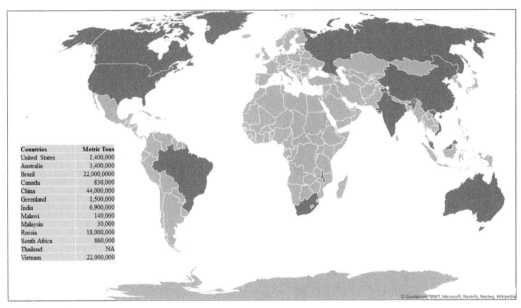

Countries	Metric Tons
United States	1,400,000
Australia	3,400,000
Brazil	22,000,0000
Canada	830,000
China	44,000,000
Greenland	1,500,000
India	6,900,000
Malawi	140,000
Malaysia	30,000
Russia	18,000,000
South Africa	860,000
Thailand	NA
Vietnam	22,000,000

Fig. 2. Worldwide distribution of REE reserves according to the USGS Mineral Commodity Summaries 2018 (USGS 2018).

0.05 ng g^{-1} for Ir and Os (Peucker-Ehrenbrink and Jahn 2001, Ravindra et al. 2004). The highest PGE concentrations are estimated to be found in the earth's core, up to 5000 ng g^{-1} (Lorand et al. 2008), as a result of primordial material accretion.

The main terrestrial natural sources of REE and PGE include volcanic emissions and weathering of local parent rock caused by rain and watercourses, whereas hydrothermal crusts and vents and Fe-Mn nodules may represent the major natural contribution of these metals to the deep-sea environment (Fig. 3). Alluvial deposits may contain PGE as native elemental metals or as alloys, typically with iridium (e.g., platiniridium), and also in nickel or copper deposits as arsenides (e.g., sperrylite, PtAs$_2$), sulphides (e.g., cooperite, (Pt, Pd, Ni)S), tellurides (e.g., PtBiTe) and antimonides (e.g., PdSb) (Fortin et al. 2011). Pyrite minerals are known to contain PGE, and they naturally occur in a variety of environments. In addition, it is important to consider that cosmic dust may contribute small amounts of REE and PGE to both terrestrial and aquatic environments (Almécija et al. 2017).

In its 2018 report, the United States Geological Survey describes REE resources as found mainly in four geological environments: carbonatites, alkaline igneous systems, ion-adsorption clay deposits, and monazite-xenotime-bearing placer deposits. Carbonatites and placer deposits are the leading sources of production of LREE elements, while ion-adsorption clays are the leading source of production of HREE (USGS 2018). The two largest sources of REE in the last five decades were Mountain Pass (California, USA) and Bayan Obo (Baotou, Inner Mongolia, China). The Mountain Pass mine ceased operations in 2015 and no REE have been explored since. Its last production in 2015 was 5,900 tons. Rare earth mining in China has seen significant growth, with 105,000 tons of REE ore reported in 2017, accounting for 81% of global production. The second largest share belonged to Australia, with 20,000 tons (USGS 2018). With the crisis of 2010, due to the decline in China's REE export quotas, countries dependent on REE-based technology have awakened to the urgent need to find other rare earth sources, either through mining (on land and from the bottom of the oceans), or through the recycling of end-of-life products. New rare earth mining sites have emerged in recent decades, such as Mount Weld in western Australia (Lynas Corporation Ltd 2017), Kvanefjeld in southern Greenland (Greenland Minerals and Energy Ltd 2017), and even a promising reserve in the deep Pacific Ocean (Takaya et al. 2018).

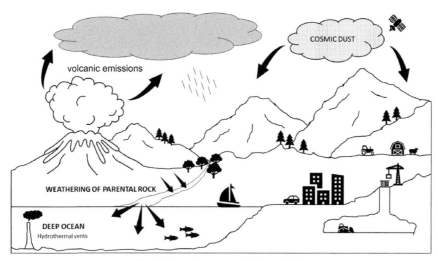

Fig. 3. Natural sources of REE and PGE.

Likewise, PGE are also critical elements and their exploitation worldwide is limited to a significantly lower number of sites. In fact, the European Union is highly dependent on the imports of those materials, particularly from China and South Africa, since it does not have any economically viable mining areas for the exploitation of technology-critical elements (Cobelo-García et al. 2015). The different PGE ores are distributed along the world in singular geological settings that allowed the formation of those unique deposits, such as the Bushveld Igneous Complex (South Africa), Noril'sk (Russia), Stillwater Complex (Montana, USA) or Great Dyke (Zimbabwe) (Holwell and McDonald 2010). The largest ore, the Bushveld Igneous Complex, contains about 75%, 54%, and 82% of Pt, Pd, and Rh world resources, respectively; it is estimated that its mineral reserve is large enough to supply the anthropogenic demand for decades (Cawthorn 2010, Glaister and Mudd 2010).

Regardless of the natural occurrence of those mantle plumes entailed in the earth's upper continental crust containing REE and PGE, the supply of those elements has almost doubled over the last decade, to more than 500 tons per year in the case of the PGE, and it is expected to increase in the upcoming years. Accordingly, mining activities for the extraction of REE and PGE, often associated with other metals of economic interest, will transform those regions in risky anthropogenic sources of those elements to the environment.

5. Anthropogenic Sources and Uses

The main anthropogenic sources of REE and PGE derive from the mining, extraction and purification activities of these metals, and in part are attributed to their use in the automobile industry (Fig. 4). In addition to mining and processing of these elements, worldwide industrial, agricultural and domestic activities also raise environmental concerns related to wastewater contaminated with REE and PGE, increasing concentrations of which have impacts on the environment. The growing global demand and uses of both REE and PGE in several technology-based industries are intimately linked to localized high contamination, which is then disseminated worldwide.

Rare earth mining and processing activities constitute a huge exploitation of the earth's resources, producing high impacts on terrestrial and marine ecosystems. One of the major concerns of rare earth mining is radioactive stockpiles resulting from primary ore deposits such as bastnäsite and monazite. The direct negative effects of rare earth mining include severe soil erosion, loss of biodiversity, alteration of land use, flooding, and air and water pollution (Carpenter et al. 2015). Mining and

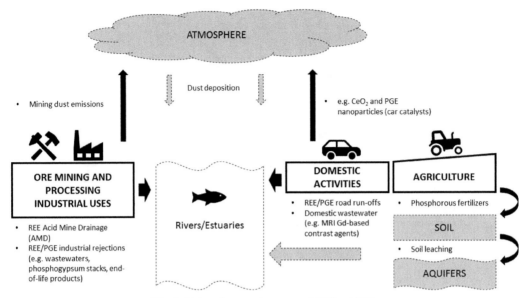

Fig. 4. Main anthropogenic sources of REE and PGE.

processing of rare earth ores is performed by complex methods including rare earth mining from ore deposits, ore grinding for rare earth oxide (REO) extraction, and separation and purification to obtain REE of industrial grade (Kumari et al. 2015, Sinha et al. 2016). Mining and production activities have also been associated with the emission of PGE into the environment, in particular in Russia and South Africa, often resulting in by-products from the smelt and extraction of other metals.

The REE are increasingly used in numerous high-technology products and processes (Table 1). They are important in a growing number of critical technologies because of their unique physical and chemical properties. Some elements of this group are vital components in much modern technology. Wind energy and electric vehicles will be considered as part of the solution for a more sustainable future, and the use of REE in this field is essential. For example, present technologies for electric vehicles and wind turbines rely heavily on Dy (dysprosium) and Nd (neodymium) for rare-earth magnets (Alonso et al. 2012). These two elements and also praseodymium (Pr) are used in wind turbines as high-strength magnets, in hard disk drives and engines in electric cars; europium (Eu), terbium (Tb), lanthanum (La), and cerium (Ce) are used for phosphor-based fluorescent lighting, smartphone screens and batteries (Guyonnet et al. 2015), and yttrium (Y) is a promising raw material for superconductors and laser technology (Nassar et al. 2015). In these products, REE are currently almost irreplaceable by other materials (Stone 2009). The industrial applications of REE include electronic devices, clean energy, aerospace, automotive and defence.

Because of their luminescent and catalytic properties, REE are widely used in high-technology and green products (lasers, camera lenses, computer memory modules, X-ray machines) (e.g., Humphries 2012), energy (batteries, lamps, superconductors) (e.g., Du and Graedel 2011, 2013), renewable energy (hybrid automobiles, wind turbines, next generation rechargeable batteries, biofuel catalysts) (e.g., Guyonnet et al. 2015, Smith Stegen 2015, Dutta et al. 2016, Grandell et al. 2016, Apergis and Apergis 2017), industrial (aerospace, caustic cleaning agents, specialized glass) (e.g., Massari and Ruberti 2013, van Gosen et al. 2014), medical science (magnetic resonance imaging (MRI) contrast agents, nuclear medicine imaging, cancer treatment applications, genetic screening tests, medical and dental lasers) (e.g., Salem and Hunter 2006, Van De Wiele et al. 2009, Lau et al. 2011, Memon et al. 2013) and defence (night-vision goggles, laser range-finders, guidance systems, communications, fluorescents and phosphors in lamps and monitors, amplifiers in fibre-optic data transmission, precision-guided weapons, "white noise" production in stealth technology)

Table 1. Rare earth elements (REE), atomic number (z), names and symbols, group and their uses.

z	Element	Element group	Uses
21	Scandium (Sc)	Associated elements	Aerospace framework, high-intensity street lamps, high-performance equipment
39	Yttrium (Y)		TV sets, cancer treatment drugs, enhancing strength of alloys
57	Lanthanum (La)	LREE	Camera lenses, battery electrodes, hydrogen storage, diesel fuel additive, catalytic converters
58	Cerium (Ce)		Catalytic converters, hybrid NiMH battery, coloured glass, LCD screens, steel production
59	Praseodymium (Pr)		Super-strong magnets, welding goggles, lasers
60	Neodymium (Nd)		Extremely strong permanent magnets, microphones, electric motors of hybrid automobiles, laser
61	Promethium (Pm)		Not usually found in nature
62	Samarium (Sm)		Cancer treatment, nuclear reactor control rods, X-ray lasers
63	Europium (Eu)	HREE	Colour TV screens, fluorescent glass, genetic screening tests
64	Gadolinium (Gd)		Shielding in neutron radiography and in nuclear reactors, nuclear marine propulsion, increases durability of alloys, contrast agents in medical MRI
65	Terbium (Tb)		TV sets, fuel cells, sonar systems
66	Dysprosium (Dy)		Commercial lighting, hard disk devices, transducers
67	Holmium (Ho)		Lasers, glass colouring, high-strength magnets
68	Erbium (Er)		Glass colourant, signal amplification for fibre optic cables, metallurgical uses
69	Thulium™		High-efficiency lasers, portable X-ray machines, high-temperature superconductor
70	Ytterbium (Yb)		Improves stainless steel, lasers, ground monitoring devices
71	Lutetium (Lu)		Refining petroleum, LED light bulbs, integrated circuit manufacturing

Adapted from webpage www.namibiararecearths.com

(e.g., Hedrick 2004, Humphries 2012, Grasso 2013, Swift 2014). More recently, medical and biological properties of REE were also evaluated and studied in the context of the geophagia phenomenon in animals and humans, considering their reasons, evaluation and relation with mineralized water consumption (Panichev 2015). Biological functions of lanthanide elements arise from the idea that such elements, although considered inessential for life, are certainly biologically active, playing an unusually important role in structure and functioning of any biological system. As such, REE have numerous applications of importance in biological analysis and in both diagnosis and therapy in medicine (Cotton and Harrowfield 2012). Bouzigues et al. (2011) also discussed different applications of rare earth-based nanoparticles, which have optical properties and low cytotoxicity that make them promising for biological applications such as biomolecule detection and imaging *in vitro*, in living cells and in small animals.

In recent years, demand for rare earth has been increasing, particularly in energy efficiency (green energy) and high technology, and the products not only are faster, lighter, smaller and more efficient, but also play an important role in developing the industries. The elevated consumption of smartphones, TVs, batteries, computers, medical equipment, and even some of the critical components in nuclear reactors depend on REE for their operation. This extensive need for REE in global manufacturing ensures constant commercial demands for these metals for the predictable future.

The increasing variety of PGE applications are also transverse to several sector-based industries of economic interest. Platinum, Rh and Pd are usually the most extensively used among the PGE and therefore among the most expensive materials. The other PGE, despite some particular applications, are often present in metal alloys or as by-product. The use of PGE in the investment sector is growing. Coins and metal bars produced from Pt and Pd are used in exchange-trade markets as investment funds in stock exchanges and bank accounts. The jewellery sector accounts for around 30% of Pt global demand, not just for ornamental objects, but also for the mechanism of watches. However, the industrial/technological sectors have the widest range of PGE applications. In electronics, PGE are used in the production of microchips, circuits, hard disks and sensors, LCD and fibreglass production. Platinum is used as a coating in airplane motor engine blades and rockets, whereas Rh is used on lighthouse reflectors and lamps. In the medical field, sensors and health devices such as pacemakers and defibrillators also contain Pt. In addition, the ability of Pt to inhibit the division of living cells turned this metal attractive in the production of Pt-based anti-cancer drugs, such as cisplatin, carboplatin and oxaliplatin. Currently, new Pt-based complexes are being explored by the pharmaceutical industry (e.g., Ndagi et al. 2017). Dental applications are also known for Pt and Pd, such as crowns and bridgeworks. The catalytic properties of PGE also have a long history in industrial production of nitric and sulphuric acids, cortisone, ammonia and fertilizers. Nevertheless, the automobile industry is the most demanding sector for PGE, accounting for 50% of Pt, 75% of Pd and 80% of Rh (Matthey 2016). Despite their use in other components of automobiles, such as fuel cells and spark plugs in combustion engines, the manufacturing of automotive catalytic converters (ACC) accounts *per se* for approximately 50% of PGE global demand. Catalytic converters were introduced in order to reduce the emissions of other pollutants such as nitrous and carbon oxides, unburned hydrocarbons and other pollutants, including particulate matter (Ravindra et al. 2004). As a result, anthropogenic emissions of PGE have grown significantly over the past decades (Zereini and Wiseman 2015). Particles released from ACC, due to degradation and abrasion, result in widespread distribution of PGE, mainly near high-traffic roads (Schäfer and Puchelt 1998), which in turn are mobilized and transported, reaching the aquatic systems.

6. Analytical Methodologies

It is of great importance to have suitable methods for the determination of REE and PGE, usually found at the ultra-trace levels in many environmental matrices. In general, most of the commonly used techniques quantify the total concentration of elements. However, sensitive and selective methods are still needed, in particular for speciation studies under relevant environmental conditions that demand sufficient selectivity and low limits of detection. Moreover, considering monitoring purposes, those techniques should be inexpensive and easy to work with. Over the last six decades, various analytical tools have been developed, in particular concerning multi-elemental capacity. For years, instrumental neutron activation analysis and isotope dilution mass spectrometry, together with other techniques such as X-ray fluorescence and inductively coupled plasma atomic emission spectrometry (ICP-AES), have been applied in the determination of REE, although with some analytical limitations and disadvantages. In addition, adsorptive cathodic stripping voltammetry (AdCSV) has been used in PGE determination. In particular, spectrophotometric methods and atomic absorption spectrometry were tentatively applied, yet also with some analytical drawbacks. More recently, the use of inductively coupled plasma mass spectrometry (ICP-MS) has been used for both REE and PGE quantification in environmental matrices. A more thorough review of the analytical methods used to determine REE and PGE would fall outside the scope of this chapter, but there are published reviews (e.g., Kantipuly and Westland 1988, Verma et al. 2002, Cobelo-García et al. 2014b, Schindl and Leopold 2015, Almecija et al. 2016, Monteiro et al. 2017).

Because of the similar physical and chemical characteristics of lanthanides, it was very difficult to separate and quantify these elements for many years. Since the 1950s, several methods have been developed in order to improve the separation and accurate determination of all REE, even when present at trace levels. With increasing worldwide demand for REE for new technological applications, the need for analytical methodologies capable of quantifying ultra-trace levels of REE has become a major area of research in analytical chemistry (Balaram 1996). With the introduction of ICP-MS in the 1980s and the development of this mass spectrometry technique in the last decades, REE analysis has changed considerably with respect to low detection limits, sensitivity, large dynamic range, and simpler procedures of sample preparation (e.g., Date and Hutchison 1987, Jarvis et al. 1989). The high sensitivity combined with relatively simple spectra in lanthanide analysis made ICP-MS the technique most used in the quantification of these elements in environmental samples.

Among the PGE studies, the greatest attention was given to Pt. More recently, Rh in environmental samples has been reckoned with, yet emphasis has been given to urban and roadside solid samples. Thus, data on its total dissolved concentrations is still lacking, as for the other PGE. The improvement of analytical techniques over the past few decades, such as electrothermal atomic absorption spectrometry and ICP-MS, allowed the individual and simultaneous determination of those elements in different environmental matrices (Zereini and Wiseman 2015). Despite the high sensitivity of ICP-MS for PGE analysis, it has some disadvantages, such as the need of matrix separation due to polyatomic interferences from other elements (Labarraque et al. 2015). Overall, this is quite an expensive and time-consuming technique, often needing a pre-concentration step of the sample. Voltammetric techniques provide a simple and sensitive way to determine Pt and Rh at ultra-trace levels, but unfortunately they are not feasible for Pd because the limits of detection are not low enough to determine it in environmental samples. Additionally, voltammetry techniques are usually less expensive and appear to be adequate for routine analysis after optimization (e.g., Locatelli 2007, Monteiro et al. 2017). Several electroanalytical procedures have been reported for PGE analysis, mostly based on the adsorptive process on a mercury electrode. Although alternative electrodes have been investigated envisaging a similar analytical performance to mercury (Silwana et al. 2014, Pérez-Ràfols et al. 2017), detection and quantification limits reported are insufficiently low to allow the study of PGE in unspoiled environments (Peucker-Ehrenbrink and Jahn 2001). More recently, the voltammetry signal transformation using the second derivative has proven to increase the accuracy of Pt and Rh determinations at the ultra-trace levels (Cobelo-García et al. 2014a, Almecija et al. 2016, Monteiro et al. 2017).

The solubility and bioavailability of REE and PGE, as with any other substance, are generally determined by speciation, depending on several variables such as pH, salinity and the presence of anions (Liang et al. 2005) and organic matter. These variables are responsible for the differences in REE and PGE speciation occurring in salt water, mostly as carbonate and chloride complexes, while in freshwater those metals are essentially complexed with humic substances and hydroxide anions. Several analytical methods have been used in the study of REE speciation in different environmental matrices, from non-nuclear to nuclear techniques such as ICP-MS and hyfenated methods (e.g., GC, LC, HPLC, HILIC), AdCSV, molecular activation analysis, proton-induced X-ray emission, and Mossbauer (Serrano et al. 2000, Tang and Johannesson 2003, Johannesson et al. 2004, Och et al. 2014). For more alkaline pH the HREE, with higher atomic number and a higher stability constant, are those that present stronger complexes with the carbonate ion, whereas in environment with low pH the most common species are REE_3^+ and $(REE)SO_4^+$. When the hydrolysis occurs at pH close to 10 the most common products are $REE(OH)_2^+$, $REE(OH)_3(aq)$ and $REE(OH)_4^-$. As to PGE in aquatic systems, Pt(II), Pt(IV), Pd(II) and Rh(III) are the most important oxidation states, depending on salinity and redox conditions. Inorganic speciation is dominated by aquo-hydroxo, aquo-chloro and mixed aquo-hydroxo-chloro complexes (Cobelo-Garcia 2013, Cobelo-García et al. 2014a). Furthermore, there is evidence of Pt and Rh interaction with organic matter, which may be a key parameter in speciation and distribution of PGE in aquatic ecosystems (Bowles et al. 1994, Cobelo-Garcia 2013). Nevertheless, there is a general lack of knowledge about PGE speciation, particularly in aquatic environments.

With the increasing demand of REE and PGE worldwide, the concern about environmental risks that these elements may pose has also grown. One of the first needs was to identify anthropogenic sources of REE and PGE and to quantify their concentrations in environmental matrices. Therefore, developed countries have devoted attention to this issue and made efforts to improve the understanding of REE and PGE speciation in the environment as well. This goal, parallel to the evolution of analytical techniques, led to outcomes more rapidly in the case of some elements rather than others. In the case of PGE, studies were devoted to Pt, whereas other PGE were more difficult to assess, namely Pd, either because of low concentrations in different environmental compartments or because of faster reaction kinetics and higher mobility in the environment. Despite the limitations of some analytical techniques at first, the scientific community was able to quantify REE and PGE in several matrices. However, efforts are still needed to improve the knowledge of their behaviour in the environment, in particular with respect to their speciation, since total concentration of an element does not reflect the potential toxicity for an organism. Higher concentrations of REE and PGE in environmental matrices are expected to pose risks in the near future and thus those potential toxic species need to be accurately determined.

7. From Urban to Transitional and Coastal Areas

The mobility and transport of REE and PGE across different environmental compartments has been a matter of debate over the past decades, owing to an increase of their concentrations especially in urban areas. The REE and PGE emitted as solutes (e.g., Gd used as contrasting agent in medical MRI or Pt used in chemotherapy) or as fine-grained size particles, even nanoparticles, may undergo chemical changes during their transport from urban areas to the nearby aquatic systems. However, studies addressing their behaviour in those compartments are still needed. One must take into account their accumulation in the different environments, their chemical reactions and changes upon different matrices, which in turn may lead to higher mobility and transport. Like other metals, REE and PGE are prone to be affected by changes in master environmental variables. These may include: the ionic strength of the involving media, such as the changes caused by the transition from freshwater to seawater; pH variations, namely from slightly acidic rainwater to alkaline seawater; and the nature of the particles, either suspended in the water column or from the sediments, the composition of which may vary, e.g., according to the Al, Si, Fe and Mn contents, and organic matter composition. Besides the essential chemical characterization of environmental samples, there is a need to understand the physical mechanisms that can drive the regional and worldwide transport of REE and PGE. Nevertheless, in this section a brief description is presented of REE and PGE occurrence in the different environmental compartments.

7.1 Rare earth elements

Atmosphere

As already mentioned, REE ore exploration and processing activities are the major anthropogenic sources of these elements into the environment. The emission of REE particles into the atmosphere has increased significantly following the worldwide large-scale exploration and use of these metals.

However, the available information on this subject is still scarce. The two main factors controlling the generation and transport of atmospheric particles are soil surface conditions and atmospheric circulation patterns (Goossens and Offer 1997, Macpherson et al. 2008). Previous studies have been carried out to investigate the long-term transport of atmospheric particles from the continents to the oceans (Arimoto et al. 1989, Sholkovitz et al. 1994, 1993, Ferrat et al. 2011), as well as REE levels in atmospheric dust in urban areas (Wang et al. 2001, 2000). Wang and Liang (2014), studying the accumulation and fractionation of REE in atmospheric particles around a mine tailing in Batou, China, found total REE concentrations up to 297 ng m^{-3}, for the suspended particulate, and 106 ng m^{-3}, for particles with an equivalent aerodynamic diameter of less than 10 μm (PM$_{10}$). These authors concluded that the REE enrichment in atmospheric particles was caused by anthropogenic sources and influenced by the strong wind.

In addition to REE ore mining and processing, the use of petroleum cracking catalysts in the oil combustion in refinery plants (e.g., Kitto et al. 1992, Pacyna and Pacyna 2001), or the emissions produced from a steel plant (e.g., Geagea et al. 2007), largely contribute to the increase of these elements into the atmosphere. Wang et al. (2000) found atmospheric particulate matter, collected in Delft (the Netherlands), with SREE concentrations between 0.22 and 33.0 ng m^{-3}. They concluded that this variation was dependent not only on anthropogenic activities, but also wind direction and associated weather conditions.

Another anthropogenic source of REE is the ACC, through the release of various metals in the form of nanoparticles. Among other elements, Ce and La are wash-coated on the surface of the monolith of the ACC, to promote thermal stability and oxygen storage (e.g., Kašpar et al. 2003); these are released into the atmosphere and contribute to the increase of pollutant emissions. On the other hand, many countries, such as the USA and UK, add CeO$_2$ nanoparticles to diesel fuels to improve combustion efficiency, as well as to reduce soot emissions (e.g., Cassee et al. 2011). Although most of the previous studies have been focused on PGE, recent studies reported higher Ce concentrations and enrichment pattern in the environment (Table 2). In addition to air transport, these Ce-enriched nanoparticles are accumulated along roadsides (in road dust), and transported either by rainwater or by road washes, through urban drainage systems to the aquatic environment. Lyubomirova et al. (2011), in a study on the fractionation of Ce and La in road dust and soils, observed that these traffic-emitted elements revealed a much higher association with the exchangeable phase, suggesting that they exist in more physicochemical forms available for biota.

Water column

The natural pattern of REE in the particulate fraction of the small rivers depends mainly on the geological characteristics of its catchment area, which can reveal enormous differences, being

Table 2. Cerium contents in road dust and soils.

Location	Sample type	Concentration (mg kg^{-1})	References
Toronto (Canada)	road dust	17–31	Wiseman et. al. 2016
	soil	37	
Toronto (Canada)	road dust	20–53	Wiseman et. al. 2013
	soil	22–119	
Several highways (Germany)	road dust	25–69	Lyubomirova et. al. 2011
Forest (North and South Bulgaria)	soil	42–89	
Perth (Australia)	road dust	35–125	Whiteley and Murray 2003
	soil	28–101	
Frankfurt am Main (Germany)	road dust	5–91	Zereini et al. 2001

practically uniform in major rivers (Goldstein and Jacobsen 1988a). The REE load in the river waters is distributed between colloidal and suspended particles, and in solution phases. Considering that these elements have a high affinity for colloids and suspended particles, they are mainly responsible for the transport and distribution of REE, with colloids being the only form of transport of the lanthanides in organic-rich rivers (Sholkovitz 1992, Tang and Johannesson 2003, Pourret et al. 2007).

Mineral phases characterized by high surface area, such as Fe- and Mn-oxyhydroxides, act as efficient adsorbents for REE, playing a key role in adsorption/desorption reactions (Johannesson and Zhou 1999, Ohta and Kawabe 2001, Quinn et al. 2004, Leybourne and Johannesson 2008).

Earlier studies on the fundamental characteristics of REE geochemistry in rivers and estuaries have shown that the concentrations and fractionation of these elements in the dissolved fraction are dependent on several parameters such as pH, alkalinity, the ionic strength or the presence of colloids (Martin et al. 1976, Sholkovitz and Elderfield 1988, Goldstein and Jacobsen 1988, Elderfield et al. 1990, Sholkovitz 1993, 1992). Therefore, the REE contribution in the transition and coastal areas depends essentially on the physicochemical properties, which, in turn, will generate different reactions, promoting the dissolution and precipitation of one or several elements and altering the REE fractionation in the water column.

Acid mine drainage (AMD) is the main environmental problem resulting from mining activities and one of the main anthropogenic sources of REE in the aquatic environment (e.g., Nieto et al. 2007, Sun et al. 2007, 2012, Burch et al. 2011, Ayora et al. 2015, Prudêncio et al. 2017). These effluents are characterized by low pH and high concentrations of dissolved elements such as heavy metals and REE causing a change in the REE fractionation patterns in the receptor medium (e.g., Pérez-López et al. 2010, Delgado et al. 2012, Lecomte et al. 2017). In a recent work on REE mobility processes in an estuary affected by AMD, Lecomte et al. (2017) found that REE contents are retained in the minerals resulting from the interaction of AMD with seawater, increasing its concentration in sediments at higher pH ($\approx$ 6), with contents higher than 1,700 μg L^{-1}. On the other hand, at low pH (3–3.5), the REE are released again into the solution. In this study, conducted in the Rio Tinto estuary (SW Spain), REE concentrations were higher (above 11,000 μg L^{-1}) at low pH values (in the river), while as pH increased (in seawater) these elements were absorbed again, particularly in the soluble salts, but also in the Fe oxyhydroxides. Similar behaviour was found by other authors in several estuarine systems affected by AMD (e.g., Burch et al. 2011, Sun et al. 2012, Sharifi et al. 2013, Soyol-Erdene et al. 2018).

As mentioned earlier, anthropogenic REE-rich particles can also be transported to the aquatic environment through urban wastewater treatment systems, which together with domestic and industrial effluents have contributed to the rising concentrations of some lanthanides and their fractionation. The best-known example is the positive anomaly of gadolinium (Gd) associated with the use of Gd-based contrast agents in MRI since the 1980s and first reported in 1996 by Bau and Dulski (1996). Since then, Gd anomalies have been observed in many rivers and lakes worldwide (e.g., Tricca et al. 1999, Nozaki et al. 2000, Elbaz-Poulichet et al. 2002, Möller et al. 2002, 2003, Knappe et al. 2005, Verplanck et al. 2005, Kulaksiz and Bau 2007, Rabiet et al. 2009, Lawrence 2010, Klaver et al. 2014), estuaries and coastal waters (Elbaz-Poulichet et al. 2002, Zhu et al. 2004, Kulaksiz and Bau 2007, Lawrence 2010) but also in groundwater (Möller et al. 2000, Knappe et al. 2005, Rabiet et al. 2009) and even in tap water (Möller et al. 2002, Kulaksiz and Bau 2011a, Tepe et al. 2014). Concentrations two orders of magnitude higher than background geological values have been also reported for other lanthanides. Kulaksiz and Bau (2011b) showed that lanthanum (La) enrichment on the Rhine River (Germany) was of anthropogenic origin from a production plant for fluid catalytic cracking catalysts. This effluent was characterized by concentrations of La up to 52 mg L^{-1}, well above the values that showed toxicological effects (e.g., Slatopolsky et al. 2005, Zhang et al. 2010). Due to these high La levels in the effluent, the positive lanthanum anomaly was still detected about 400 km downstream (Kulaksiz and Bau 2011b, Tepe et al. 2014). Later, other works showed that La is not only present in the dissolved fraction but also in the suspended particulate fraction (Klaver et al. 2014). The same authors reported the occurrence of another lanthanide on the Rhine, showing significant quantities of anthropogenic samarium (Sm) from the same industrial effluent, and also traceable along the Middle and Lower Rhine to the Dutch border (Kulaksız and Bau 2013, Klaver et al. 2014). The growing of REE anomalies in rivers and estuaries is a highly expected scenario due to the increased use of these elements in high-technology products.

Sediments

In the late 1970s and over the next two decades, several studies on river and estuarine systems showed that sediments play an important role as carriers and potential sources of REE (e.g., Martin et al. 1976, Sholkovitz and Elderfield 1988, Goldstein and Jacobsen 1988, Elderfield et al. 1990, Sholkovitz et al. 1992, Sholkovitz 1993, 1995, Ramesh et al. 1999). These studies showed that the extensive fractionation between the REE of the dissolved and particulate fraction and continental rocks resulted from the chemical weathering reactions, which occur predominantly in the solid phase during fluvial transport. The study of the geological behaviour of REE in sedimentary systems has made it possible to investigate the nature of its biogeochemical processes derived from the peculiar characteristics and properties of these elements that translate into a coherent and predictable behaviour (e.g., Leybourne et al. 2000). These properties made it possible to use the REE as input provenance markers and tracers of changes of environmental conditions, and also as tools in the study of anthropogenic sources and their effects.

One of the first reports of anthropogenic REE contamination in coastal sediments was described by Olmez et al. (1991). In this work, the authors demonstrated that the REE signature found in the upper 20 cm of sediment cores from the shelf of the San Pedro Basin, southern California, were derived from refined petroleum products (oil-refining cracking catalysts, oil-fired power plant emissions, and other products) from the Los Angeles oil refineries. These top sediments presented an enrichment in the LREE relative to the MREE and HREE, with a different signature from the crustal material.

Another study, carried out in the estuary of the Tinto and Odiel rivers (SW Spain), showed that the anomalies in the sediment REE fractionation patterns originated in the phosphogypsum residues from the fertilizer factories that used phosphorite as raw material (Borrego et al. 2004, 2005). López-González et al. (2012) also showed patterns of REE fractionation in the sediments as a consequence of AMD.

Biota

The distribution and bioavailability of lanthanides in the aquatic environment depends on their speciation, which is influenced by several physicochemical parameters such as pH or ionic strength, and the presence of different organic and inorganic elements (Elderfield et al. 1990, Moermond et al. 2001). The two dynamic processes that determine the amount of metal that can be accumulated by aquatic organisms are bioaccessibility and bioavailability. Bioaccessible compounds are potentially available but may not be within reach of the organisms. The other fraction of bioaccessible compounds, which can cross the membrane of the organism, is termed bioavailable (Semple et al. 2004). One of the anthropogenic REE characteristics is that they enter the environment in much more soluble and reactive forms than naturally occurring REE, so they are more bioavailable. So far, no essential role has been attributed to REE in relation to plants and animals. These elements become toxic when present at high concentrations because they disrupt biological processes by the displacement of calcium due to the similar size or the high affinity for the phosphate groups of the biological macromolecules.

Several laboratory studies show not only evidence of REE bioaccumulation at various taxonomic levels, but also toxicity effects on organisms, including bacteria (e.g., Balusamy et al. 2012), phytoplankton (Tai et al. 2010, Balusamy et al. 2015), aquatic plants (e.g., Yang et al. 1999, Weltje et al. 2002, Paola et al. 2007), aquatic invertebrates (Barry and Meehan 2000, Borgmann et al. 2005, Oral et al. 2010), and vertebrates (e.g., Ennevor and Beames 1993, Qiang et al. 1994, Hao et al. 1996, Hongyan et al. 2002, Cui et al. 2012, Figueiredo et al. 2018).

As previously mentioned, few studies exist on REE in biota, particularly field-based studies on their bioaccumulation. However, some studies have recently been carried out in this field because of the growing importance given to these new emerging contaminants. Squadrone et al. (2017) found mean total REE concentrations between 7.9 and 22 mg kg^{-1} (dry weight) in different seaweed species harvested at three different sites along the NW coast of the Mediterranean Sea. The highest REE concentrations were observed in the seaweeds sampled near an important commercial and tourist port, while the lowest values were observed in the samples harvested in sites of lower anthropogenic impact. Brito et al. (2018) found in two saltmarsh halophyte species (*Sarcocornia fruticosa* and *Spartina maritima*), from the Tagus estuary (SW Europe), low ability to accumulate REE. The total concentrations, although higher in the roots (6.8–68 mg kg^{-1}) than in the stems (1.2–10 mg kg^{-1}) and leaves (1.4–23 mg kg^{-1}), were much lower than the contents observed in the surrounding sediments (146–190 mg kg^{-1}). MacMillan et al. (2017) have determined REE in freshwater, marine and terrestrial ecosystems in the eastern Canadian Arctic and found high concentrations of REE at low trophic levels, especially in vegetation and aquatic invertebrates. These authors concluded that REE bioaccumulation patterns appear to be species- and tissue-specific, with some limitation for biomagnification. Akagi and Edanami (2017) determined REE in shells and soft tissues of five different bivalve species (*Ruditapes philippinarum, Mercenaria stimpsoni, Mytilus galloprovincialis, Mactra veneriformis* and *Phacosoma japonicum*) from three locations in Tokyo Bay, and found similar concentrations to those of sediment and sea water at the corresponding sites. REE concentrations in soft tissues were higher than in shells. The REE abundance patterns of both shells and soft tissues resembled those of sediments or suspended particles, rather than those dissolved in seawater, except for mussels. A study carried out by Mayfield and Fairbrother (2015) to determine REE concentrations in ten freshwater fish species in a reservoir in Washington state (USA) showed that total REE concentrations ranged from 0.014 to 3.0 mg kg^{-1} (dry weight), with a higher concentration in the organs, viscera and bones than in the muscular tissues. This work also showed that benthic species that feed on the bottom of the reservoir, and that are therefore more exposed to sediments, presented higher concentrations of REE when compared to pelagic species. The authors also reported a decrease in REE levels with increasing age, total length and weight for some species. Bosco-Santos et al. (2017), in a study carried out in a mangrove of a subtropical estuary highly impacted by activities of the fertilizer industry, found values of SREE between 0.2 and 4.6 mg kg^{-1} (dry weight) in the tissues of the crab *Ucides cordatus*. The results obtained in this study show evidence that the water absorbed during the moulting phase controls the REE chemistry in the shells. On the other hand, the authors verified that the REE chemistry in the claw muscle, in which a preferential absorption of LREE was observed, was controlled by the diet, and the fractionation of REE obtained in these tissues presented a strong correlation with the one observed in the contaminated sediment and in the materials of phosphate fertilizer production.

Herrmann et al. (2016), in a review on bioaccumulation and ecotoxicity of lanthanum in the aquatic environment, states that "no regulatory limit for REE concentrations and emissions to the environment has been defined because REE risk information is scarce." In fact, and as the authors themselves point out, REE need to be recognized as new and emerging contaminants with multiple forms of entry into the environment.

7.2 Platinum group elements

Despite the limited information available in the literature, PGE may undergo the same chemical transformations as REE and most of the heavy metals. It is hypothesized that PGE may be affected by the ionic strength and pH of the medium, therefore forming different species. Furthermore, the nature of the particles, whether in the atmospheric, terrestrial or aquatic compartments, may play a major role in adsorption, scavenging and retention mechanisms of metals. In addition, physical mechanisms, such as wind and rain events, may even force the transfer of PGE between compartments, affecting their distribution, and ultimately transferring them to the nearby aquatic systems. Along their path, many chemical changes occur, and PGE may become bioavailable. As a result, PGE concentration in sediments and aquatic vegetation often largely exceeds the natural background levels, contrasting with lower levels in waters (Zereini and Wiseman 2015).

Atmosphere

Anthropogenic activities have largely affected the natural biogeochemical cycle of PGE and the atmosphere may well be the compartment in which pollution is globally disseminated. Although activities such as mining and biomass burning contribute to the increase of PGE levels in the atmosphere, their main anthropogenic source has been largely attributed to automotive catalytic converters. Emissions of PGE from ACC vary according to the type of engine, diesel or petrol, the age of the catalyst and the velocity of the vehicles. The particles resulting from thermal degradation and mechanical abrasion from ACC during vehicle operation can be directly emitted into the atmosphere, through the exhaust fumes containing PGE in the metallic state, as fine particulate material usually in the form of micro- or nanoparticles. On the other hand, there is a consensus among scientists that PGE in those particles tend to accumulate in the pavement, as road dust or in roadside soils, therefore concentrating PGE and other metals as well. The concentrations of PGE deriving from ACC in urbanized areas worldwide are comprehensively described in the literature (e.g., Zereini and Wiseman 2015, Wiseman et al. 2016). Nevertheless, road dust emission and its resuspension continue to draw attention for the possible environmental and human health hazards in highly populated cities around the world (e.g., Zereini et al. 2004, 2012, Rauch et al. 2005, Bocca et al. 2006, Diong et al. 2016, Rinkovec et al. 2018). Amato et al. (2014) showed that road dust emissions can increase up to 35% PM10 (airborne particulate matter, PM, with aerodynamic diameter below 10 µm) at stations impacted by traffic and even in urban background sites, as well as up to 22% at industrial and rural sites. However, airborne PM of smaller dimensions is easily dispersed by the wind action, mostly depending on its intensity, which is thought to be the main mechanism responsible for regional- to long-range transport of PGE (Barbante et al. 2001, Zereini et al. 2012). In addition, those particles carrying PGE return to the earth's surface by atmospheric deposition on land or sea surface, by either wet or dry removal.

Some studies point out the presence of PGE in remote and pristine environments. In the Arctic, Barbante et al. (2001) presented the first millennial time series for PGE in ice cores from Greenland, in which three different periods were ascribed. The first period regards the past natural occurrence of PGE in ancient ice, which revealed extremely low levels of PGE. Most likely, those were derived from cosmic dust and other natural sources. A second period revealed that the atmosphere in the northern hemisphere was already contaminated with PGE, just before the introduction of ACC in the United States during the mid-1970s. The origin of PGE contamination by that period could be attributed to the large expansion of several industries, including mining and smelters of metals, chemical and petroleum, or iron and steel manufacturing, as well as some natural emissions from forest wildfires and volcanic aerosols. Finally, a third period showed a marked increase in PGE concentrations towards the recent ice layers, soon after the introduction of catalytic converters in vehicles. Similarly, the 50-year record in the Antarctic ice reported by Soyol-Erdene et al. (2011) suggests that PGE concentrations have increased in recent years most likely because of large-scale atmospheric circulation of the growing anthropogenic emissions since the 1980s.

Water column

Although PGE are not key trace elements included in the priority pollutants list to monitor in water quality programs from the European Union and United States, there is growing interest in their concentrations and distribution in waters. These include rain, riverine, estuarine, coastal and oceanic waters. Although the potential environmental effects or fate of PGE remain unclear, this interest arises from their increasing use and consequential emissions, which extend to aquatic ecosystems.

Limited data is available on PGE concentrations in rainwater. Some results point out that those concentrations are usually very low, often beneath 1 ng L⁻¹ (Eriksson 2001, Mashio et al. 2016). Mashio et al. (2016) did not find a clear relationship between Pt and the amount of precipitation. However, these authors observed a trend for higher Pt concentrations with the decrease of pH in rainwater.

Wastewaters and pluvial waters can be regarded as the main pathway of PGE transfer from urban areas to the nearby aquatic systems. One of the PGE sources to wastewaters and pluvial waters, in particular Pt, derives from the particulate matter along roadsides, which includes ACC emissions accumulated over time and washed by rainwater during storms. An additional Pt source to wastewaters is excretion products from chemotherapy patients. Either in the hospital or in home effluents, Pt enters the municipal sewage systems, which lack the suitable treatment process for those active compounds. Monteiro et al. (2017) reported the Pt and Rh dissolved concentrations in waters from two stations (inflow and outflow) in a wastewater treatment plant. These samples were collected during varying weather conditions (Fig. 5). After a period of dry weather, Pt concentrations were 16 ± 2 ng L⁻¹ and 5.0 ± 0.7 ng L⁻¹ in the inflow and outflow, respectively. Samples collected soon after the first rain depicted an increase, with Pt concentrations reaching 25 ± 2 ng L⁻¹ and 10.4 ± 0.3 ng L⁻¹, respectively in the inflow and outflow samples. In the case of Rh the concentrations found did not vary significantly yet were higher than the limit of detection of the method, averaging 0.21 ± 0.05 ng L⁻¹ in either the inflow or outflow samples. Dissolved Pt and Rh concentrations determined in those treatment plant samples reflected the anthropogenic pressure, with Pt levels intensified after the first rain, as previously reported in Europe (Laschka and Nachtwey 1997).

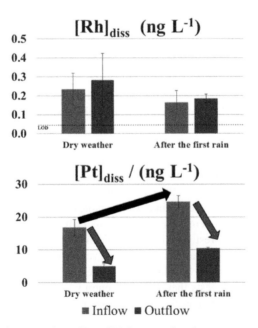

Fig. 5. Variation of the dissolved concentrations of Pt and Rh in two stations from a wastewater treatment plant before and soon after a rain event.

The concentration of Pt in rivers ranges in general up to 15 ng L^{-1} (Soyol-Erdene et al. 2011, Cobelo-Garcia 2013, Cobelo-García et al. 2014a). Data on the other PGE remain limited in the literature, despite all the efforts to understand PGE biogeochemical implications in aquatic ecosystems. As examples, dissolved Pt concentrations ranged from $\approx$ 0.02 to 0.62 pM in Lerez Estuary in Spain (Cobelo-García et al. 2013) and from 0.35 to 0.84 pM in Gironde Estuary in France (Cobelo-García et al. 2014a), while very high concentrations (between 5 and 35 pM) were reported in Tokyo Bay in Japan (Obata et al. 2006). In addition, the industries located on river and estuary banks are considered among the major and important anthropogenic sources of PGE to the waters. Many industrial processes use PGE elements as catalysts that degrade over time and release dissolvable or soluble compounds directly to superficial waters. In estuarine areas, Pt concentrations were reported to be higher than in open ocean seawater. In addition, Pt concentrations in seawater end-members exceed those of typical coastal oceanic waters (Cobelo-Garcia 2013, Cobelo-García et al. 2014a, Mashio et al. 2016, 2017), suggesting inputs of this element within the estuaries. Antonio Cobelo-García et al. (2013) observed a non-conservative behaviour of Pt during estuarine mixing, due to redox transformations along a salinity gradient.

The distribution and geochemical behaviour of PGE in oceanic waters have been studied since the early 1980s. Mainly focusing on Pt, studies in seawater have depicted different types of vertical profiles for different regions of the ocean. The literature contains descriptions of the recycled type profile in the eastern North Pacific Ocean (0.46–1.17 pmol L^{-1}; Hodge et al. 1986); the scavenged type in the Indian Ocean (0.17–1.6 pmol L^{-1}; van den Berg and Jacinto 1988); and the conservative type in the Atlantic and the western North Pacific Oceans (0.11–0.28 pmol L^{-1}; Colodner et al. 1993). From all the PGE, Rh has the second highest concentration in seawater and displays a nutrient-type depth profile in the Pacific Ocean (Bertine et al. 1993). Metals with nutrient-type or mixed profiles in the ocean are often associated with biological functions (Bruland et al. 2014); however, those were never reported for PGE.

As with the dissolved phase, data on PGE particulate concentrations in the water column, associated to suspended particulate matter (SPM), continues to be scarce. The few published studies reporting Pt concentrations in estuarine SPM pertain to the Lérez River Estuary, NW Iberian Peninsula (Cobelo-García et al. 2013), and the Gironde Estuary, SW France (Cobelo-García et al. 2014a). Concentrations of Pt in SPM were relatively low and close to the background concentration of sediments. Using SPM and dissolved Pt for the Gironde Estuary, the distribution coefficients computed were lower than those in the Lérez River Estuary. The discrepancy found may result from different hydrodynamic regimes in the two systems, and turbidity as well. In addition, different local anthropogenic pressures may affect soluble/insoluble Pt species (Cobelo-García et al. 2014a).

Sediments

Out of all the environmental compartments, sediments have been the most extensively studied. The increase of PGE concentrations in urban environments has been well documented (Zereini et al. 2004, 2007, Rauch et al. 2006, Mihaljevič et al. 2013, Wiseman et al. 2013, 2016, Rauch and Peucker-Ehrenbrink 2015, Ruchter and Sures 2015, Birke et al. 2017). However, research has focused mainly on Pt emissions, and particularly on its distribution, comprising areas under rural, urban and industrial pressures (e.g., Zereini and Wiseman 2015). It is noteworthy that studies dealing with spatially resolved distribution of PGE in worldwide aquatic environment are quite limited (e.g., Terashima et al. 1993, Wei and Morrison 1994, Zhong et al. 2012, Cobelo-García et al. 2011, 2013). For Rh, such studies do not exist and consequently the knowledge of cycling and biogeochemistry of PGE in aquatic systems remains unclear in many respects. Catalytic converters from vehicles are considered perhaps the main anthropogenic source of PGE to the environment. Yet, other anthropogenic sources cannot be disregarded and spatially resolved data from sediments are important to effectively assess local sources to the aquatic systems.

Wei and Morrison (1994) reported size-fractioned Pt concentrations in road dust and urban river sediments from Goteborg region, in Sweden, that receive pluvial runoff and sewage discharges. Despite the Pt concentrations found in sediments reflecting those of the road dust, and their source being suggested to be ACC, the authors did not find a particular spatial pattern of Pt distribution. Recently, Zhong et al. (2012) assessed the background levels and distribution of Pt and Pd in superficial sediments, and the possible sources as well, in Pearl River Estuary, China. The author's findings revealed relatively low concentrations of both elements. However, some localized high concentrations of Pt and Pd in sediments reflected other sources, such as the ones deriving from industrial processes that may have contributed to elevated PGE levels in that estuary.

When shifting from coastal to the open marine environment, data becomes scarcer, which demonstrates our present lack of knowledge on the behaviour of PGE in those environments. Earlier reports on Pt concentrations in ocean sediments point to a range up to 22 ng g^{-1} (Hodge et al. 1986). However, Terashima et al. (1993) found Pt and Pd concentrations in sediments off the Japan coast close to their crustal abundance, and both elements showed similar patterns of distribution. The authors concluded that Pt and Pd may be supplied to the deeper sediments through rivers and seawater. As dissolved species, Pt and Pd can be reduced to elemental metal species under O$_2$ minimum zones, co-precipitating with other sediment particles such as manganese oxides (Halbach et al. 1989, Terashima et al. 1993). At the continental shelf, Mashio et al. (2017) suggested that Pt in sediments could potentially have source in the coastal waters, reporting the increase of dissolved Pt concentrations with increasing depth. Nevertheless, geochemical processes at the sediment-water interface may govern the balance of PGE between compartments, changing them to species more or less bioavailable to the biota.

Biota

The biosphere is ultimately the environmental compartment of greatest concern to scientists, because of the potential hazards of these elements to humans and other terrestrial and aquatic organisms. While the free aqueous metal ions are usually regarded as the most (or only) reactive species with respect to toxicity effects towards marine organisms, one must keep in mind that different forms of a metal may be assimilated through contaminated food (Mulholland and Turner 2011). Because PGE are non-essential metals, and no biological function is known, it is assumed that they are subjected to net accumulation rather than biological regulation. In the past, research on the PGE effects in biota were focused closely on Pt, disregarding the other PGE. However, environmental and health concerns have recently grown, considering the faster reactivity of Pd when compared with Pt or Rh, and the shift to use Pd instead of Pt in ACC and industrial catalysts. Although the concentrations of PGE in aquatic environments are considerably low, ranging from pg g^{-1} to ng g^{-1} in solids and up to a small number of ng L^{-1} in waters, it is likely that those concentrations will continue to increase in the future, mostly because of the growing use of PGE in ACC and thus their becoming biologically available. The interest in aquatic organisms as potential biomonitors of PGE contamination contributed to several studies in which possible bioaccumulation and biomagnification in the food web were also addressed. While in urban areas the monitoring of PGE may use trees (Bonanno and Pavone 2015) or mosses (Ayrault et al. 2006), in aquatic systems attention is given to plants, macroinvertebrates and fish species (e.g., Moldovan et al. 2001, Zimmermann et al. 2004, Sures and Zimmermann 2007, Mulholland and Turner 2011, Ruchter and Sures 2015) from transitional areas. Recently, Ruchter and Sures (2015) showed the accumulation of Pt by the clam *Corbicula* sp. in the field; however, Pt concentrations remained relatively low. In addition, Neira et al. (2015) observed a temporal variation of Pt concentrations in mussels *Mytilus galloprovincialis* over two decades of anthropogenic emissions (i.e., ACC) and linked with the increasing concentrations in the environment. Abdou et al. (2016) reported the same trend using a historical series of wild oysters. Nevertheless, chronic exposure of organisms to the PGE in their natural environment still needs to be studied, particularly in the case of Pd.

8. Into the Deep Ocean

The major inputs of REE and PGE into the coastal regions of the ocean come through river discharges, in which the metals can derive from the weathering of continental parental rocks and major anthropogenic emissions. The atmospheric transport may also introduce REE and PGE to the oceanic surface waters (Elderfield and Greaves 1982), incorporated in both aluminosilicates and amorphous metal (predominantly Fe) oxide phases (Windom 1969). An additional source of those metals into the ocean is submarine volcanism. Although terrigenous inputs of REE and PGE apparently dominate any other inputs, those associated with marine volcanism in particular are of major importance to the deep sea and are potentially of economic interest in the near future. The main sources and sinks in the deep ocean, namely the hydrothermal systems and ferromanganese crusts and nodules, are briefly reviewed here.

The potential deep ocean metal resources are focused on three main types of deposits that contain different valuable metals. (1) The Fe-Mn crusts and nodules, consisting of mineral aggregates covering enormous areas of the seabed of the Pacific and Indian Oceans, mostly found at depths below 3,500 m. They are composed mainly of manganese, iron, copper, nickel and cobalt, along with other substances such as molybdenum, zinc and lithium. (2) The cobalt crusts, which are incrustations of minerals formed on the sides of submarine mountain ranges and seamounts and found in the western Pacific at depths of 1,000 to 3,000 m. These crusts develop as a result of the accumulation of minerals dissolved in the water and contain mainly manganese, iron, cobalt, nickel, platinum and REE. (3) Massive sulphides, which are massive ore deposits that can accumulate metals mainly at the openings of hot vents in many places on the sea floor that are or used to be volcanic. Depending on the region, they contain widely different amounts of copper, zinc, lead, gold and silver, as well as numerous important trace metals such as indium, germanium, tellurium or selenium.

The role hydrothermal systems play as sources of REE and PGE to the deep ocean has been studied since the late 1970s (e.g., Michard et al. 1983, Ruhlin and Owen 1986, Usui and Someya 1997). Ruhlin and Owen (1986) studied the REE composition of hydrothermal precipitates from the East Pacific Rise and suggested that the shale-normalized REE patterns were similar to the seawater pattern, varying only in absolute REE content. The total REE (ΣREE, including La, Ce, Nd, Sm, Eu, Tb, Yb and Lu) concentrations ranged from 14 to 467 mg kg^{-1} and the REE contents increased with the increasing distance from the paleo-rise crest, exhibiting a pronounced rise in sediments deposited below the paleolysocline. The same authors suggested that the mechanism for the REE content in hydrothermal sediments is the scavenging of REE by Fe-oxyhydroxides from seawater. The same was observed in sediments from the East Pacific Rise, where the seawater scavenging mechanism was investigated (Ruhlin and Owen 1986). German et al. (1990) also studied the hydrothermal scavenging of REE in suspended particulate samples on the Mid-Atlantic Ridge. In this study, results pointed to the REE scavenging from the water column only after the deposition of suspended hydrothermal precipitates, leading to the conclusion that deep-sea sediments should reflect the REE patterns of overlying sea water. German et al. (1990) observed the same in seawater samples from the deep Pacific near hydrothermal vents. Consequently, by transport from the vent injection point and/or by long residence time in the seawater-sediment interface, the prolonged exposure to seawater caused absolute REE content of hydrothermal sediments to approach levels found in seawater.

The study conducted by Mitra et al. (1994) evidenced the contribution of hydrothermal fluids and plumes in deep-sea REE content from the Mid-Atlantic Ridge. These deep-sea hydrothermal fluids are enriched in REE over seawater, by factors of 10 to 10^3. The same study also demonstrated that, by precipitating Fe-oxyhydroxides, scavenging processes eliminate the impact of the REE enrichments from hydrothermal fluids into seawater, such that the latter presented a net depletion of REE because of hydrothermal activity. An interesting study conducted by Hongo et al. (2007) showed that the REE enrichments in hydrothermal fluid samples from Okinawa Trough back-arc basin were in factors of 3000 in Ce, 10^3 to 10^4 in Eu, and 10^2 in other REE when compared to seawater. The authors highlighted the REE composition as a useful tool to trace the evolution of the plume during lateral transport. They also indicated that REE in the hydrothermal fluids did not act as the net source of the global REE budget; however, with the REE contents and patterns observed it was possible to trace the influence of the hydrothermal activities. More recently, several recent studies have shown the contents and behaviour of REE in hydrothermal fluids, and their contribution to the deep sea as well. Schmidt et al. (2010) indicated a total of REE (ΣREE) (including Y) concentrations in a range of 15.7 to 363 nM in filtered fluid samples from two hydrothermal vent fields at the Mid-Atlantic Ridge. This study evidenced the temporal variability of REE contents in the hydrothermal fluids mainly due to fluid temperature variations and compositional variability, concerning the elements affected by sulphate/sulphide precipitation. In hydrothermal fluids from the Manus Basin, Papua New Guinea, Craddock et al. (2010) referenced the influence of anhydrite precipitation caused by the mix between high-temperature hydrothermal fluid and locally entrained seawater. The authors indicated that ΣREE contents in acid-sulphate fluids from DESMOS and SuSu Knolls hydrothermal systems, varying from 147 to 236 nmol kg^{-1} and from 93 to 350 nmol kg^{-1}, respectively, are significantly greater than those sampled in high-temperature black smoker fluids in Vienna Woodsand and in PACMANUS hydrothermal systems, with ΣREE ranging from 2.6 to 5.8 nmol kg^{-1} and from 1.5 to 88 nmol kg^{-1}, respectively. Also, the authors suggested that although anhydrite does not affect the REE relative distributions in the hydrothermal fluids, there is an effective incorporation of REE in the anhydrite precipitate affecting the measured fluid REE contents. In another study, presented by Steinmann et al. (2012), the distribution patterns of REE were used to distinguish between metalliferous deposits precipitated directly from seawater (hydrogenous) and deposits of diagenetic or hydrothermal origin, considering the metalliferous sediments sampled near a basaltic seamount named "Dorado" on the eastern flank of the East Pacific Rise, west of Nicoya Peninsula, Costa Rica. In this study the authors indicated ΣREE concentrations in bulk sediments from 92 to 445 mg kg^{-1}, with REE PAAS-normalized patterns showing negative anomalies for Ce and positive anomalies for Y. In addition, PAAS-normalized patterns showed a slight enrichment of HREE over the LREE. Steinmann et al. (2012) also suggested that low temperature hydrothermal fluids from the present are not good indicators to explain the enrichment of REE in bulk sediments, once these fluids are strongly depleted in REE. Instead, they indicated that REE derived from Fe-Mn oxyhydroxide particles that precipitated from high temperature hydrothermal plume when the Dorado site was still located close to the East Pacific Rise. Furthermore, the study conducted by Wang et al. (2013) provides a better understanding of how REE in vent fluids could be used as tracers of sub-seafloor hydrothermal processes. The authors suggested that lower REE contents (from 813 to 1212 ng L^{-1}) in Kueishantao hydrothermal fluids, NE Taiwan, may be related to the short duration of fluid-andesite interaction under the seafloor, the boiling of fluid and the precipitation of sulphur minerals near the seafloor, when compared with those of most deep-sea hydrothermal fluids and acidic hot waters in continental geothermal fields. Additionally, among those factors, and taking into account the adsorption of small particles and the formation of REE-chloride complexes, the REE distribution patterns in the hydrothermal fluids are also affected by the pH variation, mainly at very low pH values (2.3–2.8). Ferromanganese crusts and nodules constitute an important resource of metals in the bottom of the ocean. These deep-sea deposits can be significantly enriched on several metals, such as Cu, Ni, Zn, and Co. They also include the REE (plus Y) and the PGE, in particular Pt (e.g., Muiños et al. 2013). Crusts and nodules can be classified as diagenetic, hydrogenetic, hydrothermal, and mixed-type deposits (e.g., De Baar et al. 1985, Halbach et al. 1989), depending on the mechanism of formation, occurring at the sediment-water interface and containing diagenetically reworked material (Berezhnaya et al. 2018). Hydrogenetic Fe-Mn nodules result from the precipitation of colloidal particles of Fe-Mn oxyhydroxides from seawater onto hard-rock substrates and tend to grow slowly. Several works showed the importance of Fe-Mn nodules and crusts as a source of REE and PGE in the deep ocean (Piper 1974b, Addy 1979, Elderfield et al. 1981, Gende and Quanxing 1987, Pattan and Parthiban 2011, Muiños et al. 2013, Kolesnik and Kolesnik 2015). Addy (1979) reported the patterns of REE in Mn nodules and micronodules from NW Atlantic sediments, indicating that the source of REE is probably the seawater and also that nodules had hydrogenetic origin. In the study conducted by Elderfield et al. (1981) on REE geochemistry, the relationship between REE contents in the Fe-Mn nodules and associated sediments was highlighted, providing further information on possible mechanisms that control their incorporation into the nodules. REE in nodules and sediments were closely related, pointing to an original REE seawater source, concerning the $^{43}Nd/^{44}Nd$ ratios in the nodules. However, identical ratio for sediments in combination with the REE patterns suggested that diagenetic reactions might be responsible for the transfer of elements into the nodules. In contrast, Ce appears to be added to both nodules and sediments directly from seawater and it is not involved in diagenetic reactions. Gende and Quanxing (1987) indicated higher contents of REE in Fe-Mn nodules and crusts from the South China Sea when compared with the Fe-Mn nodules and sediment from the North Pacific Ocean. They also suggested that the source of REE in those nodules derived from weathering and leaching of medium acid rock of the sea formed by volcanic activities. In addition, research concerning deep-sea Fe-Mn crusts was undertaken in the Portuguese exclusive economic zone (EEZ) in order to investigate the resources of several metals, including the REE and PGE (Muiños et al. 2013). The authors showed that all samples are highly enriched in REE and PGE relative to seawater and the earth's crust, with up to 0.29 wt.% total REEs. They also indicated that shale (PAAS) and chondrite-normalized REE patterns showed pronounced MREEs and LREEs relative to HREEs, respectively, a strong positive Ce anomaly, and a small positive Gd anomaly (only for chondrite-normalized REE patterns), which are typical of Fe-Mn crusts of hydrogenetic origin. The estimations calculated by Muiños et al. (2013) indicate the resource potential of these Fe-Mn deposits within the Portuguese EEZ, and these are comparable to parts of the central Pacific Ocean.

Recent discoveries in the French EEZ of Wallis and Futuna (SW Pacific Ocean) have brought new insight to the formation of low temperature hydrothermal Mn deposits, leading the authors to reconsider the classification and discrimination diagrams of Fe-Mn deposits and ore-forming processes and their resources of REE in the deep ocean (Josso et al. 2017). According to these authors REE concentrations in these hydrothermal samples of Fe-Mn deposits are low ($\Sigma REE = 26.1 \pm 11.5$ mg kg^{-1}), with a maximum content of 126 mg kg^{-1} REE associated with the highest concentration in P_2O_5 (0.43%) and Zr. The chondrite-normalized REE trends show two distinctive signatures. Hydrogenetic Fe-Mn crusts presented strong positive Ce anomaly, followed closely by concentrations in polymetallic nodules, whereas hydrothermal Fe-Mn precipitates showed a negative Ce anomaly. These authors highlighted the fact that the deposit-forming processes and REE concentrations are related with each type of Fe-Mn deposit.

Studies dealing with the distribution of PGE in hydrothermal systems continue to be scarce. However, in the last decades, Fe-Mn crusts and nodules have been studied mostly with respect to the economic interest in viable metals, while REE and PGE remain poorly characterized. Elements strongly enriched in crusts over abyssal Fe-Mn nodules include Fe, Co, Pt, Pb, As, Bi, V, P, Ti, Nb, Ta, Te, Sr, Th, U, Y, Zr, and the REE (Hein and Koschinsky 2014). It is also hypothesized that deep-sea systems can represent ancient volcanic-hosted massive sulphide deposits and thus provide one of the main sources of PGE in the bottom of the ocean. Despite occurring in sedimentary environments as well, layered igneous complexes are pointed out as one of the main sources of PGE (Pašava et al. 2007). Other sources at the bottom of the ocean include iron and manganese crusts and nodules (e.g., Usui and Someya 1997 and references therein, Muiños et al. 2013, Berezhnaya et al. 2018), and volcanogenic seafloor exhalates (Crocket 1990, Pašava et al. 2004). In hydrothermal deposits from the Juan de Fuca and Mid-Atlantic Ridges, Crocket (1990) reported very low Pd concentrations (3.2–3.5 µg kg^{-1}); however, in the Cu-rich samples analysed by Pašava et al. (2004) from inactive and active chimneys at Satanic Mills, from the PACMANUS hydrothermal field, the authors found larger Pd concentrations, up to 356 µg kg⁻¹. The same was observed for Rh, up to 145 µg kg⁻¹. Torokhov and Lazareva (2003) reported on Fe-Mn crusts and massive sulphides hosted in serpentinites at Logatchev in even larger concentrations than the above-mentioned studies. They identified significant enrichments of Pd in Fe-Mn crusts (900 µg kg⁻¹) and Rh (490 µg kg⁻¹) in massive sulphide ore.

Based on the five samples analysed for all the PGE, Muiños et al. (2013) found a Pt enrichment in Fe-Mn crusts from the North Atlantic, with Pt concentrations varying between 153 and 512 µg kg⁻¹, whereas the other PGE showed lower concentrations. As suggested for the REE, seawater scavenging of PGE appears to be an important mechanism in the retention of those metals in the ocean bed. Moreover, several hydrothermal fields continue to be discovered at the bottom of the ocean and those sites need subsequently to be sampled, using remotely operated vehicles, and studied in order to assess their resource potential for metals of economic interest as well as with respect to the REE and PGE (Muiños et al. 2013).

Compared to other processes, namely the exploration and mining of terrestrial resources, seafloor mining is a relatively new way to retrieve minerals and it is likely to attract mining companies to explore this growing market. Some companies have recently received permission to explore the seabed for REE and noble metals, such as Pt. Discoveries in the deep seabed of the EEZ of the Cook Islands in the Pacific Ocean have indicated potential new sources of REE and scandium (Benton 2016). In 2016, a deep-sea mining development company based in Houston Texas, The Ocean Minerals LLC, started mineral extraction with the agreement of the Cook Islands government for exclusive prospecting and exploration rights in those areas.

However, suitable technology is still needed for the comprehensive assessment and extraction process on the seafloor, despite the slowly growing engineering research undertaken in several countries (e.g., Takai et al. 2016). Moreover, environmental studies should address the regulations recommended in the United Nations Convention on the Law of the Sea and International Seabed Authority, as suggested by many researchers (e.g., Muiños et al. 2013), which demand an evaluation of risk assessment prior to any mining activity. Because anthropogenic activities in the deep ocean are likely to induce important and rapid shifts in this ecosystem, the seamounts and hydrothermal systems, and manganese nodules on abyssal plains as well, may be the endangered habitats most likely to be impacted by future mining activities. In addition, the high pressure at greater depths may strongly influence chemical reactions, thus compromising the biochemical and physiological processes in organisms (Mestre et al. 2014). This, in turn, may potentially increase the toxicity effects of pollutants or induce toxicity of the REE and PGE, for which such effect remains unknown.

9. Future Perspectives

Despite the advances in the knowledge of REE and PGE levels in transitional and marine ecosystems over the past decades, many questions continue to be posed, and further research is needed to search for the answers. There is an urgent need to elucidate the mobility, speciation and bioavailability of REE and PGE in estuaries and the adjoining coastal areas. In addition, considering the likely increase of those metals in the marine environment, the estimation of their regional and/or global fluxes needs to be evaluated. Consequently, there is also a need for new and integrative modelling tools, which will allow the prediction of "worst case scenario" impacts on regions vulnerable to anthropogenic activities. Moreover, the health issues raised by the increasing demand for those technology-critical elements still need to be elucidated, in particular with respect to species consumed by humans that may be continuously exposed to REE and PGE contamination. Thus, field studies are an urgent necessity, considering that at the earth's surface the fluxes of REE and PGE have been, and will continue to be, altered mainly by anthropogenic emissions.

Acknowledgments

The authors would like to thank the Editors for the opportunity to collaborate on this book with this chapter.

References

Abdou, M., J. Schäfer, A. Cobelo-García, P. Neira, J.C.J. Petit, D. Auger et al. 2016. Past and present platinum contamination of a major European fluvial–estuarine system: Insights from river sediments and estuarine oysters. Mar. Chem. 185: 104–110. Doi: https://doi.org/10.1016/j.marchem.2016.01.006.

Addy, S.K. 1979. Rare earth element patterns in manganese nodules and micronodules from northwest Atlantic. Geochim. Cosmochim. Acta 43: 1105–1115. Doi: 10.1016/0016-7037(79)90097-8.

Akagi, T. and K. Edanami. 2017. Sources of rare earth elements in shells and soft-tissues of bivalves from Tokyo Bay. Mar. Chem. 194. Doi: 10.1016/j.marchem.2017.02.009.

Almecija, C., A. Cobelo-Garcia and J. Santos-Echeandia. 2016. Improvement of the ultra-trace voltammetric determination of Rh in environmental samples using signal transformation. Talanta 146: 737–743. Doi: 10.1016/j.talanta.2015.06.032.

Almécija, C., A. Cobelo-García, V. Wepener and R. Prego. 2017. Platinum group elements in stream sediments of mining zones: The Hex River (Bushveld Igneous Complex, South Africa). J. African Earth Sci. 129: 934–943. Doi: https://doi.org/10.1016/j.jafrearsci.2017.02.002.

Alonso, E., A.M. Sherman, T.J. Wallington, M.P. Everson, F.R. Field, R. Roth et al. 2012. Evaluating rare earth element availability: A case with revolutionary demand from clean technologies. Environ. Sci. Technol. 46: 3406–3414. Doi: 10.1021/es203518d.

Amato, F., A. Alastuey, J. de la Rosa, Y. Gonzalez Castanedo, A.M. De la Campa, M. Pandolfi et al. 2014. Trends of road dust emissions contributions on ambient air particulate levels at rural, urban and industrial sites in southern Spain. Atmos. Chem. Phys. 14: 3533–3544. Doi: 10.5194/acp-14-3533-2014.

Apergis, E. and N. Apergis. 2017. The role of rare earth prices in renewable energy consumption: The actual driver for a renewable energy world. Energy Econ. 62: 33–42. Doi: https://doi.org/10.1016/j.eneco.2016.12.015.

Arimoto, R., R.A. Duce and B.J. Ray. 1989. Concentrations, sources and air–sea exchange of trace elements in the atmosphere over Pacific Ocean. Chem. Oceanogr. 10: 107–149.

Ayora, C., F. Macías, E. Torres and J.M. Nieto. 2015. Rare earth elements in acid mine drainage. *In*: XXXV Reunión de La Sociedad Española de Mineralogía, pp. 1–22.

Ayrault, S., C. Li and A. Gaudry. 2006. Biomonitoring of Pt and Pd with mosses. *In*: Palladium Emissions in the Environment. Springer, pp. 525–536.

Balaram, V. 1996. Recent trends in the instrumental analysis of rare earth elements in geological and industrial materials. TrAC—Trends Anal. Chem. Doi: 10.1016/S0165-9936(96)00058-1.

Balusamy, B., Y.G. Kandhasamy, A. Senthamizhan, G. Chandrasekaran, M.S. Subramanian and T.S. Kumaravel. 2012. Characterization and bacterial toxicity of lanthanum oxide bulk and nanoparticles. J. Rare Earths 30: 1298–1302. Doi: 10.1016/S1002-0721(12)60224-5.

Balusamy, B., B.E. Taştan, S.F. Ergen, T. Uyar and T. Tekinay. 2015. Toxicity of lanthanum oxide (La2O3) nanoparticles in aquatic environments. Environ. Sci. Process. Impacts 17: 1265–1270. Doi: 10.1039/C5EM00035A.

Barbante, C., A. Veysseyre, C. Ferrari, K. Van De Velde, C. Morel, G. Capodaglio et al. 2001. Greenland snow evidence of large scale atmospheric contamination for platinum, palladium, and rhodium. Environ. Sci. Technol. 35: 835–839. Doi: 10.1021/es000146y.

Barry, M.J. and B.J. Meehan. 2000. The acute and chronic toxicity of lanthanum to Daphnia carinata. Chemosphere 41: 1669–1674. Doi: 10.1016/S0045-6535(00)00091-6.

Bau, M. and P. Dulski. 1996. Anthropogenic origin of positive gadolinium anomalies in river waters. Earth Planet. Sci. Lett. 143: 245–255. Doi: 10.1016/0012-821X(96)00127-6.

Benton, D. 2016. Rare earth elements discovered in Exclusive Economic Zone of the Cook Islands deep seabed [WWW Document]. URL https://www.miningglobal.com/operations/rare-earth-elements-discovered-exclusive-economic-zone-cook-islands-deep-seabed (accessed 11.6.18).

Berezhnaya, E., A. Dubinin, M. Rimskaya-Korsakova and T. Safin. 2018. Accumulation of platinum group elements in hydrogenous Fe–Mn crust and nodules from the Southern Atlantic Ocean. Minerals 8: 275. Doi: 10.3390/min8070275.

Bertine, K.K., M. Koide and E.D. Goldberg. 1993. Aspects of rhodium marine chemistry. Mar. Chem. 4: 199–210. Doi: https://doi.org/10.1016/0304-4203(93)90012-D.

Birke, M., U. Rauch, J. Stummeyer, H. Lorenz and B. Keilert. 2017. A review of platinum group element (PGE) geochemistry and a study of the changes of PGE contents in the topsoil of Berlin, Germany, between 1992 and 2013. J. Geochemical Explor. Doi: https://doi.org/10.1016/j.gexplo.2017.09.005.

Bocca, B., S. Caimi, P. Smichowski, D. Gómez and S. Caroli. 2006. Monitoring Pt and Rh in urban aerosols from Buenos Aires, Argentina. Sci. Total Environ. 358: 255–264. Doi: https://doi.org/10.1016/j.scitotenv.2005.04.010.

Bonanno, G. and P. Pavone. 2015. Leaves of Phragmites australis as potential atmospheric biomonitors of Platinum group elements. Ecotoxicol. Environ. Saf. 114: 31–37. Doi: https://doi.org/10.1016/j.ecoenv.2015.01.005.

Borgmann, U., Y. Couillard, P. Doyle and D.G. Dixon. 2005. Toxicity of sixty-three metals and metalloids to Hyalella azteca at two levels of water hardness. Environ. Toxicol. Chem. 24: 641–652. Doi: 10.1897/04-177r.1.

Borrego, J., N. López-González, B. Carro and O. Lozano-Soria. 2005. Geochemistry of rare-earth elements in Holocene sediments of an acidic estuary: Environmental markers (Tinto River Estuary, South-Western Spain). J. Geochemical Explor. 86: 119–129. Doi: 10.1016/j.gexplo.2005.05.002.

Borrego, J., N. López-González, B. Carro and O. Lozano-Soria. 2004. Origin of the anomalies in light and middle REE in sediments of an estuary affected by phosphogypsum wastes (south-western Spain). Mar. Pollut. Bull. 49: 1045–1053. Doi: 10.1016/j.marpolbul.2004.07.009.

Bosco-Santos, A., W. Luiz-Silva, E.V. da Silva-Filho, M.D.C. de Souza, E.L. Dantas and M.S. Navarro. 2017. Fractionation of rare earth and other trace elements in crabs, Ucides cordatus, from a subtropical mangrove affected by fertilizer industry. J. Environ. Sci. (China) 54. Doi: 10.1016/j.jes.2016.05.024.

Bouzigues, C., T. Gacoin and A. Alexandrou. 2011. Biological applications of rare-earth based nanoparticles. ACS Nano. 5: 8488–8505. Doi: 10.1021/nn202378b.

Bowles, J.F.W., A.P. Gize and A. Cowden. 1994. The mobility of the platinum-group elements in the soils of the Freetown Peninsula, Sierra Leone. Can. Mineral. 32: 957LP–967.

Brito, P., M. Malvar, C. Galinha, I. Caçador, J. Canário, F. Araújo et al. 2018. Yttrium and rare earth elements fractionation in salt marsh halophyte plants. Sci. Total Environ. 643. Doi: 10.1016/j.scitotenv.2018.06.291.

Bruland, K.W., R. Middag and M.C. Lohan 2013. Controls of trace metals in seawater. In Treatise on Geochemistry: Second Edition. 8: 19–51. Elsevier Inc. https://doi.org/10.1016/B978-0-08-095975-7.00602-1.

Burch, K.R., J.B. Comer, S.F. Wolf and S.S. Brake. 2011. REE geochemistry of an acid mine drainage system in western Indiana. Abstr. with Programs—Geol. Soc. Am. 43(5): 587p.

Carpenter, D., C. Boutin, J.E. Allison, J.L. Parsons and D.M. Ellis. 2015. Uptake and effects of six rare earth elements (REEs) on selected native and crop species growing in contaminated soils. PLoS One 10. Doi: 10.1371/journal. pone.0129936.

Cassee, F.R., E.C. Van Balen, C. Singh, D. Green, H. Muijser, J. Weinstein et al. 2011. Exposure, health and ecological effects review of engineered nanoscale cerium and cerium oxide associated its use as a fuel additive. Crit. Rev. Toxicol. 41: 213–229. Doi: 10.3109/10408444.2010.529105.

Cawthorn, R.G. 2010. The platinum group element deposits of the bushveld complex in South Africa. Platin. Met. Rev. 54: 205–215. Doi: 10.1595/147106710X520222.

Cobelo-García, A., P. Neira, M. Mil-Homens and M. Caetano. 2011. Evaluation of the contamination of platinum in estuarine and coastal sediments (Tagus Estuary and Prodelta, Portugal). Mar. Pollut. Bull. 62: 646–650. Doi: 10.1016/j.marpolbul.2010.12.018.

Cobelo-Garcia, A. 2013. Kinetic effects on the interactions of Rh(III) with humic acids as determined using size-exclusion chromatography (SEC). Environ. Sci. Pollut. Res. Int. 20: 2330–2339. Doi: 10.1007/s11356-012-1113-8.

Cobelo-García, A., D.E. López-Sánchez, C. Almécija and J. Santos-Echeandía. 2013. Behavior of platinum during estuarine mixing (Pontevedra Ria, NW Iberian Peninsula). Mar. Chem. 150: 11–18. Doi: 10.1016/j. marchem.2013.01.005.

Cobelo-García, A., D.E. López-Sánchez, J. Schäfer, J.C.J. Petit, G. Blanc and A.Turner. 2014a. Behavior and fluxes of Pt in the macrotidal Gironde Estuary (SW France). Mar. Chem. 167: 93–101. Doi: 10.1016/j. marchem.2014.07.006.

Cobelo-García, A., J. Santos-Echeandía, D.E. López-Sánchez, C. Almécija and D. Omanović. 2014b. Improving the voltammetric quantification of Ill-defined peaks using second derivative signal transformation: Example of the determination of platinum in water and sediments. Anal. Chem. 86: 2308–2313. Doi: 10.1021/ac403558y.

Cobelo-García, A., M. Filella, P. Croot, C. Frazzoli, G. Du Laing, N. Ospina-Alvarez et al. 2015. COST action TD1407: network on technology-critical elements (NOTICE)—from environmental processes to human health threats. Environ. Sci. Pollut. Res. 22: 15188–15194. Doi: 10.1007/s11356-015-5221-0.

Colodner, D.C., E.A. Boyle and J.M. Edmond. 1993. Determination of rhenium and platinum in natural waters and sediments, and iridium in sediments by flow injection isotope dilution inductively coupled plasma mass spectrometry. Anal. Chem. 65: 1419–1425. Doi: 10.1021/ac00058a019.

Cotton, S.A. and J.M. Harrowfield. 2012. Lanthanides: Biological activity and medical applications. Encycl. Inorg. Bioinorg. Chem. Doi: 10.1002/9781119951438.eibc2091.

Craddock, P.R., W. Bach, J.S. Seewald, O.J. Rouxel, E. Reeves and M.K. Tivey. 2010. Rare earth element abundances in hydrothermal fluids from the Manus Basin, Papua New Guinea: Indicators of sub-seafloor hydrothermal processes in back-arc basins. Geochim. Cosmochim. Acta 74: 5494–5513. Doi: 10.1016/j.gca.2010.07.003.

Crocket, J.H. 1990. Noble metals in seafloor hydrothermal mineralization from the Juan de Fuca and Mid-Atlantic ridges; a fractionation of gold from platinum metals in hydrothermal fields. Can. Mineral. 28: 639–648.

Cui, J., Z. Zhang, W. Bai, L. Zhang, X. He, Y. Ma et al. 2012. Effects of rare earth elements La and Yb on the morphological and functional development of zebrafish embryos. J. Environ. Sci. 24: 209–213. Doi: 10.1016/S1001-0742(11)60755-9.

Date, A.R. and D. Hutchison. 1987. Determination of rare earth elements in geological samples by inductively coupled plasma source mass spectrometry. J. Anal. At. Spectrom. 2: 269. Doi: 10.1039/ja9870200269.

De Baar, H.J.W., M.P. Bacon, P.G. Brewer and K.W. Bruland. 1985. Rare earth elements in the Pacific and Atlantic Oceans. Geochim. Cosmochim. Acta 49: 1943–1959. Doi: 10.1016/0016-7037(85)90089-4.

Delgado, J., R. Pérez-López, L. Galván, J.M. Nieto and T. Boski. 2012. Enrichment of rare earth elements as environmental tracers of contamination by acid mine drainage in salt marshes: A new perspective. Mar. Pollut. Bull. 64: 1799–1808. Doi: 10.1016/j.marpolbul.2012.06.001.

Diong, H.T., R. Das, B. Khezri, B. Srivastava, X. Wang, P.K. Sikdar et al. 2016. Anthropogenic platinum group element (Pt, Pd, Rh) concentrations in PM10 and PM2.5 from Kolkata, India. Springerplus 5: 1242. Doi: 10.1186/s40064-016-2854-5.

Du, X. and T.E. Graedel. 2011. Global in-use stocks of the rare earth elements: A first estimate. Environ. Sci. Technol. 45: 4096–4101. Doi: 10.1021/es102836s.

Du, X. and T.E. Graedel. 2013. Uncovering the end uses of the rare earth elements. Sci. Total Environ. 461-462: 781–784. Doi: 10.1016/j.scitotenv.2013.02.099.

Dutta, T., K.H. Kim, M. Uchimiya, E.E. Kwon, B.H. Jeon, A. Deep et al. 2016. Global demand for rare earth resources and strategies for green mining. Environ. Res. Doi: 10.1016/j.envres.2016.05.052.

Elbaz-Poulichet, F., J.-L. Seidel and C. Othoniel. 2002. Occurrence of an anthropogenic gadolinium anomaly in river and coastal waters of Southern France. Water Res. 36: 1102–1105. Doi: 10.1016/S0043-1354(01)00370-0.

Elderfield, H., C.J. Hawkesworth, M.J. Greaves and S.E. Calvert. 1981. Rare earth element geochemistry of oceanic ferromanganese nodules and associated sediments. Geochim. Cosmochim. Acta 45: 513–528. Doi: https://doi.org/10.1016/0016-7037(81)90184-8.

Elderfield, H. and M.J. Greaves. 1982. The rare earth elements in seawater. Nature. Doi: 10.1038/296214a0.

Elderfield, H., R. Upstill-Goddard and E.R. Sholkovitz. 1990. The rare earth elements in rivers, estuaries, and coastal seawaters: processes affecting the crustal input of elements to the ocean and their significance to the composition of seawater. Geochim. Cosmochim. Acta 54: 971–991.

Ennevor, B.C. and R.M. Beames. 1993. Use of lanthanide elements to mass mark juvenile Salmonids. Can. J. Fish. Aquat. Sci. 50: 1039–1044. Doi: 10.1139/f93-120.

Eriksson, J. 2001. Concentrations of 61 trace elements in sewage sludge, farmyard manure, mineral fertiliser, precipitation and in oil and crops. Stockholm.

Ferrat, M., D.J. Weiss, S. Strekopytov, S. Dong, H. Chen, J. Najorka et al. 2011. Improved provenance tracing of Asian dust sources using rare earth elements and selected trace elements for palaeomonsoon studies on the eastern Tibetan Plateau. Geochim. Cosmochim. Acta 75: 6374–6399. Doi: 10.1016/j.gca.2011.08.025.

Figueiredo, C., T.F. Grilo, C. Lopes, P. Brito, M. Diniz, M. Caetano et al. 2018. Accumulation, elimination and neuro-oxidative damage under lanthanum exposure in glass eels (Anguilla anguilla). Chemosphere 206. Doi: 10.1016/j.chemosphere.2018.05.029.

Fortin, C., F. Wang and D. Pitre. 2011. Critical Review of Platinum Group Elements (Pd, Pt, Rh) in Aquatic Ecosystems—Research Report No R-1269.

Geagea, M.L., P. Stille, M. Millet and T. Perrone. 2007. REE characteristics and Pb, Sr and Nd isotopic compositions of steel plant emissions. Sci. Total Environ. 373: 404–419. Doi: 10.1016/j.scitotenv.2006.11.011.

Gende, B.A.O. and L.I. Quanxing. 1987. Geochemistry and their Genesis of Rare Earth Elements of Ferromanganese Nodules and Crusts From the South.

German, C.R., G.P. Klinkhammer, J.M. Edmond, A. Mura and H. Elderfield. 1990. Hydrothermal scavenging of rare-earth elements in the ocean. Nature 345: 516–518. Doi: 10.1038/345516a0.

Glaister, B.J. and G.M. Mudd. 2010. The environmental costs of platinum-PGM mining and sustainability: Is the glass half-full or half-empty? Miner. Eng. Doi: 10.1016/j.mineng.2009.12.007.

Goldstein, S.J. and S.B. Jacobsen. 1988a. Rare earth elements in river waters. Earth Planet. Sci. Lett. 89: 35–47. Doi: 10.1016/0012-821x(88)90031-3.

Goldstein, S.J. and S.B. Jacobsen. 1988b. REE in the Great Whale River estuary, northwest Quebec. Earth Planet. Sci. Lett. 88: 241–252. Doi: 10.1016/0012-821X(88)90081-7.

Goossens, D. and Z.Y. Offer. 1997. Aeolian dust erosion on different types of hills in a rocky desert: Wind tunnel simulations and field measurements. J. Arid Environ. 37: 209–229. Doi: 10.1006/jare.1997.0282.

Grandell, L., A. Lehtilä, M. Kivinen, T. Koljonen, S. Kihlman and L.S. Lauri. 2016. Role of critical metals in the future markets of clean energy technologies. Renew. Energy 95: 53–62. Doi: https://doi.org/10.1016/j.renene.2016.03.102.

Grasso, V.B. 2013. Rare earth elements in national defense: background, oversight issues, and options for congress. Congr. Res. Serv. Rep. Congr. 43.

Greenland Minerals and Energy Ltd. 2017. 2017 Annual Report—Materials for an Energy Efficient Future.

Greinacher, E. 1981. History of rare earth applications, rare earth market today. Industrial Applications of Rare Earth Elements. Doi: 10.1021/bk-1981-0164.ch001.

Griffith, W.P. 2008. The periodic table and the platinum group metals. Platin. Met. Rev. 52: 114–119. Doi: 10.1595/147106708X297486.

Gromet, L.P., L.A. Haskin, R.L. Korotev and R.F. Dymek. 1984. The North American shale composite: Its compilation, major and trace element characteristics. Geochim. Cosmochim. Acta 48: 2469–2482. Doi: 10.1016/0016-7037(84)90298-9.

Guyonnet, D., M. Planchon, A. Rollat, V. Escalon, J. Tuduri, N. Charles et al. 2015. Material flow analysis applied to rare earth elements in Europe. J. Clean. Prod. 107: 215–228. Doi: 10.1016/j.jclepro.2015.04.123.

Halbach, P., C. Kriete, B. Prause and D. Puteanus. 1989. Mechanisms to explain the platinum concentration in ferromanganese seamount crusts. Chem. Geol. 76: 95–106. Doi: https://doi.org/10.1016/0009-2541(89)90130-7.

Hans Wedepohl, K. 1995. The composition of the continental crust. Geochim. Cosmochim. Acta 59: 1217–1232. Doi: 10.1016/0016-7037(95)00038-2.

Hao, S., W. Xiaorong, H. Zhaozhe, W. Chonghua, W. Liansheng, D. Lemei et al. 1996. Bioconcentration and elimination of five light rare earth elements in carp (Cyprinus carpio L.). Chemosphere 33: 1475–1483. Doi: 10.1016/0045-6535(96)00286-X.

Hartley, F.R. (ed.). 1991. Chemistry of the Platinum Group Metals, Volume 11, 1st editio. ed. Elsevier Science.

Haskin, M.A. and L.A. Haskin. 1966. Rare earths in european shales: A redetermination. Science 154(August): 507–509. Doi: 10.1126/science.154.3748.507.

Hedrick, J.B. 2004. Rare Earths in Selected U.S. Defense Applications, 1–13.

Hein, J.R. and A. Koschinsky. 2013. Deep-ocean ferromanganese crusts and nodules. In Treatise on Geochemistry: Second Edition, 13: 273–91. Elsevier Inc. Doi: 10.1016/B978-0-08-095975-7.01111-6.

Herrmann, H., J. Nolde, S. Berger and S. Heise. 2016. Aquatic ecotoxicity of lanthanum—A review and an attempt to derive water and sediment quality criteria. Ecotoxicol. Environ. Saf. 124: 213–238. Doi: 10.1016/j.ecoenv.2015.09.033.

Hodge, V., M. Stallard, M. Koide and E.D. Goldberg. 1986. Determination of platinum and iridium in marine waters, sediments, and organisms. Anal. Chem. 58: 616–620. Doi: 10.1021/ac00294a029.

Holwell, D.A. and I. McDonald. 2010. A review of the behaviour of platinum group elements within natural magmatic sulfide ore systems. Platin. Met. Rev. 54: 26–36. Doi: 10.1595/147106709X480913.

Hongo, Y., H. Obata, T. Gamo, M. Nakaseama, J. Ishibashi, U. Konno et al. 2007. Rare earth elements in the hydrothermal system at Okinawa Trough back-arc basin. Geochem. J. 41: 1–15. Doi: 10.2343/geochemj.41.1.

Hongyan, G., C. Liang, W. Xiaorong and C. Ying. 2002. Physiological responses of Carassius auratus to ytterbium exposure. Ecotoxicol. Environ. Saf. 53: 312–316. Doi: 10.1006/eesa.2002.2223.

Humphries, M. 2012. Rare earth elements: The global supply chain. CRS—Congr. Res. Serv. 1–27.

IUPAC. 2005. Nomenclature of Inorganic Chemistry–IUPAC Recommendations 2005. Chem. Int.–Newsmag. IUPAC 27. Doi: 10.1515/ci.2005.27.6.25.

Jarvis, K.E., A.L. Gray and E. McCurdy. 1989. Avoidance of spectral interference on europium in inductively coupled plasma mass spectrometry by sensitive measurement of the doubly charged ion. J. Anal. At. Spectrom. 4: 743. Doi: 10.1039/ja9890400743.

Johannesson, K.H. and X. Zhou. 1999. Origin of middle rare earth element enrichments in acid waters of a Canadian High Arctic lake. Geochim. Cosmochim. Acta 63: 153–165.

Johannesson, K.H., J. Tang, J.M. Daniels, W.J. Bounds and D.J. Burdige. 2004. Rare earth element concentrations and speciation in organic-rich blackwaters of the Great Dismal Swamp, Virginia, USA. Chem. Geol. 209: 271–294. Doi: 10.1016/j.chemgeo.2004.06.012.

Josso, P., E. Pelleter, O. Pourret, Y. Fouquet, J. Etoubleau, S. Cheron et al. 2017. A new discrimination scheme for oceanic ferromanganese deposits using high field strength and rare earth elements. Ore Geol. Rev. 87: 3–15. Doi: 10.1016/j.oregeorev.2016.09.003.

Kantipuly, C. and A. Westland. 1988. Review of methods for the determination of lanthanides in geological samples. Talanta 35: 1–13. Doi: 10.1016/0039-9140(88)80004-3.

Kašpar, J., P. Fornasiero, N. Hickey, J. Kaspar, P. Fornasiero, N. Hickey et al. 2003. Automotive catalytic converters: current status and some perspectives. Catal. Today 77: 419–449. Doi: 10.1016/S0920-5861(02)00384-X.

Kitto, M.E., D.L. Anderson, G.E. Gordon and I. Olmez. 1992. Rare-earth distributions in catalysts and airborne particles. Environ. Sci. Technol. 26: 1368–1375. Doi: 10.1021/es00031a014.

Klaver, G., M. Verheul, I. Bakker, E. Petelet-Giraud and P. Négrel. 2014. Anthropogenic rare earth element in rivers: Gadolinium and lanthanum. Partitioning between the dissolved and particulate phases in the Rhine River and spatial propagation through the Rhine-Meuse Delta (the Netherlands). Appl. Geochemistry 47: 186–197. Doi: 10.1016/j.apgeochem.2014.05.020.

Klinger, J.M. 2015. A historical geography of rare earth elements: From discovery to the atomic age. Extr. Ind. Soc. 2: 572–580. Doi: 10.1016/j.exis.2015.05.006.

Knappe, A., P. Möller, P. Dulski and A. Pekdeger. 2005. Positive gadolinium anomaly in surface water and ground water of the urban area Berlin, Germany. Chemie der Erde—Geochemistry 65: 167–189. Doi: 10.1016/j.chemer.2004.08.004.

Kolesnik, O.N. and A.N. Kolesnik. 2015. Rare earth elements in ferromanganese nodules of the Chukchi Sea. Lithol. Miner. Resour. 50: 181–191. Doi: 10.1134/S0024490215030050.

Kulaksiz, S. and M. Bau. 2007. Contrasting behaviour of anthropogenic gadolinium and natural rare earth elements in estuaries and the gadolinium input into the North Sea. Earth Planet. Sci. Lett. 260: 361–371. Doi: 10.1016/j.epsl.2007.06.016.

Kulaksiz, S. and M. Bau. 2011a. Anthropogenic gadolinium as a microcontaminant in tap water used as drinking water in urban areas and megacities. Appl. Geochemistry 26: 1877–1885. Doi: 10.1016/j.apgeochem.2011.06.011.

Kulaksiz, S. and M. Bau. 2011b. Rare earth elements in the Rhine River, Germany: First case of anthropogenic lanthanum as a dissolved microcontaminant in the hydrosphere. Environ. Int. 37: 973–979. Doi: 10.1016/j.envint.2011.02.018.

Kulaksız, S. and M. Bau. 2013. Anthropogenic dissolved and colloid/nanoparticle-bound samarium, lanthanum and gadolinium in the Rhine River and the impending destruction of the natural rare earth element distribution in rivers. Earth Planet. Sci. Lett. 362: 43–50. Doi: 10.1016/j.epsl.2012.11.033.

Kumari, A., R. Panda, M.K. Jha, J.R. Kumar and J.Y. Lee. 2015. Process development to recover rare earth metals from monazite mineral: A review. Miner. Eng. 79: 102–115. Doi: 10.1016/J.MINENG.2015.05.003.

Labarraque, G., C. Oster, P. Fisicaro, C. Meyer, J. Vogl, J. Noordmann et al. 2015. Reference measurement procedures for the quantification of platinum-group elements (PGEs) from automotive exhaust emissions. Int. J. Environ. Anal. Chem. 95: 777–789. Doi: 10.1080/03067319.2015.1058931.

Laschka, D. and M. Nachtwey. 1997. Platinum in municipal sewage treatment plants. Chemosphere 34: 1803–1812. Doi: http://dx.doi.org/10.1016/S0045-6535(97)00036-2.

Lau, W.Y., E.C.H. Lai and T.W.T. Leung. 2011. Current role of selective internal irradiation with Yttrium-90 microspheres in the management of hepatocellular carcinoma: A systematic review. Int. J. Radiat. Oncol. 81: 460–467. Doi: https://doi.org/10.1016/j.ijrobp.2010.06.010.

Lawrence, M.G. 2010. Detection of anthropogenic gadolinium in the Brisbane River plume in Moreton Bay, Queensland, Australia. Mar. Pollut. Bull. 60: 1113–1116. Doi: 10.1016/j.marpolbul.2010.03.027.

Lecomte, K.L., A.M. Sarmiento, J. Borrego and J.M. Nieto. 2017. Rare earth elements mobility processes in an AMD-affected estuary: Huelva Estuary (SW Spain). Mar. Pollut. Bull. 121. Doi: 10.1016/j.marpolbul.2017.06.030.

Leybourne, M.I., W.D. Goodfellow, D.R. Boyle and G.M. Hall. 2000. Rapid development of negative Ce anomalies in surface waters and contrasting REE patterns in groundwaters associated with Zn-Pb massive sulphide deposits. Appl. Geochemistry 15: 695–723. Doi: 10.1016/S0883-2927(99)00096-7.

Leybourne, M.I. and K.H. Johannesson. 2008. Rare earth elements (REE) and yttrium in stream waters, stream sediments, and Fe-Mn oxyhydroxides: Fractionation, speciation, and controls over REE + Y patterns in the surface environment. Geochim. Cosmochim. Acta 72: 5962–5983. Doi: 10.1016/j.gca.2008.09.022.

Liang, T., S. Zhang, L. Wang, H.T. Kung, Y. Wang, A. Hu et al. 2005. Environmental biogeochemical behaviors of rare earth elements in soil-plant systems. Environ. Geochem. Health 27: 301–311. Doi: 10.1007/s10653-004-5734-9.

Locatelli, C. 2007. Voltammetric analysis of trace levels of platinum group metals—principles and applications. Electroanalysis 19: 2167–2175. Doi: 10.1002/elan.200704026.

Long, K.R., B.S. Van Gosen, N.K. Foley and D. Cordier. 2010. The Principal Rare Earth Elements Deposits of the United States—A Summary of Domestic Deposits and a Global Perspective.

López-González, N., J. Borrego, B. Carro, J.A. Grande, M.L. de la Torre and T. Valente. 2012. Rare-earth-element fractionation patterns in estuarine sediments as a consequence of acid mine drainage: A case study in SW Spain. Bol. Geol. y Min. 123: 55–64.

Lorand, J.P., A. Luguet and O. Alard. 2008. Platinum-group elements: A new set of key tracers for the Earth's interior. Elements 4: 47–252. Doi: 10.2113/GSELEMENTS.4.4.247.

Lynas Corporation Ltd. 2017. 2017 Annual Report.

Lyubomirova, V., R. Djingova and J.T. Van Elteren. 2011. Fractionation of traffic-emitted Ce, la and Zr in road dusts. J. Environ. Monit. 13: 1823–1830. Doi: 10.1039/c1em10187k.

MacMillan, G.A., J. Chételat, J.P. Heath, R. Mickpegak and M. Amyot. 2017. Rare earth elements in freshwater, marine, and terrestrial ecosystems in the eastern Canadian Arctic. Environ. Sci. Process. Impacts 19: 1336–1345. Doi: 10.1039/C7EM00082K.

Macpherson, T., W.G. Nickling, J.A. Gillies and V. Etyemezian. 2008. Dust emissions from undisturbed and disturbed supply-limited desert surfaces. J. Geophys. Res. Earth Surf. 113. Doi: 10.1029/2007JF000800.

Martin, J.-M., O. Høgdahl and J.C. Philippot. 1976. Rare earth element supply to the Ocean. J. Geophys. Res. 81: 3119–3124. Doi: 10.1029/JC081i018p03119.

Mashio, A.S., H. Obata, H. Tazoe, M. Tsutsumi, A. Ferrer i Santos and T. Gamo. 2016. Dissolved platinum in rainwater, river water and seawater around Tokyo Bay and Otsuchi Bay in Japan. Estuar. Coast. Shelf Sci. 180: 160–167. Doi: https://doi.org/10.1016/j.ecss.2016.07.002.

Mashio, A.S., H. Obata and T. Gamo. 2017. Dissolved platinum concentrations in coastal seawater: Boso to Sanriku Areas, Japan. Arch. Environ. Contam. Toxicol. 73: 240–246. Doi: 10.1007/s00244-017-0373-1.

Massari, S. and M. Ruberti. 2013. Rare earth elements as critical raw materials: Focus on international markets and future strategies. Resour. Policy 38: 36–43. Doi: https://doi.org/10.1016/j.resourpol.2012.07.001.

Matthey, J. 2016. Platinum Group Metals Market Report—November.

Mayfield, D.B. and A. Fairbrother. 2015. Examination of rare earth element concentration patterns in freshwater fish tissues. Chemosphere 120: 68–74. Doi: 10.1016/j.chemosphere.2014.06.010.

McLennan, S.M. 2001. Relationships between the trace element composition of sedimentary rocks and upper continental crust. Geochemistry Geophys. Geosystems. Doi: 10.1029/2000GC000109.

Memon, K., R.J. Lewandowski, A. Riaz and R. Salem. 2013. Yttrium 90 microspheres for the treatment of hepatocellular carcinoma BT—Multidisciplinary treatment of hepatocellular carcinoma. pp. 207–224. *In*: Vauthey, J.-N. and A. Brouquet (eds.). Springer Berlin Heidelberg, Berlin, Heidelberg. Doi: 10.1007/978-3-642-16037-0_14.

Mestre, N.C., R. Calado and A.M.V.M. Soares. 2014. Exploitation of deep-sea resources: The urgent need to understand the role of high pressure in the toxicity of chemical pollutants to deep-sea organisms. Environ. Pollut. 185: 369–371. Doi: https://doi.org/10.1016/j.envpol.2013.10.021.

Michard, A., F. Albarède, G. Michard, J.F. Minster and J.L. Charlou. 1983. Rare-earth elements and uranium in high-temperature solutions from East Pacific Rise hydrothermal vent field (13 °N). Nature 303: 795.

Mihaljevič, M., I. Galušková, L. Strnad and V. Majer. 2013. Distribution of platinum group elements in urban soils, comparison of historically different large cities Prague and Ostrava, Czech Republic. J. Geochemical Explor. 124: 212–217. Doi: https://doi.org/10.1016/j.gexplo.2012.10.008.

Mitra, A., H. Elderfield and M.J. Greaves. 1994. Rare earth elements in submarine hydrothermal fluids and plumes from the Mid-Atlantic Ridge. Mar. Chem. 46: 217–235. Doi: https://doi.org/10.1016/0304-4203(94)90079-5.

Moermond, C.T.A., J. Tijink, A.P. van Wezel and A.A. Koelmans. 2001. Distribution, speciation, and bioavailability of lanthanides in the Rhine-Meuse estuary, The Netherlands. Environ. Toxicol. Chem. 20: 1916–1926. Doi: 10.1002/etc.5620200909.

Moldovan, M., S. Rauch, M. Gómez, M. Antonia Palacios and G.M. Morrison. 2001. Bioaccumulation of palladium, platinum and rhodium from urban particulates and sediments by the freshwater isopod Asellus aquaticus. Water Res. 35: 4175–4183. Doi: https://doi.org/10.1016/S0043-1354(01)00136-1.

Möller, P., P. Dulski, M. Bau, A. Knappe, A. Pekdeger and C. Sommer-Von Jarmersted. 2000. Anthropogenic gadolinium as a conservative tracer in hydrology. Journal of Geochemical Exploration, 409–414. Doi: 10.1016/S0375-6742(00)00083-2.

Möller, P., T. Paces, P. Dulski and G. Morteani. 2002. Anthropogenic Gd in surface water, drainage system, and the water supply of the City of Prague, Czech Republic. Environ. Sci. Technol. 36: 2387–2394. Doi: 10.1021/es010235q.

Möller, P., G. Morteani and P. Dulski. 2003. Anomalous Gadolinium, Cerium, and Yttrium contents in the Adige and Isarco River waters and in the water of their tributaries (Provinces Trento and Bolzano/Bozen, NE Italy). Acta Hydrochim. Hydrobiol. 31: 225–239. Doi: 10.1002/aheh.200300492.

Monteiro, C.E., A. Cobelo-Garcia, M. Caetano and M.M. Correia dos Santos. 2017. Improved voltammetric method for simultaneous determination of Pt and Rh using second derivative signal transformation—application to environmental samples. Talanta 175: 1–8. Doi: https://doi.org/10.1016/j.talanta.2017.06.067.

Muiños, S.B., J.R. Hein, M. Frank, J.H. Monteiro, L. Gaspar, T. Conrad et al. 2013. Deep-sea Fe-Mn Crusts from the Northeast Atlantic Ocean: Composition and resource considerations. Mar. Georesources Geotechnol. 31: 40–70. Doi: 10.1080/1064119X.2012.661215.

Mulholland, R. and A. Turner. 2011. Accumulation of platinum group elements by the marine gastropod Littorina littorea. Environ. Pollut. 159: 977–982. Doi: https://doi.org/10.1016/j.envpol.2010.12.009.

Nassar, N.T., X. Du and T.E. Graedel. 2015. Criticality of the rare earth elements. J. Ind. Ecol. 19. Doi: 10.1111/jiec.12237.

Ndagi, U., N. Mhlongo and M.E. Soliman. 2017. Metal complexes in cancer therapy—an update from drug design perspective. Drug Des. Devel. Ther. 11: 599–616. Doi: 10.2147/DDDT.S119488.

Neira, P., A. Cobelo-García, V. Besada, J. Santos-Echeandía and J. Bellas. 2015. Evidence of increased anthropogenic emissions of platinum: Time-series analysis of mussels (1991–2011) of an urban beach. Sci. Total Environ. 514: 366–370. Doi: https://doi.org/10.1016/j.scitotenv.2015.02.016.

Nieto, J.M., A.M. Sarmiento, M. Olías, C.R. Cánovas and C. Ayora. 2007. Acid mine drainage pollution in the Tinto and Odiel rivers, SW Spain. *In*: Water-Rock Interaction—Proceedings of the 12th International Symposium on Water-Rock Interaction, WRI-12, pp. 1251–1254.

Nozaki, Y., D. Lerche, D.S. Alibo and M. Tsutsumi. 2000. Dissolved indium and rare earth elements in three Japanese rivers and Tokyo Bay: Evidence for anthropogenic Gd and In. Geochim. Cosmochim. Acta 64: 3975–3982. Doi: 10.1016/S0016-7037(00)00472-5.

Obata, H., T. Yoshida and H. Ogawa. 2006. Determination of picomolar levels of platinum in estuarine waters: A comparison of cathodic stripping voltammetry and isotope dilution-inductively coupled plasma mass spectrometry. Anal. Chim. Acta 580: 32–38. Doi: https://doi.org/10.1016/j.aca.2006.07.044.

Och, L.M., B. Müller, A. Wichser, A. Ulrich, E.G. Vologina and M. Sturm. 2014. Rare earth elements in the sediments of Lake Baikal. Chem. Geol. 376: 61–75. Doi: 10.1016/j.chemgeo.2014.03.018.

Ohta, A. and I. Kawabe. 2001. REE(III) adsorption onto Mn dioxide (δ-MnO2) and Fe oxyhydroxide: Ce(III) oxidation by δ-MnO2. Geochim. Cosmochim. Acta 65(5): 695–703. https://doi.org/10.1016/S0016-7037(00)00578-0.

Olmez, I., E.R. Sholkovitz, D. Hermann and R.P. Eganhouse. 1991. Rare earth elements in sediments off Southern California: a new anthropogenic indicator. Environ. Sci. Technol. 25: 310–316. Doi: 10.1021/es00014a015.

Oral, R., P. Bustamante, M. Warnau, A. D'Ambra, M. Guida and G. Pagano. 2010. Cytogenetic and developmental toxicity of cerium and lanthanum to sea urchin embryos. Chemosphere 81: 194–8. Doi: 10.1016/j.chemosphere.2010.06.057.

Pacyna, J.M. and E.G. Pacyna. 2001. An assessment of global and regional emissions of trace metals to the atmosphere from anthropogenic sources worldwide. Environ. Rev. 9: 269–298. Doi: 10.1139/a01-012.

Panichev, A.M. 2015. Rare earth elements: Review of medical and biological properties and their abundance in the rock materials and mineralized spring waters in the context of animal and human geophagia reasons evaluation. Achiev. Life Sci. 9: 95–103. Doi: https://doi.org/10.1016/j.als.2015.12.001.

Paola, I.M., C. Paciolla, L. d?Aquino, M. Morgana and F. Tommasi. 2007. Effect of rare earth elements on growth and antioxidant metabolism in *Lemna minor* L. Caryologia 60: 125–128. Doi: 10.1080/00087114.2007.10589559.

Pašava, J., A. Vymazalová, S. Petersen and P. Herzig. 2004. PGE distribution in massive sulfides from the PACMANUSHydrothermal field, eastern Manus basin, Papua New Guinea: implications for PGE enrichment in some ancient volcanogenic massive sulfide deposits. Miner. Depos. 39: 784–792. Doi: 10.1007/s00126-004-0442-z.

Pašava, J., A. Vymazalová and S. Petersen. 2007. PGE fractionation in seafloor hydrothermal systems: examples from mafic- and ultramafic-hosted hydrothermal fields at the slow-spreading Mid-Atlantic Ridge. Miner. Depos. 42: 423–431. Doi: 10.1007/s00126-006-0122-2.

Pattan, J.N. and G. Parthiban. 2011. Geochemistry of ferromanganese nodule-sediment pairs from Central Indian Ocean Basin. J. Asian Earth Sci. 40(2): 569–580. https://doi.org/10.1016/j.jseaes.2010.10.010.

Pérez-López, R., J. Delgado, J.M. Nieto and B. Márquez-García. 2010. Rare earth element geochemistry of sulphide weathering in the São Domingos mine area (Iberian Pyrite Belt): A proxy for fluid–rock interaction and ancient mining pollution. Chem. Geol. 276: 29–40. Doi: 10.1016/j.chemgeo.2010.05.018.

Pérez-Ràfols, C., P. Trechera, N. Serrano, J.M. Díaz-Cruz, C. Ariño and M. Esteban. 2017. Determination of Pd(II) using an antimony film coated on a screen-printed electrode by adsorptive stripping voltammetry. Talanta 167: 1–7. Doi: http://dx.doi.org/10.1016/j.talanta.2017.01.084.

Peucker-Ehrenbrink, B. and B. Jahn. 2001. Rhenium-osmium isotope systematics and platinum group element concentrations: Loess and the upper continental crust. Geochemistry, Geophys. Geosystems 2: n/a-n/a. Doi: 10.1029/2001GC000172.

Piper, D.Z. 1974a. Rare earth elements in the sedimentary cycle: A summary. Chem. Geol. 14: 285–304. Doi: 10.1016/0009-2541(74)90066-7.

Piper, D.Z. 1974b. Rare earth elements in ferromanganese nodules and other marine phases. Geochim. Cosmochim. Acta 38: 1007–1022. Doi: 10.1016/0016-7037(74)90002-7.

Piper, D.Z. and M. Bau. 2013. Normalized rare earth elements in water, sediments, and wine: Identifying sources and environmental redox conditions. Am. J. Anal. Chem. 2013: 69–83. Doi: 10.4236/ajac.2013.410A1009.

Pourret, O., M. Davranche, G. Gruau and A. Dia. 2007. Rare earth elements complexation with humic acid. Chem. Geol. 243: 128–141. Doi: 10.1016/j.chemgeo.2007.05.018.

Prudêncio, M.I., T. Valente, R. Marques, M.A.S. Braga and J. Pamplona. 2017. Rare earth elements, iron and manganese in ochre-precipitates and wetland soils of a passive treatment system for acid mine drainage. Procedia Earth Planet. Sci. 17: 932–935. Doi: 10.1016/j.proeps.2017.01.024.

Qiang, T., W. Xiao-rong, T. Li-qing and D. Le-mei. 1994. Bioaccumulation of the rare earth elements lanthanum, gadolinium and yttrium in carp (Cyprinus carpio). Environ. Pollut. 85: 345–350. Doi: 10.1016/0269-7491(94)90057-4.

Quinn, K.A., R.H. Byrne and J. Schijf. 2004. Comparative scavenging of Yttrium and the rare earth elements in seawater: Competitive influences of solution and surface chemistry. Aquat. Geochemistry 10: 59–80. Doi: 10.1023/B:AQUA.0000038959.03886.60.

Rabiet, M., F. Brissaud, J.L. Seidel, S. Pistre and F. Elbaz-Poulichet. 2009. Positive gadolinium anomalies in wastewater treatment plant effluents and aquatic environment in the Hérault watershed (South France). Chemosphere 75(8): 1057–1064. https://doi.org/10.1016/j.chemosphere.2009.01.036.

Ramesh, R., A.L. Ramanathan, R.A. James, V. Subramanian, S.B. Jacobsen and H.D. Holland. 1999. Rare earth elements and heavy metal distribution in estuarine sediments of east coast of India. Hydrobiologia 397: 89–99. Doi: 10.1023/A:1003646631589.

Rauch, S., H.F. Hemond, C. Barbante, M. Owari, G.M. Morrison, B. Peucker-Ehrenbrink et al. 2005. Importance of automobile exhaust catalyst emissions for the deposition of platinum, palladium, and rhodium in the northern hemisphere. Environ. Sci. Technol. 39: 8156–8162.

Rauch, S., B. Peucker-Ehrenbrink, L.T. Molina, M.J. Molina, R. Ramos and H.F. Hemond. 2006. Platinum group elements in airborne particles in Mexico City. Environ. Sci. Technol. 40: 7554–7560. Doi: 10.1021/es061470h.

Rauch, S. and B. Peucker-Ehrenbrink. 2015. Sources of platinum group elements in the environment. *In*: Platinum Metals in the Environment. Springer, pp. 3–17.

Ravindra, K., L. Bencs and R. Van Grieken. 2004. Platinum group elements in the environment and their health risk. Sci. Total Environ. 318: 1–43. Doi: 10.1016/S0048-9697(03)00372-3.

Rinkovec, J., G. Pehnec, R. Godec, S. Davila and I. Bešlić. 2018. Spatial and temporal distribution of platinum, palladium and rhodium in Zagreb air. Sci. Total Environ. 636: 456–463. Doi: https://doi.org/10.1016/j.scitotenv.2018.04.295.

Ruchter, N. and B. Sures. 2015. Distribution of platinum and other traffic related metals in sediments and clams (Corbicula sp.). Water Res. 70: 313–324. Doi: http://dx.doi.org/10.1016/j.watres.2014.12.011.

Ruhlin, D.E. and R.M. Owen. 1986. The rare earth element geochemistry of hydrothermal sediments from the East Pacific Rise: Examination of a seawater scavenging mechanism. Geochim. Cosmochim. Acta 50: 393–400. Doi: 10.1016/0016-7037(86)90192-4.

Salem, R. and R.D. Hunter. 2006. Yttrium-90 microspheres for the treatment of hepatocellular carcinoma: A review. Int. J. Radiat. Oncol. 66: S83–S88. Doi: https://doi.org/10.1016/j.ijrobp.2006.02.061.

Schäfer, J. and H. Puchelt. 1998. Platinum-Group-Metals (PGM) emitted from automobile catalytic converters and their distribution in roadside soils. J. Geochemical Explor. 64: 307–314. Doi: http://dx.doi.org/10.1016/S0375-6742(98)00040-5.

Schindl, R. and K. Leopold. 2015. Analysis of platinum group elements in environmental samples: A review. pp. 109–128. *In*: Zereini, F. and C.L.S. Wiseman (eds.). Platinum Metals in the Environment. Springer Berlin Heidelberg, Berlin, Heidelberg. Doi: 10.1007/978-3-662-44559-4_8.

Schmidt, K., D. Garbe-Schönberg, M. Bau and A. Koschinsky. 2010. Rare earth element distribution in > 400°C hot hydrothermal fluids from 5° S, MAR: The role of anhydrite in controlling highly variable distribution patterns. Geochim. Cosmochim. Acta 74: 4058–4077. Doi: 10.1016/j.gca.2010.04.007.

Schmitt, R.A., R.H. Smith, J.E. Lasch, A.W. Mosen, D.A. Olehy and J. Vasilevskis. 1963. Abundances of the fourteen rare-earth elements, scandium, and yttrium in meteoritic and terrestrial matter. Geochim. Cosmochim. Acta 27: 577–622. Doi: 10.1016/0016-7037(63)90014-0.

Semple, K.T., K.J. Doick, K.C. Jones, P. Burauel, A. Craven and H. Harms. 2004. Peer reviewed: Defining bioavailability and bioaccessibility of contaminated soil and sediment is complicated. Environ. Sci. Technol. 38: 228A–231A. Doi: 10.1021/es040548w.

Serrano, M.J.G., L.F.A. Sanz and D.K. Nordstrom. 2000. REE speciation in low-temperature acidic waters and the competitive effects of aluminum. Chem. Geol. 165: 167–180. Doi: 10.1016/S0009-2541(99)00166-7.

Sharifi, R., F. Moore and B. Keshavarzi. 2013. Geochemical behavior and speciation modeling of rare earth elements in acid drainages at Sarcheshmeh porphyry copper deposit, Kerman Province, Iran. Chemie der Erde 73: 509–517. Doi: 10.1016/j.chemer.2013.03.001.

Sholkovitz, E.R. and H. Elderfield. 1988. Cycling of dissolved rare earth elements in Chesapeake Bay. Global Biogeochem. Cycles. Doi: 10.1029/GB002i002p00157.

Sholkovitz, E.R. 1992. Chemical evolution of rare earth elements: fractionation between colloidal and solution phases of filtered river water. Earth Planet. Sci. Lett. 114: 77–84. Doi: 10.1016/0012-821X(92)90152-L.

Sholkovitz, E.R., T.J. Shaw and D.L. Schneider. 1992. The geochemistry of rare earth elements in the seasonally anoxic water column and porewaters of Chesapeake Bay. Geochim. Cosmochim. Acta 56: 3389–3402. Doi: 10.1016/0016-7037(92)90386-W.

Sholkovitz, E.R. 1993. The geochemistry of rare earth elements in the Amazon River estuary. Geochim. Cosmochim. Acta 57: 2181–2190. Doi: 10.1016/0016-7037(93)90559-F.

Sholkovitz, E.R., T.M. Church and R. Arimoto. 1993. Rare Earth element composition of precipitation, precipitation particles, and aerosols. J. Geophys. Res. 98: 20587. Doi: 10.1029/93JD01926.

Sholkovitz, E.R., W.M. Landing and B.L. Lewis. 1994. Ocean particle chemistry: The fractionation of rare earth elements between suspended particles and seawater. Geochim. Cosmochim. Acta. Doi: 10.1016/0016-7037(94)90559-2.

Sholkovitz, E.R. 1995. The aquatic chemistry of rare earth elements in rivers and estuaries. Aquat. Geochemistry 1: 1–34. Doi: 10.1007/BF01025229.

Silwana, B., C. van der Horst, E. Iwuoha and V. Somerset. 2014. Screen-printed carbon electrodes modified with a bismuth film for stripping voltammetric analysis of platinum group metals in environmental samples. Electrochim. Acta 128: 119–127. Doi: http://dx.doi.org/10.1016/j.electacta.2013.11.045.

Sinha, S., P. Meshram and B.D. Pandey. 2016. Metallurgical processes for the recovery and recycling of lanthanum from various resources—A review. Hydrometallurgy 160: 47–59. Doi: 10.1016/J.HYDROMET.2015.12.004.

Slatopolsky, E., H. Liapis and J. Finch. 2005. Progressive accumulation of lanthanum in the liver of normal and uremic rats. Kidney Int. 68: 2809–2813. Doi: 10.1111/j.1523-1755.2005.00753.x.

Smith Stegen, K. 2015. Heavy rare earths, permanent magnets, and renewable energies: An imminent crisis. Energy Policy 79: 1–8. Doi: 10.1016/j.enpol.2014.12.015.

Soyol-Erdene, T.-O., Y. Huh, S. Hong and S. Do Hur. 2011. A 50-year record of platinum, iridium, and rhodium in antarctic snow: Volcanic and anthropogenic sources. Environ. Sci. Technol. 45: 5929–5935. Doi: 10.1021/es2005732.

Soyol-Erdene, T.O., T. Valente, J.A. Grande and M.L. de la Torre. 2018. Mineralogical controls on mobility of rare earth elements in acid mine drainage environments. Chemosphere 205: 317–327. Doi: 10.1016/j.chemosphere.2018.04.095.

Squadrone, S., P. Brizio, M. Battuello, N. Nurra, R.M. Sartor, A. Benedetto et al. 2017. A first report of rare earth elements in northwestern Mediterranean seaweeds. Mar. Pollut. Bull. 122: 236–242. Doi: 10.1016/j.marpolbul.2017.06.048.

Steinmann, M., S. Bodeï and M. Buatier. 2012. Nd–Sr isotope and REY geochemistry of metalliferous sediments in a low-temperature off-axis hydrothermal environment (Costa Rica margin). Mar. Geol. 315-318: 132–142. Doi: https://doi.org/10.1016/j.margeo.2012.04.005.

Stone, R. 2009. As China's Rare Earth R & D becomes ever more rarefied, others tremble. Science (80-.). 325: 1336–1337.

Sun, H., F. Zhao, Y. Tang and S. Wu. 2007. Origin and behavior of Rare Earth elements in acid mine drainage. In: Water-Rock Interaction—Proceedings of the 12th International Symposium on Water-Rock Interaction, WRI-12, pp. 783–787.

Sun, H., F. Zhao, M. Zhang and J. Li. 2012. Behavior of rare earth elements in acid coal mine drainage in Shanxi Province, China. Environ. Earth Sci. 67: 205–213. Doi: 10.1007/s12665-011-1497-7.

Sures, B. and S. Zimmermann. 2007. Impact of humic substances on the aqueous solubility, uptake and bioaccumulation of platinum, palladium and rhodium in exposure studies with Dreissena polymorpha. Environ. Pollut. 146: 444–451. Doi: https://doi.org/10.1016/j.envpol.2006.07.004.

Swift, D.T.K. 2014. The economic benefits of the north american rare earths industry. Rare Earth Technol. Alliance 3.

Tai, P., Q. Zhao, D. Su, P. Li and F. Stagnitti. 2010. Biological toxicity of lanthanide elements on algae. Chemosphere 80: 1031–1035. Doi: 10.1016/j.chemosphere.2010.05.030.

Takai, K., T. Saruhashi, J. Miyazaki, I. Sawada, T. Shibuya, S. Kawaguchi et al. 2016. Method and system for recovering ocean floor hydrothermal mineral resources.

Takaya, Y., K. Yasukawa, T. Kawasaki, K. Fujinaga, J. Ohta, Y. Usui et al. 2018. The tremendous potential of deep-sea mud as a source of rare-earth elements. Scientific Reports. Doi: 10.1038/s41598-018-23948-5.

Tang, J. and K.H. Johannesson. 2003. Speciation of rare earth elements in natural terrestrial waters: assessing the role of dissolved organic matter from the modeling approach. Geochim. Cosmochim. Acta 67: 2321–2339. Doi: 10.1016/S0016-7037(02)01413-8.

Tepe, N., M. Romero and M. Bau. 2014. High-technology metals as emerging contaminants: Strong increase of anthropogenic gadolinium levels in tap water of Berlin, Germany, from 2009 to 2012. Appl. Geochemistry 45: 191–197. Doi: 10.1016/j.apgeochem.2014.04.006.

Terashima, S., H. Katayama and S. Itoh. 1993. Geochemical behavior of Pt and Pd in coastal marine sediments, southeastern margin of the Japan Sea. Appl. Geochemistry 8: 265–271. Doi: https://doi.org/10.1016/0883-2927(93)90041-E.

Torokhov, M. and L. Lazareva. 2003. PGM and PGE in Logatchev-2 ore field, MAR. InterRidge News 12(2): 33.

Tricca, A., P. Stille, M. Steinmann, B. Kiefel, J. Samuel and J. Eikenberg. 1999. Rare earth elements and Sr and Nd isotopic compositions of dissolved and suspended loads from small river systems in the Vosges mountains (France), the river Rhine and groundwater. Chem. Geol. 160: 139–158. Doi: 10.1016/S0009-2541(99)00065-0.

USGS. 2018. Mineral Commodities Summaries 2018. Doi: 10.3133/70140094.

Usui, A. and M. Someya. 1997. Distribution and composition of marine hydrogenetic and hydrothermal manganese deposits in the northwest Pacific. Geol. Soc. London, Spec. Publ. 119: 177 LP–198.

Van De Wiele, C., L. Defreyne, M. Peeters and B. Lambert. 2009. Yttrium-90 labelled resin microspheres for treatment of primary and secondary malignant liver tumors. Q. J. Nucl. Med. Mol. Imaging 53: 317–324.

van den Berg, C.M.G. and G.S. Jacinto. 1988. The determination of platinum in sea water by adsorptive cathodic stripping voltammetry. Anal. Chim. Acta 211: 129–139. Doi: http://dx.doi.org/10.1016/S0003-2670(00)83675-2.

van Gosen, B.S., D.L. Fey, A.K. Shah, P.L. Verplanck and T.M. Hoefen. 2014. Deposit model for heavy-mineral sands in coastal environments. U.S. Geological Survey Scientific Investigations Report 2010–5070–L, 51. https://doi.org/http://dx.doi.org/10.3133/sir20105070L.

Verma, S.P., E. Santoyo and F. Velasco-Tapia. 2002. Statistical evaluation of analytical methods for the determination of rare-earth elements in geological materials and implications for detection limits. Int. Geol. Rev. 44: 287–335. Doi: 10.2747/0020-6814.44.4.287.

Verplanck, P.L., H.E. Taylor, D.K. Nordstrom and L.B. Barber. 2005. Aqueous stability of gadolinium in surface waters receiving sewage treatment plant effluent Boulder Creek, Colorado. Environ. Sci. Technol. 39: 6923–6929. Doi: 10.1021/es048456u.

Wang, C., W. Zhu and R. Guicherit. 2000. Rare Earth elements and other metals in atmospheric particulate matter in the western part of theNetherlands. Water. Air. Soil Pollut. 121: 109–118.

Wang, C., W. Zhu, A. Peng and R. Guichreit. 2001. Comparative studies on the concentration of rare earth elements and heavy metals in the atmospheric particulate matter in Beijing, China, and in Delft, the Netherlands. Environ. Int. 26: 309–313. Doi: 10.1016/S0160-4120(01)00005-8.

Wang, L. and T. Liang. 2014. Accumulation and fractionation of rare earth elements in atmospheric particulates around a mine tailing in Baotou, China. Atmos. Environ. 88. Doi: 10.1016/j.atmosenv.2014.01.068.

Wang, X.Y., Z.G. Zeng, S. Chen, X.B. Yin and C.T.A. Chen. 2013. Rare earth elements in hydrothermal fluids from Kueishantao, off northeastern Taiwan: Indicators of shallow-water, sub-seafloor hydrothermal processes. Chinese Sci. Bull. 58: 4012–4020. Doi: 10.1007/s11434-013-5849-4.

Wei, C. and G.M. Morrison. 1994. Platinum in road dusts and urban river sediments. Sci. Total Environ. 146-147: 169–174. Doi: https://doi.org/10.1016/0048-9697(94)90234-8.

Weltje, L., A.H. Brouwer, T.G. Verburg, H.T. Wolterbeek and J.J.M. de Goeij. 2002. Accumulation and elimination of lanthanum by duckweed (*Lemna minor* L.) As influenced by organism growth and lanthanum sorption to glass. Environ. Toxicol. Chem. 21: 1483–1489. Doi: 10.1002/etc.5620210721.

Whiteley, J.D. and F. Murray. 2003. Anthropogenic platinum group element (Pt, Pd and Rh) concentrations in road dusts and roadside soils from Perth, Western Australia. Sci. Tot Environ. 317(1-3): 121–135.

Windom, H.L. 1969. Atmospheric dust records in permanent snowfields: Implications to marine sedimentation. GSA Bull. 80: 761–782. Doi: 10.1130/0016-7606(1969)80[761:ADRIPS]2.0.CO;2.

Wiseman, C.L.S., F. Zereini and W. Püttmann. 2013. Traffic-related trace element fate and uptake by plants cultivated in roadside soils in Toronto, Canada. Sci. Total Environ. 442: 86–95. Doi: 10.1016/j.scitotenv.2012.10.051.

Wiseman, C.L.S., Z. Hassan Pour and F. Zereini. 2016. Platinum group element and cerium concentrations in roadside environments in Toronto, Canada. Chemosphere 145: 61–67. Doi: https://doi.org/10.1016/j.chemosphere.2015.11.056.

Yang, X., D. Yin, H. Sun, X. Wang, L. Dai, Y. Chen et al. 1999. Distribution and bioavailability of rare earth elements in aquatic microcosm. Chemosphere 39: 2443–2450. Doi: 10.1016/S0045-6535(99)00172-1.

Zereini, F., C. Wiseman, F. Alt, J. Messerschmidt, J. Müller and H. Urban. 2001. Platinum and rhodium concentrations in airborne particulate matter in Germany from 1988 to 1998. Environ. Sci. Technol. 35(10): 1996–2000.

Zereini, F., F. Alt, J. Messerschmidt, A. von Bohlen, K. Liebl and W. Püttmann. 2004. Concentration and distribution of platinum group elements (Pt, Pd, Rh) in airborne particulate matter in frankfurt am main, Germany. Environ. Sci. Technol. 38: 1686–1692. Doi: 10.1021/es030127z.

Zereini, F., C. Wiseman and W. Püttmann. 2007. Changes in palladium, platinum, and rhodium concentrations, and their spatial distribution in soils along a major highway in Germany from 1994 to 2004. Environ. Sci. Technol. 41: 451–456. Doi: 10.1021/es061453s.

Zereini, F., H. Alsenz, C.L.S. Wiseman, W. Püttmann, E. Reimer, R. Schleyer et al. 2012. Platinum group elements (Pt, Pd, Rh) in airborne particulate matter in rural vs. urban areas of Germany: Concentrations and spatial patterns of distribution. Sci. Total Environ. 416: 261–268. Doi: https://doi.org/10.1016/j.scitotenv.2011.11.070.

Zereini, F. and C.L.S. Wiseman. 2015. Platinum Metals in the Environment, Environmental Science and Engineering. Springer Berlin Heidelberg, Berlin, Heidelberg. Doi: 10.1007/978-3-662-44559-4.

Zhang, H., X. He, W. Bai, X. Guo, Z. Zhang, Z. Chai et al. 2010. Ecotoxicological assessment of lanthanum with Caenorhabditis elegans in liquid medium. Metallomics, 806–810. Doi: 10.1039/c0mt00059k.

Zhong, L., W. Yan, J. Li, X. Tu, B. Liu and Z. Xia. 2012. Pt and Pd in sediments from the Pearl River Estuary, South China: background levels, distribution, and source. Environ. Sci. Pollut. Res. 19: 1305–1314. Doi: 10.1007/s11356-011-0653-7.

Zhu, Y., M. Hoshino, H. Yamada, A. Itoh and H. Haraguchi. 2004. Gadolinium anomaly in the distributions of rare earth elements observed for coastal seawater and river waters around Nagoya city. Bull. Chem. Soc. Jpn. 77: 1835–1842. Doi: 10.1246/bcsj.77.1835.

Zimmermann, S., U. Baumann, H. Taraschewski and B. Sures. 2004. Accumulation and distribution of platinum and rhodium in the European eel Anguilla anguilla following aqueous exposure to metal salts. Environ. Pollut. 127: 195–202. Doi: https://doi.org/10.1016/j.envpol.2003.08.006.

Section III
Organic Pollutants

9

Organophosphorus Compounds

*Valmir B. Silva, Renata Hellinger and Elisa S. Orth**

1. General Overview of Organophosphorus Chemistry

Organophosphorus (OP) compounds, also known as organophosphates, are a class of organic compounds that contain at least one phosphorus atom in their structure. Since the 1940s, these compounds have been developed and used as agrochemicals, chemical weapons and flame retardants. The development of organophosphorus chemistry was a search for more toxic and persistent agents than organochlorines. The OPs used as agrochemicals present several base structures, such as phosphates, phosphorotioates, phosphoroamidates, and phosphonates. These structures are derivatives of phosphoric, phosphonic, phosphinic or phosphoroamidic acids. All structures are based in the phosphorus (V), the most oxidized state, where they present tetrahedron geometry. Figure 1 shows some of these structures used nowadays as agrochemicals with the phosphorus moiety highlighted.

These agrochemicals are used to control pests, fungus or weeds in agriculture. In mammals, birds, fish, reptiles and insects, the major toxicity presented by these compounds is the inhibition of the acetyl cholinesterase enzyme (AChE) (Fukuto 1990). The AChE is responsible for hydrolyzing acetyl choline into acetic acid and choline, a crucial part of synaptic transmission in the central nervous system. The OPs that present toxicity are related to AChE inhibition and have some similarities. In addition to a central phosphorus atom, in general, the OP presents two aliphatic groups with high pK_a values (> 10) like methanol or ethanol and the third substituent is a "good" leaving group, i.e., a group with pK_a below 10. This characteristic is represented by 1,2,4 trichloropiridin-5-ol (pK_a 4.77) (Vico et al. 2009), 2-chloro-4-bromophenol (pK_a 7.92) (Jover et al. 2007), and mercapto-acetic acid methylamide (pK_a 8.05) (Barany and Merrifield 1980) in the pesticides chlorpyriphos, profenofos and dimethoate respectively. The inhibition occurs when the OP encounters the enzymatic active site and the serine alcoholic moiety attacks the central phosphorus (with auxiliary histidine moiety abstracting a serine hydrogen) displacing a leaving group (X^-), phosphorylating the alcoholic group, and in consequence deactivating the enzyme. This mechanism is shown in Fig. 2 (Delfino et al. 2009, da Silva et al. 2013). The inhibited enzyme cannot metabolize acetyl choline; hence, its concentration increases in the central nervous system, leading to collapse and death.

Department of Chemistry, Universidade Federal do Paraná; CP 19081, CEP 81531-990; Curitiba-PR (Brasil).
* Corresponding author: elisaorth@ufpr.br

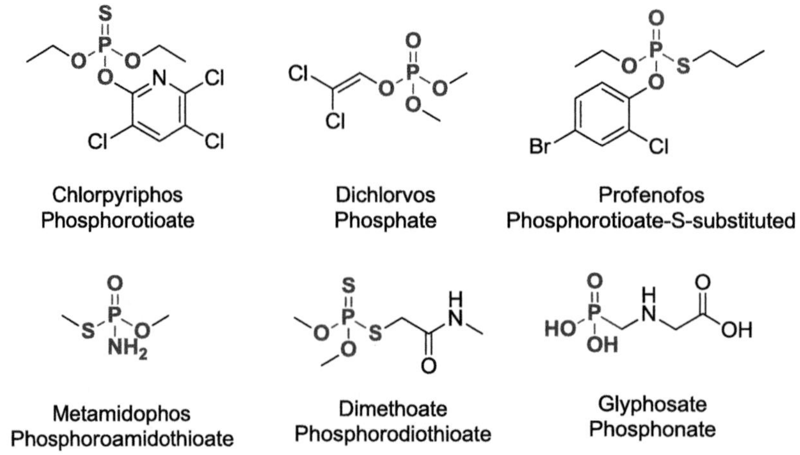

Fig. 1. Some representative agrochemicals.

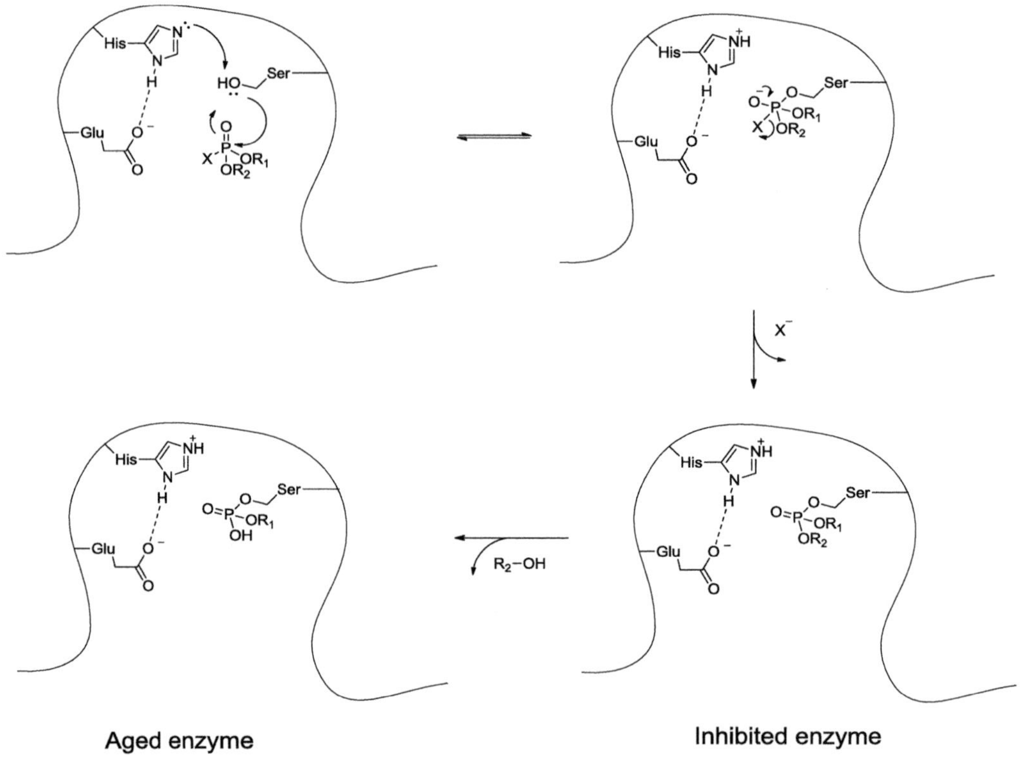

Fig. 2. Mechanism of enzymatic inhibition by organophosphates.

After the enzyme is inhibited, another slow reaction occurs called "aging", in which a second P-O bond is broken with formation of a phosphate diester. This form of OPs is known to be dramatically unreactive and in this sense the recovery of enzymatic activity in aged enzymes remains a challenge. Although the most important process related to OP toxicity is the enzymatic inhibition of AChE, some OPs (e.g., pesticides) can react before encountering the enzyme in desulfurization or isomerization process. These processes generate products that inhibit other esterases and AChE also (Myers et

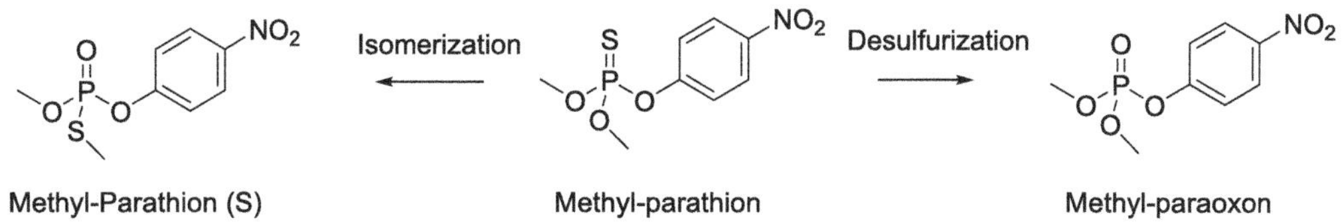

Fig. 3. Methyl-parathion reactions.

al. 1952). Figure 3 exemplifies these reactions in the case of the conventional pesticide methyl-parathion. Isomerization leads to a compound with a P=O replacing P=S and the desulfurization gives a phosphate ester: methyl-paraoxon.

Isomerization reactions are known to occur in polar solvents with high temperatures; however, *in vivo* isomerization is not yet detected, making this intoxication pathway less thoroughly evaluated. *In vitro* analysis showed that the isomerized product is more toxic than the original OP, but the compound inhibits other esterases (Myers et al. 1952). On one hand the phosphorothioates (such as methyl-parathion) are weak AChE inhibitors, but on the other hand the oxon analogs are very toxic and are activated in living organisms such as mammals and fish (Sultatos and Murphy 1983, Sultatos 1994, Ma and Chambers 1995, Straus et al. 2000). *In vivo*, the oxidation represents the most important pathway that transforms the phosphorothioate compound into a more toxic one (Myers et al. 1952), so the studies of oxon analogs of pesticides are very common (Myers et al. 1952, Eyer et al. 2009, Tsuda et al. 1997). Furthermore, in organic media, the phosphorothioates desulfurization occurs when the OP is in contact with oxidants such as molecular bromine (Lee et al. 2002), or organic peroxides (Zhu et al. 2017, Jackson et al. 1992, Swinson et al. 1988).

Organophosphates are less persistent than organochlorines and in ocean water these compounds can suffer a variety of reactions. Generally, the hydrolysis and oxidation (desulfuration) reactions are most commonly known. For example, the pesticide methyl-parathion was evaluated with water (salinity 3%) compatible with sea water in acid, basic and neutral pH. The pesticide was relatively stable in neutral and acid conditions, suffering slow dealkylation, and in alkaline media the ion hydroxide reacts with central phosphorus, promoting dearylation. Further, salinity decreases the alkaline hydrolysis rate, favoring methyl-parathion contamination in ocean waters. In the same study, other reactions promoted by microorganisms were evaluated and presented the following reaction products: dealkylated methyl-parathion, reduced methyl-parathion and dealkylation of reduced product, synthesized in Fig. 4 (Badawy and El-Dib 1984).

The liver in vertebrate organisms is the main organ of metabolization and biotransformation of chemical compounds (Williams and Iatropoulos 2002). The elimination of chemical compounds that are lipophilic comes from their metabolic transformation into hydrophilic compounds, allowing their excretion through urine or feces. This process of metabolizing the chemical compounds occurs in three phases. In phases I and II, they are related to the biotransformation of these compounds, in which they are converted into water-soluble metabolites. Phase III corresponds to the excretion of products resulting from biotransformation. During phase I, oxidation, reduction and hydrolysis reactions occur, which modifies or introduces some functional group to the contaminant molecule in order to allow the conjugated reactions of phase II. In phase II, the conjugation of the contaminant with endogenous substrates occurs, which results in a greater solubility of the chemical compound in water. From this, during phase III, it is possible to excrete the contaminant to the outside of the cells through the transmembrane channels (Stegeman and Hahn 1994).

To better understand the process by which aquatic organisms metabolize OP compounds, the metabolization of the agrochemical diazinon is exemplified and its chemical structure is presented in Fig. 5. This compound is used as an OP insecticide to control insects in soils, plants, fruits and vegetables. In addition, it can be applied in the control of some domestic pests such as flies, fleas and cockroaches (Garfitt et al. 2002).

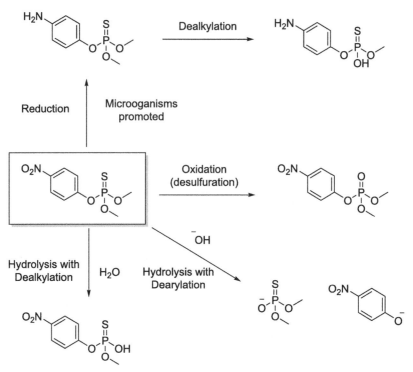

Fig. 4. Possible pathways of methyl-parathion in marine environments.

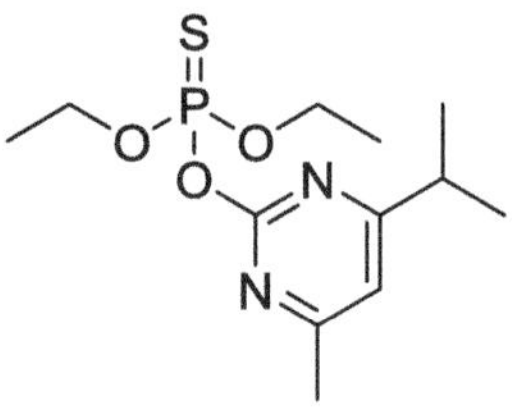

Fig. 5. Chemical structure of diazinon.

Diazinon is considered a toxic chemical for vertebrate animals (Sams et al. 2003). Some studies were carried out to verify its occurrence in the aquatic environment. A study of the tropical coastal sediments of the Klang River Estuary in Africa detected the presence of diazinon in all the sampling stations evaluated (Omar et al. 2018). In addition, its action in aquatic organisms was studied and it was verified that the reproductive system of the fish is damaged, as shown in a decrease in the egg spawning and egg laying rates. Another observed fact was that diazinon may delay the sexual maturation of aquatic invertebrates (Larkin and Tjeerdema 2000).

The diazinon molecule can be metabolized through the metabolic processes of phases I and II to diazoxon. As discussed earlier, this process is performed by cytochrome P-450 and the analog formed, diazoxon, has a much higher toxicity than diazinon, causing serious health effects in aquatic organisms. The main degradation processes include thiol ester hydrolysis. The main mechanisms of degradation of this insecticide are: hydrolysis of the thiol ester by the A-esterases; oxidation and hydroxylation of the isopropyl group carried out by mixed-function monooxidases in phase I; and conjugation of the pyrimidyl ring with the tripeptide glutathione, which occurs in phase II. Phase I products are diazinons, monohydroxyl and diazoxon (Larkin and Tjeerdema 2000). A diagram of the metabolic process described is given in Fig. 6.

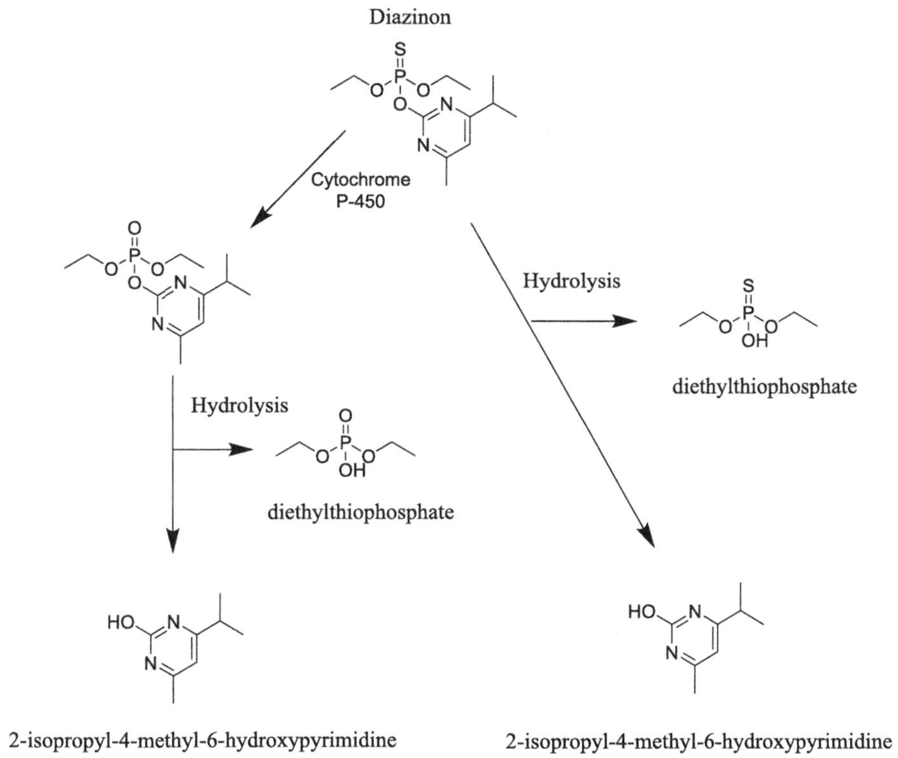

Diazinon

Fig. 6. Metabolic process of diazinon.

Land areas and fresh water in rivers are the areas most affected by pesticide contamination because of proximity of application. Moreover, by several processes these pesticides can pollute distant oceans, lakes and bays. The pesticides diazinon, iprobenfos, fenitrothion and chlorpyrifos were detected in Mizuro Bay (Japan) in water and sediment samples. Furthermore, mussels were a bioaccumulator in which three of the four pesticides were more concentrated in these organisms than in the sediments (Harino et al. 2014). In estuarine areas of Korea (Kunsan, Danghang, Sachon and Kwangyang), high concentrations of low toxic organophosphorus iprobenfos were found (between 42.1 and 528.0 ng g^{-1}). In sediments, the pesticide chlorpyrifos was the most concentrated OP detected (2.4 and 1.7 ng g^{-1}); because of its acute toxicity, it is an indication of accumulation power of these particles in marine environments (Yu et al. 2001). Another study carried out in Korea (Saemangeum Bay) showed high levels of OP iprobenfos (1840 ng L^{-1}) (Li et al. 2005). In Jiulong Estuary River in China, a 26 ton/year flux of OP pesticides was estimated, with a warning of 5.43 ton/year for the pesticide methamidophos, considered by the authors to be the most dangerous OP evaluated in that study (Zulin et al. 2002). In the Mar Menor Lagoon located in the marine coast of Spain, the pesticide concentration analyzed in 2010 was below 20 ng g^{-1}, except for chlorpyrifos (Moreno-González and León 2017). All values obtained in the lagoon are related to the Mediterranean Sea, evidencing the capacity of this water body to retain the contaminants. Another study carried out with water and sediment marine samples in Central America showed high presence of chlorpyrifos and low presence of parathion (between 0 and 1.2 ng g^{-1} for parathion and 0 and 34.2 ng g^{-1} for chlorpyrifos). The presence of parathion is argued by the authors because this pesticide is the most commonly used in the region and shows high persistence in the environment. Chlorpyrifos even in the studies described above showed high absorption in sediments due to low water solubility and high affinity to organic matter (Readman et al. 1992). In this context, some researchers evaluated *in vitro* chlorpyrifos activity with marine living organisms, showing that this compound can be very toxic (Serrano et al. 1997, Kim et al. 2004). In Mexico, where shrimp aquaculture is an important economic activity, chlorpyrifos was detected in Altata-ensanada del Pabellón. The concentration of the OP was 2.4 ng L^{-1}, considered by the authors a high contaminant level, necessitating an understanding of its toxicity in aquatic systems (Páez-Osuna et al. 1998). Chlorpyrifos was encountered in the Alaskan Arctic in concentrations close to 80 ng L^{-1} and in minor concentrations from the East Asian Sea to the high Arctic, where the pesticide can be trapped in the ice (Garbarino et al. 2002, Zhong et al. 2011). These studies evidence that some water bodies and aquatic organisms can accumulate the pesticides. Living organisms and humans are in contact with these pesticides even if the marine waters are far from cultivation areas. Hence, the persistence and toxicity should be evaluated in marine waters.

2. Monitoring

Monitoring techniques are more consolidated for freshwater and wastewater, while in ocean waters this concern is more recent (Kröger et al. 2002). As mentioned above, several data were presented regarding OPs in ocean waters. The levels were monitored by collection of water samples and consolidated analytical techniques such as gas or liquid chromatography coupled to mass spectrometer. Sample preparation usually relies on extraction techniques such as liquid-liquid extraction (Harino et al. 2014, Yu et al. 2001, Li et al. 2005, Serrano et al. 1997), ultrasonic and pressurized liquid extraction (Moreno-González and León 2017), and soxlet extraction (Readman et al. 1992, Zhong et al. 2011, Ma et al. 2017).

In relation to OP pesticides, some biosensors are being developed based on AChE inhibition (Kröger et al. 2002, Sarkar et al. 2006). AChE inhibition is one of the most characteristic symptoms of OP intoxication. In this regard, some authors evaluate the enzymatic activity of marine species to monitor contamination and impact on these species (Sturm et al. 1999). This biomarker is used in freshwater but the studies in marine species are still being consolidated. Magni et al. (2006), in a study with samples from Oristano Gulf, suggests that differences observed in AChE activity in fishes are related with the use of pesticides in well-defined stations of the year. In a study with common oyster, two cholinesterases were found, one sensitive to classical enzymatic inhibitors and another insensitive. In this way, this oyster showed an interesting capacity to be a biomarker, since the activity difference between two enzymes can be determined with standard procedures and the enzymatic activity differences are attributed to OPs in the environment (Bocquené et al. 1997). Using AChE as biomarker, Matozzo et al. (2005) observed differences in enzymatic activity between the clams living in a highly polluted lagoon (Lagoon of Venice) and reference clams. The authors noticed two factors determining seasonal differences in enzymatic activity during the year: a variation of degree of pollutant in the lagoon and animal physiology itself. Fulton and Key analyzed some results about estuarine fish species and invertebrate intoxication by OP pesticides to evaluate the capacity of these animals to be biomarkers of OP intoxication. The authors observed different relations between species determined to be intoxicated with several OP pesticides. They defend the potential of this tool but reiterate that more studies should be developed in this field (Hu et al. 2014).

One way of analyzing the cytotoxicity of aquatic organisms is by using biomarkers of aquatic contamination. Biomarkers are a biochemical system that makes it possible to evaluate the state of degradation and contamination of an environment by analyzing the health of the organisms inserted in this ecosystem. Alteration in health may occur at various levels of biological organization, such as biochemical changes, molecular damage, and changes in the physiological pattern (Timbrell 2008). There are several types of aquatic contamination biomarkers and herein we will focus on two models: oxidative stress biomarkers and morphological biomarkers.

The oxidative stress biomarkers serve to delimit the levels of aquatic contamination through the formation of free radicals (Valavanidis et al. 2006). The cytotoxicity of some pesticides, such as OPs, can be determined from the production and accumulation of H_2O_2 within the cells (Abdollahi et al. 2004). H_2O_2 gives rise to the formation of the hydroxyl radical ($\bullet OH$), which, because of its high reactivity, is capable of oxidizing DNA biomolecules and membranes (Bagchi et al. 1995). There are studies that show the potential of some pesticides, among them organophosphates, to induce oxidative stress using free radicals (Banerjee et al. 1999). In addition, research has been done to assess this damage in aquatic organisms (Yan et al. 2017).

Another system that can be used consists of morphological biomarkers, in which cellular, tissue and anatomical morphological patterns are analyzed when there is exposure of fish or other aquatic organisms to contaminants (Van der Oost et al. 2003) (Murty 1986). There are studies that report the potential of some organophosphates to weaken the health of various aquatic organisms through morphological changes. A survey of the Java medaka fish found that glyphosate, one of the most widely used OP compounds in global agriculture, causes changes in heart rate and morphological deficiencies. Irregular shape of abdomen and cellular rupture in the fin and head were observed (Yusof et al. 2014). These studies show the potential of organophosphates in changing physiological patterns of some aquatic organisms.

3. Cytotoxicity Assessment in Marine Species

Anthropogenic activities have led to an increase in the level of xenobiotic compounds in aquatic environments, causing serious problems in the health of living beings that inhabit them (Cajaraville et al. 2000). Among the various types of contaminants, several types of pesticides stand out, including OP compounds. These compounds in nature present several problems because they are potentially cytotoxic and/or carcinogenic and are produced on a large scale (Livingstone et al. 1993).

The uncontrolled and excessive use of organophosphates in the control of organisms that are undesirable for agriculture causes diverse impacts on the environment and is extremely harmful when these residues are present in soils, water, air, plants and animals (dos Santos and da Silva 2007). The pollution of aquatic environments by organophosphates occurs through incorrect agricultural practices, which use pesticides in an excessive and inadequate way, and these particles arrive in seawater through rain, atmospheric dust, or flow of rivers (Edwards 1973).

In the oceans, organophosphates are mostly found coupled with other particles, such as plasticizers (Wei et al. 2015), since it is extremely difficult to measure their nearly infinitesimal concentration in the oceans (Edwards 1973). The pesticides present in the oceans can penetrate aquatic organisms in different ways depending on their physico-chemical characteristics (Rand and Petrocelli 1985). In most cases, water-soluble compounds can penetrate an organism through the entire surface of the body, gills and mouth. Since the lipophilic compounds are absorbed in suspended particles, organic matter or biological systems, they are absorbed by ingestion through the gastrointestinal tract (Rand and Petrocelli 1985). Therefore, aquatic organisms may accumulate concentrations of pesticides much higher than those found in the water they contain, since there is a portion of pesticide absorbed by the surface of the body and another that is ingested.

In addition, there are several studies that analyze the health consequences of aquatic organisms when they come into direct contact with some OP compounds and what this entails throughout the aquatic ecosystem. A study conducted in Hawaii has verified the consequences of exposure of sea turtles to OP glyphosate. The results obtained were several adverse effects on the digestive system and on the general health of these animals due to the decrease of the mixed bacterial communities of the intestinal tract when exposed to glyphosate (Kittle et al. 2018). In China, organophosphate residues were detected in the tissues of tilapia, eel and turbot species (Sun et al. 2011). Another study conducted in Japan with coastal fish (Kuda et al. 2011) showed that p-nitrophenol, which is toxic to marine animals (Howe et al. 1994) and is an intermediate in the degradation of certain organophosphates (Spain and Gibson 1991), was present in 17 samples from the 19 analyzed.

Organophosphates can be found in various forms in the ocean. One of the most common forms is found in microplastics (Browne et al. 2010, 2011, Endo et al. 2005, Thompson et al. 2004); it is estimated that there are 35,540 tons of microplastic particles floating in the oceans (Eriksen et al.

2014). Microplastics are recognized as a global threat to marine and coastal environments (Arthur et al. 2009) because they have several toxic additives in their manufacture, such as organophosphorus esters (Browne et al. 2013, Faure et al. 2015, Net et al. 2015, Wei et al. 2015) and, in addition, their surface area with high specificity makes it possible for them to absorb various organic pollutants (Fisner et al. 2013) (Wang et al. 2015).

There are several microplastics produced with organophosphorus esters coupled in their structure that are found in aquatic environments, some examples being polyethylene fragments with tris(2-chloroethyl) phosphate, polystyrene foams with tris(1-chloro-2-propyl) phosphate, and polyethylene pellets with tri-butyl phosphate (Zhang et al. 2018). These structures are represented in Fig. 7.

Organophosphorus flame retardants are physically bonded to the materials and not chemically bonded, so they easily leach into the aquatic environment (Tanaka et al. 2013). They readily pass through volatilization, abrasion and dissolution processes (Wei et al. 2015). These contaminants are transported through atmospheric air to the aquatic environment (Möller et al. 2012). Studies showed that this contribution of atmospheric air is quite significant; in the Mediterranean Sea there are 13 to 260 tons of OP flame retardant residues per year coming from the atmosphere (Castro-Jiménez et al. 2014). Because of their small size, they can be ingested by zooplankton, mussels, fish, seabirds and whales (Browne et al. 2008, Cole et al. 2013, Li et al. 2015, Syberg et al. 2015, Tanaka et al. 2013). These OPs are used as additives to modify the surface of certain materials, such as plastics, furniture, textiles, electronics and vehicles, protecting them against fire (Marklund et al. 2003). Organophosphorus flame retardants have been applied in several industries in recent decades (Marklund et al. 2003, Reemtsma et al. 2008), and their global use in 2011 reached 500,000 tons (Yuxiang 2011). Hazard assessments of OP flame retardants are still ongoing (Wei et al. 2015), but there are already studies demonstrating their toxicity and carcinogenicity (Van der Veen and de Boer 2012). Figure 8 shows the main organophosphates used as retardant films.

This fact results in a high contribution to the pollution of marine biota, since these OPs are transferred to aquatic organisms by ingestion and and contaminate the intestinal tissues, causing diverse biological effects (Browne et al. 2013). The toxicity of these compounds is different from that of OP pesticides. Some of them can stimulate cancer, tumors and other illness (Zhong et al. 2017). The OP flame retardants presented in Fig. 8 were evaluated in an expedition from the North Pacific to the Arctic Ocean and those OPs were quantified in range of 159–4658 pg g^{-1} (dry sediment). The author noticed influence in structure of OPs (represented by the constant K_{ow}, a property that expresses the water affinity of the compound) and the range of detection. Compounds such as TiBP and TnBP with high values of K_{ow} (i.e., low water affinity) present difficulties in being transported across the ocean to the Arctic, showing a tendency to be adsorbed; OPs such as TCCP and TCEP with low K_{ow} value (i.e., high water affinity) are transported more easily across the ocean (Ma et al. 2017).

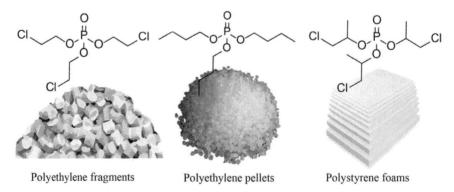

Polyethylene fragments   Polyethylene pellets   Polystyrene foams

Fig. 7. Common OP structures used as plasticizers.

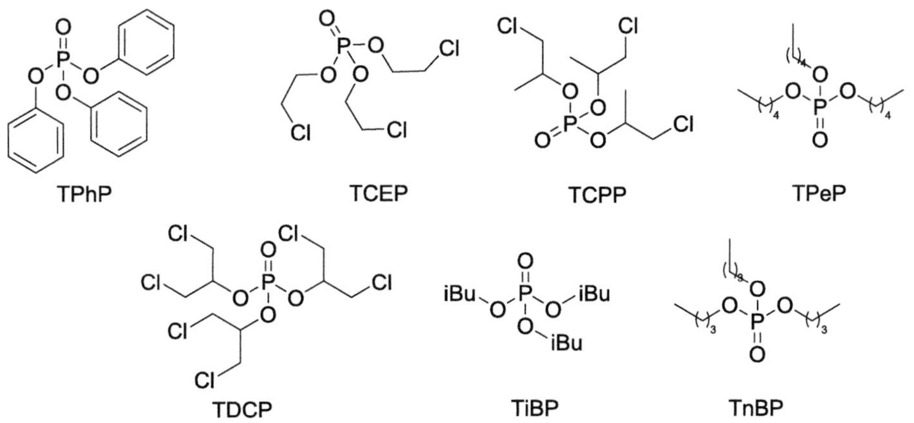

Fig. 8. Common OP structures used as flame retardants.

Nowadays China is one of the largest consumers and manufacturers of flame retardants, and some researchers are developing studies to measure contamination in the China Sea (Zhong et al. 2017). In the Yellow and Bohai Seas (China), the OP concentration varied between 0 and 31 ng L^{-1}, with chlorinated OPs more persistent than alkyl and aryl OPs (Zhong et al. 2017). A study carried out in an expedition from Asia to the Arctic seas evidenced between 230–2900 pg m^{-3} and 120–170 pg m^{-3} (in 2010 and 2011, respectively) in airborne particles of eight different OPs used as flame retardants and plasticizers (Möller et al. 2012). The most persistent are the chlorinated OPs, since those compounds are normally more concentrated than alkyl and aryl OPs. Another study carried out in Chinese seas showed high chlorinated OP levels with mean of 424.6 ng L^{-1} (sum of four different OPs). The seawaters near Lianyungang are the most polluted in this evaluation, with levels higher than 1000 ng L^{-1} (Hu et al. 2014).

In conclusion, the OPs present a minor persistence grade in relation to former pesticides based on organochlorine compounds. The studies discussed here show the scientific community's concern regarding the impact of OPs in the marine environment. Hence, the monitoring and evaluation of toxicity in aquatic systems must continue systematically around the globe (Browne et al. 2013).

4. Food Web Bioaccumulation

Aquatic organisms are divided into trophic levels in food chains and are separated according to the role they play in the system. There are basically three divisions in which aquatic organisms are classified: producers, consumers and decomposers (Van der Oost et al. 2003). The first trophic level is composed of the producers, who are usually the basis of the food chain. The main producer in the aquatic environment is phytoplankton (Ghosal et al. 2002), which does not consume other organisms, its energy source being photosynthesis (Behrenfeld et al. 2001). Consumers are those organisms that are not able to make their own food and therefore need to consume other organisms, and they are divided into three groups (Hogg 2013): primary consumers, which are herbivorous organisms, for example, zooplankton; secondary consumers, carnivores that feed on herbivorous organisms; and tertiary consumers, carnivores that feed on other carnivores. Finally, decomposers break down dead plants and animal waste and release the energy back into the system; bacteria are the main components of the decomposition system (Wardle and Yeates 1993).

Ocean pollution and the poisoning of marine biota by organophosphates is directly linked to the biomagnification process, which is defined by the transfer of a chemical along the food chain, resulting in the bioaccumulation and bioconcentration of pollutants in aquatic animal organisms in successive trophic levels (Yu et al. 1998).

Plankton are the basis of the food chain of many aquatic systems and have a high potential for accumulating contaminants present in water (Chiuchiolo et al. 2004) (Siriwong et al. 2008). Thus, it is extremely important to know the dynamics of the aquatic food chain (Hardy 1924), as the pollution and intoxication of plankton have serious consequences for the rest of the marine biota, such as the transfer of contaminants and biomagnification to higher trophic levels (Andrady 2011). The zooplankton found in coastal waters and estuaries are in constant contact with pesticides and other organic pollutants. A study of the zooplankton *Mysidopsis bahia* showed bioaccumulation of methyl-parathion and forate (Morgan 2012). Another study conducted in the US Virgin Islands National Park found more than 10 different types of organophosphates, including tris(1,3-dichloro-2-propyl) phosphate and chlorpyrifos, which showed the highest values of plankton concentrations (Bargar et al. 2013).

As already discussed, organophosphates are widely found in microplastics that are dispersed in aquatic environments and pollute them (Faure et al. 2015, Zhang et al. 2018). Because microplastics are extremely small, plankton often ingest them because they recognize them as food (Farrell and Nelson 2013). Microplastics then transfer along the food chain, which generates an increase in the bioaccumulation of these pollutants in organisms (Setälä et al. 2014). Moreover, such pollution has an impact on predator/prey relationships (Wright et al. 2013).

Flame retardants are another source of OPs in marine waters. Since the 1970s, when OP flame retardants were recognized as having potential to accumulate in aquatic organisms, there have been research and data on the biodegradability and bioaccumulation of these compounds in the aquatic environment (Saeger et al. 1979). Knowing the bioaccumulation and biomagnification potentials of these pollutants is very important in assessing their ecological risk.

To evaluate the bioaccumulation potential, the agency REACH (Registration, Evaluation, Authorisation and Restriction of Chemicals) analyzes the log K_{ow} value for each OP compound. K_{ow} is the partition coefficient of n-octanol/water and is defined as the ratio of the concentration of a chemical in n-octanol to the equilibrium water at a certain temperature. Chemicals as well as organophosphates with high log K_{ow} values are of utmost concern to nature as they have the potential to affect living organisms (Hou et al. 2016). The log K_{ow} values of the main organophosphates found in the environment are described in Table 1 (Hou et al. 2016).

According to the REACH scale (Hou et al. 2016), OPs that are classified as bioaccumulative and pose the greatest risk to nature are tricresyl phosphate, resorcinol bis(diphenylphosphate), trixylenyl phosphate, tris(2-ethylhexyl) phosphate, and 2-ethylhexyldiphenyl phosphate. The structures of the two compounds that present the greatest bioaccumulative potential are represented in Fig. 9.

The bioaccumulation of OP flame retardants is evident at several trophic levels. A study carried out in several European coastal areas with different species of fish and bivalves shows that tris(2-chloroethyl) phosphate and tris(2-butoxyethyl) phosphate flame retardants present high concentrations in these animals, whereas phosphate of tris(2-butoxyethyl) reached 98.4 ng g^{-1} in shark mullet (Álvarez-Muñoz et al. 2015). In Sweden a study investigated the concentration of 11 different flame retardants and OP plasticizers in the muscular tissues of fish. Eleven different species of fish from coastal areas were studied and all presented concentrations of 11 OPs analyzed in the organism (Sundkvist et al. 2010).

One method to determine whether biomagnification occurs along a food chain is by analyzing the trophic magnification factor (TMF), which is the slope of the trophic level curve versus the logarithm of the body waste. It is considered that a trophic enlargement of the food chain has occurred when the TMFs are greater than 1 (Borgå et al. 2012).

There are two types of food chains that can be analyzed in the marine environment: benthic and pelagic. The benthic food chain encompasses organisms that inhabit the deep ocean, and these distribute themselves within or on sediments such as coral (Tagliapietra and Sigovini 2010). In the pelagic food chain aquatic organisms are able to swim and move around and they do not depend on the seabed (Sommer et al. 2002).

Table 1. Log K_{ow} values for some organophosphates.

Name	log K_{ow}
Tricresyl phosphate	5.48
Tributyl phosphate	4.00
Tri-iso-butyl phosphate	3.60
Triethyl phosphate	0.87
Tris(2-ethylhexyl) phosphate	9.49
Triphenyl phosphate	4.70
Tris(2-chloroethyl) phosphate	1.63
Tris(chloroisopropyl) phosphate	2.89
Tris(1,3-dichloro-2-propyl) phosphate	3.65
Tris(2-butoxyethyl) phosphate	3.00
Trixylenyl phosphate	7.98
2-ethylhexyldiphenyl phosphate	6.30
Tetrekis(2-chlorethyl) dichloroisopentyl diphosphate	1.9
Resorcinol bis(diphenylphosphate)	5.82

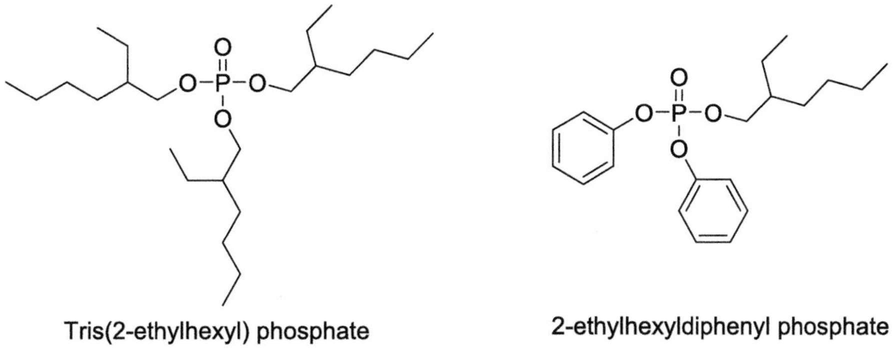

Tris(2-ethylhexyl) phosphate   2-ethylhexyldiphenyl phosphate

Fig. 9. Molecular structure of organophosphates with higher bioaccumulation potential.

In the case of OP flame retardants, biomagnification in benthic food chains is more likely than in pelagic food chains. This is because organophosphates are adsorbed to particles, which are more abundant in the deep ocean (Brandsma et al. 2015). In a study in the Western Scheldt estuary in the Netherlands, nine OP flame retardants were analyzed in two food chains, one pelagic and one benthic. For the benthic food web, invertebrate species such as cockle, crab, shrimp and some species of fish such as sculpin, plaice, goby and sole were collected. For the pelagic food web, suspended algae, jellyfish, lanctopago fish and a species of terrestrial piscivorous bird were collected. In addition, zooplankton of *Mysis* sp. were collected in both food chains. This study showed that the organophosphates tris(2-chloroethyl) phosphate, tris(2-chloroisopropyl) phosphate and tris(2-butoxyethyl) phosphate had TMFs greater than 1: the values were 2.6, 2.2 and 3.5 respectively, showing that these OP flame retardants cause a trophic increase in the food chain of this system. In the case of the pelagic food chain, there was a dilution of OP levels, which was probably due to the metabolism of the organisms studied (Brandsma et al. 2015).

Other studies challenge the idea that there is biomagnification of OPs along the food chains in the aquatic environment. A survey of 20 fish belonging to the same food chain from Manila Bay in the Philippines showed that organophosphates were detected in virtually all samples. In some cases the concentrations were 1000 ng g^{-1} lipid weight. However, no relationship was found between OP compounds and fish weight or length, indicating that these pollutants are not biomagnified through the food chain of this system (Kim et al. 2011).

Uncontrolled use of these organic pollutants is a significant threat to marine biota because there is a high potential for their transfer between trophic levels. In addition, they can concentrate in food webs (Zulin et al. 2002). These pesticides affect the structure and diversity of aquatic communities (Schulz and Liess 1999). Studies of their impact on the aquatic environment and the relationship between producers and consumers are extremely important for the preservation of the ecosystem (Price et al. 1980).

References

Abdollahi, Mohammad, Akram Ranjbar, Shahin Shadnia, Shekoufeh Nikfar and Ali Rezaiee. 2004. Pesticides and oxidative stress: a review. Med. Sci. Monit. 10(6): RA141–RA147.

Álvarez-Muñoz, Diana, Sara Rodríguez-Mozaz, Ana Luísa Maulvault, Alice Tediosi, Margarita Fernández-Tejedor, Freek Van den Heuvel et al. 2015. Occurrence of pharmaceuticals and endocrine disrupting compounds in macroalgaes, bivalves, and fish from coastal areas in Europe. Environ. Res. 143: 56–64.

Andrady, Anthony L. 2011. Microplastics in the marine environment. Mar. Pollut. Bull. 62(8): 1596–1605.

Arthur, Courtney, Joel E. Baker and Holly A. Bamford. 2009. Proceedings of the International Research Workshop on the Occurrence, Effects, and Fate of Microplastic Marine Debris, September 9–11, 2008, University of Washington Tacoma, Tacoma, WA, USA.

Badawy, Mohamed I. and Mohamed A. El-Dib. 1984. Persistence and fate of methyl parathion in sea water. Bull. Environ. Contam. Toxicol. 33(1): 40–49.

Bagchi, D., M. Bagchi, E.A. Hassoun and S.J. Stohs. 1995. *In vitro* and *in vivo* generation of reactive oxygen species, DNA damage and lactate dehydrogenase leakage by selected pesticides. Toxicology 104(1-3): 129–140.

Banerjee, B.D., V. Seth, A. Bhattacharya, S.T. Pasha and A.K. Chakraborty. 1999. Biochemical effects of some pesticides on lipid peroxidation and free-radical scavengers. Toxicol. Lett. 107(1-3): 33–47.

Barany, George and R.B. Merrifield. 1980. Kinetics and mechanism of the thiolytic removal of the dithiasuccinoyl (Dts) amino protecting group. J. Am. Chem. Soc. 102(9): 3084–3095.

Bargar, Timothy A., Virginia H. Garrison, David A. Alvarez and Kathy R. Echols. 2013. Contaminants assessment in the coral reefs of Virgin Islands National Park and Virgin Islands Coral Reef National Monument. Mar. Pollut. Bull. 70(1-2): 281–288.

Behrenfeld, Michael J., James T. Randerson, Charles R. McClain, Gene C. Feldman, Sietse O. Los, Compton J. Tucker et al. 2001. Biospheric primary production during an ENSO transition. Science 291(5513): 2594–2597.

Bocquené, Gilles, Anne Roig and Didier Fournier. 1997. Cholinesterases from the common oyster (Crassostrea gigas) Evidence for the presence of a soluble acetylcholinesterase insensitive to organophosphate and carbamate inhibitors. FEBS Lett. 407(3): 261–266.

Borgå, Katrine, Karen A. Kidd, Derek C.G. Muir, Olof Berglund, Jason M. Conder, Frank A.P.C. Gobas et al. 2012. Trophic magnification factors: considerations of ecology, ecosystems, and study design. Integr. Environ. Assess. Manage. 8(1): 64–84.

Brandsma, Sicco H., Pim E.G. Leonards, Heather A. Leslie and Jacob de Boer. 2015. Tracing organophosphorus and brominated flame retardants and plasticizers in an estuarine food web. Sci. Total Environ. 505: 22–31.

Browne, Mark A., Awantha Dissanayake, Tamara S. Galloway, David M. Lowe and Richard C. Thompson. 2008. Ingested microscopic plastic translocates to the circulatory system of the mussel, Mytilus edulis (L.). Environ. Sci. Technol. 42(13): 5026–5031.

Browne, Mark A., Tamara S. Galloway and Richard C. Thompson. 2010. Spatial patterns of plastic debris along estuarine shorelines. Environ. Sci. Technol. 44(9): 3404–3409.

Browne Mark A., Phillip Crump, Stewart J. Niven, Emma Teuten, Andrew Tonkin, Tamara Galloway et al. 2011. Accumulation of microplastic on shorelines worldwide: sources and sinks. Environ. Sci. Technol. 45(21): 9175–9179.

Browne, Mark Anthony, Stewart J. Niven, Tamara S. Galloway, Steve J. Rowland and Richard C. Thompson. 2013. Microplastic moves pollutants and additives to worms, reducing functions linked to health and biodiversity. Curr. Biol. 23(23): 2388–2392.

Cajaraville, Miren P., Maria J. Bebianno, Julián Blasco, Cinta Porte, Carmen Sarasquete and Aldo Viarengo. 2000. The use of biomarkers to assess the impact of pollution in coastal environments of the Iberian Peninsula: a practical approach. Sci. Total Environ. 247(2-3): 295–311.

Castro-Jiménez, Javier, Naiara Berrojalbiz, Mariana Pizarro and Jordi Dachs. 2014. Organophosphate ester (OPE) flame retardants and plasticizers in the open Mediterranean and Black Seas atmosphere. Environ. Sci. Technol. 48(6): 3203–3209.

Chiuchiolo, Amy L., Rebecca M. Dickhut, Michele A. Cochran and Hugh W. Ducklow. 2004. Persistent organic pollutants at the base of the Antarctic marine food web. Environ. Sci. Technol. 38(13): 3551–3557.

Cole, Matthew, Pennie Lindeque, Elaine Fileman, Claudia Halsband, Rhys Goodhead, Julian Moger et al. 2013. Microplastic ingestion by zooplankton. Environ. Sci. Technol. 47(12): 6646–6655.

da Silva, Natália Alvarenga, Willian Garcia Birolli, Mirna Helena Regali Seleghim and André Luiz Meleiro Porto. 2013. Biodegradation of the organophosphate pesticide profenofos by marine fungi.

Delfino, Reinaldo T., Tatiana S. Ribeiro and José D. Figueroa-Villar. 2009. Organophosphorus compounds as chemical warfare agents: A review. J. Braz. Chem. Soc. 20(3): 407–428.

dos Santos, Josilane Rodrigues and Joelmir Marques da Silva. 2007. Toxicologia de agrotóxicos em ambientes aquáticos. Oecol. bras. 11(4): 565–573.

Edwards, Clive Arthur. 1973. Persistent pesticides in the environment. Crit. Rev. Environ. Control (Ed. 2).

Endo, Satoshi, Reiko Takizawa, Keiji Okuda, Hideshige Takada, Kazuhiro Chiba, Haruyuki Kanehiro et al. 2005. Concentration of polychlorinated biphenyls (PCBs) in beached resin pellets: variability among individual particles and regional differences. Mar. Pollut. Bull. 50(10): 1103–1114.

Eriksen, Marcus, Laurent C.M. Lebreton, Henry S. Carson, Martin Thiel, Charles J. Moore, Jose C. Borerro et al. 2014. Plastic pollution in the world's oceans: more than 5 trillion plastic pieces weighing over 250,000 tons afloat at sea. PloS One 9(12): e111913.

Eyer, Florian, Darren M. Roberts, Nicholas A. Buckley et al. 2009. Extreme variability in the formation of chlorpyrifos oxon (CPO) in patients poisoned by chlorpyrifos (CPF). Biochem. Pharmacol. 78(5): 531–537.

Farrell, Paul and Kathryn Nelson. 2013. Trophic level transfer of microplastic: Mytilus edulis (L.) to Carcinus maenas (L.). Environ. Pollut. 177: 1–3.

Faure, Florian, Colin Demars, Olivier Wieser, Manuel Kunz and Luiz Felippe De Alencastro. 2015. Plastic pollution in Swiss surface waters: nature and concentrations, interaction with pollutants. Environ. Chem. 12(5): 582–591.

Fisner, Mara, Satie Taniguchi, Fabiana Moreira, Márcia C. Bícego and Alexander Turra. 2013. Polycyclic aromatic hydrocarbons (PAHs) in plastic pellets: Variability in the concentration and composition at different sediment depths in a sandy beach. Mar. Pollut. Bull. 70(1-2): 219–226.

Fukuto, T. Roy. 1990. Mechanism of action of organophosphorus and carbamate insecticides. Environ. Health Perspect. 87: 245–254.

Garbarino, John R., Elaine Snyder-Conn, Thomas J. Leiker and Gerald L. Hoffman. 2002. Contaminants in Arctic snow collected over northwest Alaskan sea ice. Water, Air, Soil Pollut. 139(1-4): 183–214.

Garfitt, S.J., K. Jones, H.J. Mason and J. Cocker. 2002. Exposure to the organophosphate diazinon: data from a human volunteer study with oral and dermal doses. Toxicol. Lett. 134(1-3): 105–113.

Ghosal, S., M. Rogers and A. Wray. 2002. The effects of turbulence on phytoplankton.

Hardy, A.C. 1924. The herring in relation to its animate environment I. The food and feeding habits of the herring with special reference to the east coast of England. Fish. Invest. Lond. 2(3): 53.

Harino, Hiroya, Emi Yatsuzuka, Chiaki Yamao, Masaaki Ueno and Madoka Ohji. 2014. Current status of organophosphorus compounds contamination in Maizuru Bay, Japan. J. Mar. Biol. Assoc. U.K. 94(1): 43–49.

Hogg, Stuart. 2013. Essential Microbiology: John Wiley & Sons.

Hou, Rui, Yiping Xu and Zijian Wang. 2016. Review of OPFRs in animals and humans: Absorption, bioaccumulation, metabolism, and internal exposure research. Chemosphere 153: 78–90.

Howe, George E., Leif L. Marking, Terry D. Bills, Michael A. Boogaard and Foster L. Mayer Jr. 1994. Effects of water temperature on the toxicity of 4-nitrophenol and 2, 4-dinitrophenol to developing rainbow trout (Oncorhynchus mykiss). Environ. Toxicol. Chem. 13(1): 79–84.

Hu, Mengyang, Jun Li, Beibei Zhang, Qinglan Cui, Si Wei and Hongxia Yu. 2014. Regional distribution of halogenated organophosphate flame retardants in seawater samples from three coastal cities in China. Mar. Pollut. Bull. 86(1-2): 569–574.

Jackson, John A., Clifford E. Berkman and Charles M. Thompson. 1992. Stereoselective and chemoselective oxidation of phosphorothionates using MMPP. Tetrahedron Lett. 33(41): 6061–6064.

Jover, Jesús, Ramón Bosque and Joaquim Sales. 2007. Neural network based QSPR study for predicting pKa of phenols in different solvents. QSAR Comb. Sci. 26(3): 385–397.

Kim, Joon-Woo, Tomohiko Isobe, Kwang-Hyeon Chang, Atsuko Amano, Rommel H. Maneja, Peter B. Zamora et al. 2011. Levels and distribution of organophosphorus flame retardants and plasticizers in fishes from Manila Bay, the Philippines. Environ. Pollut. 159(12): 3653–3659.

Kim, Wan-Soo, Seong-Jin Yoon and Dong-Beom Yang. 2004. Effects of chlorpyrifos on the endogenous rhythm of the Manila clam, Ruditapes philippinarum (Bivalvia: Veneridae). Mar. Pollut. Bull. 48(1-2): 182–187.

Kittle, Ronald P., Karla J. McDermid, Lisa Muehlstein and George H. Balazs. 2018. Effects of glyphosate herbicide on the gastrointestinal microflora of Hawaiian green turtles (Chelonia mydas) Linnaeus. Marine Pollution Bulletin 127: 170–174.

Kröger, Silke, Sergey Piletsky and Anthony P.F. Turner. 2002. Biosensors for marine pollution research, monitoring and control. Mar. Pollut. Bull. 45(1-12): 24–34.

Kuda, Takashi, Daisuke Kyoi, Hajime Takahashi, Kazuhiro Obama and Bon Kimura. 2011. Detection and isolation of p-nitrophenol-lowering bacteria from intestine of marine fishes caught in Japanese waters. Mar. Pollut. Bull. 62(8): 1622–1627.

Larkin, Daniel J. and Ronald S. Tjeerdema. 2000. Fate and effects of diazinon. Rev. Environ. Contam. Toxicol. 166: 49–82.

Lee, Hye-Sung, Young Ah Kim, Young Ae Cho and Yong Tae Lee. 2002. Oxidation of organophosphorus pesticides for the sensitive detection by a cholinesterase-based biosensor. Chemosphere 46(4): 571–576.

Li, Donghao, Meihua Dong, Won Joon Shim, Sang Hee Hong, Jae-Ryoung Oh, Un Hyuk Yim et al. 2005. Seasonal and spatial distribution of nonylphenol and IBP in Saemangeum Bay, Korea. Mar. Pollut. Bull. 51(8-12): 966–974.

Li, Jiana, Dongqi Yang, Lan Li, Khalida Jabeen and Huahong Shi. 2015. Microplastics in commercial bivalves from China. Environ. Pollut. 207: 190–195.

Livingstone, David R., Philippe Lemaire, Anne Matthews, Laurence Peters, David Bucke and Robin J. Law. 1993. Pro-oxidant, antioxidant and 7-ethoxyresorufin O-deethylase (EROD) activity responses in liver of dab (Limanda limanda) exposed to sediment contaminated with hydrocarbons and other chemicals. Mar. Pollut. Bull. 26(11): 602–606.

Ma, Tangeng and Janice E. Chambers. 1995. A kinetic analysis of hepatic microsomal activation of parathion and chlorpyrifos in control and phenobarbital-treated rats. J. Biochem. Toxicol. 10(2): 63–68.

Ma, Yuxin, Zhiyong Xie, Rainer Lohmann, Wenying Mi and Guoping Gao. 2017. Organophosphate ester flame retardants and plasticizers in ocean sediments from the North Pacific to the Arctic Ocean. Environ. Sci. Technol. 51(7): 3809–3815.

Magni, Paolo, Giovanni De Falco, Carla Falugi, M. Franzoni, Martino Monteverde, Emanuela Perrone et al. 2006. Genotoxicity biomarkers and acetylcholinesterase activity in natural populations of Mytilus galloprovincialis along a pollution gradient in the Gulf of Oristano (Sardinia, western Mediterranean). Environ. Pollut. 142(1): 65–72.

Marklund, Anneli, Barbro Andersson and Peter Haglund. 2003. Screening of organophosphorus compounds and their distribution in various indoor environments. Chemosphere 53(9): 1137–1146.

Matozzo, Valerio, Andrea Tomei and Maria Gabriella Marin. 2005. Acetylcholinesterase as a biomarker of exposure to neurotoxic compounds in the clam Tapes philippinarum from the Lagoon of Venice. Mar. Pollut. Bull. 50(12): 1686–1693.

Möller, Axel, Renate Sturm, Zhiyong Xie, Minghong Cai, Jianfeng He and Ralf Ebinghaus. 2012. Organophosphorus flame retardants and plasticizers in airborne particles over the Northern Pacific and Indian Ocean toward the polar regions: Evidence for global occurrence. Environ. Sci. Technol. 46(6): 3127–3134.

Moreno-González, R. and V.M. León. 2017. Presence and distribution of current-use pesticides in surface marine sediments from a Mediterranean coastal lagoon (SE Spain). Environ. Sci. Pollut. Res. 24(9): 8033–8048.

Morgan, Mark D. 2012. Ecology of Mysidacea. Vol. 10: Springer Science & Business Media.

Murty, Ayyagari S. 1986. Toxicity of Pesticides to Fish. Vol. 2: CRC press Boca Raton, FL.

Myers, D.K., B. Mendel, H.R. Gersmann and J.A.A. Ketelaar. 1952. Oxidation of thiophosphate insecticides in the rat. Nature 170(4332): 805.

Net, Sopheak, Richard Sempéré, Anne Delmont, Andrea Paluselli and Baghdad Ouddane. 2015. Occurrence, fate, behavior and ecotoxicological state of phthalates in different environmental matrices. Environ. Sci. Technol. 49(7): 4019–4035.

Omar, Tuan Fauzan Tuan, Ahmad Zaharin Aris, Fatimah Md Yusoff and Shuhaimi Mustafa. 2018. Occurrence, distribution, and sources of emerging organic contaminants in tropical coastal sediments of anthropogenically impacted Klang River estuary, Malaysia. Mar. Pollut. Bull. 131: 284–293.

Páez-Osuna, Federico, Saúl R. Guerrero-Galván and Ana C. Ruiz-Fernández. 1998. The environmental impact of shrimp aquaculture and the coastal pollution in Mexico. Mar. Pollut. Bull. 36(1): 65–75.

Price, Peter W., Carl E. Bouton, Paul Gross, Bruce A. McPheron, John N. Thompson and Arthur E. Weis. 1980. Interactions among three trophic levels: influence of plants on interactions between insect herbivores and natural enemies. Annu. Rev. Ecol. Syst. 11(1): 41–65.

Rand, Gary M. and Sam R. Petrocelli. 1985. Fundamentals of Aquatic Toxicology: Methods and Applications. FMC Corp., Princeton, NJ.

Readman, James W., Laval Liong Wee Kwong, Laurence D. Mee, Jean Bartocci, Göran Nilvé, J.A. Rodriguez-Solano et al. 1992. Persistent organophosphorus pesticides in tropical marine environments. Mar. Pollut. Bull. 24(8): 398–402.

Reemtsma, Thorsten, José Benito Quintana, Rosario Rodil, Mónica Garcı and Isaac Rodrı. 2008. Organophosphorus flame retardants and plasticizers in water and air I. Occurrence and fate. TrAC, Trends Anal. Chem. 27(9): 727–737.

Saeger, Victor W., Orville Hicks, Robert G. Kaley, Paul R. Michael, James P. Mieure and E. Scott Tucker. 1979. Environmental fate of selected phosphate esters. Environ. Sci. Technol. 13(7): 840–844.

Sams, C., J. Cocker and M.S. Lennard. 2003. 544 Metabolism of chlorpyrifos and diazinon by human liver microsomes. Toxicol. Lett. 144: s146.

Sarkar, A., D. Ray, Amulya N. Shrivastava and Subhodeep Sarker. 2006. Molecular biomarkers: their significance and application in marine pollution monitoring. Ecotoxicology 15(4): 333–340.

Schulz, Ralf and Mathias Liess. 1999. A field study of the effects of agriculturally derived insecticide input on stream macroinvertebrate dynamics. Aquat. Toxicol. 46(3-4): 155–176.

Serrano, R., F. Hernández, F.J. López and J.B. Pena. 1997. Bioconcentration and depuration of chlorpyrifos in the marine mollusc Mytilus edulis. Arch. Environ. Contam. Toxicol. 33(1): 47–52.

Setälä, Outi, Vivi Fleming-Lehtinen and Maiju Lehtiniemi. 2014. Ingestion and transfer of microplastics in the planktonic food web. Environ. Pollut. 185: 77–83.

Siriwong, W., K. Thirakhupt, D. Sitticharoenchai, M. Borjan and M. Robson. 2008. Organochlorine pesticide residues in plankton, Rangsit agricultural area, Central Thailand. Bull. Environ. Contam. Toxicol. 81(6): 608–612.

Sommer, Ulrich, Herwig Stibor, Alexis Katechakis, Frank Sommer and Thomas Hansen. 2002. Pelagic food web configurations at different levels of nutrient richness and their implications for the ratio fish production: primary production. In Sustainable increase of marine harvesting: fundamental mechanisms and new concepts: Springer.

Spain, Jim C. and David T. Gibson. 1991. Pathway for biodegradation of p-nitrophenol in a Moraxella sp. Appl. Environ. Microbiol. 57(3): 812–819.

Stegeman, John J. and Mark E. Hahn. 1994. Biochemistry and molecular biology of monooxygenases: current perspectives on forms, functions, and regulation of cytochrome P450 in aquatic species. In Aquatic Toxicology: Molecular, Biochemical, and Cellular Perspectives.

Straus, David L., Daniel Schlenk and Janice E. Chambers. 2000. Hepatic microsomal desulfuration and dearylation of chlorpyrifos and parathion in fingerling channel catfish: lack of effect from Aroclor 1254. Aquat. Toxicol. 50(1-2): 141–151.

Sturm, A., H.C. Da Silva De Assis and P.-D. Hansen. 1999. Cholinesterases of marine teleost fish: enzymological characterization and potential use in the monitoring of neurotoxic contamination. Mar. Environ. Res. 47(4): 389–398.

Sultatos, Lester G. and Sheldon D. Murphy. 1983. Kinetic analyses of the microsomal biotransformation of the phosphorothioate insecticides chlorpyrifos and parathion. Toxicol. Sci. 3(1): 16–21.

Sultatos, Lester G. 1994. Mammalian toxicology of organophosphorus pesticides. J. Toxicol. Environ. Health, Part A 43(3): 271–289.

Sun, Xiaojin, Fang Zhu, Jiabin Xi, Tongbu Lu, Hong Liu, Yexiang Tong et al. 2011. Hollow fiber liquid-phase microextraction as clean-up step for the determination of organophosphorus pesticides residues in fish tissue by gas chromatography coupled with mass spectrometry. Mar. Pollut. Bull. 63(5-12): 102–107.

Sundkvist, Anneli Marklund, Ulrika Olofsson and Peter Haglund. 2010. Organophosphorus flame retardants and plasticizers in marine and fresh water biota and in human milk. J. Environ. Monit. 12(4): 943–951.

Swinson, Joel, Lamar Field, Norman E. Heimer, Danuta Michalska, Donald D. Muccio, Prasad L. Polavarapu et al. 1988. Thiono compounds. 9. Use of spectra to study intermediates in the oxidation of thiono phosphorus, compounds. Phosphorus Sulfur Relat. Elem. 35(1-2): 159–172.

Syberg, Kristian, Farhan R. Khan, Henriette Selck et al. 2015. Microplastics: addressing ecological risk through lessons learned. Environ. Toxicol. Chem. 34(5): 945–953.

Tagliapietra, Davide and Marco Sigovini. 2010. Benthic fauna: collection and identification of macrobenthic invertebrates. Terre et Environnement 88: 253–261.

Tanaka, Kosuke, Hideshige Takada, Rei Yamashita, Kaoruko Mizukawa, Masa-aki Fukuwaka and Yutaka Watanuki. 2013. Accumulation of plastic-derived chemicals in tissues of seabirds ingesting marine plastics. Mar. Pollut. Bull. 69(1-2): 219–222.

Thompson, Richard C., Ylva Olsen, Richard P. Mitchell, Anthony Davis, Steven J. Rowland, Anthony W.G. John et al. 2004. Lost at sea: where is all the plastic? Science 304(5672): 838–838.

Timbrell, John A. 2008. Principles of Biochemical Toxicology: CRC Press.

Tsuda, Taizo, Mihoko Kojima, Hiroyuki Harada, Atsuko Nakajima and Shigeru Aoki. 1997. Acute toxicity, accumulation and excretion of organophosphorous insecticides and their oxidation products in killifish. Chemosphere 35(5): 939–949.

Valavanidis, Athanasios, Thomais Vlahogianni, Manos Dassenakis and Michael Scoullos. 2006. Molecular biomarkers of oxidative stress in aquatic organisms in relation to toxic environmental pollutants. Ecotox. Environ. Safe. 64(2): 178–189.

Van der Oost, Ron, Jonny Beyer and Nico P.E. Vermeulen. 2003. Fish bioaccumulation and biomarkers in environmental risk assessment: a review. Environmental Toxicology and Pharmacology 13(2): 57–149.

Van der Veen, Ike and Jacob de Boer. 2012. Phosphorus flame retardants: properties, production, environmental occurrence, toxicity and analysis. Chemosphere 88(10): 1119–1153.

Vico, Raquel V., Rita H. de Rossi and Elba I. Buján. 2009. Reactivity of the insecticide chlorpyrifos-methyl toward hydroxyl and perhydroxyl ion. Effect of cyclodextrins. J. Phys. Org. Chem. 22(7): 691–702.

Wang, Fei, Kai Min Shih and Xiao Yan Li. 2015. The partition behavior of perfluorooctanesulfonate (PFOS) and perfluorooctanesulfonamide (FOSA) on microplastics. Chemosphere 119: 841–847.

Wardle, D.A. and G.W. Yeates. 1993. The dual importance of competition and predation as regulatory forces in terrestrial ecosystems: evidence from decomposer food-webs. Oecologia 93(2): 303–306.

Wei, Gao-Ling, Ding-Qiang Lia, Mu-Ning Zhuo, Yi-Shan Liao, Zhen-Yue Xie, Tai-Long Guo et al. 2015. Organophosphorus flame retardants and plasticizers: sources, occurrence, toxicity and human exposure. Environ. Pollut. 196: 29–46.

Williams, Gary M. and Michael J. Iatropoulos. 2002. Alteration of liver cell function and proliferation: differentiation between adaptation and toxicity. Toxicol. Pathol. 30(1): 41–53.

Wright, Stephanie L., Darren Rowe, Richard C. Thompson and Tamara S. Galloway. 2013. Microplastic ingestion decreases energy reserves in marine worms. Curr. Biol. 23(23): R1031–R1033.

Yan, Saihong, Huimin Wu, Jianhui Qin, Jinmiao Zha and Zijian Wang. 2017. Halogen-free organophosphorus flame retardants caused oxidative stress and multixenobiotic resistance in Asian freshwater clams (Corbicula fluminea). Environ. Pollut. 225: 559–568.

Yu, Jun, Dong Ho Lee, Kyung Tae Kim, Dong Beom Yang and Jae Sam Yang. 2001. Distribution of organophosphorus pesticides in some estuarine environments in Korea. Fish. Aquat. Sci. 4(4): 201–207.

Yu, Ming-Ho, Wayne Landis and Ruth Sofield. 1998. Introduction to Environmental Toxicology: Molecular Substructures to Ecological Landscapes: CRC Press.

Yusof, Shahrizad, Ahmad Ismail and Mohamad Shafiq Alias. 2014. Effect of glyphosate-based herbicide on early life stages of Java medaka (Oryzias javanicus): A potential tropical test fish. Marine Pollution Bulletin 85(2): 494–498.

Yuxiang, O.U. 2011. Developments of organic phosphorus flame retardant industry in China [J]. Chem. Ind. Eng. Prog. 1: 033.

Zhang, Haibo, Qian Zhou, Zhiyong Xie, Yang Zhoub, Chen Tu, Chuancheng Fu et al. 2018. Occurrences of organophosphorus esters and phthalates in the microplastics from the coastal beaches in north China. Sci. Total Environ. 616: 1505–1512.

Zhong, Guangcai, Zhiyong Xie, Minghong Cai, Axel Möller, Renate Sturm, Jianhui Tang et al. 2011. Distribution and air–sea exchange of current-use pesticides (CUPs) from East Asia to the high Arctic Ocean. Environ. Sci. Technol. 46(1): 259–267.

Zhong, Mingyu, Jianhui Tang, Lijie Mi et al. 2017. Occurrence and spatial distribution of organophosphorus flame retardants and plasticizers in the Bohai and Yellow Seas, China. Mar. Pollut. Bull. 121(1-2): 331–338.

Zhu, Yong-Zhe, Min Fu, In-Hong Jeong, Jeong-Han Kim and Chuan-Jie Zhang. 2017. Metabolism of an Insecticide Fenitrothion by Cunninghamella elegans ATCC36112. J. Agric. Food Chem. 65(49): 10711–10718.

Zulin, Zhang, Hong Huasheng, Wang Xinhong, Lin Jianqing, Chen Weiqi and Xu Li. 2002. Determination and load of organophosphorus and organochlorine pesticides at water from Jiulong River Estuary, China. Mar. Pollut. Bull. 45(1-12): 397–402.

10

Halogenated Pollutants in Marine and Coastal Environments

Natalia E. Cappelletti

1. Introduction

This chapter examines the partitioning behavior of halogenated pollutants in the coastal and marine environment. We outline this behavior in quantitative terms by identifying the phases or compartments of interest, and the expected halogenated pollutant concentrations that can be found in these compartments.

Halogenated pollutants—organic compounds containing one or more carbon-chlorine, carbon-bromine, carbon-iodine, or carbon-fluorine bonds—are a group of compounds that includes such well-known persistent substances as 1,1,1-trichloro-2,2-bis(*p*-chlorophenyl) ethane (DDT) and the polychlorinated biphenyls (PCBs). Because of their relative stability, some of these chemicals find a variety of industrial uses, the PCBs having been used for many years in electrical transformers and condensers. Other compounds, such the short-chain chlorofluorocarbons, have found applications in aerosol dispensers. Earlier studies focused on chlorinated pollutants because of their toxicity and high stability in the environment. In the last three decades, concern has been expanded also to brominated, fluorinated and iodine-containing compounds. There are many organohalogen compounds that are highly stable and lipophilic, and therefore accumulate in the environment. The lipophilic nature of these compounds also permits rapid penetration of cell membranes and uptake by tissues, and it is not surprising that highly lipophilic toxicants are distributed and concentrated in body fat (Klaassen 2008). This tendency is indicated by the octanol-water partition coefficient > 10^3 (Table 1). The environmental accumulation and potential toxicological significance of long-term storage of numerous organohalogen compounds are proved by the dominant occurrence of halogenated compounds in lists of priority pollutants, e.g., the Dirty Dozen of the Stockholm Convention. Organohalogen compounds, like PCBs and organochlorine pesticides, have a very low water solubility (Table 1) and

Consejo Nacional de Investigaciones Científicas y Técnicas, Argentina, Grupo de investigación en Geología Ambiental, Universidad Nacional de Avellaneda, Comisión investigaciones científicas, Provincia de Buenos Aires. M. Bravo 1460 (1870). Avellaneda, Argentina.
Email: ncappelletti@undav.edu.ar

Table 1. Water solubility (mg l^{-1}) at 25°C, and log octanol-water partitioning coefficients (log K_{ow}) of persistent organohalogen compounds (from Mackay et al. 2006).

Organohalogen compounds	Water solubility	Log K_{ow}
Aldrin	2.7×10^{-2}	6.2
DDT	5.5×10^{-3}	6.3
4,4'-DDE	2.0×10^{-3}	6.9
Dieldrin	0.20	5.4
Endrin	0.23	5.2
Endosulfan	0.28	4.8
Heptachlor	5.6×10^{-2}	6.1
gamma-HCH	7.0	3.7
Mirex	2.0×10^{-2}	7.1
2,2',5,5'Tetrachlorobiphenyl	1.5×10^{-2}	5.9
2,2',4,5,5'Pentachlorobiphenyl	2.6×10^{-2}	6.3
2,3',4,4',5 Pentachlorobiphenyl	1.9×10^{-2}	6.8
Tetrachlorinated dibenzo-*p*-dioxins	1.5×10^{-5}	7.1
Tetrachlorinated dibenzofurans	4.2×10^{-4}	6.5

a strong affinity for phase surface of particles; therefore, in natural waters, the typical particle-water distribution coefficient falls between 10^3 and 10^5 (Olsen et al. 1982). Consequently, sediments are an important sink, and estuarine and coastal marine sediments act as short- or long-term reservoirs. Thus, the distribution of hydrophobic compounds in the aquatic ecosystem depends on sorption reactions over sediment particles, digenetic processes such as biological and chemical transformation reactions, diffusion, advection, and mixing and resuspension. The relative importance of each of these processes depends on the molecular structure, biological activity and physical-chemical properties of individual organic compounds, as well as properties of the sedimentary environment (Brownawell and Farrington 1985). The photolysis and biodegradation reactions determine the overall half-life time of persistent organohalogen compounds in aquatic environments, whereas hydrolysis is neglected because in environmental conditions it is extremely slow (Sinkkonen and Paasivirta 2000). The photodegradation takes place in water layers as long as the light is available and rates fall directly with the depth of the water column (Sinkkonen and Paasivirta 2000). Biodegradation rate in the real aquatic environment depends on characteristics of the aquatic system, concentrations of inorganic and organic nutrients, presence of particulate matter, temperature, oxygen concentration, redox potential, adaptation of the microbial population and the concentration of the chemical studied (Howard and Banerjee 1984).

The potential dangers of certain organohalogens resulted in measures to regulate their input into the seas as well as monitoring at the local and regional levels. Several monitoring programs are being and/or have been carried out to identify areas or regions of concern, estimate the hazards caused by persistent organohalogens to humans and the marine environment, and assess the effectiveness of the measures taken (Roose and Brinkman 2005). Because of hydrophobicity of several organohalogen compounds, marine sediments and biota have been used as sample types for such analytes. Particularly, marine mammals have been widely used as bioindicators of organochlorinated compounds, because they are long-lived and they are at the top of the aquatic food chain. Besides, they can transfer several organohalogen compounds, such as PCBs and chlorinated pesticides, through lactation over generations and most have a low capacity for xenobiotic degradation (Tanabe et al. 1988).

Following this outline of some basic features of organohalogen compounds, a more detailed discussion is given of recent observations regarding the distribution of PCBs, organochlorine pesticides and polychlorinated dibenzo-*p*-dioxins and polychlorinated dibenzofurans in the coastal and marine environment. Finally, we summarize each section with a conceptual model to rationalize how organohalogen compounds can be distributed in a background and contaminated marine ecosystem.

Throughout the chapter, the concentrations of selected organohalogen compounds are expressed as mass per gram of dry weight (pg g^{-1}, ng g^{-1} or µg g^{-1}), or mass per gram of lipid weight (pg g^{-1} l.w., ng g^{-1} l.w. or µg g^{-1} l.w.), or it is indicated if they are expressed in another way.

2. Polychlorinated Biphenyls

Polychlorinated biphenyls (PCBs) form a group of 209 possible congeners, in which the biphenyl can be substituted by one to ten chlorine atoms at different positions (Fig. 1). PCBs have been linked to a wide range of human health and ecosystem effects, in particular neurodevelopmental, reproductive and endocrine-related effects, even at relatively low PCB doses (ATSDR 2000). Commercial synthesis of PCBs started in 1929 and they were used as heat transfer fluids, hydraulic lubricants, and dielectric fluids. Although the production and use of PCBs were prohibited by the Stockholm Convention on Persistent Organic Pollutants in 2004 worldwide, they are still present in materials with a long use phase as sealants of concrete buildings and paints (Jartun et al. 2009, Robson et al. 2010), and they can also be formed as by-products of certain industrial processes or waste incineration (Yu et al. 2006, Nakao et al. 2006). Thus, substantial amounts of PCBs are still emitted from primary and secondary sources in cities (Diamond et al. 2010, Diefenbacher et al. 2016). Among PCBs, the most toxic subset of 12 non-ortho and mono-ortho dioxin-like PCBs congeners have been included in an equivalent toxicity scheme that allows the calculation of total equivalent toxicity (TEQ; Van den Berg et al. 2006).

PCBs enter the aquatic environment mainly via industrial effluent, discharge of untreated wastes, and treated effluent or as a result of local events (e.g., leakage from equipment using transformer oil with added PCBs). Eganhouse and Sherblom (2001) evaluated the impact of combined sewer overflows effluent that discharges into the Dorchester Bay, in the northwestern portion of Boston Harbor (USA). They estimated annual mass emission rates of 67 kg of PCBs to the harbor. PCB concentrations in the effluent ranged from 19 to 22 ng l^{-1} in low-flow, dry weather conditions, and 26 to 730 µg l^{-1} during periods of variable rainfall. Their results suggested that PCBs were derived from sewage inputs during wet weather, where inflows could exceed the capacity of the primary wastewater treatment plant, and raw sewage and runoff was directed to relief drain. Consistently, the cities of Vancouver and Victoria on Canada's west coast discharge wastewater in which PCB concentrations averaged 13 ± 4 ng l^{-1} and 10 ± 2 ng l^{-1}, respectively (Dinn et al. 2012). On the other hand, Sánchez-Avila et al. (2009) did not detect PCBs in wastewater samples collected directly from sewers of secondary treatment plants from the area of Maresme (Catalonia, Spain), which ultimately discharge into the Mediterranean Sea.

Atmospheric deposition can also be a significant source of PCBs to surface waters, especially downwind of urban and industrialized areas (Park et al. 2001). Dickhut and Gustafson (1995) reported

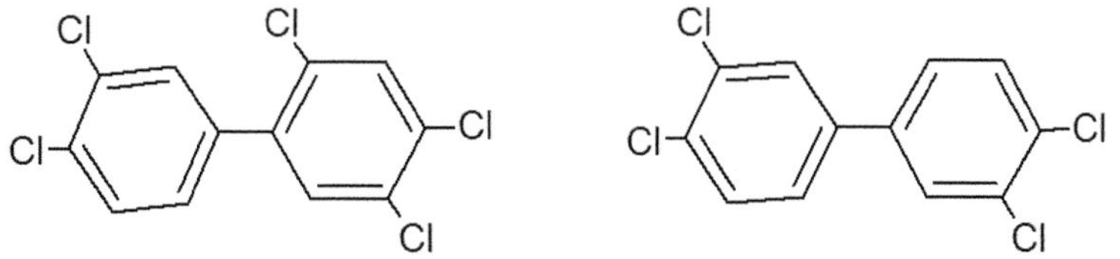

Fig. 1. Structures of 2,3',4,4',5 pentachlorobiphenyl (left) and 3,3',4,4'tetrachlorobiphenyl (right).

atmospheric deposition (dry plus wet) fluxes of PCB to Chesapeake Bay (Virginia, USA), which ranged from 1.6 to 2.3 µg m^{-2} yr^{-1}. Park et al. (2001) also studied the atmospheric deposition for PCBs in Galveston Bay (Texas, USA). In that study, the direct deposition rate (dry plus wet) and loading of PCBs to Galveston Bay were 6.40 µg m^{-2} yr^{-1} and 9.15 kg yr^{-1}, respectively. The authors found that the yearly dry deposition rate of PCBs was over three times the wet deposition rate and the higher rates of dry deposition are likely related to the drier and hotter climate.

On the other hand, paint may constitute another major source of PCBs coming from urban environments because chips of paint may easily fall from surfaces during heavy rainfall and wind, facilitating a particle-bound dispersion through the stormwater sewage system (Jartun et al. 2009).

Once PCBs entered aquatic environments, the half-life of congeners ranged from < 1 to 67 yr in water, and from < 1 to 23 yr in sediments (Paasivirta and Sinkkonen 2009). PCBs have been shown to undergo biodegradation via aerobic oxidative processes and anaerobic reductive processes. Several microorganisms have been isolated that can aerobically degrade PCBs, preferentially degrading the more lightly chlorinated congeners. Anaerobic bacteria attack more highly chlorinated PCB congeners through reductive dechlorination. In general, this microbial process removes preferentially the *meta* and *para* chlorines, resulting in a depletion of highly chlorinated PCB congeners with corresponding increases in lightly chlorinated, *ortho*-substituted PCB congeners. The altered congener distribution of residual PCB contamination observed in several aquatic sediments was the earliest evidence of the anaerobic dechlorination of PCBs (Abramowicz 1995). Øfjord et al. (1994) studied reductive dechlorination of PCBs by methanogenic and sulfate-amended laboratory cultures enriched from marine sediments (Puget Sound, USA) in sea-salts media. Under culture conditions the extent of dechlorination was 7–8% over a period of 1 yr, 14–15% *meta* and 9–10% of *para* chlorines were removed with the accumulation of tetrachlorobiphenyls. Matturro et al. (2016) followed PCB biodegradation in microcosms containing marine sediments collected from Mar Piccolo (Taranto, Italy) chronically contaminated by PCBs. They found that concentrations of the most representative congeners detected in the original sediment decreased between 24% and 47% after 70 d of anaerobic incubation. On the other hand, evidence for aerobic degradation of PCBs has also been observed in the field based on the detection of characteristic metabolites in sediment cores (Field and Sierra-Alvarez 2008).

Measurements of PCBs in seawater are difficult because of the low levels and the potential for sample contamination. However, Gioia et al. (2008) presented one of the first comprehensive and reliable data sets of PCB concentrations in Atlantic seawater. In that study, concentrations of PCBs ranged from 0.24 to 5.7 pg l^{-1} in the Atlantic Ocean (from Germany to South Africa). Comparable PCB concentration (2.7 pg l^{-1}) was reported by Maldonado and Bayona (2002) in open sea water of the Black Sea. In that study, the highest values of dissolved PCBs were detected in the Danube prodelta (77.7–102 pg l^{-1}) and Odessa depression (59.7 pg l^{-1}), showing the relevance of rivers to the transport of dissolved organochlorines to the Black Sea. Other studies reported comparable concentrations, such as PCB concentrations in water samples collected in the Arkona Sea, a region of the southern Baltic Sea, which ranged from 8 to 17 pg l^{-1} (Bruhn et al. 2003), and seawater concentrations in the western Mediterranean, Italy, which ranged from non-detectable levels to 65 pg l^{-1} (Marrucci et al. 2013). On the other hand, concentrations of dissolved PCB were higher than previous data in samples collected from Singapore's coastal marine environment, which ranged from 0.05 to 1.8 ng l^{-1} (Wurl and Obbard 2005).

When PCBs are released into the aquatic environment, a considerable proportion becomes adsorbed to sediments, as might be expected from its lipophilicity. Pavlou and Dexter (1979) reported that approximately 80% of PCBs added to surface waters is retained in the sediment. Pozo et al. (2009) reported PCB concentrations in sediments from 15 Italian Marine Protected Areas. In that study, with the exception of a site linked to a local "hot spot", mean PCB concentrations in sediments were low and ranged from 0.1 to 8 ng g^{-1}. Those results are lower than PCB concentrations reported for surface sediment (0.8 to 19 ng g^{-1}; Sobek et al. 2014) and sedimentation material (3.02 to 11 ng g^{-1}; Korhonen et al. 2013) of the Gulf of Bothnia and the Gulf of Finland. Tolosa et al. (2010) also reported comparable levels of PCBs in sediments from Caribbean coastal areas of Cienfuegos Bay, situated in the southern central part of Cuba (1.9 to 16 ng g^{-1}). However, in environments affected by wastewater discharges, PCB levels are higher than those reported in the aforementioned sites. For example, Castells et al. (2008) found PCB concentrations ranged from 7 to 44 ng g^{-1} in coastal marine sediment samples collected near the mouth of Beso's River (Barcelona, NE Spain), at the site of wastewater treatment plant discharge. Similarly, in several industrialized coastal locations in Bahia Blanca Estuary (Argentina), Arias et al. (2013) reported that PCB levels in sediment averaged 24 ± 10 ng g^{-1}.

Hoekstra et al. (2002) studied the distribution of PCB in water and zooplankton samples from locations in the Alaskan Beaufort Sea coast and the Canadian Arctic archipelago. They reported PCB concentrations in zooplankton in the same range than previous studies (from 46 to 207 ng g^{-1} l.w.). On the other hand, Knickmeyer and Steinhart 1989 reported that plankton from British and Continental coastal in the North Sea contained higher concentrations of PCBs (3–28 ug g^{-1} l.w.) compared to less polluted central North Sea (< 1–7 ug g^{-1} l.w.).

The soft tissue of mussels can provide information of greater environmental relevance because of the ability to accumulate levels of toxic compounds far in excess of those in the surrounding water or sediments (Muncaster et al. 1990). Thompson et al. (1999) presented the results of PCB concentrations in bivalves from the Arcachon Bay on the Atlantic coast of France. They reported very low PCB concentrations in bivalves, ranging from 2.15 to 6.28 ng g^{-1} l.w. (13–34 ng g^{-1} d.w.). On the other hand, Vorkamp et al. (2010) compared and assessed PCB levels in bivalves from marine environments on a global scale. In that study, the concentration of PCBs in samples from the Solomon Islands, Tasmania, and the Chilean archipelago were below detection limits. The highest PCB concentrations were found in samples from Boston Harbor and Sydney estuary, ranging from 147 to 358 ng g^{-1} d.w. If normalized to lipid content (about 2%), the PCB concentrations increased to 6,000–15,000 ng g^{-1} l.w. Comparable PCB concentrations (8,200–15,300 ng g^{-1} l.w.) were reported in bivalves from the coastal zone of Ushuaia (Argentina), where anthropogenic activities in the city of Ushuaia and the surrounding areas had affected the environmental quality of the bay (Amin et al. 2011).

Bodin et al. (2007a) obtained data for the concentration ranges of PCBs in the hepatopancreas of large crustaceans from the French coasts. They reported that the concentration of PCBs was about 6- to 40-fold higher in crustaceans from the area exposed to large contaminant discharge from the Seine River (1776–6332 ng g^{-1} l.w.) than in those measured in non-impacted areas (154–970 ng g^{-1} l.w.).

PCB concentration and pattern in fish tissue have been shown to vary with the feeding types of fish species (Weber and Goerke 2003). For example, PCBs were quantified in bluefin tuna collected from the southern Tyrrhenian Sea (Corsolini et al. 2007). Bluefin tuna feed on diverse food items depending on their age; thus, they occupy different trophic levels during their lifespan. Concentrations ranged from 13 to 16,839 ng g^{-1} l.w., increasing significantly with size and trophic levels. Froescheis et al. (2000) studied the levels and patterns of organochlorines in deep-sea or bottom-dwelling biota in comparison to surface-living species. In that study, PCB concentrations in fish living in the upper layers of the North Atlantic and the South Atlantic and at the continental shelf of California were compared to the levels in deep-sea or bottom-dwelling fish within the same geographic area. The deep-sea biota showed significantly higher burdens (480–2080 ng g^{-1} l.w.) than surface-living species of the same region (108–220 ng g^{-1} l.w.), suggesting that biota of the deep-sea act also as a global sink for PCBs.

Regard fish collected from wastewater-impacted coastal waters, Lewis et al. (2002) reported that fish from northwestern Florida and southwestern Alabama exhibited average PCB concentrations from 1 to 69 ng g^{-1} w.w. Similarly, Greenfield et al. (2005) reported that PCB concentrations in fish collected in urbanized water bodies of San Francisco Bay ranged from 13 to 191 ng g^{-1} w.w. (3250 to 6143 ng g^{-1} l.w.).

Marine birds and mammals have served as useful environmental contamination indicators because many species are top-level predators and have a high potential to biomagnify pollutants. For example, Magellanic penguins, *Spheniscus magellanicus*, were good indicators of environmental pollution in temperate regions of South America. Baldassin et al. (2016) studied penguins from the Pacific Ocean (Chile) and Atlantic Ocean (Brazil and Uruguay), and they found that PCB concentrations in liver of penguins on the Chilean coast (178 ng g^{-1} w.w.) were significantly higher than in specimens from the other areas (21 and 104 ng g^{-1} w.w.). The authors related those levels to the fact that all penguins on the Chilean coast were found near Concepcion, which is considered the most industrialized region in the country. Similarly, Leonel et al. (2010) studied PCB concentrations in the blubber of Franciscana (*Pontoporia blainvillei*), a small cetacean with a distribution restricted to the southwest Atlantic Ocean. They found a steady state of PCB levels in Franciscana between 1995 and 2005, ranging from 1,135 to 10,555 ng g^{-1} l.w., which was higher than those reported in common and Fraser's dolphins (1977–7541 ng g^{-1} l.w.) from Argentina in the South Atlantic (Durante et al. 2016). Notably, García-Alvarez et al. (2014) reported PCB concentrations two-fold greater than the threshold for adverse biological effects (17,000 ng g^{-1} l.w.; Jepson et al. 2005) in bottlenose dolphins from the Canary Islands located in the Atlantic Ocean. Comparable results were reported for bottlenose dolphins from locations along the US East Coast, the Gulf of Mexico, and Bermuda (33–450 ug g^{-1} l.w.; Kucklick et al. 2011). Those studies show that legacy persistent organic compounds such as PCBs, while banned or phased out of production, are still pervasive threats to marine animal health.

Based on the previously reported data, a conceptual model of PCB concentrations in a background and an impacted marine ecosystem is shown in Fig. 2.

3. Organochlorine Pesticides

Organochlorine pesticides (OCPs) were used extensively for both agricultural and sanitary purposes. The first organochlorine insecticide to be widely used was 1,1,1-trichloro-2,2-bis(p-chlorophenyl) ethane (DDT; Fig. 3). After World War II, DDT gained ground as an insecticide for many purposes. Then, hexachlorocyclohexane (HCH), and particularly the gamma-isomer (lindane; Fig. 3), began to be intensively used as a contact insecticide, especially in agriculture and forestry. In the next decade, a group of OCPs called cyclodienes were introduced in North America and Europe. These include the compounds aldrin, dieldrin, endrin, heptachlor, endosulfan, and chlordane (Fig. 3). Although substantial portions of applied pesticide were dissipated at the site of application through chemical and biological degradation processes, still a reasonable fraction of the OCP residues reached the oceans through agricultural run-off, atmospheric transport, and sewerage discharge. Additionally, OCPs have environmental significance because of their stability, toxicity, and tendency to concentrate on the lipids of organisms.

DDT and some of its relative compounds act as nerve poisons. They are able to bind tightly to a site located on certain sodium channels that pass through the nerve membrane. When a nerve is affected by DDT, the pore stays open for a long time, with the consequence that the passage of action potential along the nerve is disrupted (Walker 2014). The metabolic transformations of DDT take place slowly, and particularly significant is the transformation into the highly persistent metabolite DDE. In birds, the metabolite DDE can inhibit the transport of calcium ions from blood to developing eggshell, so that the shell becomes thin and fragile.

The OCP compounds used throughout mid-latitudes and the tropics over the past decades have been introduced into high latitudes by long-range transport via the atmosphere, river discharge, and surface ocean current. As long ago as 1966, Sladen et al. reported DDT contamination in the Antarctic ecosystems, and since then there has been an increasing interest in studying and monitoring

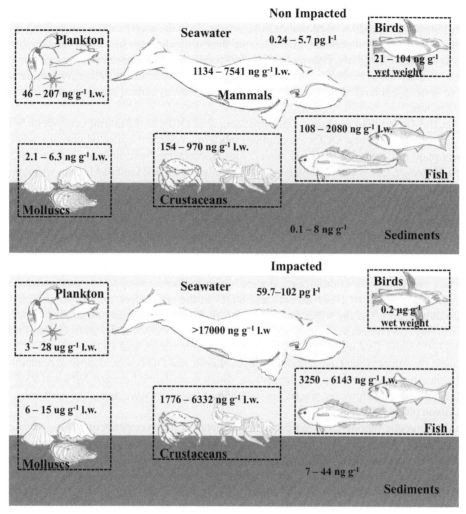

Fig. 2. A conceptual model of PCB concentrations in a background (upper) and an impacted marine ecosystem (bottom).

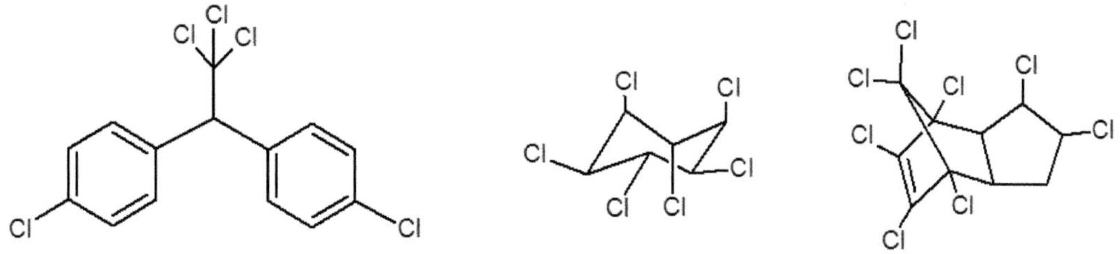

Fig. 3. Structures of DDT (left), lindane (center), and chlordane (right).

the presence of pollutants in pristine areas of the world. Actually, except the DDT for sanitary use, OCPs have been phased out in most countries.

Transport of OCPs by rivers and streams passing from the main agricultural and urban areas is the main source of pollution in the coastal and marine environment. In some regions, heavy annual floods cause a high transport of contaminants from agricultural areas to the sea. For example, in

India, OCPs, especially DDT and HCH, were used extensively till recently for agricultural and sanitary purposes and OCP concentrations differ markedly in the eastern and western coast of India, reflecting differing agricultural and other uses and their ultimate input into the coastal environment (Sarkar et al. 2008). Similarly, Haynes et al. (2000) found that OCP concentrations in sediment and seagrass collected along the Great Barrier Reef and greater Queensland coastline are associated with intensive agricultural land use (primarily sugarcane production) carried out along the coast in high rainfall regions.

Also, the discharge of sewage is a potential source of OCPs to the marine ecosystem. Syakti et al. (2012) found the highest OCP concentration in marine sediment collected near the wastewater treatment plant of Cortiou-Marseille (France), with a decreasing trend in total OCP concentrations seaward. However, Lewis et al. (2000) have demonstrated that in the Gulf of Mexico, which receives wastewaters from about 3,700 discharge points, chlorinated pesticides were minor contaminants in the sediments and were usually below the method detection limit values.

On a regional scale, the atmospheric input has a wider scope for impact over the sea, because the riverine flux may influence only the nearshore areas, since much of the river load of OCPs will be deposited near the mouth of the river (GESAMP 1989). Wurl and Obbard (2005) studied the wet deposition of OCPs in Singapore's coastal marine environment. OCP concentrations in the sea-surface microlayer after rainfall period were higher by factors of up to 50; thus, wet deposition was a significant source of HCH isomers and DDTs to the sea-surface microlayer in Singapore's coastal environment. On the other hand, Park et al. (2001) estimated annual flux of OCPs to Galveston Bay (US) waters via atmospheric deposition. In that study, the dry and wet deposition was 0.14, 1.73, 0.75 and 1.94 $\mu g\ m^{-2}\ yr^{-1}$ for HCB, HCHs, chlordanes, and DDTs, respectively. In the same study, the authors found that chlordanes, cyclodienes, and DDTs were deposited predominantly in particulate phase via wet and dry deposition, which contributed between 84% and 93% of the total direct deposition. However, the HCHs with higher solubility were deposited predominantly in the dissolved phase (89%).

In the Antarctic marine environment, two independent measurements of DDTs indicated that 1–4 $kg\ yr^{-1}$ DDTs were currently being transported to the sea because of glacier ablation (Chiuchiolo et al. 2004, Geisz et al. 2008). Glaciers accumulate atmospherically deposited contaminants, and melting glacial ice provides a source of contaminants to surface waters during the spring and summer seasons. Coincidently, Dickhut et al. (2005) demonstrated that atmospheric influx of HCHs was still occurring in Antarctic coastal waters on the basis of water/air fugacity ratios.

It appears that plastic debris adsorbs, accumulates and transports persistent organic pollutants such as OCPs. Plastic materials constitute one of the most persistent macroscopic pollutants in oceanic waters and beaches in the world. The only pesticide that we found associated with the plastic debris was DDT and its metabolites DDD and DDE, and the concentration of DDTs ranged from 22 to 7100 ng/g (Rios et al. 2007).

The detection of OCPs today is explained mainly by their persistence, long half-life and lipophilicity of the parent compound as well as of the metabolic forms into which they transform. It is plausible that significant partitioning to sediments may increase the aquatic lifetime of certain pesticides by suppressing the rate of hydrolysis and photochemical degradation (Weber et al. 2010). The persistence of OCPs in water and sediment has a wide half-life range. For example, the persistent half-life of DDT in aquatic environments has been suggested to be approximately 4 yr (Carvalho et al. 1992). Half-lives for DDE in tropical marine sediment microcosms ranged from 35 wk under methanogenic conditions at room temperature to 495 wk under sulfidogenic conditions at 10°C (Quensen et al. 2001).

The concentration of OCPs in dissolved phase has been reported mainly for HCHs, a more soluble compound. The HCHs were the only organochlorine pesticides routinely detected in dissolved phase of the Antarctic seawater (Dickhut et al. 2005). Concentrations of α-HCH ranged from 1.65 to 4.54 pg l^{-1}, and concentrations of γ-HCH ranged from 0.90 to 10.6 pg l^{-1} in seawater to coastal

Antarctic seas. Lohmann et al. (2009) reported the comparable concentration of HCHs in surface water samples from across the North Atlantic and "European" Arctic seawater (0.85–9.61 pg l^{-1}). In this last study, the concentration of DDTs in dissolved phase ranged from 0.24 to 0.74 pg l^{-1}. In a study in the coastal regions of Mumbai, the concentration of OCPs in samples of seawater was three orders of magnitude higher than those of the Arctic and Antarctic Sea (Pandit et al. 2002). The high concentrations of HCH isomers (0.16–16 ng l^{-1}) and DDT and its metabolites (3.0–33 ng l^{-1}) were attributed to the use of these insecticides in agricultural as well as anti-malaria sanitary activities throughout India.

In the marine environment, OCPs can be dissolved in the water phase or associated with particles depending on their hydrophobicity and organic carbon-water partition coefficients. In a study conducted in the Arcachon Bay, in the Atlantic coast of France, six DDT-related compounds were reported in sediments (Thompson et al. 1999). The total concentration was 1.1 ng g^{-1} for the DDTs, and the profile was dominated by DDT and DDD. Strandberg et al. (1998) reported slightly higher levels of OCPs in sediments from the Bothnian Sea. In that study, DDT and HCH concentrations ranged from 1.9 to 6.9 ng g^{-1} and from 5.0 to 7.0 ng g^{-1}, respectively. In the more polluted area, in the Mediterranean coast around the sewage outfall of the wastewater treatment plant of Cortiou Marseille (France), the concentrations of DDTs were 0.7 to 114.3 ng g^{-1} in sediment samples (Syakti et al. 2012). Similarly, sediments from the northern Queensland (Australia) nearshore are contaminated with a range of pesticides because of intensive agricultural land uses (primarily sugarcane production). Haynes et al. (2000) reported lindane (0.08 ± 0.19 ng g^{-1}), dieldrin (0.05 ± 0.37 ng g^{-1}), DDT (0.05 ± 0.26 ng g^{-1}), and DDE (0.05 ± 0.26 ng g^{-1}) concentrations in nearshore sediments from the Great Barrier Reef and greater Queensland coastline. For the Caribbean area of Cuba, concentrations of DDTs in the sediments of Cienfuegos Bay varied from 0.26 to 12.8 ng g^{-1}. The high values of DDT residue levels were accounted for by the discharge of untreated effluents from the Cienfuegos Hospital and the cities of Junco Sur and Tulipán. DDT was used in Cuba since 1950 on a variety of agricultural crops and for the control of human disease vectors. The largest agricultural use of DDT was on rice production before its ban in 1989 (Tolosa et al. 2010).

The bioaccumulation of OCPs in the Alaskan and Canadian Arctic planktonic copepods was studied by Hoekstra et al. (2002). In general, the most common OCPs in both water and zooplankton samples were HCHs, the more water-soluble, less hydrophobic compounds. In planktonic copepods, HCHs (9.6–30 ng g^{-1} l.w.) was the dominant analyte group, followed by chlordanes (2.6–17 ng g^{-1} l.w.) and DDTs (4.3–21 ng g^{-1} l.w.).

Knickmeyer and Steinhart (1989) determined HCB, HCHs, and DDE in the North Sea plankton during the declining spring plankton bloom. The amounts of HCB range from 1 to 49 ng g^{-1} l.w. in zooplankton and from 21 to 380 ng g^{-1} l.w. in phytoplankton. The concentrations of DDE range from 4 to 115 ng g^{-1} l.w. in zooplankton and from 80 to 620 ng g^{-1} l.w. in phytoplankton. Samples from the northern North Sea, with the exception of the Scottish coast and the Shetlands, show lower concentrations than those from the southern North Sea. Sources of DDE pollution are Humber, Thames, Rhine, Ems and the waters of the Baltic Sea. The concentrations of lindane range from 2 to 130 ng g^{-1} l.w. in zooplankton and from 60 to 810 ng g^{-1} l.w. in phytoplankton. Samples influenced by North Atlantic currents between Scotland and Shetland as well as samples from the central North Sea show high concentrations.

Mussels are used in many marine pollution monitoring and OCP assessment studies. The OCP levels in mussel on a global scale were studied by Vorkamp et al. (2010). In that study, comparing the lowest, but detectable, concentrations with the highest ones, the total DDT concentrations span three orders of magnitude (0.19–138 ng g^{-1}). In the low range of concentrations, Thompson et al. (1999) reported very low DDT concentrations in mussels of Arcachon Bay, in the Atlantic coast of France (4.32 ng g^{-1} l.w.). On the other hand, Kurt and Ozkoc (2004) conducted a mussel monitoring survey along the mid-Black Sea coast of Turkey in order to assess concentrations of OCPs. In that study, concentrations of DDTs in mussels ranged from 45 to 184 ng g^{-1} l.w. (considering 3% of lipid content), higher than obtained from similar studies. Similarly, Naso et al. (2005) reported DDT concentration in blue mussel (177 ng g^{-1} l.w.) from the Gulf of Naples, which is subject to OCP pollution from the intensively cultivated areas, the highly populated urban centers, and the large industrial complexes clustered along the coast.

Bodin et al. (2007b) evaluated the concentration ranges of OCPs in different edible crustaceans from the French coasts. In spider crab hepatopancreas, DDE levels varied between 4.5 and 52 ng g^{-1} l.w. The specimens from a site localized in the Seine Bay were the most contaminated, which confirmed the exposure to large contaminant discharges from the Seine. Magalhães et al. (2012) reported OCP levels in crabs from two different places inside the Santos Bay and Moela Island near one of the most economically important metropolitan areas in southern Brazil. Total DDT in the crabs exhibited the highest concentrations, ranging from 154 to 410 ng g^{-1} l.w. Among DDT metabolites, only the reduced (DDD) or oxidized form (DDE) was observed in crab tissues.

In a global assessment of residue levels of OCPs in fish, skipjack tuna collected from Asian, Seychelles, and Brazilian offshore waters and open seas from the southern hemisphere were significantly lower than those collected from the northern hemisphere (Ueno et al. 2003). In skipjack tuna from the southern hemisphere, the concentration of DDTs, HCHs and chlordanes ranged from 31 to 92, < 0.3 to 2.7, and 7.8 to 63 ng g^{-1} l.w., respectively. On the other hand, in skipjack tuna from the northern hemisphere, the concentration of DDTs, HCHs, and chlordanes ranged from 30 to 670, 0.8 to 28 and 13 to 150 ng g^{-1} l.w., respectively. Considerably higher concentrations of DDTs were found in the muscle of fish collected in the Gulf of Lions (NW Mediterranean), which ranged from 2242 to 4423 ng g^{-1} l.w. (Solé et al. 2001). Consistently, the concentration of DDTs ranged from 1275 to 2091 ng g^{-1} l.w. in sport fish collected in San Francisco Bay in the summer (Greenfield et al. 2005). In the latest study, Greenfield et al. (2005) documented changes in contamination of OCPs over time at seasonal, interannual, and decadal time scales. It is a well-known fact that trophic habits can influence the levels of organochlorine compounds in fish tissue. In this regard, Corsolini et al. (2005) reported that in bluefin tuna, DDE concentrations (9.1–375 ng g^{-1} l.w.) were a function of size and trophic level, while no correlations were observed for HCB (0.003–5.4 ng g^{-1} l.w.).

In marine birds, Geisz et al. (2008) evaluated the levels of DDTs in a resident Antarctic apex predator, the Adélie penguin. In the Antarctic food web, the majority of higher trophic level organisms rely heavily on krill (*Euphausia superba* and *E. crystallorophias*). The DDTs in the fat of Adélie penguins from the Palmer Archipelago and Cape Crozier ranged from 101 to 450 ng g^{-1} l.w. On the other hand, Ruus et al. (2002) reported levels of OCPs in the herring gull (*Larus argentatus*) collected in the Hvaler and Torbjørnskjær Archipelago, located in the outer Oslofjord in southeastern Norway, close to the Swedish border. The area receives water from rivers running through rich agricultural areas and with several industrial activities. In the herring gull, DDTs had the highest concentration, averaging 6285 ± 6667 ng g^{-1} l.w.

Aguilar et al. (2002) reviewed data about levels of OCPs in selected species of marine mammals with a wide geographical distribution and spatial patterns of variation. Marine mammals from the temperate fringe of the northern hemisphere, particularly fish-eating species inhabiting the mid-latitudes of Europe and North America, display the greatest DDT loads. Similarly, Muir et al. (2000) evaluated circumpolar trends in levels and proportions of persistent OCPs in ringed seal. The ringed seal is the most abundant Arctic pinniped with a circumpolar distribution, making this species an ideal candidate for examining spatial trends of persistent organic pollutants in the Arctic. In that study, highest concentrations of DDT were present in samples from Yenisey Gulf and Jarfjord in northern Norway (1184–6840 ng g^{-1} l.w). Lowest levels of these OCs were found in samples from NW Greenland and Arctic Bay (297–1832 ng g^{-1} l.w). Kucklick et al. (2011) sampled blubber of live bottlenose dolphins along the US Atlantic coast, which was used to identify geographic variations in OCP concentrations and patterns. DDTs constituted the compound class with the next highest concentration ranging from 8.03 µg g^{-1} l.w. in Biscayne Bay-south up to 51.0 µg g^{-1} l.w. in dolphins sampled near Cape May. In order of decreasing rank, concentrations of DDTs were followed by

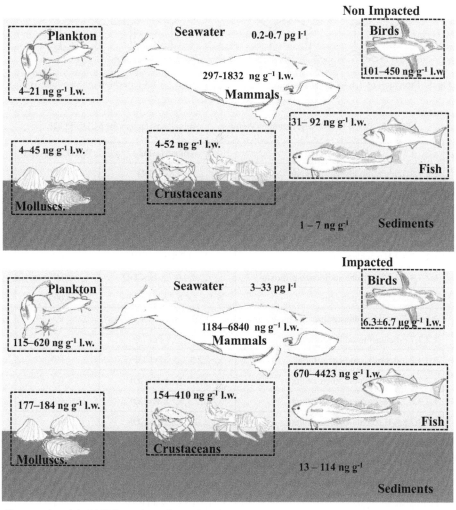

Fig. 4. Conceptual model of OCP in marine and coastal environments, which compares DDT concentrations in a background (upper) and an impacted marine ecosystem (bottom).

chlordane, mirex, dieldrin, and HCB. On the other hand, Leonel et al. (2010) reported the relatively low concentration of OCPs in blubber samples of *Pontoporia blainvillei* (Franciscana dolphin) from southern Brazil, especially when compared to concentrations in the northern hemisphere. The concentrations of DDTs in Franciscana dolphin blubber ranged from 230 to 3447 ng g^{-1} l.w.

4. Polychlorinated Dibenzo-*p*-dioxins and Polychlorinated Dibenzofurans

Polychlorinated dibenzo-*p*-dioxins (PCDDs) and polychlorinated dibenzofurans (PCDFs) are persistent organic pollutants collectively termed "dioxins". There are 210 chemically different PCDD/Fs depending on the number and position of chlorine atoms around the two benzene rings (Fig. 5). PCDD/Fs have received considerable public and scientific attention because of the acute toxicity of 2,3,7,8-tetrachlorodibenzo-p-dioxin (2378-TCDD), which has one of the lowest known lethal dose values (Klaassen 2008). The toxicity of the 17 most dangerous congeners is scaled against the most toxic congener, 2378-TCDD, through toxic equivalence factors, and overall equivalent toxicity

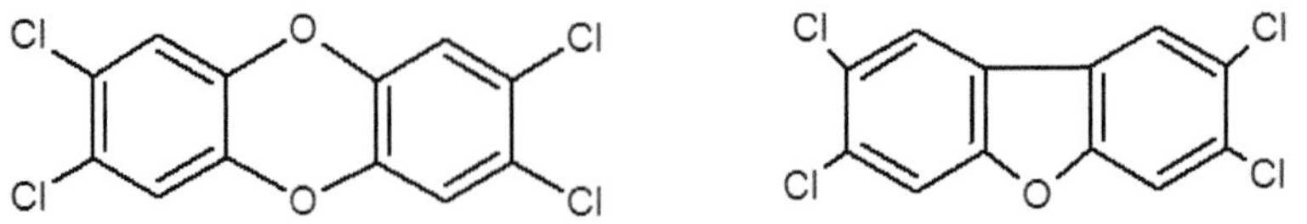

Fig. 5. Structures of 2,3,7,8-tetrachlorodibenzo-*p*-dioxin (left) and 2,3,7,8-tetrachlorodibenzo-furan (right).

Table 2. World Health Organization (WHO) toxic equivalency factors for PCDD/PCDF.

IUPAC name	Abbreviation	WHO 2005 TEF
2,3,7,8-Tetrachlorodibenzo-p-dioxin	2,3,7,8-TCDD	1
1,2,3,7,8-Pentachlorodibenzo-p-dioxin	1,2,3,7,8-PeCDD	1
1,2,3,4,7,8-Hexachlorodibenzo-p-dioxin	1,2,3,4,7,8-HxCDD	0.1
1,2,3,6,7,8-Hexachlorodibenzo-p-dioxin	1,2,3,6,7,8-HxCDD	0.1
1,2,3,7,8,9-Hexachlorodibenzo-p-dioxin	1,2,3,7,8,9-HxCDD	0.1
1,2,3,4,6,7,8-Heptachlorodibenzo-p-dioxin	1,2,3,4,6,7,8-HpCDD	0.01
Octachlorodibenzo-p-dioxin	1,2,3,4,6,7,8,9-OCDD	0.0003
2,3,7,8-Tetrachlorodibenzofuran	2,3,7,8-TCDF	0.1
1,2,3,7,8-Pentachlorodibenzofuran	1,2,3,7,8-PeCDF	0.03
2,3,4,7,8-Pentachlorodibenzofuran	2,3,4,7,8-PeCDF	0.3
1,2,3,4,7,8-Hexachlorodibenzofuran	1,2,3,4,7,8-HxCDF	0.1
1,2,3,6,7,8-Hexachlorodibenzofuran	1,2,3,6,7,8-HxCDF	0.1
1,2,3,7,8,9-Hexachlorodibenzofuran	1,2,3,7,8,9-HxCDF	0.1
2,3,4,6,7,8-Hexachlorodibenzofuran	2,3,4,6,7,8-HxCDF	0.1
1,2,3,4,6,7,8-Heptachlorodibenzofuran	1,2,3,4,6,7,8-HpCDF	0.01
1,2,3,4,7,8,9-Heptachlorodibenzofuran	1,2,3,4,7,8,9-HpCDF	0.01
Octachlorodibenzofuran	1,2,3,4,6,7,8,9-OCDF	0.0003

of complex mixtures (TEQs) can be obtained to assess health risks (Table 2; Van den Berg et al. 2006). PCDD/F compounds are generated as unwanted by-products from different processes. These involve combustion, municipal waste incineration (Tuppurainen et al. 1998, Fiedler 1998, Fiedler et al. 2000), metal industries (Antunes et al. 2012), bushfires (Gullett et al. 2008), and prescribed and uncontrolled burning (Lemieux 2000, Lundin et al. 2013). PCDD/Fs are formed through homogeneous and heterogeneous pathways (Fiedler 1998). The homogeneous pathway involves the reaction of structurally related precursors in the gas phase, especially chlorinated phenols (Altarawneh et al. 2009b). The ratio of PCDD/Fs in the gas phase depends strongly on the temperature, between 400 and 800°C, oxygen concentration and the chlorination patterns of the chlorophenols (Altarawneh et al. 2009b). Other compounds can serve as potential precursors for the formation of PCDD/Fs, including chlorobenzene (Sommeling et al. 1994), hydroquinone (Truong et al. 2008) and permethrin (Altarawneh et al. 2009a). The heterogeneous pathways take place between 250 and 450°C by catalytic-assisted coupling of precursors, and by the so-called *de novo* process, which proceeds through burnoff of a carbonaceous matrix with simultaneous oxidation and chlorination in the presence of oxygen (Altarawneh et al. 2009b). It is widely accepted that transition-metal species, especially copper compounds, are the most efficient to catalyze the formation of PCDD/Fs through the two heterogeneous pathways (Fiedler 1998).

There is growing evidence that PCDD/Fs are extremely harmful to marine and freshwater ecosystems, especially when they bioaccumulate through aquatic food webs (Sundqvist 2009). The sources of PCDD/F in the marine coastal environment are atmospheric deposition, river discharge, and direct runoff. The total estimated dry aerosol deposition of PCDD/Fs to the Atlantic Ocean ranged from 200 (Baker and Hites 1999) to 500 kg yr^{-1} (Jurado et al. 2005). Coincident dry deposition fluxes of PCDD/F were reported in the Black and Mediterranean Seas, ranging from 4 to 210 kg yr^{-1} (Castro-Jiménez et al. 2010). These results are background deposition fluxes but PCDD/Fs atmospheric inputs are expected to be higher in coastal areas and during large combustion events. Salamanca et al. (2016) found increases in PCDD/F concentration in Chilean coastal waters before large forest fires near the coastal zone, reflecting the change of atmospheric flux of PCDD/Fs due to particular events. With regard to point discharges to the coast, industrial wastewater is also an important source of PCDD/F to marine ecosystems. PCDD/F concentration in sediments of Swedish coastal areas of the Baltic Sea showed that pulp, paper, and other wood industry activities have strongly affected PCDD/F levels and patterns in the surface sediments, and previous use of chlorophenols appear to still be playing a major role (Sundqvist et al. 2009). In the Norwegian coast, NE North Sea wastewater emission of PCDD/Fs from a magnesium plant was estimated to be approximately 500 g TEQ yr^{-1} in 1988; according to regulations, annual TEQ emission to water had to be reduced to 5 g (Musdalslien et al. 1995). As a result of improper wastewater treatment practices associated with vinyl chloride monomer production in the 1970s, PCDD/Fs leached into the Gulf of Finland in the Baltic Sea (11.4 kg of PCDD/s or 32.0 g of TEQ) resulted in about 26 km^2 polluted area near the plant (Isosaari et al. 2000). Thompson et al. (1992) presented levels of dioxins and furans associated with sewer outfalls in Australia and its impact on deep water. In the Melbourne sewerage system, in the Port Phillip Bay, Australia, concentrations of total PCDD/Fs in the treatment complex varied from 0.7 to 18 pg l^{-1}. Consequently, in the sediments from the Port Phillip Bay, the PCDD/F concentrations varied from 9 to 62,300 pg g^{-1} for PCDDs and from 0.2 to 1700 pg g^{-1} for PCDFs (Thompson et al. 1992).

Reports of PCDD/F levels in the open sea or coastal waters are scarce in the literature. Concentrations reported by Castro-Jiménez et al. (2010) in open Mediterranean waters ranged from 42 to 64 pg m^{-3} (3–5 pg TEQ m^{-3}). Similarly, Ishaq et al. (2009) reported concentrations of PCDD/F in surface water from fjords in southern Norway. PCDD/F ranged from 0.35 to 0.67 ng m^{-3} (5.4–8.3 pg TEQ m^{-3}) in a reference area in Skagerrak Sea, and 7.4 to 160 ng m^{-3} (16–2857 pg TEQ m^{-3}) in coastal waters in Greenland fjords in the south of Norway, which have since the 1950s been substantially polluted by PCDD/F discharges from a magnesium production plant that operated from 1951 until 2002.

The major sink for PCDD/Fs in the marine environment is sediment burial, and once PCDD/Fs enter the marine environment they will remain available in the marine environment for a long time (Armitage et al. 2009). Sinkkonen and Paasivirta (2000) reported an average degradation half-life— photodegradation and biodegradation—of PCDD/Fs in marine sediments of 75 yr based on work done in the Baltic, where the average temperature is 7°C; a rough estimate of temperature dependence on environmental half-life suggests a 2.2- to 3-fold increase/decrease per 10°C temperature change (Sinkkonen and Paasivirta 2000).

Castro Jimenez et al. (2013) measured dioxins for the first time in deep-sea biota and sediment samples in the Blanes canyon, the largest submarine canyon in the Catalan Sea (northwestern Mediterranean Sea). PCDD/F concentrations in bottom sediment from the Blanes submarine canyon ranged from 102 to 681 pg g^{-1} (1.34 to 5.59 pg TEQ g^{-1}), and they were generally in the range of those reported for surface sediments in the Mediterranean Sea coastal areas and in aquatic ecosystems worldwide that have low to moderate anthropogenic impact. Similarly, Guidotti et al. (2013) reported that the concentration of PCDD/F ranged from 12 to 193 pg g^{-1} (0.12–1.25 pg TEQ g^{-1}) in twelve marine sediments collected during November 2009 in six different coastal areas in Italy. Dioxin levels in the sediment have been reported to be higher in the Gulf of Finland (731–76,900 pg g^{-1}), and it is assumed that the Kymijoki River with high levels of PCDD/F in sediments—up to

350,000 TEQ pg g^{-1}—is the main contributor of dioxins to the Gulf of Finland (Korhonen et al. 2013). Similarly, in their aforementioned work, Ishaq et al. (2009) reported concentrations of PCDD/F between 25 and 730 ng g^{-1} (2.2–10 TEQ ng g^{-1}) in sediments from Greenland fjords coastal impacted area.

In aquatic systems, PCDD/F adsorb easily to particles (Zhao et al. 2011), accumulate in aquatic organisms, and are biomagnified through the food chain to humans (Kim et al. 2009). However, differences in the PCDD/F composition in the food affect their assimilation. Naito et al. (2003) assessed PCDD/Fs concentration in a food web from the Tokyo Bay, which comprised phytoplankton, zooplankton, debris of surface sediment, shellfish, benthic invertebrates, and fish. Their results showed that bioaccumulation patterns of PCDD/Fs vary greatly among congeners. The patterns of total PCDD/Fs exhibited decreasing concentrations of higher chlorinated congeners with an increase in trophic level, while the 2,3,7,8-substituted PCDDs/DFs congeners with a lower degree of chlorination showed increasing concentrations with an increase in trophic level. Similarly, Ruus et al. (2006) evaluated the PCDD/F concentration through marine food webs in two neighboring fjords with different exposure history in the Skagerrak area, southern Norway. Their results indicated that octachlorinated dioxins and furans constituted the highest percentages in the pollution source and they found a limited bioaccumulation of PCDD/Fs through the food webs because of lower bioaccumulation of higher chlorinated congeners. The low uptake of the higher chlorinated congeners due to reduced membrane permeability, and possibly slow transport through intestinal aqueous phases because of low aqueous solubility, best explains those findings.

The studies about PCDD/Fs in the lowest marine trophic levels (e.g., plankton) show highly variable concentrations, indicating site-specific pollution and other factors such as plankton biomass (Morales et al. 2015) and species-specific differences (Naito et al. 2003). For example, Morales et al. (2015) investigated PCDD/F levels in plankton from the Atlantic, Pacific, and Indian Oceans. They found highly variable concentrations of PCDD/Fs ranging from 0.01 to 6.6 pg TEQ g^{-1} d.w. (1–66 pg TEQ g^{-1} l.w.). On the other hand, Naito et al. (2003) reported that PCDD/F concentration in plankton from Tokyo Bay was about 98 ng g^{-1} l.w. (370 pg TEQ g^{-1} l.w.). Broman et al. (1992) reported comparable PCDD/F levels in plankton collected in the Stockholm archipelago, between 1,233 and 11,316 pg g^{-1} l.w. (22–149 pg TEQ g^{-1} l.w). In general, the PCDD/F pattern in plankton showed a predominance of octa- and heptachlorinated dioxins and furans.

Recently, Carro et al. (2018) collected species of bivalve mollusk from several points of Galician littoral; these coastal zones are productive and receive many contributions of contaminants that become a potential risk for marine species. During the period from 2006 to 2014, PCDD/Fs in mollusk ranged from 0.03 to 0.62 pg TEQ g^{-1} w.w. (0.4 to 9 pg TEQ g^{-1} l.w.), below the level considered safe for human consumption (4 pg TEQ g^{-1} w.w.; European Commission 2006), and consistent with levels reported in clam and mussel of the Adriatic Sea (0.07–0.24 pg WHO-TEQ g^{-1} w.w.; Bayarri et al. 2001). As part of the NOAA National Status and Trends Program (Wade et al. 2014), concentrations of PCDD/F in bivalves from coastal locations along the US East, West, and Gulf Coasts, Hawaii and Alaska were reported, and they ranged from 4.6 to 78 pg g^{-1} w.w. (12 to 122 pg TEQ g^{-1} l.w.). In crustaceans from northwest of the Mediterranean Sea, Castro-Jiménez et al. (2013) reported PCDD/F concentrations of 220 to 795 pg g^{-1} l.w. (13 to 90 pg TEQ g^{-1} l.w), two orders of magnitude lower than those reported for crustaceans from Norwegian fjords (about 1300 to 3200 pg TEQ g^{-1} l.w.; Ruus et al. 2006).

Ueno et al. (2005), in order to elucidate the global distribution of dioxins in offshore waters and open seas, used skipjack tuna (*Katsuwonus pelamis*) as a bioindicator. Skipjack tuna were collected from various regions in the world (offshore waters of Japan, Taiwan, Philippines, Indonesia, Seychelles and Brazil, and the Japan Sea, the East China Sea, the South China Sea, the Indian Ocean, and the North Pacific Ocean) between 1997 and 2001. Residue levels of PCDD/Fs in skipjack tuna from the northern hemisphere (ND-72 pg g^{-1} l.w.) were found to be apparently higher than those from the southern hemisphere (ND-4.2 pg g^{-1} l.w.). Among the regions surveyed in that study, higher contamination levels of PCDD/Fs were found in offshore waters around temperate Asian regions. The estimated toxic equivalents were in the range of 0.02–0.57 pg TEQs g^{-1} w.w. (0.43–15 pg TEQs g^{-1} l.w.). Similarly, Brown et al. (2006) collected marine fish from California and reported levels to fall within < 0.3 to 1.5 pg TEQ g^{-1} w.w. (3.8–193 pg TEQs g^{-1} l.w.), and they found the highest levels in the San Francisco Bay (enclosed bay, highly urbanized) and the Port of Oakland (port in an industrialized area). Consistently, Bayarri et al. (2001) reported PCDD/F levels to fall within 0.07 and 1.07 pg TEQ g^{-1} w.w. in fish from Italian marine environment in the Adriatic Sea. In a more impacted environment, Isosaari et al. (2000) reported that in several wild fish species (perch, pike, sculpin, and flounder) caught near old vinylchloride monomer industry, in the Gulf of Finland in the Baltic Sea, the concentration of PCDD/F (0.71–8.72 ng TEQ g^{-1} w.w.) was 2–9 times higher than the levels measured in fish from background sites of the Gulf of Finland (0.48–2.73 pg TEQ g^{-1} w.w.).

Marine mammals are top-level predators in the marine ecosystem. Because of their long life spans and the lower metabolic capacity of toxic organic contaminants, marine mammals have been used as bioindicators of contamination by persistent organic pollutants including PCDD/Fs. Brazilian coastal cetaceans (Franciscana, Guiana and rough-toothed dolphins) from five Brazilian states were used as indicators of micropollutant trophic flow (Dorneles et al. 2013). The PCDD/F concentrations in hepatic tissue and blubber ranged from 27 to 448 pg g^{-1} l.w. (about 4.0–14 pg TEQ g^{-1} l.w.). As expected, concentrations found in the blubber of rough-toothed dolphins from Brazil that feed on top-predator fish were among the highest ever reported for marine mammals (PCDD at 747 pg g^{-1} l.w.; PCDF at 1313 pg g^{-1} l.w.). Pinzone et al. (2015) reported PCDD/F levels in large cetaceans permanently inhabiting the Mediterranean Sea (sperm whale, *Physeter macrocephalus*, and the long-finned whale, *Balaenoptera physalus*). In blubber biopsies collected between 2006 and 2013 from a relatively confined area (off the southern coast of France), they found concentrations ranging from 0.14 to 0.52 ng g^{-1} l.w. (12–57 pg TEQ g^{-1} l.w.). Jeong et al. (2016) reported temporal trends of PCDD/Fs in the blubber of finless porpoises (*Neophocaena asiaeorientalis*) collected in Korean waters between 2003 and 2010. They reported concentrations lower than threshold values for marine mammals, ranging from 0.02 to 38 pg TEQ g^{-1} l.w. They found also a significant decreasing trend between the sampling years (reduction rates of 57% and 54% for PCDDs and PCDFs respectively). Their results suggest that the regulations on dioxin-like contaminants have been effective for marine mammals in Korea.

Based on the previously reported data, a conceptual model of PCDD/F concentrations in a background and an impacted marine ecosystem is shown in Fig. 6.

5. Conclusion

The scientific knowledge and understanding of sources, exposure pathways, environmental levels, trends, and health effects of organohalogen compounds have improved considerably since the 1990s. However, there are still research gaps that should be filled. For example, data of organohalogen compounds in the marine environment are lacking for many developing countries in which the problems of inadequate waste management infrastructure and high volume of waste emissions are growing larger. On the other hand, it is necessary to understand both primary emissions and reemissions to the atmosphere from reservoirs in the global environment. As this chapter shows, atmospheric deposition is an important source of organohalogen compounds to the sea, even of compounds currently banned. Thus, improvements in sampling and analytical methodologies, monitoring programs, and environmental models are necessary to close the knowledge gap and understand how human activities impact our marine environment. The improvements must take into account data quality (standardization of sampling, storage, and analytical protocols and harmonization of analytic methods), since they are crucial to decision-making leading to resource management and contaminant regulation.

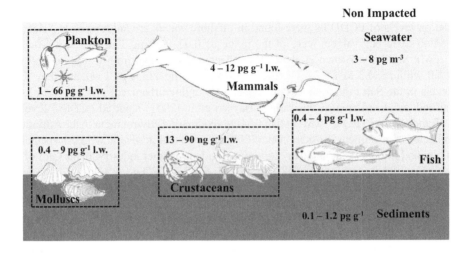

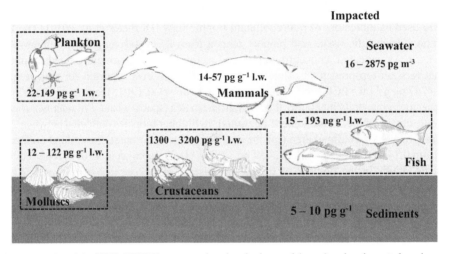

Fig. 6. A conceptual model of TEQ-PCDD/F concentrations in a background (upper) and an impacted marine ecosystem (bottom).

References

Abramowicz, D.A. 1995. Aerobic and anaerobic PCB biodegradation in the environment. Environ. Health Perspect. 103: 97–99.

Aguilar, A., A. Borrell and P.J.H. Reijnders. 2002. Patterns in geographical and temporal variation of organochlorine pollutant concentrations in marine mammals. Mar. Environ. Res. 53: 425–452.

Altarawneh, M., D. Carrizo, A. Ziolkowski, E.M. Kennedy, B.Z. Dlugogorski, J.C. Mackie. 2009a. Pyrolysis of permethrin and formation of precursors of polychlorinated dibenzo-p-dioxins and dibenzofurans (PCDD/F) under non-oxidative conditions. Chemosphere 74: 1435–1443.

Altarawneh, M., B.Z. Dlugogorski, E.M. Kennedy and J.C. Mackie. 2009b. Mechanisms for formation, chlorination, dechlorination and destruction of polychlorinated dibenzo-p-dioxins and dibenzofurans (PCDD/Fs). Prog. Energy Combust. Sci. 35: 245–274.

Amin, O.A., L.I. Comoglio and J.L. Sericano. 2011. Polynuclear aromatic and chlorinated hydrocarbons in mussels from the coastal zone of Ushuaia, Tierra del Fuego, Argentina. Environ. Toxicol. Chem. 30: 521–529.

Antunes, P., P. Viana, T. Vinhas, J. Rivera and E.M.S.M. Gaspar. 2012. Emission profiles of polychlorinated dibenzodioxins, polychlorinated dibenzofurans (PCDD/Fs), dioxin-like PCBs and hexachlorobenzene (HCB) from secondary metallurgy industries in Portugal. Chemosphere 88: 1332–1339.

Arias, A.H., A. Vazquez-Botello, G. Diaz and J.E. Marcovecchio. 2013. Accumulation of polychlorinated biphenyls (PCBs) in navigation channels, harbors and industrial areas of the Bahia Blanca estuary, Argentina. Int. J. Environ. Res. 7: 925–936.

Armitage, J.M., M.S. McLachlan, K. Wiberg and P. Jonsson. 2009. A model assessment of polychlorinated dibenzo-p-dioxin and dibenzofuran sources and fate in the Baltic Sea. Sci. Total Environ. 407: 3784–3792.

ATSDR. 2000. Toxicological profile for polychlorinated biphenyls (PCBs).

Baker, J.I. and R.A. Hites. 1999. Polychlorinated Dibenzo-p-dioxins and dibenzofurans in the remote North Atlantic marine atmosphere. Environ. Sci. Technol. 33: 14–20.

Baldassin, P., S. Taniguchi, H. Gallo, A. Maranho, C. Kolesnikovas, D.B. Amorim et al. 2016. Persistent organic pollutants in juvenile Magellanic Penguins (Spheniscus magellanicus) in South America. Chemosphere 149: 391–399.

Bayarri, S., L.T. Baldassarri, N. Iacovella, F. Ferrara and A. di Domenico. 2001. PCDDs, PCDFs, PCBs and DDE in edible marine species from the Adriatic Sea. Chemosphere 43: 601–610.

Bodin, N., A. Abarnou, D. Fraisse, S. Defour, V. Loizeau, A.M. Le Guellec et al. 2007a. PCB, PCDD/F and PBDE levels and profiles in crustaceans from the coastal waters of Brittany and Normandy (France). Mar. Pollut. Bull. 54: 657–668.

Bodin, N., A. Abarnou, A.M. Le Guellec, V. Loizeau and X. Philippon. 2007b. Organochlorinated contaminants in decapod crustaceans from the coasts of Brittany and Normandy (France). Chemosphere 67: S37–S47.

Broman, D., C. Rolff, C. Näf, Y. Zebühr, B. Fry and J. Hobbie. 1992. Using ratios of stable nitrogen isotopes to estimate bioaccumulation and flux of polychlorinated dibenzo-p-dioxins (PCDDs) and dibenzofurans (PCDFs) in two food chains from the Northern Baltic. Environ. Toxicol. Chem. 11: 331–345.

Brown, F.R., J. Winkler, P. Visita, J. Dhaliwal and M. Petreas. 2006. Levels of PBDEs, PCDDs, PCDFs, and coplanar PCBs in edible fish from California coastal waters. Lipids 64: 276–286.

Brownawell, B.J. and J.W. Farrington. 1985. Biogeochemistry of PCBs in marine sediments and interstitial waters. Mar. Estuar. Geochemistry 30: 97–120.

Bruhn, R., S. Lakaschus and M.S. McLachlan. 2003. Air/sea gas exchange of PCBs in the southern Baltic Sea. Atmos. Environ. 37: 3445–3454.

Carro, N., I. García, M. Ignacio and A. Mouteira. 2018. Polychlorinated dibenzo-P-dioxins and dibenzofurans (PCDD/Fs) and dioxin-like polychlorinated biphenyls (dl-PCBS) in bivalve mollusk from Galician Rías (N. W., SPAIN). Chemosphere 197: 782–792.

Carvalho, F., S. Fowler, J. Readman and L. Mee. 1992. Pesicide residues in tropical coastal lagoons. *In*: Applications of Isotopes and Radiation in Conservation of the Environment. International Atomic Energy Agency, 699 p.

Castells, P., J. Parera, F.J. Santos and M.T. Galceran. 2008. Occurrence of polychlorinated naphthalenes, polychlorinated biphenyls and short-chain chlorinated paraffins in marine sediments from Barcelona (Spain). Chemosphere 70: 1552–1562.

Castro-Jiménez, J., S.J. Eisenreich, M. Ghiani, G. Mariani, H. Skejo, G. Umlauf et al. 2010. Atmospheric occurrence and deposition of polychlorinated dibenzo-p-dioxins and dibenzofurans (PCDD/Fs) in the open Mediterranean Sea. Environ. Sci. Technol. 44: 5456–5463.

Castro-Jiménez, J., G. Rotllant, M. Ábalos, J. Parera, J. Dachs, J.B. Company et al. 2013. Accumulation of dioxins in deep-sea crustaceans, fish and sediments from a submarine canyon (NW Mediterranean). Prog. Oceanogr. 118: 260–272.

Chiuchiolo, A., R. Dickhut, M. Cochran and H. Ducklow. 2004. Persistent organic pollutants at the base of the Antarctic marine food web. Environ. Sci. Technol. 38: 3553–3557.

Corsolini, S., N. Ademollo, T. Romeo, S. Greco and S. Focardi. 2005. Persistent organic pollutants in edible fish: A human and environmental health problem. Microchem. J. 79: 115–123.

Corsolini, S., G. Sara, N. Borghesi and S. Focardi. 2007. Transfer and magnification in Bluefin Tuna (Thunnus thynnus) from the Mediterranean Sea. Environ. Sci. Technol. 41: 4227–4233.

Diamond, M., L. Melymuk, S.A. Csiszar and M. Robson. 2010. Estimation of PCB stocks, emissions, and urban fate: Will our policies reduce concentrations and exposure? Environ. Sci. Technol. 44: 2777–2783.

Dickhut, R.M. and K.E. Gustafson. 1995. Atmospheric inputs of selected polycyclic aromatic hydrocarbons and polychlorinated biphenyls to southern Chesapeake Bay. Mar. Pollut. Bull. 30: 385–396.

Dickhut, R.M., A. Cincinelli, M. Cochran and H.W. Ducklow. 2005. Atmospheric concentrations and air-water flux of organochlorine pesticides along the Western Antarctic Peninsula. Environ. Sci. Technol. 39: 465–470.

Diefenbacher, P.S., A.C. Gerecke, C. Bogdal and K. Hungerbu. 2016. Spatial distribution of atmospheric PCBs in Zurich, Switzerland: Do joint sealants still matter? Environ. Sci. Technol. 50: 232–239.

Dinn, P.M., S.C. Johannessen, P.S. Ross, R.W. MacDonald, M.J. Whiticar, C.J. Lowe et al. 2012. PBDE and PCB accumulation in benthos near marine wastewater outfalls: The role of sediment organic carbon. Environ. Pollut. 171: 241–248.

Dorneles, P.R., P. Sanz, G. Eppe, A.F. Azevedo, C.P. Bertozzi, M.A. Martínez et al. 2013. High accumulation of PCDD, PCDF, and PCB congeners in marine mammals from Brazil: A serious PCB problem. Sci. Total Environ. 463-464: 309–318.

Durante, C.A., E.B. Santos-Neto, A. Azevedo, E.A. Crespo and J. Lailson-Brito. 2016. POPs in the South Latin America: Bioaccumulation of DDT, PCB, HCB, HCH and Mirex in blubber of common dolphin (Delphinus delphis) and Fraser's dolphin (Lagenodelphis hosei) from Argentina. Sci. Total Environ. 572: 352–360.

Eganhouse, R.P. and P.M. Sherblom. 2001. Anthropogenic organic contaminants in the effluent of a combined sewer over flow: impact on Boston Harbor. Mar. Environ. Res. 51: 51–74.

European Commission. 2006. Commission Regulation (EC) No. 1881/2006, Official Journal of the European Union.

Fiedler, H. 1998. Thermal formation of PCDD/PCDF: a survey. Environ. Eng. Sci. 15: 49–58.

Fiedler, H., C. Lau and G. Eduljee. 2000. Statistical analysis of patterns of PCDDs and PCDFs in stack emission samples and identification of a marker congener. Waste Manag. Res. 18: 283–292.

Field, J.A. and R. Sierra-Alvarez. 2008. Microbial transformation and degradation of polychlorinated biphenyls. Environ. Pollut. 155: 1–12.

Froescheis, O., R. Looser, G.M. Cailliet, W.M. Jarman and K. Ballschmiter. 2000. The deep sea as a final global sink of semivolatile persistant organic pollutants? Part I: PCBs in surface and deep-sea dwelling fish in the North and South Atlantic and the Monteray Bay Canyon (California). Chemosphere 40: 661–670.

García-Alvarez, N., V. Martín, A. Fernández, J. Almunia, A. Xuriach, M. Arbelo et al. 2014. Levels and profiles of POPs (organochlorine pesticides, PCBs, and PAHs) in free-ranging common bottlenose dolphins of the Canary Islands, Spain. Sci. Total Environ. 493: 22–31.

Geisz, H.N., R.M. Dickhut, M.A. Cochran, W.R. Fraser and H.W. Ducklow. 2008. Melting glaciers: A probable source of DDT to the Antarctic marine ecosystem. Environ. Sci. Technol. 42: 3958–3962.

GESAMP (IMO/FAO/UNESCO-IOC/WMO/IAEA/UNEP/UNDP Joint Group of Experts on the Scientific Aspects of Marine Environmental Protection). 1989. Atmospheric Input of Trace Species to the World Oceans.

Gioia, R., L. Nizzetto, R. Lohmann, J. Dachs, C. Temme and K.C. Jones. 2008. Polychlorinated biphenyls (PCBs) in air and seawater of the Atlantic Ocean: sources, trends and processes. Environ. Sci. Technol. 42: 1416–1422.

Greenfield, B.K., J.A. Davis, R. Fairey, C. Roberts, D. Crane and G. Ichikawa. 2005. Seasonal, interannual, and long-term variation in sport fish contamination, San Francisco Bay. Water Pollut. Control 336: 25–43.

Guidotti, M., C. Protano, C. Dominici, S. Chiavarini, N. Cimino and M. Vitali. 2013. Determination of selected polychlorinated dibenzo-p-dioxins/furans in marine sediments by the application of gas-chromatography-triple quadrupole mass spectrometry. Bull. Environ. Contam. Toxicol. 90: 525–530.

Gullett, B., A. Touati and L. Oudejans. 2008. PCDD/F and aromatic emissions from simulated forest and grassland fires. Atmos. Environ. 42: 7997–8006.

Haynes, D., J. Müller and S. Carter. 2000. Pesticide and herbicide residues in sediments and seagrasses from the Great barrier reef World Heritage Area and Queensland Coast. Mar. Pollut. Bull. 41: 279–287.

Hoekstra, P.F., T.M. O'Hara, C. Teixeira, S. Backus, A.T. Fisk and D.C.G. Muir. 2002. Spatial trends and bioaccumulation of organochlorine pollutants in marine zooplankton from the Alaskan and Canadian Arctic. Environ. Toxicol. Chem. 21: 575–583.

Howard, P.H. and S. Banerjee. 1984. Interpreting results from biodegradability tests of chemicals in water and soil. Environ. Toxicol. Chem. 3: 551–562. Doi: 10.1002/etc.5620030405.

Ishaq, R., N.J. Persson, Y. Zebühr, D. Broman and K. Næs. 2009. PCNs, PCDD/Fs, and non-orthoPCBs, in water and bottom sediments from the industrialized Norwegian Grenlandsfjords. Environ. Sci. Technol. 43: 3442–3447.

Isosaari, P., T. Kohonen, H. Kiviranta, J. Tuomisto and T. Vartiainen. 2000. Assessment of levels, distribution, and risks of polychlorinated dibenzo-p-dioxins and dibenzofurans in the vicinity of a vinyl chloride monomer production plant. Environ. Sci. Technol. 34: 2684–2689.

Jartun, M., R.T. Ottesen, E. Steinnes and T. Volden. 2009. Painted surfaces–important sources of polychlorinated biphenyls (PCBs) contamination to the urban and marine environment. Environ. Pollut. 157: 295–302.

Jeong, Y., S.J. Kim, K.H. Shin, S.Y. Hwang, Y.R. An and H.B. Moon. 2016. Accumulation and temporal changes of PCDD/Fs and dioxin-like PCBs in finless porpoises (Neophocaena asiaeorientalis) from Korean coastal waters: Tracking the effectiveness of regulation. Mar. Pollut. Bull. 105: 30–36.

Jepson, P.D., P.M. Bennett, R. Deaville, C.R. Allchin, J.R. Baker and R. Law. 2005. Relationships between Polychlorinated Biphenyls and health status in harbor porpoises (Phocoena phocoena) stranded in the United Kingdom. Environ. Toxicol. Chem. 24: 238–248.

Jurado, E., F.M. Jaward, R. Lohmann, K. Jones, R. Simo and J. Dachs. 2005. Atmospheric dry deposition of persistent organic pollutants to the global oceans. Environ. Sci. Technol. 39: 2426–2435.

Kim, K.-S., S.C. Lee, K.-H. Kim, W.J. Shim, S.H. Hong, K.H. Choi et al. 2009. Survey on organochlorine pesticides, PCDD/Fs, dioxin-like PCBs and HCB in sediments from the Han river, Korea. Chemosphere 75: 580–7.

Klaassen, C.D. 2008. Casarett and Doull's Toxicology. The Basic Science of Poisons. Seventh Edition. McGraw-Hill. New York, 1331 pp.

Knickmeyer, R. and H. Steinhart. 1989. Cyclic organochlorines in plankton from the North Sea in spring. Estuar. Coast. Shelf Sci. 28: 117–127.

Korhonen, M., S. Salo, H. Kankaanpää, H. Kiviranta, P. Ruokojärvi and M. Verta. 2013. Chemosphere sedimentation of PCDD/Fs and PCBs in the Gulf of Finland and the Gulf of Bothnia, the Baltic Sea. Chemosphere, 1–7.

Kucklick, J., L. Schwacke, R. Wells, A. Hohn, A. Guichard, J. Yordy et al. 2011. Bottlenose dolphins as indicators of persistent organic pollutants in the western North Atlantic Ocean and northern Gulf of Mexico 3036. Environ. Sci. Technol. 45: 4270–4277.

Kurt, P.B. and H.B. Ozkoc. 2004. A survey to determine levels of chlorinated pesticides and PCBs in mussels and seawater from the Mid-Black Sea Coast of Turkey. Mar. Pollut. Bull. 48: 1076–1083.

Lemieux, P. 2000. Emissions of polychlorinated dibenzo-p-dioxins and polychlorinated dibenzoforuans from the open burning of household waste in Barrells. Environ. Sci. Technology 34: 7.

Leonel, J., J.L. Sericano, G. Fillmann, E. Secchi and R.C. Montone. 2010. Long-term trends of polychlorinated biphenyls and chlorinated pesticides in franciscana dolphin (Pontoporia blainvillei) from Southern Brazil. Mar. Pollut. Bull. 60: 412–8.

Lewis, M., D. Weber, R. Stanley and B. Albrecht. 2000. Treated wastewater as a source of sediment contamination in Gulf of Mexico near-coastal areas: A survey. Environ. Toxicol. Chem. 19: 192–203.

Lewis, M.A., G.I. Scott, D.W. Bearden, R.L. Quarles, J. Moore, E.D. Strozier et al. 2002. Fish tissue quality in near-coastal areas of the Gulf of Mexico receiving point source discharges. Sci. Total Environ. 284: 249–261.

Lohmann, R., R. Gioia, K.C. Jones, L. Nizzetto, C. Temme, Z. Xie et al. 2009. Organochlorine pesticides and PAHs in the surface water and atmosphere of the North Atlantic and Arctic Ocean. Environ. Sci. Technol. 43: 5633–5639.

Lundin, L., B. Gullett, W.F. Carroll, A. Touati, S. Marklund and H. Fiedler. 2013. The effect of developing nations' municipal waste composition on PCDD/PCDF emissions from open burning. Atmos. Environ. 79: 433–441.

Mackay, D., W.Y. Shiu, K. Ma and S.C. Lee. 2006. Properties and environmental fate second edition introduction and hydrocarbons. Chemphyschem A European Journal of Chemical Physics and Physical Chemistry.

Magalhães, C.A., S. Taniguchi, M.J. Cascaes and R.C. Montone. 2012. PCBs, PBDEs and organochlorine pesticides in crabs Hepatus pudibundus and Callinectes danae from Santos Bay, State of São Paulo, Brazil. Mar. Pollut. Bull. 64: 662–667.

Maldonado, C. and J. Bayona. 2002. Organochlorine compounds in the North-western Black Sea Water: Distribution and water column process. Estuar. Coast. Shelf Sci. 54: 527–540.

Marrucci, A., B. Marras, S.S. Campisi and M. Schintu. 2013. Using SPMDs to monitor the seawater concentrations of PAHs and PCBs in marine protected areas (Western Mediterranean). Mar. Pollut. Bull. 75: 69–75.

Matturro, B., C. Ubaldi, P. Grenni, A.B. Caracciolo and S. Rossetti. 2016. Polychlorinated biphenyl (PCB) anaerobic degradation in marine sediments: microcosm study and role of autochthonous microbial communities. Environ. Sci. Pollut. Res. 23: 12613–12623.

Morales, L., J. Dachs, M.C. Fernández-Pinos, N. Berrojalbiz, C. Mompean, B. González-Gaya et al. 2015. Oceanic sink and biogeochemical controls on the accumulation of polychlorinated dibenzo-p-dioxins, dibenzofurans, and biphenyls in plankton. Environ. Sci. Technol. 49: 13853–13861.

Muir, D., F. Riget, M. Cleemann, J. Skaare, L. Kleivane, H. Nakata et al. 2000. Circumpolar trends of PCBs and organochlorine pesticides in the Arctic marine environment inferred from levels in ringed seals. Environ. Sci. Technol. 34: 2431–2438.

Muncaster, B.W., P.D.N. Hebert and R. Lazar. 1990. Biological and physical factors affecting the body burden of organic contaminants in freshwater mussels. Arch. Environ. Contam. Toxicol. 19: 25–34.

Musdalslien, U., N. Standal and J. Johansen. 1995. Pilot plant test with a wet electrostatic precipitator for reducing PCDD/PCDF in corrosive off gas from magnesium production. Chemosphere 23: 1097–1108.

Naito, W., J. Jin, Y.S. Kang, M. Yamamuro, S. Masunaga and J. Nakanishi. 2003. Dynamics of PCDDs/DFs and coplanar-PCBs in an aquatic food chain of Tokyo Bay. Chemosphere 53: 347–362.

Nakao, T., O. Aozasa, S. Ohta and H. Miyata. 2006. Formation of toxic chemicals including dioxin-related compounds by combustion from a small home waste incinerator 62: 459–468.

Naso, B., D. Perrone, M.C. Ferrante, M. Bilancione and A. Lucisano. 2005. Persistent organic pollutants in edible marine species from the Gulf of Naples, Southern Italy. Sci. Total Environ. 343: 83–95.

Øfjord, G.D., J.A. Puhakka and J.F. Ferguson. 1994. Reductive dechlorination of Aroclor 1254 by marine sediment cultures. Environ. Sci. Technol. 28: 2286–2294.

Olsen, C.R., N.H. Cutshall and I.L. Larsen. 1982. Pollutant-Particle associations and dynamics in coastal marine environments: a review. Mar. Chem. 11: 501–533.

Paasivirta, J. and S.I. Sinkkonen. 2009. Environmentally relevant properties of All 209 polychlorinated biphenyl congeners for modeling their fate in different natural and climatic conditions. J. Chem. Eng. Data 54: 1189–1213.

Pandit, G.G., S.K. Sahu and S. Sadasivan. 2002. Distribution of HCH and DDT in the coastal marine environment of Mumbai, India. J. Environ. Monit. 4: 431–434.

Park, J.-S., T.L. Wade and S. Sweet. 2001. Atmospheric deposition of organochlorine contaminants to Galveston Bay, Texas. Atmos. Environ. 35: 3315–3324.

Pavlou, S.P. and R.N. Dexter. 1979. Distribution of Polychlorinated Biphenyls (PCB) in Estuarine Ecosystems. Testing the Concept of Equilibrium Partitioning in the Marine Environment. Environ. Sci. and Technol. 13(1): 65–71.

Pinzone, M., H. Budzinski, A. Tasciotti, D. Ody, G. Lepoint, J. Schnitzler et al. 2015. POPs in free-ranging pilot whales, sperm whales and fin whales from the Mediterranean Sea: Influence of biological and ecological factors. Environ. Res. 142: 185–196.

Pozo, K., D. Lazzerini, G. Perra, V. Volpi, S. Corsolini and S. Focardi. 2009. Levels and spatial distribution of polychlorinated biphenyls (PCBs) in superficial sediment from 15 Italian Marine Protected Areas (MPA). Mar. Pollut. Bull. 58: 773–776.

Quensen, J.F., J.M. Tiedje, M.K. Jain and S.A. Mueller. 2001. Factors controlling the rate of DDE dechlorination to DDMU in Palos Verdes margin sediments under anaerobic conditions. Environ. Sci. Technol. 35: 286–291.

Rios, L.M., C. Moore and P.R. Jones. 2007. Persistent organic pollutants carried by synthetic polymers in the ocean environment. Mar. Pollut. Bull. 54: 1230–1237.

Robson, M., L. Melymuk, S.A. Csiszar, A. Giang, M.L. Diamond and P.A. Helm. 2010. Continuing sources of PCBs: The significance of building sealants. Environ. Int. 36: 506–513.

Roose, P. and U.A.T. Brinkman. 2005. Monitoring organic microcontaminants in the marine environment: Principles, programmes and progress. TrAC-Trends Anal. Chem. 24: 897–926.

Ruus, A., K. Ugland and J. Skaare. 2002. Influence of trophic position on organochlorine concentrations and compositional patterns in a marine food web. Environ. Toxicol. Chem. 21: 2356–2364.

Ruus, A., J.A. Berge, O.A. Bergstad, J.A. Knutsen and K. Hylland. 2006. Disposition of polychlorinated dibenzo-p-dioxins (PCDDs) and polychlorinated dibenzofurans (PCDFs) in two Norwegian epibenthic marine food webs. Chemosphere 62: 1856–1868.

Salamanca, M., C. Chandía and A. Hernández. 2016. Impact of forest fires on the concentrations of polychlorinated dibenzo-p-dioxin and dibenzofurans in coastal waters of central Chile. Sci. Total Environ. 573: 1397–1405.

Sánchez-Avila, J., J. Bonet, G. Velasco and S. Lacorte. 2009. Determination and occurrence of phthalates, alkylphenols, bisphenol A, PBDEs, PCBs and PAHs in an industrial sewage grid discharging to a municipal wastewater treatment plant. Sci. Total Environ. 407: 4157–4167.

Sarkar, S.K., B.D. Bhattacharya, A. Bhattacharya, M. Chatterjee, A. Alam, K.K. Satpathy et al. 2008. Occurrence, distribution and possible sources of organochlorine pesticide residues in tropical coastal environment of India: An overview. Environ. Int. 34: 1062–1071.

Sinkkonen, S. and J. Paasivirta. 2000. Degradation half-life times of PCDDs, PCDFs and PCBs for environmental fate modeling. Chemosphere 40: 943–9.

Sladen, W., C. Menzie and W. Reichel. 1966. © 1966 Nature Publishing Group. Nature 210: 670–673.

Sobek, A., K. Wiberg, K.L. Sundqvist, P. Haglund, P. Jonsson and G. Cornelissen. 2014. Coastal sediments in the Gulf of Bothnia as a source of dissolved PCDD/Fs and PCBs to water and fish. Sci. Total Environ. 487: 463–470.

Solé, M., C. Porte and J. Albaigés. 2001. Hydrocarbons, PCBs and DDT in the NW Mediterranean deep-sea fish Mora moro. Deep. Res. Part I Oceanogr. Res. Pap. 48: 495–513.

Sommeling, P.M., P. Mulder and R. Louw. 1994. Formation of PCDFs during chlorination and oxidation of chlorobenzene in chlorine/oxygen mixtures around 340°C. Chemosphere 29: 2015–2018.

Strandberg, B., B. van Bavel, P. Bergqvist, D. Broman, R. Ishaq, C. Naf et al. 1998. Occurrence, sedimentation, and spatial variations of organochlorine contaminants in settling particulate matter and sediments in the northern part of the Baltic Sea. Environ. Sci. Technol. 32: 1754–1759.

Sundqvist, K. 2009. Sources of dioxins and other POPs to the marine environment. Thesis, 85 pp.

Sundqvist, K.L., M. Tysklind, P. Geladi, I. Cato and K. Wiberg. 2009. Congener fingerprints of tetra- through octa-chlorinated dibenzo-p-dioxins and dibenzofurans in Baltic surface sediments and their relations to potential sources. Chemosphere 77: 612–620.

Syakti, A.D., L. Asia, F. Kanzari, H. Umasangadji, L. Malleret, Y. Ternois et al. 2012. Distribution of organochlorine pesticides (OCs) and polychlorinated biphenyls (PCBs) in marine sediments directly exposed to wastewater from Cortiou, Marseille. Environ. Sci. Pollut. Res. 19: 1524–1535.

Tanabe, S., S. Watanabe, H. Kan and R. Tatsukawa. 1988. Capacity and mode of Pcb metabolism in small cetaceans. Mar. Mammal. Sci. 4: 103–124.

Thompson, G.B., J.C. Chapman and B.J. Richardson. 1992. Disposal of hazardous wastes in Australia: implications for marine pollution. Mar. Pollut. Bull. 25: 155–162.

Thompson, S., H. Budzinski, P. Garrigues and J.F. Narbonne. 1999. Comparison of PCB and DDT distribution between water-column and sediment-dwelling bivalves in Arcachon Bay, France. Mar. Pollut. Bull. 38: 655–662.

Tolosa, I., M. Mesa-Albernas and C.M. Alonso-Hernandez. 2010. Organochlorine contamination (PCBs, DDTs, HCB, HCHs) in sediments from Cienfuegos bay, Cuba. Mar. Pollut. Bull. 60: 1619–1624.

Truong, H., S. Lomnicki and B. Dellinger. 2008. Mechanisms of molecular product and persistent radical formation from the pyrolysis of hydroquinone. Chemosphere 71: 107–113.

Tuppurainen, K., I. Halonen, P. Ruokojärvi, J. Tarhanen and J. Ruuskanen. 1998. Formation of PCDDs and PCDFs in municipal waste incineration and its inhibition mechanisms: A review. Chemosphere 36: 1493–1511.

Ueno, D., S. Takahashi, H. Tanaka, A.N. Subramanian, G. Fillmann, H. Nakata et al. 2003. Global pollution monitoring of PCBs and organochlorine pesticides using skipjack tuna as a bioindicator. Arch. Environ. Contam. Toxicol. 45: 378–389.

Ueno, D., M. Watanabe, A. Subramanian, H. Tanaka, G. Fillmann, P.K.S. Lam et al. 2005. Global pollution monitoring of polychlorinated dibenzo-p-dioxins (PCDDs), furans (PCDFs) and coplanar polychlorinated biphenyls (coplanar PCBs) using skipjack tuna as bioindicator. Environ. Pollut. 136: 303–13.

Van den Berg, M., L.S. Birnbaum, M. Denison, M. De Vito, W. Farland, M. Feeley et al. 2006. The 2005 World Health Organization reevaluation of human and mammalian toxic equivalency factors for dioxins and dioxin-like compounds. Toxicol. Sci. 93: 223–241.

Vorkamp, K., J. Strand, J.H. Christensen, T.C. Svendsen, P. Lassen, A.B. Hansen et al. 2010. Polychlorinated biphenyls, organochlorine pesticides and polycyclic aromatic hydrocarbons in a one-off global survey of bivalves. J. Environ. Monit. 12: 1141–1152.

Wade, T.L., S.T. Sweet, J.L. Sericano, D.A. DeFreitas and G.G. Lauenstein. 2014. Polychlorinated dibenzo-p-dioxins and dibenzofurans detected in bivalve samples from the NOAA National Status and Trends Program. Mar. Pollut. Bull. 81: 317–324.

Walker, C. 2014. Ecotoxicology. Effects of Pollutants on the Natural Environment. Taylor & Francis Group, Boca Raton.

Weber, J., C.J. Halsall, D. Muir, C. Teixeira, J. Small, K. Solomon et al. 2010. Endosulfan, a global pesticide: A review of its fate in the environment and occurrence in the Arctic. Sci. Total Environ. 408: 2966–2984.

Weber, K. and H. Goerke. 2003. Persistent organic pollutants (POPs) in antarctic fish: Levels, patterns, changes. Chemosphere 53: 667–678.

Wurl, O. and J.P. Obbard. 2005. Chlorinated pesticides and PCBs in the sea-surface microlayer and seawater samples of Singapore. Mar. Pollut. Bull. 50: 1233–1243.

Yu, B.-W., G.-Z. Jin, Y.-H. Moon, M.-K. Kim, J.-D. Kyoung and Y.-S. Chang. 2006. Emission of PCDD/Fs and dioxin-like PCBs from metallurgy industries in S. Korea. Chemosphere 62: 494–501.

Zhao, X., H. Zhang, J. Fan, D. Guan, H. Zhao, Y. Ni et al. 2011. Dioxin-like compounds in sediments from the Daliao River Estuary of Bohai Sea: Distribution and their influencing factors. Mar. Pollut. Bull. 62: 918–925.

11

Polycyclic Aromatic Hydrocarbons
Sources, Occurrence, Levels, Distribution and Ecotoxicological Fate at Coastal and Deep Ocean

Ana L. Oliva,[1,*] *Ana C. Ronda,*[1,2] *Lautaro Girones,*[1] *Melina M. Orazi,*[1]
Tatiana Recabarren-Villalón,[1] *Jorge E. Marcovecchio*[1,3,4,5]
and Andrés H. Arias[1,6]

1. Introduction

Oceans cover 75% of the earth's surface, and they contain 90% of the planet's water. It is estimated that more than half of the human population (65%) lives along the coasts (Small and Nichols 2003), with a general tendency of people to move from continental regions towards the coasts (Costanza 1994, Marcovecchio et al. 2014). As a consequence, marine environments support an intensive and increasing anthropogenic pressure including inputs from agriculture, industrial and municipal wastewaters, oil spills and direct industrial discharges.

Environmental pollution by persistent organic pollutants (POPs) has triggered an increasing concern all over the world. POPs can persist in the environment for a long time and travel a long distance from their original source. POPs can be released in polluted regions far from their origin, crossing boundaries, and their control requires international efforts (Tanabe 1991, Wania and Mackay 1996, Fu et al. 2003). Moreover, POPs are able to accumulate in the environment (water, air, soils, sediments and biota) and exert biological effects on humans and wildlife. Among POPs, polycyclic aromatic hydrocarbons (PAHs) are ubiquitous contaminants in marine and coastal environments.

[1] Instituto Argentino de Oceanografía (CONICET/UNS), Camino La Carrindanga km 7.5, 8000 Bahía Blanca, Argentina.
[2] Departamento de Biología, Bioquímica y Farmacia, Universidad Nacional del Sur, Av Alem 1253, 8000 Bahía Blanca, Argentina.
[3] Universidad de la Fraternidad de Agrupaciones Santo Tomás de Aquino, Gascón 3145, 7600 Mar del Plata, Argentina.
[4] Universidad Tecnológica Nacional – FRBB, 11 de Abril 445, 8000 Bahía Blanca, Argentina.
[5] Academia Nacional de Ciencias Exactas, Físicas y Naturales, Av. Alvear 1711, 4to piso, 1014 Buenos Aires, Argentina.
[6] Departamento de Química, Area III, Química Analítica. Universidad Nacional del Sur, Av Alem 1253, 8000 Bahía Blanca, Argentina.
* Corresponding author: anaoliva@criba.edu.ar

Hundreds of different combinations of PAHs exist, but only a small group of these compounds are considered pollutants of environmental importance. This chapter presents general aspects of PAHs as environmental pollutants and synthesizes information on the presence and levels of PAHs in the coasts and deep sea throughout the world.

2. Chemical Characteristics of PAHs

PAHs are a large group of compounds formed of two or more fused benzene rings or, in some cases, a pentagonal ring. Additionally, many families of PAHs present functional groups that can include diverse elements. Based on the molecular structure, PAHs can be divided into two categories, namely low molecular weight PAHs (LMW-PAHs) with two or three aromatic rings, and high molecular weight PAHs (HMW-PAHs) with four or more rings.

PAHs are solid at room temperature and characterized by low steam pressures and low water solubility, as well as high melting and boiling points. In general, as molecular weight increases, their hydrophobicity, adsorption to particulate matter, toxicity and persistence in the environment tend to increase. It was established that the aqueous solubility of PAHs decreases with each additional ring in their structure (Masih et al. 2010). PAH compounds also have high octanol-water coefficients (K_{ow}). This aspect of PAHs is extremely important, since their low solubility in water and high affinity for organic compounds determine their behavior and distribution in the environment, facilitating their bioaccumulation in organisms and concentration in different environmental matrices. Other characteristics of these compounds are light sensitivity, heat resistance, conductivity, emittability, corrosion resistance, and physiological action (Masih et al. 2012). Table 1 shows the molecular structure and physicochemical properties of a few selected PAHs.

3. Sources of PAHs

PAHs can originate from three possible sources: petrogenic, pyrolytic and natural. Although PAHs can occur naturally, anthropogenic activities are generally considered the major source of PAHs released into the environment.

Petrogenic PAHs are formed during geological processes such as the formation of oil and coal. PAHs are found naturally in fossil fuels such as crude oil and coal, being the result of a slow process that suffer the organic matter at low temperatures and high pressures. In this process, the formation of derivative products such as alkylated PAHs is favored (Neff 1979). Examples of petrogenic PAH sources include oceanic and freshwater oil spills, leaks of underground storage tanks, and small emissions of gasoline, motor oil and related substances associated with marine shipping (Tolosa et al. 1996, Abdel-Shafy and Mansour 2016, Hussain et al. 2018).

Pyrolytic PAHs result from an incomplete, short-duration combustion of organic matter at high temperatures (Neff 1979, Burgess et al. 2003). Although the combustion may have natural origins, such as forest fires or volcanic eruptions, these classes of PAHs are frequently associated with anthropogenic activities and urban development. Incomplete combustion, derived from both natural and anthropogenic origin, has been identified as the single largest contributor of PAHs in the environment (Zhang and Tao 2009, Abdel-Shafy and Mansour 2016). On the one hand, pyrolytic PAHs can be intentionally formed, during the destructive distillation of coal into coke and coal tar or the thermal cracking of petroleum residuals into lighter hydrocarbons. On the other hand, they can be unintentionally formed during the incomplete combustion of motor fuels in cars and trucks and the incomplete combustion of fuel oils in heating systems.

Natural PAHs can be formed by the transformation of organic compounds present in soils or sediments under conditions of anoxia or hypoxia (*diagenetic PAHs*), or they can be produced by direct

Table 1. Molecular structure and physicochemical properties of PAHs.

Compound	Molecular weight	Molecular structure	Melting point (°C)	Boiling point (°C)	Solubility in water (mg/L)	Log K_{ow}
Naphthalene	128		80	218	31	3.37
Acenaphthylene	152		92	270	16.1	4
Acenaphthrene	154		96	279	3.8	3.92
Fluorene	166		116	295	1.9	4.18
Phenanthrene	178		101	339	1.1	4.57
Anthracene	178		216	340	0.045	4.54
Fluoranthene	202		111	375	0.26	5.22
Pyrene	202		156	360	0.132	5.18
Benzo[a]anthracene	228		160	435	0.011	5.91
Chrysene	228		255	448	0.002	5.86
Benzo[b]fluoranthene	252		168	481	0.0015	5.8
Benzo[k]fluoranthene	252		217	481	0.0008	6

Table 1 contd. ...

...Table 1 contd.

Compound	Molecular weight	Molecular structure	Melting point (°C)	Boiling point (°C)	Solubility in water (mg/L)	Log K_{ow}
Benzo[a]pyrene	252		179	495	0.0038	6.04
Indeno[1,2,3,c-d]pyrene	276		164	536	Insoluble	6.58
Dibenzo[a,h]anthracene	278		267	524	0.006	6.75
Benzo[gh]perylene	276		277	500	0.00026	6.5

biosynthesis from organisms such as bacteria, fungus and some insects (*biogenic PAHs*) (Hites et al. 1980, Burgess et al. 2003, Kovács et al. 2008, Pietzsch et al. 2010, Nakata et al. 2014). For example, phenanthrene can be synthesized by precursors such as alkyl-phenanthrenes present in plant tissues (Sims and Overcash 1983, Kovács et al. 2008), while perylene can be produced in marine sediments under anoxic conditions (Venkatesan 1988, Wilcke et al. 2003).

There are important differences in the chemical composition of PAH mixtures depending on the sources of emission. In general, pyrolytic PAHs present dominance of high molecular mass compounds (corresponding to compounds of four to six rings), while petrogenic PAHs present dominance of compounds of two to three rings (Sanders et al. 2002, Dahle et al. 2003). Moreover, the pyrolysis process produces PAHs associated with soot carbon. This is a much stronger association and persists when PAHs are deposited in aquatic systems, affecting their partitioning and bioavailability. Moreover, PAHs associated with soot carbon appear to be more resistant to environmental degradation (Burgess et al. 2003, Thorsen et al. 2004).

4. Distribution and Fate of PAHs in the Marine Environment

PAHs can be introduced into marine environments by different ways: spillage of fossil fuels, ship traffic, atmospheric depositions, urban runoff, and municipal and industrial wastewater discharge. Their final fate in the environment will be influenced by many physicochemical processes and the characteristics of the media in which PAHs reside (Fig. 1).

Atmospheric transport is the most important pathway for environmental distribution of PAHs (Birgül et al. 2011, Abdel-Shafy and Mansour 2016). Particle-bound PAHs can be transported long distances from the emission source and removed from the atmosphere through precipitation and dry deposition to remote areas. Consequently, PAHs have been detected around the world even at sites located far from industrial activity, such us the tropics and polar regions (Wania and Mackay 1996, Wilcke 2007).

In addition to the chemical nature of the PAHs, it must be taken into account that in surface water PAHs can be transformed by photooxidation, chemical oxidation, and microbial metabolism

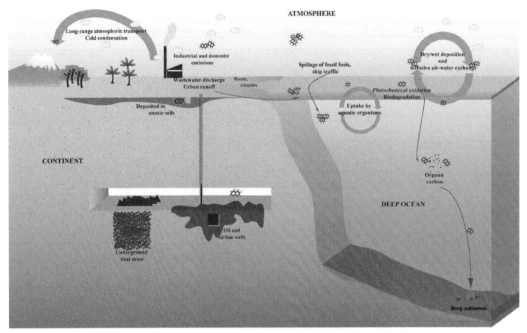

Fig. 1. Distribution and fate of PAHs in the marine environment.

Color version at the end of the book

(Patrolecco et al. 2010), changing their original chemical nature. Owing to their hydrophobic nature in aquatic environments, PAHs are transported from surface waters by volatilization and sorption to settling particles and, finally, deposited in sediments, which are the most important PAH reservoir in the marine environment (Soclo et al. 2000, Culotta et al. 2006, Qiao et al. 2006, Chen and Chen 2011, Acquavita et al. 2014). Sediments can act as stores of PAH, which under specific hydrodynamic conditions or changes in the aquatic systems (i.e., bioturbation, dredging, acidification) can be re-suspended or re-dissolved, becoming bioavailable for aquatic organisms (Katayama et al. 2010, Maletić et al. 2019).

5. PAHs in Seawater

Inputs of PAHs in seawater could be related to different sources such as untreated wastewater discharge, urban runoff, refinery effluents, vessel discharge and/or spills, vehicular emission, and atmospheric deposition as well as seasonal hydrological variation (Neff 1979, Suman et al. 2016, Han and Currell 2017). PAH distribution between dissolved and particulate fractions depends on both the intrinsic solubility of each PAH and the availability of binding substrates, such as suspended particles and organic matter content (Latimer and Zheng 2003).

An extensive revision of the available literature of the last ten years shows a wide range of PAH levels in seawater. Figure 2 shows PAH distribution in several seawater samples worldwide. Concentration of these compounds varied from lower than the limit detection to more than 6000 ng/L. The largest numbers of studies were conducted in Asia (especially in China) and Europe. Conversely, there are few studies that were conducted to assess PAHs in America.

The highest PAH concentrations are generally found associated with anthropogenic coastal activities, such as shipping harbors and industrial cores (Lipiatou and Saliot 1991, Cincinelli et al.

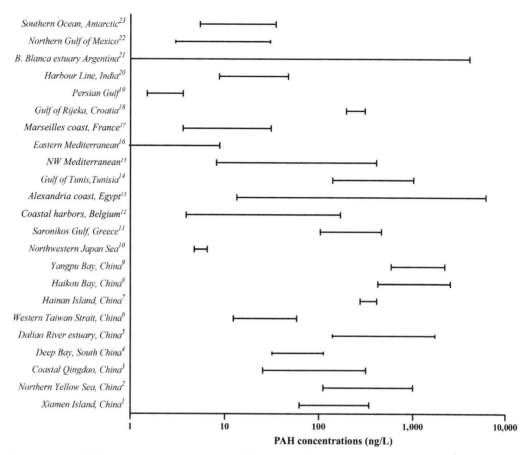

Fig. 2. Range of PAH concentration in seawater (ng PAHs/L) from different surface water bodies of the world. (1) Ya et al. 2014, (2) Zong et al. 2014, (3) Jin et al. 2014, (4) Qiu et al. 2009, (5) Guo et al. 2007, (6) Wu et al. 2011, (7) Xiang et al. 2018, (8) Li et al. 2015a, (9) Li et al. 2015b, (10) Chizhova et al. 2013, (11) Valavanidis et al. 2008, (12) Monteyne et al. 2013, (13) El-Naggar et al. 2018, (14) Mzoughi and Chouba 2011, (15) Guigue et al. 2014, (16) Berrojalbiz et al. 2011, (17) Guigue et al. 2011, (18) Bihari et al. 2007, (19) Mirza et al. 2012, (20) Dhananjayan et al. 2011, (21) Arias et al. 2009, (22) Botello et al. 2015, (23) Cao et al. 2018.

2001, Wurl and Obbard 2004, Guitart et al. 2007, Louvado et al. 2015). In terms of PAH profiles, LMW-PAHs are the predominant compounds in most sampling sites (Guo et al. 2007, Valavanidis et al. 2008, Wu et al. 2011, Ya et al. 2014, Guigue et al. 2014, Jin et al. 2014, Zong et al. 2014, Han and Currell 2017), while HMW-PAHs are in a much smaller proportion. This trend could be explained by the chemical characteristics of PAHs. On the one hand, PAH solubility decreases as mass weight increases. On the other hand, as the octanol-water partition coefficient (Kow) increases, the HMW-PAHs tend to be absorbed onto or associated with organic particles present in the water column or sediments. Therefore, HMW-PAHs are more readily enriched in sediments than in waters, due to their lower aqueous solubility and hydrophobic nature (Patrolecco et al. 2010, Han and Currell 2017).

6. PAHs in Marine Sediments

Once in the aquatic media, PAHs are distributed between dissolved compounds, colloids, suspended particulate matter, surface sediments and biota (Readman et al. 1987, Zhou et al. 1998, King et al. 2004).

Despite the great variety of matrices in which PAHs can be associated, their distribution is influenced by intrinsic physicochemical properties such as solubility and hydrophobicity, making sediments their primary site of storage (Latimer and Zheng 2003). Moreover, environmental behavior and bioavailability of PAHs are source dependent. Burgess et al. (2003) concluded that, in general, petrogenic PAHs tend to be associated with local or point sources (e.g., refineries, close to road and navigation channels), while pyrolytic PAHs are widely geographically distributed and often dominate aquatic environments in terms of concentrations and geographical distributions.

PAHs associated with sediments are considered to be less bioavailable to organisms and therefore less toxic for aquatic biota than PAHs dissolved in seawater. However, because of re-suspension or bioturbation, storms, dredging and tides, PAHs can be removed from sediments to the water columns (Eggleton and Thomas 2004, Maletić et al. 2019).

In addition to the contaminant physicochemical properties, sedimentation is influenced by a number of physicochemical parameters of sediments, such as their adsorption capabilities, their partitioning constant at the water–sediment interface, the total organic carbon content, and size particle distribution (Viguri et al. 2002, Culotta et al. 2006, Fei et al. 2011, Chen et al. 2013, Bigus et al. 2014, He et al. 2014). Generally, smaller particles exhibit a higher ratio of surface area to volume than larger ones and thus have higher organic carbon content. Consequently, the content of the adsorbed contaminants in smaller particles is often higher (Chiou et al. 1998, He et al. 2014).

As mentioned before, marine sediments represent the most important reservoir of pollutants (Gagnon and Fisher 1997, Li et al. 2015a, Abdollahi et al. 2013) and are considered to be a permanent source of these pollutants as dissolved and particle-bound contaminants to the overlying waters (Zhang et al. 2017). Accordingly, sediments represent important archives of environmental information and a source of knowledge about the possible fate of pollutants over the long term (Tuncel and Topal 2015, Yuan et al. 2017). Moreover, the vertical profiles of PAHs in dated sediment cores can be employed to examine pollution mechanisms, which are significant for predicting future pollution tendencies and assessing potential environmental risks (Bigus et al. 2014).

The scientific interest in evaluating the presence of PAHs in sediment is reflected in the large number of papers on this topic worldwide. Table 2 provides an exhaustive review from data available of PAH concentrations in the last ten years grouped by continents. Since the focus is on seawater, the largest number of studies were carried out in Asia, especially in diverse marine water bodies of China, followed by America and Europe. A great variance of PAH concentrations occurred in sampling sites, with a range from lower than the limit detection (LOD) to 1,099,410 ng/g dry wet (d.w.) from Asia, < 0.01 to 23,400 ng/g d.w. from Oceania, < LOD to > 32,000 ng/g d.w. from America, < LOD to 30,100 ng/g d.w. from Europe, < LOD to 14,100 ng/g d.w. from Africa and 0.54 to 1908 ng/g d.w. from Antarctica. The highest concentrations were found at the Gulf of Kutch, Gujarat, India (118,280 to 1,099,410 ng/g dw), while the lowest were found in Antarctic sites.

In general, PAH levels were higher in harbors and urban/industrial core and tended to decrease as the distance from these sites increased (Lipiatou and Saliot 1991, Soclo et al. 2000, Latimer and Zheng 2003). According to the data available, the composition of PAHs in sediments is dominated by mixed sources (petrogenic/pyrolytic) (e.g., Sprovieri et al. 2007, Johnson-Restrepo et al. 2008, Liu et al. 2012, Chen et al. 2013), with a high proportion of HMW compounds (four to six rings), in contrast to the general distributions in dissolved or vapor phases, which tend to be dominated by compounds of two to three rings.

7. Sources, Fate and Occurrence of PAHs in Deep Ocean

PAH analysis in deep ocean sediments presents additional challenges compared to coastal sediments because of the sampling difficulty and high cost. In order to obtain deep sediments, researchers use corer box, corer gravity, Van Veen dredges and unmanned vehicles. In addition to the complex logistics, another factor is the size area and the sampling intensity needed to obtain representative

Table 2. Range and media of PAH concentration (ppb dry weight) in sediments from worldwide locations.

Continent	Sampling site	Country	Nº PAHs	Range (ppb)	Media (ppb)	Sources	Reference
	Mediterranean coast	Egypt	16	3.5–14,100	1,524.49	mainly petrogenic	Barakat et al. 2011
	Mediterranean coast		16	208.69–1,020.02	530.68	mainly petrogenic	El Nemr et al. 2013
	Abu Qir Bay		16	<LOD-2,660	478	mainly pyrolytic	Khairy et al. 2009
	Abu Qir Bay, Eastern Harbors, Western Harbors and El Max Bay		15	4.2–886.08	169.37	mixed (pyrolytic/petrogenic)	Shreadah et al. 2011
	Northern Red Sea		16	0.74–456.91	32.94	mainly pyrolytic	Salem et al. 2014
	Gulf of Suez		16	1,667.02–2,671.27	2,072.68	mainly pyrolytic	Younis et al. 2018
	Mediterranean coastline of Morocco (Tangier Nador)	Morocco	–	2,000–198,000	50.850	biogenic and petrogenic inputs	Er-Raioui et al. 2009
	Togo coastline	Togo	21	4–257	92	mainly pyrolytic	Gnandi et al. 2011
	Gulf of Gabes	Tunisia	17	174.50–10,769	1,707.99	mixed (pyrolytic/petrogenic)	Zaghden et al. 2014
	Khniss Tunisian coast, Jarzouna coast in Bizerte		17	916.40–4,541.10	2,254.44	mixed (pyrolytic/petrogenic)	Zrafi-Nouira et al. 2010
	West coast of the Gulf of Tunis		24	530.60–3,736	1,401.1	mixed (pyrolytic/petrogenic)	Mzoughi and Chouba 2011
	Gulf of Aden	Yemen	16	2.10–199	26.59	mixed (pyrolytic/petrogenic)	Mostafa et al. 2009
	Zanzibar archipelago	Zanzibar	11	25.04–76.70	33.25	mixed (pyrolytic/petrogenic)	Mahugija et al. 2017
South America	San Jorge Gulf, Patagonia Argentina	Argentina	16	<LOD-94.90	45.47	mainly pyrolytic	Commendatore et al. 2015
	Bahía Blanca estuary		16	15–10,260	1,500	mainly pyrolytic	Arias et al. 2010
	Bahía Blanca Estuary		17	19.70–30,054.50	1,798.50	mixed (pyrolytic/petrogenic)	Oliva et al. 2015
	Ushuaia Bay		17	<LOD-360	163.14	mixed (pyrolytic/petrogenic)	Commendatore et al. 2012
	Mar del Plata Port		16	180–17,094	4,292.24	mainly pyrolytic	Laitano et al. 2015

Table 2 contd. ...

...Table 2 contd.

Continent	Sampling site	Country	N° PAHs	Range (ppb)	Media (ppb)	Sources	Reference
	Paranagua Bay	Brazil	16	36.9–89.10	64.31	–	Rizzi et al. 2016
	Babitonga Bay		16	8.70–5,489	565.80	mixed (pyrolytic/petrogenic)	Dauner et al. 2018
	SE coastal region of Brazil		37	30.7–8,064	727	mixed (pyrolytic/petrogenic)	Pinheiro et al. 2017
	Mucuripe and Pecém Harbors		33	<1–292	77.95	mainly pyrolytic	Buruaem et al. 2016
	Capibaribe estuarine system		16	<LOD-497.6	222.72	mixed (pyrolytic/petrogenic)	Maciel et al. 2015
	Itajaí-Açu estuarine system,		16	28.40–1,237.20	217.90	mixed (pyrolytic/petrogenic)	Frena et al. 2017
	São Sebastião Channel		23	<LOD-370	117.38	mainly petrogenic	Da Silva and Bícego 2010
	Paranagua estuarine system and Santos estuarine system		16	9.59–1,406.13	258.17	mixed (pyrolytic/petrogenic)	Torres et al. 2012
	Paranagua Bay		16	26.33–406.76	157.30	mixed (pyrolytic/petrogenic)	Froehner et al. 2010
	São Paulo River estuary		16	11.45–1,825.35	392.48	mainly pyrolytic	Nascimento et al. 2017
	Santos–São Vicente estuarine system		–	193.02–4,803.62	1,499.05	–	Buruaem et al. 2013
	The Santos estuary		38	60–8,680	-	mixed (pyrolytic/petrogenic)	Martins et al. 2011
	Gulf of Cariaco	Venezuela	13	<LOD-5,090	511.33	mainly petrogenic	Romero et al. 2013
	Colombian shoreline	Colombia	16	<LOD-89.9	26.14	mainly pyrolytic	Caballero-Gallardo et al. 2015
	Cartagena Bay, Totumo marsh, and Caimanera marsh		16	98.4–3,210	1,248	mixed (pyrolytic/petrogenic)	Johnson-Restrepo et al. 2008
	Montevideo harbor	Uruguay	16	303.80–3,261.60	1,428.66	mainly petrogenic	Muniz et al. 2015
	Inner sea of Chiloé Island	Chile	16	49.2–821	210.40	mainly pyrolytic	Montory et al. 2008
	San Vicente Bay		16	290–6,118	2,025	mainly pyrolytic	Pozo et al. 2011
	Chilean coast		10	125–38,734	14,459	mixed (pyrolytic/petrogenic)	Aguirre-Martínez et al. 2009

North America	Estero de Urias	Mexico	11	27–418	158.19	mixed (pyrolytic/petrogenic)	Jaward et al. 2012
	Pacific coast, in the Gulf of California, Gulf of Tehuan tepec and southern part of the Gulf of Mexico		16	14.90–195	62.24	mixed (pyrolytic/petrogenic)	Piazza et al. 2008
	South Texas coast of the Gulf of Mexico		16	1–328 (ppm)	92.64 (ppm)	mainly pyrolytic	Sabourin et al. 2012
	Northern Gulf of Mexico		43	44–160	–	mainly petrogenic	Adhikari et al. 2016
	Gulf of Mexico		16	175–856	–	mixed (pyrolytic/petrogenic)	Wang et al. 2014
	Gulf of Guacanayabo	Cuba	–	120–140	124	–	Arencibia-Carballo et al. 2017
	Cienfuegos bay		40	450–10,500	3,956.47	mixed (pyrolytic/petrogenic)	Tolosa et al. 2009
	Coast of British Columbia	Canada	10	43–665	173.53	mainly petrogenic	Yunker et al. 2014
	Coast of British Columbia		16	3.70–19,400	2,333.03	mainly petrogenic	Harris et al. 2011
	San Diego Bay, California	EEUU	14	<LOD–911	146.85	mainly pyrolytic	Neira et al. 2017
	Southern California coast		28	77–11,076	6,768	–	Sabin et al. 2010
Europe	Protected natural reserve of estuary of Urdaibai, Bay of Biscay	Basque Country	16	0.70–140	–	mainly pyrolytic	Cortazar et al. 2008
	Rijeka Bay, northern Adriatic	Croatia	11	53–12,532	–	mixed (pyrolytic/petrogenic)	Alebic-Juretic 2011
	Gemi Konagi, Girne and Gazi Magusa	Cyprus	16	4.90–76	49.13	mainly pyrolytic	Darılmaz et al. 2013
	Eastern Mediterranean	Greece	14	2.10–180.70	67.35	–	Tsapakis et al. 2010
	Hellenic coastal zone, eastern Mediterranean		21	108–26,633	2,374.02	mixed (pyrolytic/petrogenic)	Botsou and Hatzianestis 2012

Table 2 contd. ...

...Table 2 contd.

Continent	Sampling site	Country	N° PAHs	Range (ppb)	Media (ppb)	Sources	Reference
	Italian marine protected areas	Italy	16	0.71–1,550	155.26	pyrolytic	Perra et al. 2011
	Gulf of Gela		16	2.40–434	–	–	Orecchio et al. 2010
	Northern Adriatic		16	310–1,550	6,806.75	pyrolytic	Guerra 2012
	Porto Marghera, Venice		15	–	54,000	–	Colacicco et al. 2010
	Svalbard	Norwegian Arctic	15	25–38	30.50	petrogenic	Jiao et al. 2009
	Porto region	Portugal	16	51.98–54.79	–	–	Rocha et al. 2011
	Spanish northern continental shelf	Spain	13	22–47,528	–	mixed (pyrolytic/petrogenic)	Viñas et al. 2010
	Estuary of Urdaibai, Bay of Biscay		10	856–3,495	1,638.45	combustion origin (both coal and fuel)	Bustamante et al. 2010
	Ukrainian estuaries, Crimean Peninsula	Ukraine	16	<LOD–82,700	–	mixed (pyrolytic/petrogenic)	Burgess et al. 2011
Asia	Persian Gulf and the Gulf of Oman	China	43	0.30–3,540	163.50	mixed (pyrolytic/petrogenic)	de Mora et al. 2010
	Bohai Bay		16	37.20–206.60	134.20	mixed (pyrolytic/petrogenic)	Qian et al. 2016
	Daya Bay		16	42.50–158.20	126.20	mixed (pyrolytic/petrogenic)	Yan et al. 2009
	Yellow Sea		18	27–224	82	pyrolytic	Liu et al. 2012
	Beibu Gulf		15	3.01–388	95. 50	mixed (pyrolytic/petrogenic)	Li et al. 2015c
	Yantai coast		16	450–4,299	2,492.90	pyrolytic	Lang and Yang 2014
	Rizhao		16	76.40–27,512	2,622.60	mixed (pyrolytic/petrogenic)	Lang et al. 2013
	Zhanjiang Bay and Leizhou Bay		16	21.72–933.90	210.47	mixed (pyrolytic/petrogenic)	Huang et al. 2012a
	Dalian, Liaodong Peninsula		46	172–4,700 (winter; 71.10–1,090 (summer)	1,140 (winter); 196 (summer)	mixed (pyrolytic/petrogenic)	Hong et al. 2016

	Location	Country	n	Range	Mean	Source	Reference
	Bohai Sea		10	24.70–2,079.40	130.60	mainly pyrolytic	Zhang et al. 2009
	Bohai Bay and adjacent shelf		16	140.60–300.70	188	mainly pyrogenic	Hu et al. 2010
	East China Sea		16	257–1,381	592	mainly from mixed combustion and petroleum	Deng et al. 2013
	Nantong coast, China		16	1.40–87.10	19.90	petrogenic	Liu et al. 2017
	Bhavnagar, Gujarat	India	16	5,020–981,180	345,000	mixed (pyrolytic/petrogenic)	Dudhagara et al. 2016
	Gulf of Kutch, Gujarat		16	118,280–1,099,410	321,635	petrogenic	Rajpara et al. 2017
	Imam Khomeini Port, Persian Gulf	Iran	16	2,885.8–5,482.23	4,031.66	pyrolytic	Abdollahi et al. 2013
	Taean Peninsula	Korea	nr	2–530,000	38,003	Hebei Spirit oil spill	Lee et al. 2013
	Korean coast		16	15.58–1,082.26	302.95	mainly pyrolytic	Choi et al. 2011
	Kuwait Bay	Kuwait	30	12.90–1,286	120.14	combustible sources	Lyons et al. 2015
	Klang Strait	Malaysia	16	100.30–3,446.90	994.02	mainly pyrolytic sources with petroleum inputs	Sany et al. 2014
	Prai and Malacca rivers		16	716–7,938	2,689.67	residues of biomass, coal, and petroleum	Keshavarzifard et al. 2014
	Gulf War	Saudi Arabia	43	<LOD–1,556,888	21,189.4	oil spill	Bejarano and Michel 2010
	Qatar, Arabian Gulf		34	2.60–1,025	117.30	mixture of petroleum and combustion-derived sources	Soliman et al. 2014
	Kaohsiung Harbor	Taiwan	17	472–16,201	5,764	mainly pyrolytic	Chen and Chen 2011
Asia-Oceania	Segara Anakan Nature Reserve	Indonesia	16	375–29,517	–	petrogenic and biogenic	Syakti et al. 2013
	Losari Beach and adjacent areas, South Sulawesi		16	3,555–9947	4,967.67	mixed (pyrolytic/petrogenic)	La Nafie et al. 2016

Table 2 contd. ...

...*Table 2 contd.*

Continent	Sampling site	Country	N° PAHs	Range (ppb)	Media (ppb)	Sources	Reference
Oceania	Sydney Harbor	Australia	16	54–23,440	–	mixed (pyrolytic/petrogenic)	Aryal et al. 2013
	Kogarah		16	490–5,190	1,760	mainly pyrolytic	Nguyen et al. 2014
Antarctic	Fildes Peninsula, King George Island (Antarctica)	Antarctic	16	0.54–228	34.64	mixed (pyrolytic/petrogenic)	Préndez et al. 2011
	Admiralty Bay, King George Island, Antarctica		11	< 30–454.90	–	mainly pyrolytic	Martins et al. 2010
	Potter Cove		–	12.1–210	35,925	mainly petrogenic	Dauner et al. 2015
	Potter Peninsula, King George Island, South Shetlands Islands		25	36.50–1,908.40	595.60	mixed (pyrolytic/petrogenic)	Curtosi et al. 2009

conclusions. These limitations restrict study of pollutants in deep ocean sediments and are reflected in the scarcity of scientific literature on the subject. Most of the works were carried out by research groups from developed countries; the Mediterranean and Baltic seas and the Arctic Ocean are more often studied. Moreover, a determining factor in the viability of ocean sediment sampling is the depth. It has been possible to obtain sediments of the oceanic bottom to more than 11,000 m depth (Kato et al. 1997).

The major inputs of PAHs into open sea come from the atmosphere, such as dry/wet deposition and diffusive air-water exchange, direct discharges and continental runoffs into coastal margins and subsequent off-shelf export (Leister and Baker 1994, Lipiatou et al. 1997, Yunker et al. 2002, Parinos and Gogou 2016, Ke et al. 2017). Continental shelves and marginal seas are the crucial areas for terrigenous pollutant exchange between the coast and open sea, acting as a global sink for organic compounds such as PAHs (Ya et al. 2017, Cai et al. 2018). Consequently, because of the relatively low aqueous solubility of PAHs, these compounds are sorbed onto soil and sediments. Then, they sink to the seafloor (Lipiatou et al. 1997, Dachs et al. 2002, Adhikari et al. 2015, 2016). The half-lives of PAHs in deep sea environments increase because of the high salinity, low temperature, and absence of light (Tansel et al. 2011, Louvado et al. 2015). It has been documented that the augmented salinity increases PAH sediment adsorption and consequently reduces PAH bioavailability (Marini and Frapiccini 2013, Louvado et al. 2015). Moreover, as temperature decreases, PAH solubility is reduced and thus the biodegradation decreases (Eriksson et al. 2003, Marini and Frapiccini 2013), while in absence of light PAH photodegradation is slightly lower (Marini and Frapiccini 2013, Louvado et al. 2015). Geochemical processes such as hydrothermal activity and cold seeps could be another important source of PAH distribution in offshore sites such as the mid-Atlantic ridge (Cui et al. 2008, Konn et al. 2009, Louvado et al. 2015). PAHs are also transported through oceanic circulation and long-range atmospheric transport (Dachs et al. 1996, Jurado et al. 2004, Garcia-Flor et al. 2005, Huang et al. 2012b). Global scale studies of the relationship between atmospheric transport and air-sea exchange of organic compounds such as PAHs may make it possible to understand their mobility and impact on remote areas (Dachs et al. 2002, Pozo et al. 2009, Luek et al. 2017). According to the cold condensation hypothesis proposed by Wania and Mackay (1996), semi-volatile organic pollutants are constantly cycled from the gaseous phase in warmer latitudes to the cooler polar regions and are deposited or bioaccumulated in aquatic and terrestrial environments where revolatilization is minimal (Luek et al. 2017).

According to the literature available, PAH levels in deep ocean sediments increased towards the north and proximately to the coast. The research by Ohkouchi et al. (1999) in the Pacific Ocean excellently reflects this pattern: low PAH concentrations are at low latitudes, between 0.81 and 3.67 ng/g (15° S to 24° N) and an exponential increase is observed towards 48° N from 3.47 ng/g at 27° N to 60.6 ng/g at 48° N. Some authors point out that this increase could be due to the lower temperature and luminosity, typical of environments close to the poles, which would produce a decrease in the rate of biological and physicochemical degradation and water solubility (Tansel et al. 2011, Marini and Frapiccini 2013). In the Mediterranean Sea the highest values were found in the west, in the Gulf of Lion (1513 ng/g) at 1050 m depth (Domine et al. 1994). This zone is between Spain, France and Sardinia (Italy) and has a highly populated and industrialized area with important tributaries such as the Ebro River and the Rhone River. In the Baltic Sea, PAH level distribution was dominated by depth, the highest concentrations being found at the deepest points, and 3624 ng/g being the maximum in Eastern Gotland Basin at 237 m depth (Pikkarainen 2004). In the Arctic Ocean, PAH levels were further dominated by depth and geographical position of sampling points, with higher levels towards the Barents Sea (northern Norway, Finland and Greenland) and lower levels towards the Bering Sea (Alaska and Siberia). The highest concentration of PAHs (1426 ng/g) was found in the western Barents Sea at 200 m depth (Boitsov et al. 2009), while the maximum mean was found exactly at the north pole at 4230 m depth (Yunker et al. 2011). In the South China Sea, the highest PAH concentrations were observed in the south and at 665 m depth,

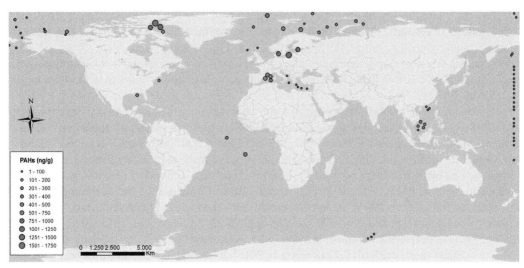

Fig. 3. Levels of PAHs in deep oceans (Domine et al. 1994, Takada et al. 1994, Tolosa et al 1996, Yunker et al. 1996, 2011, Guzzella and De Paolis 1994, Ohkouchi et al. 1999, Yang 2000, Pikkarainen 2004, Bouloubassi et al 2006, 2012, Cui et al. 2008, Boitsov et al. 2009, Shao et al. 2010, Zaborska et al. 2011, Parinos et al. 2013, Mandalakis et al. 2014, Foster et al. 2015, Romero et al. 2015, Xue et al. 2016, Zhao et al. 2016, Ma et al. 2017, Kaiser et al. 2018, Webster et al. 2018).

Color version at the end of the book

near Vietnam (Gui-Peng Yang 2000). In the Pacific Ocean, PAH levels tended to be higher toward the north pole and the highest concentration (60.6 ng/g) was found at 5334 m depth (Ohkouchi et al. 1999) (175° E, 48.10° N). In relation to the Atlantic and Antarctic oceans, it is not possible to determine trends or behavior of PAHs because studies carried out were punctual or did not have the purpose of evaluating environmental quality. However, relatively high concentrations have been found in the Mid-Atlantic Ridge (260 and 445 ng/g) (Shao et al. 2010, Cui et al. 2008) and lower concentrations were found in the eastern Antarctic Ocean (30.93 ng/g) (Xue et al. 2016).

Finally, a special case is the Gulf of Mexico, part of the Caribbean Sea, which has become one of the most extensively studied semi-enclosed seas in the world since the Deepwater Horizon oil spill in 2010. The Deepwater Horizon spill is considered to be the largest marine oil spill in the history of the petroleum industry and released ~ 2.1×10^{10} g of PAHs into the northern Gulf of Mexico waters (Reddy et al. 2012, Adhikari et al. 2016). Therefore, most of the preexisting studies in this zone no longer reflect the current state of the environment and high concentrations with high temporal variability are found. The post-spill studies also fail to reflect the reality, being simply a picture of a particular moment in the history and the natural and induced remediation process. Along with temporal variations there are also important spatial variations, with concentrations ranging from 11,600 ng/g next to the spill site to 28 ng/g a few kilometers away (Romero et al. 2015).

8. Ecotoxicological Fate of PAHs in the Marine Environment

It has been documented that chimney cleaners in England and German coal tar workers developed skin cancer in the late 19th century (Pott 1775). However, only in 1930 was it demonstrated that soot, tar and oils were carcinogenic substances and that such mixtures contained PAH compounds. The carcinogenic potential of PAHs was acknowledged and reported by Cook et al., who isolated a cancer-producing hydrocarbon from coal tar (Cook et al. 1933). With this recognition, a new biological era of research started that aimed to determine the cause and nature of carcinogenic effects that these compounds provoke in organisms, including marine organisms.

PAHs are considered priority pollutants of environmental concern because of their toxic, carcinogenic and mutagenic characteristics, endocrine disruption, reproductive impairments and effects of immune function on growth and development in lower and higher trophic levels (USEPA 2002, IARC 2010). On the basis of all existing information on the harmful effects and high toxicity of PAHs in living organisms and the environment, many PAHs (and the mixtures that contain them) have been considered priority pollutants by several international agencies. For example, 16 PAHs have been included among the 129 priority pollutants listed by the United States Environmental Protection Agency (USEPA). Furthermore, this agency has classified as Group B2, probable human carcinogens, the following seven PAHs: benzo[a]anthracene, chrysene, benzo[b]fluoranthene, benzo[k] fluoranthene, benzo[a]pyrene, indeno[1,2,3cd]pyrene and dibenzo[a,h]anthracene. Naphthalene is classified as a possible human carcinogen (group C), with limited evidence in animals and lack of information in humans (USEPA 2002). The International Agency for Research on Cancer, which is part of the World Health Organization, has classified benzo[a]pyrene as human carcinogenic and dibenzo[a,h]anthracene as probably carcinogenic, while naphthalene, benzo[a]anthracene, chrysene, benzo[b]fluoranthene, benzo[k]fluoranthene and indene[1,2,3,cd]pyrene are considered probably carcinogenic to humans (IARC 2010).

The toxicity mechanism of PAHs on organisms is mainly due to their lipophilicity characteristics. The lipophilicity properties of PAHs enable them to readily penetrate cell membranes (Yu 2005). Marine organisms have different routes of PAH entrance: feeding on contaminated prey, direct contact with the skin, and uptake through respiration via gills (Meador et al. 1995, Streit 1998). Some marine organisms have the metabolic capacity to biotransform PAHs. Once PAHs enter the body, they are metabolized in a number of tissues such as liver, kidney and lungs. The main route of PAH biotransformation is through the mitochondrial cytochrome P450 (CYP1A1) enzyme and mechanisms that involve the activation of glutathione S-transferase. Subsequent metabolism renders them more water-soluble, making the original PAH precursor easier for the organism to remove. Therefore, PAH metabolites can be excreted by bile or urine. However, certain organisms are not able to biotransform PAHs or remove their products and therefore they bioaccumulate these compounds to a limited degree, mainly in adipose tissue. Despite the metabolic mechanisms unleashed to eliminate PAHs from the body in some organisms, a portion of PAH metabolites can be more toxic than their original precursor (Varanasi et al. 1993, Johnson-Restrepo et al. 2008). Actually, the most important way of toxic action of these compounds is through the reactive intermediates that are produced during PAH biotransformation. Xue and Warshawsky described the principal metabolic pathways that yield reactive PAH intermediates (Xue and Warshawsky 2005). They demonstrated two pathways to produce electrophiles that may covalently bind to DNA, generating DNA adducts that subsequently have teratogenic, mutagenic and carcinogenic consequences. On the one hand, PAHs can be oxidized by the CYP1A1, leading to formation of electrophilic diol-epoxides that are hydrolyzed by epoxide hydrolase, resulting in arene oxides. On the other hand, successive CYP1A-catalyzed oxidations take place and yield a radical anion.

In addition to their genotoxic effects, PAHs can act as endocrine disruptors, causing adverse effects on reproduction, including in marine organisms (Vignet et al. 2016, Collier et al. 2013, Isobe et al. 2007). However, the exact mechanisms by which PAHs have this effect is a subject of study at present, given that it depends on the characteristics of each species in question.

All marine organisms are exposed to PAH compounds from the environment, since these chemicals are easily associated with the sediment or with suspended particulate matter. Biomagnification, an increase in tissue concentration over two or more trophic levels, is generally not expected to occur for PAHs, except possibly in species from the lower trophic levels that are not able to effectively metabolize these compounds (Meador et al. 1995). However, many studies have shown that marine organisms can bioaccumulate PAHs from the environment. PAH effects and bioaccumulation occur in a wide range of tissues depending on variations in the environmental concentrations, time of exposure, and species ability to metabolize them. There are two important variables that have a major

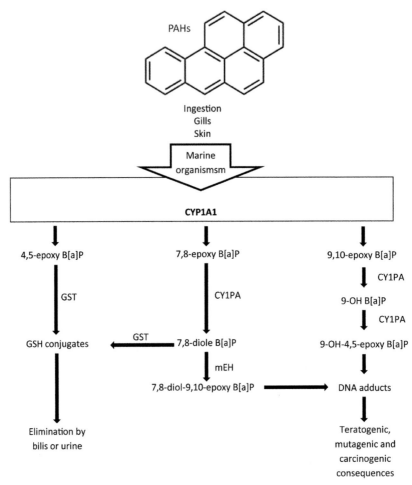

Fig. 4. Metabolic activation pathways of PAHs (adapted from Lodovici et al. 2004). GSH: glutathione. GST: glutathione S-transferase.

influence on contaminant body residues. One is PAH bioavailability, which is mainly affected by a few environmental variables, such as organic carbon, contact time (aging), desorption rate, PAH source (petrogenic vs. pyrogenic), and sediment surface area. The other is the organism's physiology, including lipid levels, rates of uptake, biotransformation and elimination capacities. Taking into account the lipid levels in organisms, a bioaccumulation model has been proposed based on lipid content that accounts for variations in toxicity due to differing species sensitivities and chemical differences (Di Toro et al. 2000, Mathew et al. 2008). Different concentrations of PAHs have been evidenced in a great variety of marine organisms such as invertebrates, fish, mussels, plants, and microorganisms, and the use of these species has been recommended for testing sediment ecotoxicity (Walker et al. 2013, Keshavarzifard et al. 2017, Barthe et al. 2008, Lotufo 1998, Walker and MacAskill 2014, Simpson et al. 2016, Barron et al. 2000, Di Toro et al. 2000, Mathew et al. 2008). As is well documented, these organisms have the capacity for bioaccumulation or biomagnification (in some cases) in lipid tissues. Therefore, in addition to sediment toxicity, they can be used to examine PAH uptake. In order to assess PAH accumulation in organisms and evaluate their toxicity potential, several factors have been determined and widely used. The biota-sediment accumulation factor (BSAF) defined from paired observations of chemical concentrations in biota and sediment have been defined in an EPA document (USEPA 2009). This parameter describes bioaccumulation of sediment-associated organic compounds or metals in tissues of ecological receptors. BSAFs are appropriate for describing bioaccumulation of sediment contaminants in aquatic food webs with non-equilibrium conditions between both the sediment and fish, and the sediment and its overlying water. Equilibrium is regarded as a reference condition for describing degrees of disequilibrium and thus is not a requirement for measurement, prediction, or application of BSAFs. Moreover, the use of sediment quality guidelines has been proposed and widely reviewed (Burton 2002, Den Besten et al 2002, Kothe 2003, De Deckere et al. 2011, Maletić et al. 2019). These reviews have documented that sediment quality guidelines are among the remaining challenges for the better protection of aquatic biodiversity, particularly sediment-dwelling organisms (Maletić et al. 2019). Moreover, calculation of other parameters has been formulated in order to evaluate PAH toxicology in human and benthic organisms: risk quotient approach, toxicity equivalent, and mutagen equivalent (Ranjbar Jafarabadi et al. 2017).

Despite the fact that mechanisms of PAH entry into marine organisms are similar, compared to pelagic species, it is believed that benthic species can accumulate higher concentrations of PAHs since they often ingest sediment, detritus, or other benthic species that may be exposed to higher concentrations of PAHs (Abdolahpur Monikh et al. 2014). The principal form of toxicity elicited by PAHs to benthic invertebrates is narcosis. Narcotic toxicants frequently demonstrate additive toxicity; thus, narcotic toxicant effects can be added together to summarize the total amount of toxicity present in a mixture of chemicals (as occurs in sediments) (USEPA 2009). Narcosis results in the alteration of cell membrane function (from mild toxic effects to mortality, depending upon the exposure) (USEPA 2009). Moreover, an agonism of the Aryl hydrocarbon receptor has been proposed as another mechanism of PAH chronic toxicity for marine organisms (Barron et al. 2000).

In addition to organisms that inhabit marine environments, humans can also be affected by the presence of PAHs in these environments. Even when humans are not directly exposed to PAH-contaminated marine sediments, these compounds may enter the food chain because of their properties of bioaccumulation principally and to a lesser extent biomagnification, and therefore have adverse effects on human health. The decisive factors that affect potential effects on human are status, age, exposure time and concentration of PAHs. Long-term exposure to PAHs can arise as kidney and liver damage, lung malfunction, skin inflammation, embryotoxic effects during pregnancy, and genotoxic and carcinogenic effects on humans. It is important to note that PAH exposure is never due to a single compound, so it is necessary to understand the dynamics of PAH metabolism in organisms (Maletić et al. 2019).

References

Abdel-Shafy, H.I. and M.S. Mansour. 2016. A review on polycyclic aromatic hydrocarbons: source, environmental impact, effect on human health and remediation. Egypt. J. Petrol. 25: 107–123.

Abdolahpur Monikh, F., M. Hosseini, J. Kazemzadeh Khoei and A.F. Ghasemi. 2014. Polycyclic aromatic hydrocarbons levels in sediment, benthic, benthopelagic and pelagic fish species from the Persian Gulf. Int. J. Environ. Res. 8: 839–848.

Abdollahi, S., Z. Raoufi, I. Faghiri, A. Savari, Y. Nikpour and A. Mansouri. 2013. Contamination levels and spatial distributions of heavy metals and PAHs in surface sediment of Imam Khomeini Port, Persian Gulf, Iran. Mar. Pollut. Bull. 71: 336–345.

Acquavita, A., J. Falomo, S. Predonzani, F. Tamberlich, N. Bettoso and G. Mattassi. 2014. The PAH level, distribution and composition in surface sediments from a Mediterranean Lagoon: The Marano and Grado Lagoon (Northern Adriatic Sea, Italy). Mar. Pollut. Bull. 81: 234–241.

Adhikari, P.L., K. Maiti and E.B. Overton. 2015. Vertical fluxes of polycyclic aromatic hydrocarbons in the northern Gulf of Mexico. Mar. Chem. 168: 60–68.

Adhikari, P.L., K. Maiti, E.B. Overton, B.E. Rosenheim and B.D. Marx. 2016. Distributions and accumulation rates of polycyclic aromatic hydrocarbons in the northern Gulf of Mexico sediments. Environ. Pollut. 212: 413–423.

Aguirre-Martínez, G., A. Rudolph, R. Ahumada, R. Loyola and V. Medina. 2009. Non-specific toxicity in port sediments: an approach to the content of critical pollutants. Rev. Biol. Mar. Oceanogr. 44: 725–735.

Alebic-Juretic, A. 2011. Polycyclic aromatic hydrocarbons in marine sediments from the Rijeka Bay area, Northern Adriatic, Croatia, 1998–2006. Mar. Pollut. Bull. 62: 863–869.

Arencibia-Carballo, G., A. Betanzos-Vega, M. Pérez Prohenza, C. Ocano Busia, A. Rodríguez Gil and A. Tripp-Quesada. 2017. Polycyclic aromatic hydrocarbons in superficial sediments of the coastal zone of Campechuela–Niquero, Cuba. Revista Cubana de Investigaciones Pesqueras 34: 68–73.

Arias, A.H., C.V. Spetter, R.H. Freije and J.E. Marcovecchio. 2009. Polycyclic aromatic hydrocarbons in water, mussels (*Brachidontes* sp., *Tagelus* sp.) and fish (*Odontesthes* sp.) from Bahía Blanca Estuary, Argentina. Estuar. Coast. Shelf. Sci. 85: 67–81.

Arias, A.H., A. Vazquez-Botello, N. Tombesi, G. Ponce-Vélez, H. Freije and J.E. Marcovecchio. 2010. Presence, distribution, and origins of polycyclic aromatic hydrocarbons (PAHs) in sediments from Bahía Blanca estuary, Argentina. Environ. Monit. Assess. 160: 301–314.

Aryal, R., S. Beecham, S. Vigneswaran, J. Kandasamy and R. Naidu. 2013. Spatial variation of polycyclic aromatic hydrocarbons and equivalent toxicity in Sydney Harbour, Australia. J. Water Clim. Chang. 4: 364–372.

Barakat, A.O., A. Mostafa, T.L. Wade, S.T. Sweet and N.B. El Sayed. 2011. Distribution and characteristics of PAHs in sediments from the Mediterranean coastal environment of Egypt. Mar. Pollut. Bull. 62: 1969–1978.

Barron, M.G., M.G. Carls, R. Heintz and S.D. Rice. 2000. Evaluation of fish early life-stage toxicity models of chronic embryonic exposures to complex polycyclic aromatic hydrocarbon mixtures. Environ. Toxicol. Chem. 19: 1951–1970.

Barthe, M., E. Pelletier, G.D. Breedveld and G. Cornelissen. 2008. Passive samplers versus surfactant extraction for the evaluation of PAH availability in sediments with variable levels of contamination. Chemosphere 71: 1486–1493.

Bejarano, A.C. and J. Michel. 2010. Large-scale risk assessment of polycyclic aromatic hydrocarbons in shoreline sediments from Saudi Arabia: environmental legacy after twelve years of the Gulf war oil spill. Environ. Pollut. 158: 1561–1569.

Berrojalbiz, N., J. Dachs, M.J. Ojeda, M.C. Valle, J. Castro-Jiménez, J. Wollgast et al. 2011. Biogeochemical and physical controls on concentrations of polycyclic aromatic hydrocarbons in water and plankton of the Mediterranean and Black Seas. Global Biogeochemical Cycles 25(4).

Bigus, P., M. Tobiszewski and J. Namieśnik. 2014. Historical records of organic pollutants in sediment cores. Mar. Pollut. Bull. 78: 26–42.

Bihari, N., M. Fafanđel and V. Piškur. 2007. Polycyclic aromatic hydrocarbons and ecotoxicological characterization of seawater, sediment, and mussel Mytilus galloprovincialis from the Gulf of Rijeka, the Adriatic Sea, Croatia. Arch. Environ. Contam. Toxicol. 52: 379–387.

Birgül, A., Y. Tasdemir and S.S. Cindoruk. 2011. Atmospheric wet and dry deposition of polycyclic aromatic hydrocarbons (PAHs) determined using a modified sampler. Atmospheric Research 101: 341–353.

Boitsov, S., H.K.B. Jensen and J. Klungsøyr. 2009. Geographical variations in hydrocarbon levels in sediments from the Western Barents Sea. Nor. J. Geol. 89: 91–10.

Botello, A.V., L.A. Soto, G. Ponce-Vélez and S. Villanueva. 2015. Baseline for PAHs and metals in NW Gulf of Mexico related to the Deepwater Horizon oil spill. Estuar. Coast. Shelf. Sci. 156: 124–133.

Botsou, F. and I. Hatzianestis. 2012. Polycyclic aromatic hydrocarbons (PAHs) in marine sediments of the Hellenic coastal zone, eastern Mediterranean: levels, sources and toxicological significance. J. Soils Sediments 12: 265–277.

Bouloubassi, I., L. Me'janelle, R. Pete, J. Fillaux, A. Lorre and V. Point. 2006. PAH transport by sinking particles in the open Mediterranean Sea: A 1 year sediment trap study. Mar. Pollut. Bull. 52: 560–571.

Bouloubassi, I., V. Roussiez, M. Azzoug and A. Lorre. 2012. Sources, dispersal pathways and mass budget of sedimentary polycyclic aromatic hydrocarbons (PAH) in the NW Mediterranean margin, Gulf of Lions. Mar. Chem. 142-144: 18–28.

Burgess, R.M., M.J. Ahrens, C.W. Hickey, P.J. den Besten, D. ten Hulscher, B. van Hattum et al. 2003. Geochemistry of PAHs in aquatic environment: Source, persistence and distribution. pp 35–47. *In*: Douben, P.E.T. (ed.). PAHs: An Ecotoxicological Perspective, Wiley, England.

Burgess, R.M., I.M. Konovets, L.S. Kipnis, A.V. Lyashenko, V.A. Grintsov, A.N. Petrov et al. 2011. Distribution, magnitude and characterization of the toxicity of Ukrainian estuarine sediments. Mar. Pollut. Bull. 62: 2442–2462.

Burton, G.A. 2002. Sediment quality criteria in use around the world. Limnology. 3: 65–76.

Buruaem, L.M., I. Braga de Castro, M.A. Hortellani, S. Taniguchi, G. Fillmann, S.T. Sasaki et al. 2013. Integrated quality assessment of sediments from harbour areas in Santos-São Vicente Estuarine System, Southern Brazil. Estuar. Coast. Shelf. Sci. 130: 179–189.

Buruaem, L.M., S. Taniguchi, S. Tarou Sasaki, M. Caruso Bícego, L. Costa-Lotufo and D. Souza Abessa. 2016. Hydrocarbons in surface sediments of harbor areas in a tropical region (Ceará state, northeast Brazil). Environ. Earth. Sci. 75: 642.

Bustamante, J., A. Albisu, L. Bartolomé, A. Prieto, A. Atutxa, S. Arrasate et al. 2010. Levels of polycyclic aromatic hydrocarbons, polychlorinated byphenyls, methylmercury and butyltins in the natural UNESCO reserve of the biosphere of Urdaibai (Bay of Biscay, Spain). Int. J. Environ. Anal. Chem. 90: 722–736.

Caballero-Gallardo, K., A. Guerrero-Castilla, B. Johnson-Restrepo, J. de la Rosa and J. Olivero-Verbel. 2015. Chemical and toxicological characterization of sediments along a Colombian shoreline impacted by coal export terminals. Chemosphere 138: 837–846.

Cai, M., M. Duan, J. Guo, M. Liu, A. Qi, Y. Lin et al. 2018. PAHs in the Northern South China Sea: Horizontal transport and downward export on the continental shelf. Mar. Chem. 202: 121–129.

Cao, S., G. Na, R. Li, L. Ge, H. Gao, S. Jin et al. 2018. Fate and deposition of polycyclic aromatic hydrocarbons in the Bransfield Strait, Antarctica. Mar. Poll. Bull. 137: 533–541.

Chen, C.F., C.W. Chen, C.D. Dong and C.M. Kao. 2013. Assessment of toxicity of polycyclic aromatic hydrocarbons in sediments of Kaohsiung Harbor, Taiwan. Sci. Total Environ. 463: 1174–1181.

Chen, C.W. and C.F. Chen. 2011. Distribution, origin, and potential toxicological significance of polycyclic aromatic hydrocarbons (PAHs) in sediments of Kaohsiung Harbor, Taiwan. Mar. Pollut. Bull. 63: 417–423.

Chiou, C.T., S.E. McGroddy and D.E. Kile. 1998. Partition characteristics of polycyclic aromatic hydrocarbons on soils and sediments. Environ. Sci. Technol. 32: 264–269.

Chizhova, T., K. Hayakawa, P. Tishchenko, H. Nakase and Y. Koudryashova. 2013. Distribution of PAHs in the northwestern part of the Japan Sea. Deep Sea Res. Part II Top. Stud. Oceanogr. 86: 19–24.

Choi, H.G., H.B. Moon, M. Choi and J. Yu. 2011. Monitoring of organic contaminants in sediments from the Korean coast: spatial distribution and temporal trends (2001–2007). Mar. Pollut. Bull. 62: 1352–1361.

Cincinelli, A., A.M. Stortini, M. Perugini, L. Checchini and L. Lepri. 2001. Organic pollutants in sea-surface microlayer and aerosol in the coastal environment of Leghorn—(Tyrrhenian Sea). Mar. Chem. 76: 77–98.

Colacicco, A., G. De Gioannis, A. Muntoni, E. Pettinao, A. Polettini and R. Pomi. 2010. Enhanced electrokinetic treatment of marine sediments contaminated by heavy metals and PAHs. Chemosphere 81: 46–56.

Collier, T.K., B.F. Anulacion, M.R. Arkoosh, J.P. Dietrich, J.P. Incardona, L.L Johnson et al. 2013. Effects on fish of polycyclic aromatic hydrocarbons (PAHs) and naphthenic acid exposures. pp 195–255. *In*: Fish physiology, Academic Press.

Commendatore, M.G., M.L. Nievas, O. Amin and J.L. Esteves. 2012. Sources and distribution of aliphatic and polyaromatic hydrocarbons in coastal sediments from the Ushuaia Bay (Tierra del Fuego, Patagonia, Argentina). Mar. Environ. Res. 74: 20–31.

Commendatore, M.G., M. Franco, P. Gomes Costa, I. Castro, G. Fillmann, G. Bigatti et al. 2015. BTS, PAHS, OCPS and PCBS in sediments and bivalve mollusks in a midlatitude environment from the patagonian coastal zone. Environ. Toxicol. Chem. 34: 2750–2763.

Cook, J.W., C.L. Hewett and I. Hieger. 1933. The isolation of a cancer producing hydrocarbon from coal tar. Part I, II and III. J. Chem. Soc. 395405.

Cortazar, E., L. Bartolomé, S. Arrasate, A. Usobiaga, J.C. Raposo, O. Zuloaga et al. 2008. Distribution and bioaccumulation of PAHs in the UNESCO protected natural reserve of Urdaibai, Bay of Biscay. Chemosphere 72: 1467–1474.

Costanza, R. 1994. Ecological economics and the management of coastal and estuarine ecosystems. pp. 485 *In*: Dyer, Orth (ed.). Changes in Fluxes in Estuaries: Implications from Science to Management. ECSA/ERF Symposium, Univ. of Plymouth.

Cui, Z., Q. Lai, C. Dong and Z. Shao. 2008. Biodiversity of polycyclic aromatic hydrocarbondegrading bacteria from deep sea sediments of the Middle Atlantic Ridge. Environ. Microbiol. 10: 2138–2149.

Culotta, L., C. De Stefano, A. Gianguzza, M.R. Mannino and S. Orecchio. 2006. The PAH composition of surface sediments from Stagnone coastal lagoon, Marsala (Italy). Mar. Chem. 99: 117–127.

Curtosi, A., E. Pelletier, C.L. Vodopivez and W.P. Mac Cormack. 2009. Distribution of PAHs in the water column, sediments and biota of Potter Cove, South Shetland Islands, Antarctica. Antarct. Sci. 21: 329–339.

Da Silva, D. and M.C. Bícego. 2010. Polycyclic aromatic hydrocarbons and petroleum biomarkers in São Sebastião Channel, Brazil: Assessment of petroleum contamination. Mar. Environ. Res. 69: 277–286.

Dachs, J., J.M. Bayona, S.W. Fowler, J.C. Miquel and J. Albaigés. 1996. Vertical fluxes of polycyclic aromatic hydrocarbons and organochlorine compounds in the western Alboran Sea (southwestern Mediterranean). Mar. Chem. 52: 75–86.

Dachs, J., R. Lohmann, W.A. Ockenden, L. Méjanelle, S.J. Eisenreich and K.C. Jones. 2002. Oceanic biogeochemical controls on global dynamics of persistent organic pollutants. Environ. Sci. Technol. 36: 4229–4237.

Dahle, S., V.M. Savinov, G.G. Matishov, A. Evenset and K. Næs. 2003. Polycyclic aromatic hydrocarbons (PAHs) in bottom sediments of the Kara Sea shelf, Gulf of Ob and Yenisei Bay. Sci. Total Environ. 306: 57–71.

Darılmaz, E., A. Kontaş, E. Uluturhan, I Akçalı and O. Altay. 2013. Spatial variations in polycyclic aromatic hydrocarbons concentrations at surface sediments from the Cyprus (Eastern Mediterranean): relation to ecological risk assessment. Mar. Pollut. Bull. 75: 174–181.

Dauner, A., T.H. Dias, F.K. Ishiia, B.G. Libardoni, R.A. Parizzi and C.C. Martins. 2018. Ecological risk assessment of sedimentary hydrocarbons in a subtropical estuary as tools to select priority areas for environmental management. J. Environ. Manag. 223: 417–425.

Dauner, A.L.L., E.A. Hernández, W.P. MacCormack and C.C. Martins. 2015. Molecular characterisation of anthropogenic sources of sedimentary organic matter from Potter Cove, King George Island, Antarctica. Sci. Total Environ. 502: 408–416.

De Deckere, E., W. de Cooman, V. Leloup, P. Meire, C. Schmitt and P.C. von der Ohe. 2011. Development of sediment quality guidelines for freshwater ecosystems. J. Soils Sediments 11: 504–517.

de Mora, S., I. Tolosa, S.W. Fowler, J.P. Villeneuve, R. Cassi and C. Cattini. 2010. Distribution of petroleum hydrocarbons and organochlorinated contaminants in marine biota and coastal sediments from the ROPME Sea Area during 2005. Mar. Pollut. Bull. 60: 2323–2349.

Den Besten, P.J., E. de Deckere, M.E. Babut, B. Power, T.A. DelValls, C. Zago et al. 2002. Biological effects-based sediment quality in ecological risk assessment for european waters. J. Soils Sediments 3: 144–162.

Deng, W., X.G. Li, S.Y. Li, Y.Y. Ma and D.H. Zhang. 2013. Source apportionment of polycyclic aromatic hydrocarbons in surface sediment of mud areas in the East China Sea using diagnostic ratios and factor analysis. Mar. Pollut. Bull. 70: 266–273.

Dhananjayan, V., S. Muralidharan and V.R. Peter. 2012. Occurrence and distribution of polycyclic aromatic hydrocarbons in water and sediment collected along the Harbour Line, Mumbai, India. Int. J. Oceanogr. 2012.

Di Toro, D.M., J.A. McGrath and D.J. Hansen. 2000. Technical basis for narcotic chemicals and polycyclic aromatic hydrocarbon criteria. I. Water and tissue. Environ. Toxicol. Chem. 19: 1951–1970.

Domine, D., J. Devillers, P. Garrigues, H. Budzinski, M. Chastrette and W. Karcher. 1994. Chemometrical evaluation of the PAH contamination in the sediments of the Gulf of Lion (France). Sci. Total Environ. 155: 9–24.

Dudhagara, D.R., R.K. Rajpara, J.K. Bhatt, H.B. Gosai, B.K. Sachaniya and B.P. Dave. 2016. Distribution, sources and ecological risk assessment of PAHs in historically contaminated surface sediments at Bhavnagar coast, Gujarat, India. Environ. Pollut. 213: 338–346.

Eggleton, J. and K.V. Thomas. 2004. A review of factors affecting the release and bioavailability of contaminants during sediment disturbance events. Environ. Int. 30: 973–980.

El Nemr, A., M.M. El-Sadaawy, A. Khaled and S.O. Draz. 2013. Aliphatic and polycyclic aromatic hydrocarbons in the surface sediments of the Mediterranean: assessment and source recognition of petroleum hydrocarbons. Environ. Monit. Assess. 185: 4571–4589.

El-Naggar, N.A., H.I. Emara, M.N. Moawad, Y.A. Soliman and A.A. El-Sayed. 2018. Detection of polycyclic aromatic hydrocarbons along Alexandria's coastal water, Egyptian Mediterranean Sea. Egypt J. Aquat. Res. 44: 9–14.

Eriksson, M., E. Sodersten, Z. Yu, G. Dalhammar and W.W. Mohn. 2003. Degradation of polycyclic aromatic hydrocarbons at low temperature under aerobic and nitrate-reducing conditions in enrichment cultures from northern soils. Appl. Environ. Microbiol. 69: 275–284.

Er-Raioui, H., S. Bouzid, M. Marhraoui and A. Saliot. 2009. Hydrocarbon pollution of the Mediterranean coastline of Morocco. Ocean Coast Manag. 52: 124–129.

Fei, Y.H., X.D. Li and X.Y Li. 2011. Organic diagenesis in sediment and its impact on the adsorption of bisphenol A and nonylphenol onto marine sediment. Mar. Pollut. Bull. 63: 578–582.

Foster, K., G.A. Stern, J. Carrie, J.N.-L. Bailey, P.M. Outridge, H. Sanei et al. 2015. Spatial, temporal, and source variations of hydrocarbons in marine sediments from Baffin Bay, Eastern Canadian Arctic. Sci. Total Environ. 506-507: 430–443.

Frena, M., G.A. Bataglion, S.S. Sandini, K.N. Kuroshima, M.N. Eberlin and L.A.S. Madureira. 2017. Distribution and sources of aliphatic and polycyclic aromatic hydrocarbons in surface sediments of Itajaí-Açu estuarine system in Brazil. J. Braz. Chem. 28: 603–614.

Froehner, S., M. Maceno, E. Cardoso Da Luz, D. Botelho Souza and K. Scurupa Machado. 2010. Distribution of polycyclic aromatic hydrocarbons in marine sediments and their potential toxic effects. Environ. Monit. Assess. 168: 205–13.

Fu, J., B. Mai, G. Sheng, G. Zhang, X. Wang, P.A. Peng et al. 2003. Persistent organic pollutants in environment of the Pearl River Delta, China: an overview. Chemosphere 52: 1411–1422.

Gagnon, C. and N.S. Fisher. 1997. Bioavailability of sediment-bound methyl and inorganic mercury to a marine bivalve. Environ. Sci. Technol. 31: 993–998.

García-Flor, N., C. Guitart, M. Ábalos, J. Dachs, J.M. Bayona and J. Albaigés. 2005. Enrichment of organochlorine contaminants in the sea surface microlayer: An organic carbon-driven process. Mar. Chem. 96: 331–345.

Gnandi, K., B.A.M. Bandowe, D.D. Deheyn, M. Porrachia, M. Kersten and W. Wilcke. 2011. Polycyclic aromatic hydrocarbons and trace metal contamination of coastal sediment and biota from Togo. J. Environ. Monit. 13: 2033–2041.

Guerra, R. 2012. Polycyclic aromatic hydrocarbons, polychlorinated biphenyls and trace metals in sediments from a coastal lagoon (Northern Adriatic, Italy). Water, Air and Soil Pollution 223: 85–98.

Guigue, C., M.M. Tedetti, S. Giorgi and M. Goutx. 2011. Occurrence and distribution of hydrocarbons in the surface microlayer and subsurface water from the urban coastal marine area off Marseilles, Northwestern Mediterranean Sea. Mar. Pollut. Bull. 62: 2741–2752.

Guigue, C., M. Tedetti, N. Ferretto, N. Garcia, L. Méjanelle and M. Goutx. 2014. Spatial and seasonal variabilities of dissolved hydrocarbons in surface waters from the Northwestern Mediterranean Sea: Results from one year intensive sampling. Sci. Total Environ. 466-467: 650–662.

Gui-Peng Yang. 2000. Polycyclic aromatic hydrocarbons in the sediments of the South China Sea. Environ. Pollut. 108: 163–171.

Guitart, C., N. García-Flor, J.M. Bayona and J. Albaigés. 2007. Occurrence and fate of polycyclic aromatic hydrocarbons in the coastal surface microlayer. Mar. Pollut. Bull. 54: 186–194.

Guo, W., M. He, Z. Yang, C. Lin, X. Quan and H. Wang. 2007. Distribution of polycyclic aromatic hydrocarbons in water, suspended particulate matter and sediment from Daliao River watershed, China. Chemosphere 68: 93–104.

Guzzella, L. and A. De Paolis. 1994. Polycyclic aromatic hydrocarbons in sediments of the Adriatic Sea. Mar. Pollut. Bull. 28: 159–169.

Han, D. and M.J. Currell. 2017. Persistent organic pollutants in China's surface water systems. Sci. Total Environ. 580: 602–625.

Harris, K., M.B. Yunker, N. Dangerfield and P.S. Ross. 2011. Sediment-associated aliphatic and aromatic hydrocarbons in coastal British Columbia, Canada: Concentrations, composition, and associated risks to protected sea otters. Environ. Pollut. 159: 2665–2674.

He, X., Y. Pang, X. Song, B. Chen, Z. Feng and Y. Ma. 2014. Distribution, sources and ecological risk assessment of PAHs in surface sediments from Guan River Estuary, China. Mar. Pollut. Bull. 80: 52–58.

Hites, R.A., R.E. Laflamme, J.G. Windsor, J.W. Farrington and W.G. Deuser. 1980. Polycyclic aromatic hydrocarbons in an anoxic sediment core from the Pettaquamscutt River (Rhode Island, USA). Geochimica et Cosmochimica Acta 44: 873–878.

Hong, W.J., H. Jia, Y.F. Li, Y. Sun, X. Liu and L. Wang. 2016. Polycyclic aromatic hydrocarbons (PAHs) and alkylated PAHs in the coastal seawater, surface sediment and oyster from Dalian, Northeast China. Ecotox. Environ. Safe. 128: 11–20.

Hu, N., X. Shi, J. Liu, P. Huang, Y. Liu and Y. Liu. 2010. Concentrations and possible sources of PAHs in sediments from Bohai Bay and adjacent shelf. Environ. Earth Sci. 60: 1771–1782.

Huang, W., Z. Wang and W. Yan. 2012a. Distribution and sources of polycyclic aromatic hydrocarbons (PAHs) in sediments from Zhanjiang Bay and Leizhou Bay, South China. Mar. Pollut. Bull. 64: 1962–1969.

Huang, H.C., C.L. Lee, C.H. Lai, M.D Fang and I.C. 2012b. Transboundary movement of polycyclic aromatic hydrocarbons (PAHs) in the Kuroshio Sphere of the western Pacific Ocean. Atmos. Environ. 54: 470–479.

Hussain, K., R.R. Hoque, S. Balachandran, S. Medhi, M.G. Idris, M. Rahman et al. 2018. Monitoring and risk analysis of PAHs in the environment. Handbook of Environmental Materials Management, 1–35.

IARC. 2010. International Agency for Research on Cancer. Some Non-heterocyclic Polycyclic Aromatic Hydrocarbons and Some Related Exposures, Vol 92. http://monographs.iarc.fr/ENG/Monographs/vol92/mono92.pdf.

Isobe, T., H. Takada, M. Kanai, S. Tsutsumi, K.O. Isobe, R. Boonyatumanond et al. 2007. Distribution of polycyclic aromatic hydrocarbons (PAHs) and phenolic endocrine disrupting chemicals in South and Southeast Asian mussels. Environ. Monit. Assess. 135: 423–440.

Jaward, F.M., H.A. Alegria, J.G. Galindo Reyes and A. Hoare. 2012. Levels of PAHs in the waters, sediments, and shrimps of Estero de Urias, an estuary in Mexico, and their toxicological effects. Sci. World J. 9.

Jiao, L., G.J. Zheng, T.B. Minh, B. Richardson, L. Chen, Y. Zhang et al. 2009. Persistent toxic substances in remote lake and coastal sediments from Svalbard, Norwegian Arctic: levels, sources and fluxes. Environ. Pollut. 157: 1342–1351.

Jin, Q., L. Pan, D. Liu, F. Hu and M. Xiu. 2014. Assessing PAHs pollution in Qingdao coastal area (China) by the combination of chemical and biochemical responses in scallops, *Chlamys farreri*. Mar. Pollut. Bull. 89: 473–480.

Johnson-Restrepo, B., J. Olivero-Verbel, S. Lu, J. Guette-Fernandez, R. Baldiris-Avila, I. O'Byrne-Hoyos et al. 2008. Polycyclic aromatic hydrocarbons and their hydroxylated metabolites in fish bile and sediments from coastal waters of Colombia. Environ. Pollut. 151: 452–459.

Jurado, E., F.M. Jaward, R. Lohmann, K.C. Jones, R. Simó and J. Dachs. 2004. Atmospheric dry deposition of persistent organic pollutants to the Atlantic and inferences for the global oceans. Environ. Sci. Technol. 38: 5505–5513.

Kaiser, D., D.E. Schulz-Bull and J.J. Waniek. 2018. Polycyclic and organochlorine hydrocarbons in sediments of the northern South China Sea. Mar. Pollut. Bull. 137: 668–676.

Katayama, A., R. Bhula, G.R. Burns, E. Carazo, A. Felsot, D. Hamilton et al. 2010. Bioavailability of xenobiotics in the soil environment. pp 1–86. *In*: Ware, G.W. (ed.). Reviews of Environmental Contamination and Toxicology. Springer, New York, NY.

Kato, C., L. Li, J. Tamaoka and K. Horikoshi. 1997. Molecular analyses of the sediment of the 11,000-m deep Mariana Trench. Extremophiles 1: 117–23.

Ke, H., M. Chen, M. Liu, M. Chen, M. Duan, P. Huang et al. 2017. Fate of polycyclic aromatic hydrocarbons from the north pacific to the arctic: Field measurements and fugacity model simulation. Chemosphere 184: 916–923.

Keshavarzifard, M., M.P. Zakaria, T.S. Hwai, F.M. Yusuff, S. Mustafa, V. Vaezzadeh et al. 2014. Baseline distributions and sources of polycyclic aromatic hydrocarbons (PAHs) in the surface sediments from the Prai and Malacca Rivers, Peninsular Malaysia. Mar. Pollut. Bull. 88: 366–372.

Keshavarzifard, M., F. Moore, B. Keshavarzi and R. Sharifi. 2017. Polycyclic aromatic hydrocarbons (PAHs) in sediment and sea urchin (Echinometra mathaei) from the intertidal ecosystem of the northern Persian Gulf: distribution, sources, and bioavailability. Mar. Pollut. Bull. 123: 373–380.

Khairy, M.A., M. Kolb, A.R. Mostafa, E.F. Anwar, E.F. and M. Bahadir. 2009. Risk assessment of polycyclic aromatic hydrocarbons in a Mediterranean semi-enclosed basin affected by human activities (Abu Qir Bay, Egypt). J. Hazard. Mater. 170: 389–397.

King, A.J., J.W. Readman and J.L. Zhou. 2004. Dynamic behaviour of polycyclic aromatic hydrocarbons in Brighton marina, UK. Mar. Pollut. Bull. 48: 229–239.

Konn, C., J.L. Charlou, J.P. Donval, N.G. Holm, F. Dehairs and S. Bouillon. 2009. Hydrocarbons and oxidized organic compounds in hydrothermal fluids from Rainbow and Lost City ultramafic-hosted vents. Chem. Geol. 258: 299–314.

Kothe, H. 2003. Existing sediment management guidelines: an overview, what will happen with the sediment/dredged material? J. Soils Sediments 3: 139–143.

Kovács, A., A. Vasas and J. Hohmann. 2008. Natural phenanthrenes and their biological activity. Phytochemistry 69: 1084–1110.

La Nafie, N., M.R. Asmawati and A. Arief. 2016. Distribution of Polycyclic Aromatic Hydrocarbons (PAHs) in Sediments of Losari Beach and adjacent areas, South Sulawesi, Indonesia. Int. J. of Appl. Chem. 12: 675–682.

Laitano, M.V., I.B. Castro, P.G. Costa, G. Fillmann and M. Cledón. 2015. Butyltin and PAH Contamination of Mar del Plata Port (Argentina) sediments and their influence on adjacent coastal regions. Bull. Environ. Contam. Toxicol. 95: 513–520.

Lang, Y.H., X. Yang, H. Wang, W. Yang and G.L. Li. 2013. Diagnostic ratios and positive matrix factorization to identify potential sources of PAHS in sediments of the Rizhao Offshore, China. Polycycl. Aromat. Compd. 33: 161–172.

Lang, Y.H. and W. Yang. 2014. Source apportionment of PAHs using Unmix model for Yantai coastal surface sediments, China. Bull. Environ. Contam. Toxicol. 92: 30–35.

Latimer, J.S. and J. Zheng. 2003. The sources, transport, and Fate of PAHs in the marine environment. pp. 9–35. *In*: Douben, P.E.T. (ed.). PAHs: An Ecotoxicological Perspective, Wiley, England.

Lee, C.H., J.H. Lee, C.G. Sung, S.D. Moon, S.K. Kang, J.H. Lee et al. 2013. Monitoring toxicity of polycyclic aromatic hydrocarbons in intertidal sediments for five years after the Hebei Spirit oil spill in Taean, Republic of Korea. Mar. Pollut. Bull. 76: 241–249.

Leister, D.L. and J.E. Baker. 1994. Atmospheric deposition of organic contaminants to the Chesapeake Bay. Atmos. Environ. 28: 1499–1520.

Li, P., J. Cao, X. Diao, B. Wang, B. Zhou, Q. Han et al. 2015b. Spatial distribution, sources and ecological risk assessment of polycyclic aromatic hydrocarbons in surface seawater from Yangpu Bay, China. Mar. Pollut. Bull. 93: 53–60.

Li, P., R. Xue, Y. Wang, R. Zhang and G. Zhang. 2015c. Influence of anthropogenic activities on PAHs in sediments in a significant gulf of low-latitude developing regions, the Beibu Gulf, South China Sea: distribution, sources, inventory and probability risk. Mar. Pollut. Bull. 90: 218–226.

Li, Y.H., P. Li, W.D. Ma, Q.Q Song, H.L. Zhou, H. Qin et al. 2015a. Spatial and temporal distribution and risk assessment of polycyclic aromatic hydrocarbons in Surface seawater from the Haikou Bay, China. Mar. Pollut. Bull. 92: 244–251.

Lipiatou, E. and A. Saliot. 1991. Fluxes and transport of anthropogenic and natural polycyclic aromatic hydrocarbons in the western Mediterranean Sea. Mar. Chem. 32: 51–71.

Lipiatou, E., I. Tolosa, R. Simo, I. Bouloubassi, J. Dachs, S. Marti et al. 1997. Mass budget and dynamics of polycyclic aromatic hydrocarbons in the Mediterranean Sea. Deep Sea Res. Part II Top. Stud. Oceanogr. 44: 881–905.

Liu, L.Y., J.Z. Wang, G.L. Wei, Y.F. Guan and E.Y. Zeng. 2012. Polycyclic aromatic hydrocarbons (PAHs) in continental shelf sediment of China: implications for anthropogenic influences on coastal marine environment. Environ. Pollut. 167: 155–162.

Liu, N., X. Li, D. Zhang, Q. Liu, L. Xiang, K. Liu et al. 2017. Distribution, sources, and ecological risk assessment of polycyclic aromatic hydrocarbons in surface sediments from the Nantong Coast, China. Mar. Pollut. Bull. 114: 571–576.

Lodovici, M., V. Akpan, C. Evangelisti and P. Dolara. 2004. Sidestream tobacco smoke as the main predictor of exposure to polycyclic aromatic hydrocarbons. J. Appl. Toxicol. 24: 277–281.

Lotufo, G.R. 1998. Lethal and sublethal toxicity of sediment-associated fluoranthene to benthic copepods: application of the critical-body-residue approach. Aquat. Toxicol. 44: 17–30.

Louvado, A., N.C.M. Gomes, M.M.Q. Simões, A. Almeida, D.F.R. Cleary and A. Cunha. 2015. Polycyclic aromatic hydrocarbons in deep sea sediments: microbe–pollutant interactions in a remote environment. Sci. Total Environ. 526: 312–328.

Luek, J.L., R.M. Dickhut, M.A. Cochran, R.L. Falconer and H. Kylin. 2017. Persistent organic pollutants in the Atlantic and southern oceans and oceanic atmosphere. Sci. Total Environ. 583: 64–71.

Lyons, B.P., J.L. Barber, H.S. Rumney, T.P.C. Bolam, P. Bersuder, R.J. Law et al. 2015. Baseline survey of marine sediments collected from the State of Kuwait: PAHs, PCBs, brominated flame retardants and metal contamination. Mar. Pollut. Bull. 100: 629–636.

Ma, Y., C.J. Halsall, Z. Xie, D. Koetke, W. Mi, R. Ebinghaus et al. 2017. Polycyclic aromatic hydrocarbons in ocean sediments from the North Pacific to the Arctic Ocean. Environ. Pollut. 227: 498–504.

Maciel, D., J.R. Botelho de Souza, S. Taniguchi, M. Caruso Bícego and E. Zanardi-Lamardo. 2015. Sources and distribution of polycyclic aromatic hydrocarbons in an urbanized tropical estuary and adjacent shelf, Northeast of Brazil. Mar. Pollut. Bull. 101: 429–433.

Mahugija, J.A., K.N. Ahmed and Y.M. Makame. 2017. Polycyclic aromatic hydrocarbons (PAHs) contamination in coastal mangrove ecosystems of the Zanzibar archipelago. Western Indian Ocean Journal of Marine Science 16: 25–34.

Maletić, S.P., J.M. Beljin, S.D. Rončević, M.G. Grgić and B.D. Dalmacija. 2019. State of the art and future challenges for polycyclic aromatic hydrocarbons is sediments: sources, fate, bioavailability and remediation techniques. J. Hazard. Mater. 365: 467–482.

Mandalakis, M., P.N. Polymenakou, A. Tselepides and N. Lampadariou. 2014. Distribution of aliphatic hydrocarbons, polycyclic aromatic hydrocarbons and organochlorinated pollutants in deep-sea sediments of the southern Cretan margin, eastern Mediterranean Sea: a baseline assessment. Chemosphere 106: 28–35.

Marcovecchio, J.E., R.H. Freije and A. Vazquez Botello. 2014. Introducción general. pp. 13–18. *In*: Marcovecchio, J.E. and H.R. Freije (eds.). Procesos Químicos en estuarios. Editorial de la Universidad Tecnológica Nacional, Buenos Aires.

Marini, M. and E. Frapiccini. 2013. Persistence of polycyclic aromatic hydrocarbons in sediments in the deeper area of the Northern Adriatic Sea (Mediterranean Sea). Chemosphere 90: 1839–1846.

Martins, C.C., M.C. Bícego, N.L. Rose, S. Taniguchi, R.A. Lourenço, R.C.L. Figueira et al. 2010. Historical record of polycyclic aromatic hydrocarbons (PAHs) and spheroidal carbonaceous particles (SCPs) in marine sediment cores from Admiralty Bay, King George Island, Antarctica. Environ. Pollut. 158: 192–200.

Martins, C.C., M.C. Bícego, M.M. Mahiques, R.C. Figueira, M.G. Tessler and R.C. Montone. 2011. Polycyclic aromatic hydrocarbons (PAHs) in a large South American industrial coastal area (Santos Estuary, Southeastern Brazil): sources and depositional history. Mar. Pollut. Bull. 63: 452–8.

Masih, J., A. Masih, A. Kulshrestha, R. Singhvi and A. Taneja. 2010. Characteristics of polycyclic aromatic hydrocarbons in indoor and outdoor atmosphere in the North central part of India. J. Hazard. Mater. 177: 190–198.

Masih, J., R. Singhvi, K. Kumar, V.K. Jain and A. Taneja. 2012. Seasonal variation and sources of polycyclic aromatic hydrocarbons (PAHs) in indoor and outdoor air in a semi arid tract of northern India. Aerosol Air Qual Res. 12: 515–525.

Mathew, R., J.A. McGrath and D.M. Di Toro. 2008. Modeling polycyclic aromatic hydrocarbon bioaccumulation and metabolism in time-variable early life-stage exposures. Environ. Toxicol. Chem. 27: 1515–1525.

Meador, J.P., E. Casillas, C.A. Sloan and U. Varanasi. 1995. Comparative bioaccumulation of polycyclic aromatic hydrocarbons from sediment by two infaunal invertebrates. Mar. Ecol. Prog. Ser. 123: 107–124.

Mirza, R.M., A.D. Mohammadi, A. Sohrab, A. Safahieh, A. Savari and P. Hajeb. 2012. Polycyclic aromatic hydrocarbons in seawater, sediment, and rock oyster *Saccostrea cucullata* from the northern part of the Persian Gulf (Bushehr Province). Water, Air and Soil Pollution 223: 189–198.

Monteyne, E., P. Roose and C.R. Janssen. 2013. Application of a silicone rubber passive sampling technique for monitoring PAHs and PCBs at three Belgian coastal harbours. Chemosphere 91: 390–398.

Montory, M., G. Chiang, D. Fuentes-Ríos, H. Palma-Fleming and R. Barra. 2008. Polychlorinated Biphenyls (PCBs) and Polycyclic Aromatic Hidrocarbons in sediments from the inner sea of Chiloé island. Results from the cimar 10 cruise. Cienc. Tecnol. Mar. 31: 67–81.

Mostafa, A.R., T.L. Wade, S.T. Sweet, A.K.A. Al-Alimi and A.O. Barakat. 2009. Distribution and characteristics of polycyclic aromatic hydrocarbons (PAHs) in sediments of Hadhramout coastal area, Gulf of Aden, Yemen. J. Mar. Syst. 78: 1–8.

Muniz, P., N. Venturini, C.C. Martins, A.B. Munshi, F. García-Rodríguez, E. Brugnoli et al. 2015. Integrated assessment of contaminants and monitoring of an urbanized temperate harbor (Montevideo, Uruguay): a 12-year comparison. Braz. J. Oceanogr. 63: 311–330.

Mzoughi, N. and L. Chouba. 2011. Distribution and partitioning of aliphatic hydrocarbons and polycyclic aromatic hydrocarbons between water, suspended particulate matter, and sediment in harbours of the West coastal of the Gulf of Tunis (Tunisia). J. Environ. Monit. 13: 689–698.

Nakata, H., K. Uehara, Y. Goto, M. Fukumura, H. Shimasaki, K. Takikawa et al. 2014. Polycyclic aromatic hydrocarbons in oysters and sediments from the Yatsushiro Sea, Japan: Comparison of potential risks among PAHs, dioxins and dioxin-like compounds in benthic organisms. Ecotox. Environ. Safe. 99: 61–68.

Nascimento, R.A., M. de Almeida, N.C.F. Escobar, S.L.C. Ferreira, J. Mortatti and A.F.S. Queiroz. 2017. Sources and distribution of polycyclic aromatic hydrocarbons (PAHs) and organic matter in surface sediments of an estuary under petroleum activity influence, Todos os Santos Bay, Brazil. Mar. Pollut. Bull. 119: 223–230.

Neff, J.M. 1979. Polycyclic aromatic hydrocarbons in the aquatic environment.

Neira, C., J. Cossaboon, G. Mendoza, E. Hoh and L.A. Levin. 2017. Occurrence and distribution of polycyclic aromatic hydrocarbons in surface sediments of San Diego Bay marinas. Mar. Pollut. Bull. 114: 466–479.

Nguyen, T.C., P. Loganathan, T.V. Nguyen, S. Vigneswaran, J. Kandasamy, D. Slee et al. 2014. Polycyclic aromatic hydrocarbons in road-deposited sediments, water sediments, and soils in Sydney, Australia: comparisons of concentration distribution, sources and potential toxicity. Ecotox. Environ. Safe. 104: 339–348.

Ohkouchi, N., K. Kawamura and H. Kawahata. 1999. Distributions of three- to seven-ring polynuclear aromatic hydrocarbons on the deep sea floor in the Central Pacific. Environ. Sci. Technol. 33: 3086–3090.

Oliva, A.L., P.Y. Quintas, N.S. La Colla, A.H. Arias and J.E. Marcovecchio. 2015. Distribution, sources, and potential ecotoxicological risk of polycyclic aromatic hydrocarbons in surface sediments from Bahía Blanca Estuary, Argentina. Arch. Environ. Contam. Toxicol. 69: 163–72.

Orecchio, S., S. Cannata and L. Culotta. 2010. How building an underwater pipeline connecting Libya to Sicilian coast is affecting environment: polycyclic aromatic hydrocarbons (PAHs) in sediments; monitoring the evolution of the shore approach area of the Gulf of Gela (Italy). J. Hazard. Mater. 181: 647–658.

Parinos, C., A. Gogou, I. Bouloubassi, R. Pedrosa-Pàmies, I. Hatzianestis, A. Sanchez-Vidal et al. 2013. Occurrence, sources and transport pathways of natural and anthropogenic hydrocarbons in deep-sea sediments of the eastern Mediterranean Sea. Biogeosciences 10: 6069–6089.

Parinos, C. and A. Gogou. 2016. Suspended particle-associated PAHs in the open eastern Mediterranean Sea: occurrence, sources and processes affecting their distribution patterns. Mar. Chem. 180: 42–50.

Patrolecco, L., N. Ademollo, S. Capri, R. Pagnotta and S. Polesello. 2010. Occurrence of priority hazardous PAHs in water, suspended particulate matter, sediment and common eels (*Anguilla anguilla*) in the urban stretch of the River Tiber (Italy). Chemosphere 81: 1386–1392.

Perra, G., K. Pozo, C. Guerranti, D. Lazzeri, V. Volpi, S. Corsolini et al. 2011. Levels and spatial distribution of polycyclic aromatic hydrocarbons (PAHs) in superficial sediment from 15 Italian marine protected areas (MPA). Mar. Pollut. Bull. 62: 874–877.

Piazza, R., A.C. Ruiz-Fernández, M. Frignani, R. Zangrando, L.G. Bellucci, I. Moret et al. 2008. PCBs and PAHs in surficial sediments from aquatic environments of Mexico City and the coastal states of Sonora, Sinaloa, Oaxaca and Veracruz (Mexico). Environ. Geol. 54: 1537–1545.

Pietzsch, R., S.R. Patchineelam and J.P. Torres. 2010. Polycyclic aromatic hydrocarbons in recent sediments from a subtropical estuary in Brazil. Mar. Chem. 118: 56–66.

Pikkarainen, A.L. 2004. Polycyclic aromatic hydrocarbons in baltic sea sediments. Polycycl. Aromat. Compd. 24: 667–679.

Pinheiro, P., C.G. Massone and R.S. Carreira. 2017. Distribution, sources and toxicity potential of hydrocarbons in harbour sediments: A regional assessment in SE Brazil. Mar. Pollut. Bull. 120: 6–17.

Pott, P. 1775. Chirurgical observations relative to the cataract, the polypus of the nose, the cancer of the scrotum, the different kinds of ruptures and the mortification of the toes and feet. *In*: Hawes, L., W. Clarke and R. Collins, London, pages 6348, 1775.

Pozo, K., T. Harner, S.C. Lee, F. Wania, D.C.G. Muir and K.C. Jones. 2009. Seasonally resolved concentrations of persistent organic pollutants in the global atmosphere from the first year of the GAPS study. Environ. Sci. Technol. 43: 796–803.

Pozo, K., G. Perra, V. Menchi, R. Urrutia, O. Parra, A. Rudolph et al. 2011. Levels and spatial distribution of polycyclic aromatic hydrocarbons (PAHs) in sediments from Lenga Estuary, central Chile. Mar. Pollut. Bull. 62: 1572–1576.

Préndez, M., C. Barra, C. Toledo and P. Richter. 2011. Alkanes and polycyclic aromatic hydrocarbons in marine surficial sediment near Antarctic stations at Fildes Peninsula, King George Island. Antarct. Sci. 23: 578–588.

Qian, X., B. Liang, W. Fu, X. Liu and B. Cui. 2016. Polycyclic aromatic hydrocarbons (PAHs) in surface sediments from the intertidal zone of Bohai Bay, Northeast China: Spatial distribution, composition, sources and ecological risk assessment. Mar. Pollut. Bull. 112: 349–358.

Qiao, M., C. Wang, S. Huang, D. Wang and Z. Wang. 2006. Composition, sources, and potential toxicological significance of PAHs in the surface sediments of the Meiliang Bay, Taihu Lake, China. Environ. Int. 32: 28–33.

Qiu, Y., G. Zhang, G. Liu, L. Guo, X. Li and O. Wai. 2009. Polycyclic aromatic hydrocarbons (PAHs) in the water column and sediment core of Deep Bay, South China. Estuar. Coast. Shelf. Sci. 83: 60–66.

Rajpara, R.K., D.R. Dudhagara, J.K. Bhatt, H.B. Gosai and B.P. Dave. 2017. Polycyclic aromatic hydrocarbons (PAHs) at the Gulf of Kutch, Gujarat, India: Occurrence, source apportionment, and toxicity of PAHs as an emerging issue. Mar. Pollut. Bull. 119: 231–238.

Ranjbar Jafarabadi, A., A. Riyahi Bakhtiari, M. Aliabadian and A. Shadmehri Toosi. 2017. Spatial distribution and composition of aliphatic hydrocarbons, polycyclic aromatic hydrocarbons and hopanes in superficial sediments of the coral reefs of the Persian Gulf, Iran, Environ. Pollut. 224: 195–223.

Readman, J.W., R.F.C. Mantoura and M.M. Rhead. 1987. A record of polycyclic aromatic hydrocarbon (PAH) pollution obtained from accreting sediments of the Tamar estuary, UK: evidence for non-equilibrium behaviour of PAH. Sci. Total Environ. 66: 73–94.

Reddy, C.M., J.S. Arey, J.S. Seewald, S.P. Sylva, K.L. Lemkau, R.K. et al. 2012. Composition and fate of gas and oil released to the water column during the deepwater horizon oil spill. Proc. Natl. Acad. Sci. U.S.A. 109: 20229–20234.

Rizzi, J., E. Pérez-Albaladejo, D. Fernandes, J. Contreras, S. Froehner and C. Porte. 2016. Characterization of quality of sediments from Paranagua Bay (Brazil) combined *in vitro* bioassays and chemical analyses. Environ. Toxicol. Chem. 36: 1811–1819.

Rocha, M.J., E. Rocha, C. Cruzeiro, P.C. Ferreira and P.A. Reis. 2011. Determination of polycyclic aromatic hydrocarbons in coastal sediments from the Porto region (Portugal) by microwave-assisted extraction, followed by SPME and GC-MS. J. Chromatogr. Sci. 49: 695–701.

Romero, D., G. Martínez, F. Brito and E. Rodríguez. 2013. Estudio de línea base en la determinación de hidrocarburos aromáticos policíclicos totales en sedimentos superficiales del sector oriental del Golfo de Cariaco, Venezuela. Avances en Química 8: 47–54.

Romero, I.C., P.T. Schwing, G.R. Brooks, R.A. Larson, D.W. Hastings, G. Ellis et al. 2015. Hydrocarbons in deep-Sdea dediments following the 2010 deepwater horizon blowout in the Northeast Gulf of Mexico. PLoS ONE 10: 5.

Sabin, L.D., K.A. Maruya, W. Lao, D.W. Diehl, D. Tsukada, K.D. Stolzenbach et al. 2010. Exchange of polycyclic aromatic hydrocarbons among the atmosphere, water, and sediment in coastal embayments of southern California, USA. Environ. Toxicol. Chem. 29: 265–274.

Sabourin, D.T., J.E. Silliman and K.B. Strychar. 2012. Polycyclic aromatic hydrocarbon contents of coral and surface sediments off the South Texas coast of the Gulf of Mexico. Int. J. Bio. 5: 1.

Salem, D.M.A., F.A.E.M. Morsy, A. El Nemr, A. El-Sikaily and A. Khaled. 2014. The monitoring and risk assessment of aliphatic and aromatic hydrocarbons in sediments of the Red Sea, Egypt. Egypt. J. Aquat. Res. 40: 333–348.

Sanders, M., S. Sivertsen and G. Scott. 2002. Origin and distribution of polycyclic aromatic hydrocarbons in surficial sediments from the Savannah River. Arch. Environ. Contam. Toxicol. 43: 438–48.

Sany, S.B.T., R. Hashim, A. Salleh, M. Rezayi, A. Mehdinia and O. Safari. 2014. Polycyclic aromatic hydrocarbons in coastal sediment of Klang Strait, Malaysia: distribution pattern, risk assessment and sources. PloS One 9: e94907.

Shao, Z., Z. Cui, C. Dong, Q. Lai and L. Chen. 2010. Analysis of a PAH-degrading bacterial population in subsurface sediments on the Mid-Atlantic Ridge. Deep Sea Res. Part I Oceanogr. Res. Pap. 57: 724–730.

Shreadah, M.A., T.O. Said, M.I.A. El Monem, E.M. Fathallah and M.E. Mahmoud. 2011. PAHs in sediments along the semi-closed areas of Alexandria, Egypt. J. Environ. Prot. 2(06): 700.

Simpson, S.L., O. Campana and K.T. Ho. 2016. Sediment toxicity testing. pp. 199–237. *In*: Blasco, J., P.M. Chapman, O. Campana and M. Hampel (eds.). Marine Ecotoxicology: Current Knowledge and Future Issues, Academic Press Incorporated, Orlando.

Sims, R.C. and M.R. Overcash. 1983. Fate of polynuclear aromatic compounds (PNAs) in soil-plant systems. pp. 1–68. *In*: Gunter, F.A. (ed.). Residue Reviews. Springer New York.

Small, C. and R.J. Nicholls. 2003. A global analysis of human settlement in coastal zones. J. Coast. Res. 19: 584–599.

Soclo, H.H., P.H. Garrigues and M. Ewald. 2000. Origin of polycyclic aromatic hydrocarbons (PAHs) in coastal marine sediments: case studies in Cotonou (Benin) and Aquitaine (France) areas. Mar. Pollut. Bull. 40: 387–396.

Soliman, Y.S., E.M.S. Al Ansari and T.L. Wade. 2014. Concentration, composition and sources of PAHs in the coastal sediments of the exclusive economic zone (EEZ) of Qatar, Arabian Gulf. Mar. Pollut. Bull. 85: 542–548.

Sprovieri, M., M.L. Feo, L. Prevedello, D.S. Manta, S. Sammartino, S. Tamburrino et al. 2007. Heavy metals, polycyclic aromatic hydrocarbons and polychlorinated biphenyls in surface sediments of the Naples harbour (southern Italy). Chemosphere 67: 998–1009.

Streit, B. 1998. Bioaccumulation of contaminants in fish. pp. 353–387. *In*: Braunbeck, T., D.E. Hinton and B. Streit. (eds.). Fish Ecotoxicology. Birkhäuser Basel.

Suman, S., A. Sinha and A. Tarafdar. 2016. Polycyclic aromatic hydrocarbons (PAHs) concentration levels, pattern, source identification and soil toxicity assessment in urban traffic soil of Dhanbad, India. Sci. Total Environ. 545: 353–360.

Syakti, A.D., N.V. Hidayati, E. Hilmi, A. Piram and P. Doumenq. 2013. Source apportionment of sedimentary hydrocarbons in the Segara Anakan Nature Reserve, Indonesia. Mar. Pollut. Bull. 74: 141–148.

Takada, H., J.W. Farrington, M.H. Bothner, C.G. Johnson and B.W. Tripp. 1994. Transport of sludge-derived organic pollutants to deep-sea sediments at deep water dump site 106. Environ. Sci. Technol. 28: 1062–1072.

Tanabe, S. 1991. Fate of toxic chemicals in the tropics. Mar. Pollut. Bull. 22: 259–260.

Tansel, B., C. Fuentes, M. Sanchez, K. Predoi and M. Acevedo. 2011. Persistence profile of polyaromatic hydrocarbons in shallow and deep Gulf waters and sediments: effect of water temperature and sediment–water partitioning characteristics. Mar. Pollut. Bull. 62: 2659–2665.

Thorsen, W.A., W.G. Cope and D. Shea. 2004. Bioavailability of PAHs: Effects of soot carbon and PAH source. Environ. Sci. Technol. 38: 2029–2037.

Tolosa, I., J.M. Bayona and J. Albaigés. 1996. Aliphatic and polycyclic aromatic hydrocarbons and sulfur/oxygen derivatives in northwestern Mediterranean sediments: spatial and temporal variability, fluxes, and budgets. Environ. Sci. Technol. 30: 2495–2503.

Tolosa, I., M. Mesa-Albernas and C.M. Alonso-Hernandez. 2009. Imputs and sources of hydrocarbons in sediment from Cienfuegos bay, Cuba. Mar. Pollut. Bull. 58: 1624–1634.

Torres, R.J., A. Cesar, C.D.S. Pereira, R.B. Choueri, D.M.S. Abessa, M.R.L. do Nascimento et al. 2012. Bioaccumulation of polycyclic aromatic hydrocarbons and mercury in oysters (*Crassostrea rhizophorae*) from Two Brazilian Estuarine Zones. International Journal of Oceanography 8 pages.

Tsapakis, M., E. Dakanali, E.G. Stephanou and I. Karakassis. 2010. PAHs and n-alkanes in Mediterranean coastal marine sediments: aquaculture as a significant point source. J. Environ. Monit. 12: 958–963.

Tuncel, S.G. and T. Topal. 2015. Polycyclic aromatic hydrocarbons (PAHs) in sea sediments of the Turkish Mediterranean coast, composition and sources. Environ. Sci. Pollut. Res. 22: 4213–4221.

USEPA. 2002. Polycyclic Organic Matter. US Environmental Protection Agency. Avalible from: https://www.epa.gov/sites/production/files/2016-09/documents/polycyclic-organic-matter.pdf.

USEPA. 2009. Estimation of Biota sediment Accumulation Factor (BSAF) From Paired Observations of Chemical Concentrations in Biota and Sediment, U.S. Environmental Protection Agency, Ecological Risk Assessment Support Center, Cincinnati, OH, EPA/600/R-06/047.

Valavanidis, A., T. Vlachogianni, S. Triantafillaki, M. Dassenakis, F. Androutsos and M. Scoullos. 2008. Polycyclic aromatic hydrocarbons in surface seawater and in indigenous mussels (Mytilus galloprovincialis) from coastal areas of the Saronikos Gulf (Greece). Estuar. Coast. Shelf. Sci. 79: 733–739.

Varanasi, U., D.W. Brown, T. Hom, D.G. Burrows, C.A. Sloan, L.J. Field et al. 1993. Survey of Alaskan subsistence fish, marine mammal, and invertebrate samples collected 1989–91 for exposure to oil spilledfrom the Exxon Valdez, Vol 1. NOAA Technical Memorandum NMFS-NWFSC-12.

Venkatesan, M.I. 1988. Occurrence and possible sources of perylene in marine sediments-a review. Mar. Chem. 25: 1–27.

Vignet, C., T. Larcher, B. Davail, L. Joassard, K. Le Menach, T. Guionnet et al. 2016. Fish reproduction is disrupted upon lifelong exposure to environmental PAHs fractions revealing different modes of action. Toxics 4: 26.

Viguri, J., J. Verde and A. Irabien. 2002. Environmental assessment of polycyclic aromatic hydrocarbons (PAHs) in surface sediments of the Santander Bay, Northern Spain. Chemosphere 48: 157–165.

Viñas, L., M.A. Franco, J.A. Soriano, J.J. González, J. Pon and J. Albaigés. 2010. Sources and distribution of polycyclic aromatic hydrocarbons in sediments from the Spanish northern continental shelf. Assessment of spatial and temporal trends. Environ. Pollut. 158: 1551–1560.

Walker, T.R., D. MacAskill and P. Weaver. 2013. Legacy contaminant bioaccumulation in rock crabs in Sydney Harbour during remediation of the Sydney Tar Ponds, Nova Scotia, Canada. Mar. Pollut. Bull. 77: 412–417.

Walker, T.R. and D. MacAskill. 2014 Monitoring water quality in Sydney Harbour using blue mussels during remediation of the Sydney Tar Ponds, Nova Scotia, Canada. Environ. Monit. Assess. 186: 1623–1638.

Wang, Z., Z. Liu, K. Xu, L.M. Mayer, Z. Zhang, A.S. Kolker et al. 2014. Concentrations and sources of polycyclic aromatic hydrocarbons in surface coastal sediments of the northern Gulf of Mexico. Geochem. Trans. 15: 2.

Wania, F. and D. Mackay. 1996. Peer reviewed: tracking the distribution of persistent organic pollutants. Environ. Sci. Technol. 30: 390A–396A.

Webster, L., M. Russell, N. Shepherd, G. Packer, E.J. Dalgarno and F. Neat. 2018. Monitoring of Polycyclic Aromatic Hydrocarbons (PAHs) in Scottish Deepwater environments. Mar. Pollut. Bull. 128: 456–459.

Wilcke, W., W. Amelung, M. Krauss, C. Martius, A. Bandeira and M. Garcia. 2003. Polycyclic Aromatic Hydrocarbon (PAH) patterns in climatically different ecological zones of Brazil. Org. Geochem. 34: 1405–1417.

Wilcke, W. 2007. Global patterns of polycyclic aromatic hydrocarbons (PAHs) in soil. Geoderma 141: 157–166.

Wu, Y.L., X.H. Wang, Y.Y. Li and H.S. Hong. 2011. Occurrence of Polycyclic Aromatic Hydrocarbons (PAHs) in seawater from the Western Taiwan Strait, China. Mar. Pollut. Bull. 63: 459–463.

Wurl, O. and J.P. Obbard. 2004. A review of pollutants in the sea-surface microlayer (SML): a unique habitat for marine organisms. Mar. Pollut. Bull. 48: 1016–1030.

Xiang, N., C. Jiang, T. Yang, P. Li, H. Wang, Y. Xie et al. 2018. Occurrence and distribution of Polycyclic Aromatic Hydrocarbons (PAHs) in seawater, sediments and corals from Hainan Island, China. Ecotox. Environ. Safe. 152: 8–15.

Xue, W. and D. Warshawsky. 2005. Metabolic activation of polycyclic and heterocyclic aromatic hydrocarbons and DNA damage: a review. Toxicol. Appl. Pharmacol. 206: 73–93.

Xue, R., L. Chen, Z. Lu, J. Wang, H. Yang, J. Zhang et al. 2016. Spatial distribution and source apportionment of PAHs in marine surface sediments of Prydz Bay, East Antarctica. Environ. Pollut. 219: 528–536.

Ya, M., X. Wang, Y. Wu, C. Ye and Y. Li. 2014. Enrichment and partitioning of polycyclic aromatic hydrocarbons in the sea surface microlayer and subsurface water along the coast of Xiamen Island, China. Mar. Pollut. Bull. 78: 110–117.

Ya, M.L., Y.L. Wu, Y.Y. Li and X.H. Wang. 2017. Transport of terrigenous polycyclic aromatic hydrocarbons affected by the coastal upwelling in the northwestern coast of South China Sea. Environ. Pollut. 229: 60–68.

Yan, W., J. Chi, Z. Wang, W. Huang and G. Zhang. 2009. Spatial and temporal distribution of polycyclic aromatic hydrocarbons (PAHs) in sediments from Daya Bay, South China. Environ. Pollut. 157: 1823–1830.

Yang, G.P. 2000. Polycyclic aromatic hydrocarbons in the sediments of the South China Sea. Environ. Pollut. 108: 163–171.

Younis, A.M., Y.A. Soliman, E.M. Elkady and M.H. El-Naggar. 2018. Assessment of polycyclic aromatic hydrocarbons in surface sediments and some fish species from the Gulf of Suez, Egypt. Egypt. J. Aquat Res. 22: 49–59.

Yu, M.H. 2005. Environmental Toxicology: Biological and Health Effects of Pollutants (2nd), CRC Press, 156670670X (alk. paper), Boca Raton.

Yuan, H., E. Liu, E. Zhang, W. Luo, L. Chen, C. Wang et al. 2017. Historical records and sources of polycyclic aromatic hydrocarbons (PAHs) and organochlorine pesticides (OCPs) in sediment from a representative plateau lake, China. Chemosphere 173: 78–88.

Yunker, M., F.A. McLaughlin, M.G. Fowler and B.R. Fowler. 2014. Source apportionment of the hydrocarbon background in sediment cores from Hecate Strait, a pristine sea on the west coast of British Columbia, Canada. Org. Geochem. 76: 235–258.

Yunker, M.B., L.R. Snowdon, R.W. Macdonald, J.N. Smith, M.G. Fowler, D.N. Skibo et al. 1996. Polycyclic aromatic hydrocarbon composition and potential sources for sediment samples from the beaufort and barents Seas. Environ. Sci. Technol. 30: 1310–1320.

Yunker, M.B., R.W. Macdonald, R. Vingarzan, R.H. Mitchell, D. Goyette and S. Sylvestre. 2002. PAHs in the Fraser River basin: a critical appraisal of PAH ratios as indicators of PAH source and composition. Org. Geochem. 33: 489–515.

Yunker, M.B., R.W. Macdonald, L.R. Snowdon and B.R. Fowler. 2011. Alkane and PAH biomarkers as tracers of terrigenous organic carbon in Arctic Ocean sediments. Org. Geochem. 42: 1109–11.

Zaborska, A., J. Carroll, K. Pazdro and J.Pempkowiak. 2011. Spatio-temporal patterns of PAHs, PCBs and HCB in sediments of the western Barents Sea. Oceanologia 53: 1005–1026.

Zaghden, H., M. Kallel, B. Elleuch, J. Oudot, A. Saliot and S. Sayadi. 2014. Evaluation of hydrocarbon pollution in marine sediments of Sfax coastal areas from the Gabes Gulf of Tunisia, Mediterranean Sea. Environ. Earth Sci. 72: 1073–1082.

Zhang, P., J. Song and H. Yuan. 2009. Persistent organic pollutant residues in the sediments and mollusks from the Bohai Sea coastal areas, North China: an overview. Environ. Int. 35: 632–646.

Zhang, Y. and S. Tao. 2009. Global atmospheric emission inventory of polycyclic aromatic hydrocarbons (PAHs) for 2004. Atmos. Environ. 43: 812–819.

Zhang, Y., Y. Han, J. Yang, L. Zhu and W. Zhong. 2017. Toxicities and risk assessment of heavy metals in sediments of Taihu Lake, China, based on sediment quality guidelines. J. Environ. Sci. 62: 31–38.

Zhao, M., W. Wang, Y. Liu, L. Dong, L. Jiao, L. Hu et al. 2016. Distribution and sources of polycyclic aromatic hydrocarbons in surface sediments from the Bering Sea and western Arctic Ocean. Mar. Pollut. Bull. 104: 379–385.

Zhou, J.L., T.W. Fileman, S. Evans, P. Donkin, C. Llewellyn, J.W. Readman et al. 1998. Fluoranthene and pyrene in the suspended particulate matter and surface sediments of the Humber Estuary, UK. Mar. Pollut. Bull. 36: 587–597.

Zong, H., X. Ma, G. Na, C. Huo, X. Yuan and Z. Zhang. 2014. Polycyclic aromatic hydrocarbons (PAHs) in the mariculture zones of China's northern Yellow Sea. Mar. Pollut. Bull. 85: 172–178.

Zrafi-Nouira, I., D.S. Nimerm, R. Bahri, N. Mzoughi, A. Aissi, H. Ben Abdennebi et al. 2010. Distribution and sources of polycyclic aromatic hydrocarbons around a petroleum refinery rejection area in Jarzouna-Bizerte (Coastal Tunisia). Soil Sediment. Contam. 19: 292–306.

12

Brominated Flame Retardants

Norma Tombesi,[1,]* *Karla Pozo,*[2,3] *Mónica Alvarez*[1]
and *Andrés H. Arias*[1,4]

1. Introduction

Fire is considered one of the most destructive forces on the planet. Throughout history, fire has caused losses of human lives, injury, and extensive damage to property (Innes and Innes 2012). So the search for ways to protect structures from fire began a long time ago. The term "flame retardant" (FR) is used to identify any substance applied or added to a wide range of products to inhibit, suppress, or delay ignition and to prevent the spread of fire (Aschberger et al. 2017).

Probably asbestos was the first product used by humans for the purpose of delaying fire expansion, and its first known use in textiles dates back to Ancient Egypt (2000–3000 BCE). This mineral was used for a long time and throughout the world, but its use began to be limited when harmful effects were detected in people who handled asbestos in some way. In about 450 BC, Herodotus reported the use of alum by Egyptians to reduce the flammability of wood, and the use of a mixture of alum and vinegar on wood by the Romans (~ 200 BCE) (Hindersinn 1990).

The first known patent for a flame retardant was granted in Great Britain for a mixture of alum, sulfate, iron, and borax (Wyld 1735). Later, Gay-Lussac (1821) carried out the first systematized study of FRs using cellulosic materials. He concluded that the most effective flame-retarding salts had low melting points and also produced glassy deposits on the fiber surface, or decomposed into non-flammable vapors, diluting cellulose-derived flammable gases. Thus, numerous studies and compounds emerged as FRs, which gave rise to numerous patents throughout the first half of the 20th century.

[1] INQUISUR, Departamento de Química, Universidad Nacional del Sur-CONICET, Bahía Blanca, Argentina.
[2] Faculty of Science, Research Center for Toxic Compounds in the Environment, Masaryk University, Brno, Czech Republic.
[3] Facultad de Ingeniería y Tecnología, Universidad San Sebastián, Concepción, Chile.
[4] Instituto Argentino de Oceanografía, CCT-CONICET, Bahía Blanca, Argentina.
* Corresponding author: ustombes@uns.edu.ar

Roman ships were coating with alum and vinegar to prevent fire.

Joseph Louis Gay-Lussac: carried out the first systematized study of FRs using cellulosic materials.

At the same time, the effectiveness of halogenated compounds in the treatment of materials such as textiles and paper as a method of flame proofing was recorded in several patents published in the early 20th century, for example, Carleton (1932), Dryfus (1933), and Hanson (1931), who specifically mentions a preference for the use of bromine compounds because they are more efficacious as FRs than chlorine compounds.

2. Use of FRs

The use of FRs has drastically increased with industrialization (Guerra et al. 2010). Potential fire hazards have been dramatically augmented with technological advances and widespread use of synthetic polymers. Thus, FRs are often added or applied to a great variety of products such as furnishings (foam, upholstery, mattresses, carpets, curtains, and fabric blinds); electronics and electrical

WHERE FLAME RETARDANTS ARE FOUND

In home insulation

In upholstered furniture containing polyurethane foam—manufacturers add it to meet flammability standards enacted by California but followed nationwide

In the plastic casing of some electronics

In dust—children are exposed to higher doses of flame retardants than adults because they spend more time on the floor and put things in their mouths

In carpet padding made with recycled foam

In some baby products containing polyurethane foam, including highchairs and diaper-changing pads

SOURCES: EPA, Tribune reporting

KATIE NIELAND/TRIBUNE

Fig. 1. Flame retardants hard to avoid. Sam Roe, Michael Hawthorne, David Eads, Katie Nieland, and Joe Germuska, May 5, 2012. *Chicago Tribune Reporter*. http://media.apps.chicagotribune.com/flames/flame-retardants-hard-to-avoid.html.

devices (computers, laptops, phones, televisions, and household appliances, plus wires and cables); building and construction materials (including electrical wires and cables, and insulation materials, such as polystyrene and polyurethane insulation foams); and transportation products (seats, seat covers and fillings, bumpers, overhead compartments, and other parts of automobiles, airplanes, and trains) (https://www.niehs.nih.gov/health/topics/agents/flame_retardants/index.cfm). Figure 1 shows only a few examples of places where FRs can be found in a domestic environment.

3. FR Classifications

A great variety of chemicals with different properties and molecular structures act as FRs and can be classified by diverse criteria.

FRs may be mixed with the base material or chemically bonded to it (Speight 2017). **Additive** FRs are incorporated into the polymer either prior to, during or (most frequently) following polymerization, and **reactive** FRs are reactive components chemically built into a polymer molecule (INCHEM 1997). Additive FRs are of particular concern for the environment since they are most likely to leak out of the products and migrate from the polymeric materials.

FRs can act by **physical** or **chemical action** on one of the stages of the combustion process (Guerra et al. 2010). Fundamentally, four processes are involved in polymer flammability: preheating, decomposition, ignition and combustion/propagation. The combustion process can be retarded by **physical action** by *cooling* (endothermic process of FR decomposition) or *diluting* the substrate in the gas phase (i.e., formation of water) and the solid phase (alumina trihydrate and magnesium hydroxide), or by *coating* the substrate (shielding it against the attack of oxygen and heat) with phosphorous and nitrogen compounds. With regard to **chemical action**, the most significant

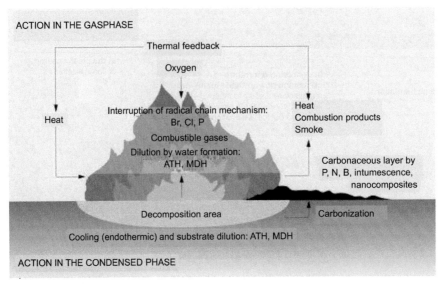

Fig. 2. Different modes of action of FRs (https://www.flameretardants-online.com/flame-retardants/mode-of-action).

reactions interfering with the combustion process take place in the solid-gas phases (Troitzsch 1990). Thus, the free radical mechanism of the combustion process that takes place *in the gas phase* is interrupted by the flame retardant. The exothermic processes are thus stopped, the system cools down, and the supply of flammable gases is reduced and eventually completely suppressed. The reaction *in the solid phase* can take place by two forms of action (Fig. 2). (1) The FR can cause a layer of carbon to form on the polymer surface, e.g., by means of the dehydrating action that generates double bonds in the polymer, and forming the carbonaceous layer by cyclizing and cross-linking. (2) The other mode of action on the solid phase can be acceleration of polymer decomposition, causing marked flow of the polymer and, consequently, its removal from the area of influence of the flame (INCHEM 1997).

Finally, according to chemical composition, the FRs can be classified as **organic** or **inorganic based**. By this classification, halogenated FRs are principally of organic composition, and phosphorous and non-phosphorous compounds.

4. Brominated Flame Retardants

Most halogenated organic FRs are based on chlorine and bromine. Halogen-containing FRs rise in efficiency in the order of F < Cl < Br < I (Boryniec and Przygocki 2001). While bromine and chlorine compounds are the only halogen compounds having commercial importance as FR chemicals, fluorine compounds are practically not used since they are costly and ineffective because the C–F bond is too strong. Iodine compounds, although effective, are expensive and have an excessively loose bond to carbon (Pettigrew 1993). Consequently, only organochlorine and organobromine compounds are suitable as FRs. Thus, with greater trapping efficiency and lower decomposing temperature, organobromine compounds have become more widespread as FRs than their organochlorine counterparts.

Bromine is currently used in a large number of products as pesticides, gasoline additives, drilling fluids, and biocides, but currently the largest application for bromine-based compounds is in FR production. Figure 3 shows the global consumption of FRs in 2010 by region and chemical composition, where brominated FRs represent 21% of the FR market.

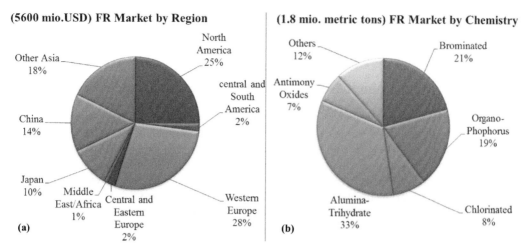

(5600 mio.USD) FR Market by Region

- North America 25%
- central and South America 2%
- Western Europe 28%
- Central and Eastern Europe 2%
- Middle East/Africa 1%
- Japan 10%
- China 14%
- Other Asia 18%

(a)

(1.8 mio. metric tons) FR Market by Chemistry

- Brominated 21%
- Organo-Phophorus 19%
- Chlorinated 8%
- Alumina-Trihydrate 33%
- Antimony Oxides 7%
- Others 12%

(b)

Fig. 3. Global consumption of flame retardants (2010): FR market by region (a) and chemistry (b). Graphic built with data extracted and adapted from https://www.flameretardants-online.com/images/itempics/6/4/1/item_18146_pdf_1.pdf.

Table 1. Five main classes of BFRs and their common uses.

Name (abbreviation)	Common uses
Polybrominated diphenyl ethers (PBDEs)	plastics, textiles, electronic castings, circuitry
Hexabromocyclododecanes (HBCDDs)	thermal insulation in the building industry
Tetrabromobisphenol A (TBBP-A) and other phenols	printed circuit boards, thermoplastics (mainly in TVs)
Polybrominated biphenyls (PBBs)	consumer appliances, textiles, plastic foams
Other brominated flame retardants	

Source: https://www.efsa.europa.eu/en/topics/topic/brominated-flame-retardants

More than 80 different aliphatic, cyclo-aliphatic, aromatic, and polymeric compounds have been registered as brominated FRs. Table 1 shows the main classes of BFRs and their common uses.

5. BFRs and the Environment

An accident in 1973, referred to as the "Michigan disaster", first raised the alarm about the harmful effects of BFRs on human health and animals. In a plant in Michigan (USA), polybrominated biphenyls (PBBs) were inadvertently added to animal feed that was distributed to farms. The plant was dismantled, and the site was declared a hazardous waste site and eventually remediated. The consequences of this accident continue to this day (Venier et al. 2015).

Not all FRs present concerns, but the halogenated compounds that contain chlorine or bromine bonded to carbon often are associated with health and environmental alarms. Consistently, in recent years, due to growing recognition of their toxicity, potential bioaccumulation, and high environmental persistence, stricter policies have been placed on some BFRs. The Stockholm Convention, an international treaty to protect human health and the environment from persistent organic pollutants (POPs), in its fourth meeting in 2008 decided to amend the regulation of organobromines hexabromobiphenyl, hexabromodiphenyl ether and heptabromodiphenyl ether, and tetrabromodiphenyl ether and pentabromodiphenyl ether in the Annex A (Elimination). At the Sixth

Conference of Parties in 2013, hexabromocyclododecane (HBCDD) was added (UNEP 2018). At the eighth meeting held in 2017, decabromodiphenyl ether (commercial mixture, c-DecaBDE) was added to this list.

5.1 Scientific studies about BFRs in coastal and deep ocean environment

An exhaustive search was carried out on all studies published until 31/12/2017 in Scopus with the general search criteria shown in Fig. 4 to evaluate the existing information on BFRs in coastal and oceanic environments. Subsequently, the articles were evaluated individually to define their relevance to the objective of this chapter, and revisions were excluded to avoid duplication of information. Additional searches were made to find those articles that might have been lost, for example, with the subject "mussel", "tissue", "bivalve", "shell" or "fish", to the biota matrix.

On the other hand, Fig. 5 shows the evolution of scientific research about BFRs in coastal and ocean environment in the time and according to the search strategy described above. The increase in the number of studies is consistent with the increase in environmental concern with respect to BFRs.

The first studies were recorded in the 1980s and were carried out mainly on the coasts of Japan (Watanabe et al. 1983, 1985, 1986, 1987) and in sites associated with seas of northern Europe (Anderson and Blomkvist 1981, Jansson et al. 1987). Andersson and Blomkvist (1981) found a level mean of 1.4 mg kg^{-1} fat of PBDEs in sea trout catch (average fat content of 1.1%) in the Klosterfjorden Bay from Sweden in 1979; in sediments of Osaka Bay (Japan), Watanabe et al. (1983) reported levels of TBBP-A from 0.5 to 4.5 ppb, and of hexabromobenzene from undetected levels to 6.2 ppb Watanabe et al. (1986). In the 1990s, new studies on BFRs included various environmental matrices from coastal areas and seas of northern Europe, including the Swedish coast (Nylund et al. 1992, Andersson and Wartanian 1992, Sellström et al. 1993, Haglund et al. 1997) and Latvian coast (Olsson et al. 1999), but also reached other marine areas like the Atlantic coast of the United States (Kuehl et al. 1991, Kannan et al. 1998) and other areas corresponding to the North Atlantic (de Boer

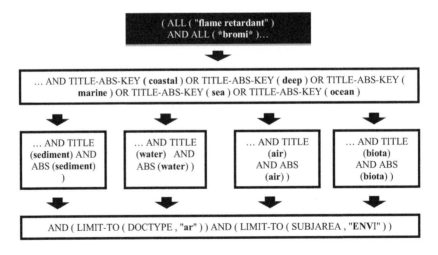

ALL: All fields
TITLE-ABS-KEY: Article title, Abstract, Keywords
TITLE: Title
ABS: Abstract
LIMIT TO DOCTYPE "ar": Document-type Article
LIMIT-TO (SUBJAREA) "ENVI": Subject Area: ENVIRONMENTAL SCIENCE

Fig. 4. General search criteria carried out on all studies published until 12/31/2017 in Scopus.

et al. 1998, Bottaro et al. 1999, Lindström et al. 1999). In the 2000s, the number of scientific articles, although oscillating, increased markedly and new regions were added, although the majority also corresponded to the northern hemisphere, e.g., the Mediterranean Sea (Pettersson et al. 2004, Eljarrat et al. 2005, Johansson et al. 2006), and various regions of the Asian continent (Bayen et al. 2003, Liu et al. 2005, Takigami et al. 2005, Mai et al. 2005, Kajiwara et al. 2006, Wurl et al. 2006, Gevao et al. 2006, Chen et al. 2006, Ramu et al. 2007a, b, Isobe et al. 2007, Moon et al. 2007a, b, c, Pan et al. 2007, Xiang et al. 2007, Minh et al. 2007, Binelli et al. 2007, Luo et al. 2007, Zhang et al. 2008, Choi et al. 2008, Tanabe, 2008, Jin et al. 2008, Moon et al. 2008, Terauchi et al. 2009, Yu et al. 2009 and Wang et al. 2009). This trend can be observed in Fig. 6, which shows spatial distribution of studies about BFRs (number of articles) related to coastal and oceanic environments and corresponding to the search strategy described above.

With a few exceptions, the first surveys on levels of BFRs in the southern hemisphere were made in the current decade. Thus, the articles found included studies in the Antarctic (Dickhut et al. 2012, Strobel et al. 2016, Fromant et al. 2016), South America (Dias et al. 2013, Barón et al. 2013a, b, Cappelletti et al. 2015, Pozo et al. 2015, Wang et al. 2017, Tombesi et al. 2017), South

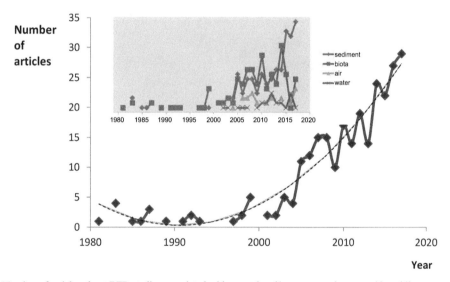

Fig. 5. Number of articles about BFR studies associated with coastal and/or ocean environment (dotted line: second order polynomial adjustment). Inset: number of articles by matrix in the same time period.

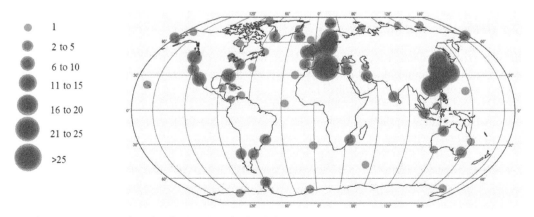

Fig. 6. Spatial distribution of studies (number of articles) about BFRs related to coastal and oceanic environments.

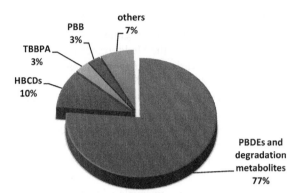

Fig. 7. Percentage distribution about the main BFRs studied in coastal and deep ocean environment according to chemical type.

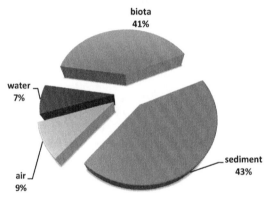

Fig. 8. Percentage contribution by environmental matrix in articles about BFRs in coastal and deep sea around the world.

Atlantic (Rochman et al. 2014, Pegoraro et al. 2016), South Africa (La Guardia et al. 2013), Indian Ocean (Cheng et al. 2015), and Australia (Drage et al. 2015, Anim et al. 2017).

Finally, taking into account the type of chemical compound and the matrix studied, the percentage distribution of the studies carried out is represented in Figs. 7 and 8 respectively. The BFRs more often studied in marine and coast environments were PBDEs and their metabolites (77%), followed by HBCDs (10%), TBBPA (3%), and PBB (3%); meanwhile, the matrices were sediment (43%) and biota (41%), followed by air (9%) and water (7%).

5.2 BFR levels and global trends

In order to have a global view of the current levels and trends of BFRs on coastal and oceanic marine environments, some recent studies were selected among the most representative, and their results are reproduced and/or briefly described below. Taking into account the trend observed in the present review, mainly studies on PBDEs (the BFR most studied, Fig. 7) are considered. PBDEs have been principally marketed in three formulations: deca-BDE, octa-BDE and penta-BDE. La Guardia et al. (2006) characterized composition of six common technical FR mixtures (two formulations of each), and established the following principal contributions of PBDE congeners: BDE-209 to deca-BDE; BDE-99, BDE-47 and BDE-100 to penta-BDE; and BDE-209 or BDE-183 (depending on the formulation) to octa-BDE. This information is relevant in assessing possible sources and environmental fate, always considering other aspects such as the selective congener biomagnification, degradation, and transport.

5.3 BFRs in marine water

According to its low water solubility, Lee and Kim (2015) observed that little information has been reported about PBDEs in water and these studies are usually associated with particulate phases. This trend was also observed in our review (Fig. 8), where the studies in water matrix presented the lowest contribution (7%). Among the most recent articles on PBDEs in water are studies by Salvadó et al. (2016), Ju et al. (2016), Möller et al. (2011), Lee et al. (2018).

PBDE concentrations (sum of 14 congeners) from 0.3 to 11.2 pg L^{-1} in surface seawater (dissolved and particulate) were measured by Salvadó et al. (2016) in the Arctic Ocean, and with higher concentrations in the pan-Arctic shelf seas and lower levels in the interior basin. BDE-209 was the dominant congener in most of the pan-Arctic areas except for the ones close to North America, where penta-BDE and tetra-BDE congeners predominate, according to Möller et al. (2011), who found dominance of BDE-47 and BDE-99 congeners in the East Greenland Sea in 2009, and PBDE concentration (sum of 10 congeners) in seawater ranged from 0.005 to 0.64 pg L^{-1}. Thus, regional studies show different trends in relation to the proximity to possible sources of contamination. Lee et al. (2018) found in Korea a total PBDE concentration (sum of 19 congeners) in seawater from 1.58 to 6.94 ng L^{-1}, and with highest levels at a station adjacent to a sewage outfall of the treatment plant and predominance of BDE-209 congener. Ju et al. (2016) observed that PBDE concentrations (sum of 14 congeners) were higher in the inner and mouth regions than in the outer and middle region of seawater from Jiaozhou Bay, northern China (strongly affected by urbanization and anthropogenic activities, including industries, agriculture, aquaculture, and ship traffic) and also BDE-209 dominance. Moreover, Ju et al. (2016) reported PBDE concentrations in dissolved phase decreased with water depth in the water column from 0.24 to 1.35 ng L^{-1} in the surface water and from 0.09 to 0.38 ng L^{-1} in the bottom water; for suspended particulate matter, the concentrations were 2.27 to 5.45 ng L^{-1} in the surface water and 1.17 to 2.23 ng L^{-1} in the bottom water (Fig. 9).

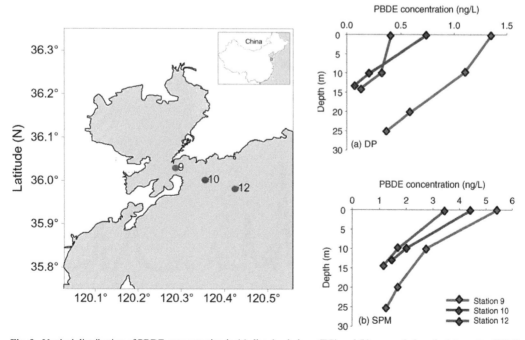

Fig. 9. Vertical distribution of PBDE concentration in (a) dissolved phase (DP) and (b) suspended particulate matter (SPM) of seawater from Jiaozhou Bay, northern China (Ju et al. 2016).

5.4 BFRs in air

The review on studies of BFR levels in air in marine environment showed low contribution (9%), similar to levels in water matrix (7%) (Fig. 8). In order to compare the PBDE levels in the oceanic atmosphere many factors must be taken into account: geographic (e.g., longitude, latitude, ocean current) and climatic parameters (e.g., temperature, pressure, humidity, wind speed and direction, monsoon), and instruments for sampling and analysis as described by Chao et al. (2014). Also, different number of individual congeners and the atmospheric phase sampled (vapor, particle or both—Al-Omran 2018), besides structural chemical aspects (e.g., chiral characteristics) and the possibility of different photodegradation rates among others, must be taken into account. Among the articles on PBDEs in air were those by Chao et al. (2014), Pegoraro et al. (2016), Wang et al. (2017) and Castro-Jiménez et al. (2017).

Levels of atmospheric PBDEs (14 congeners) between 0.14 and 58.3 pg m^{-3} over the Pacific Ocean near southern Taiwan and the northern Philippines (sampling by ~ 40 h, with PS-1 sampler Graseby Andersen, USA) were found by Chao et al. (2014), who observed that the highest PBDE concentrations in the global oceanic atmosphere were over the marginal seas. Likewise, Pegoraro et al. (2016) found concentrations of PBDEs (14 congeners) from 0.69 to 2.58 pg m^{-3} (0.07 to 0.49 pg m^{-3} and 0.58 to 1.81 pg m^{-3} for PBDE-47 and PBDE-209 respectively) in air samples collected using a high-volume air sampler (48 h of sampling) on board the Argentinean research vessel Puerto Deseado (CONICET) in the south Atlantic. Levels remained elevated in the near-shore environment and then dropped off substantially beyond a distance of about 400 km. Meanwhile, Castro-Jiménez et al. (2017) found PBDE concentrations (27 congeners) from 2.2 to 17.6 pg m^{-3} (~ 9.0 pg m^{-3}, median) for Marseille (France) and from 1.0 to 54.0 pg m^{-3} (~ 6.0 pg m^{-3}, median) for Bizerte (Tunisia) in a comparative study of aerosol samples during 2015–2016 in two coastal cities at both the African and European coasts of the western Mediterranean, showing the PBDE-209 congener to be the major contributor for both sites.

On the other hand, concentrations of PBDEs (sum of 27 congeners) in air (gas + particle) at the Chinese Great Wall Station, West Antarctica (January 2011 to 2014), from 0.60 to 16.1 pg m^{-3} (using a high-volume air sampler, for 7 d) were reported by Wang et al. (2017), including BDE-209 concentration range from 0.18 to 6.35 pg m^{-3} and BDE-47 from 0.006 to 0.37 pg m^{-3}. Additionally, Wang et al. (2017) reported temporal variation of PBDEs over the whole sampling period (Fig. 10) and observed that the results were relatively higher than the available data in the Antarctic atmosphere and slightly lower than those from the Arctic regions.

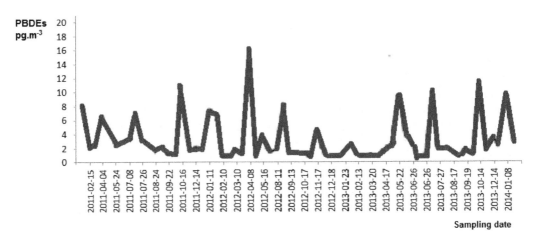

Fig. 10. Temporal variation of PBDEs (sum of 27 congeners) over the sampling period 2011–2014 at the Chinese Great Wall Station, West Antarctica (Wang et al. 2017).

Finally, from a global point of view, Rauert et al. (2018) observed that the levels of PBDEs in air found in samples taken during 2014 in the Global Atmospheric Passive Sampling (GAPS) Network were similar to those previously reported from samples collected in 2005 at GAPS sites, suggesting global background atmospheric concentrations of PBDEs have not declined since regulatory measures were implemented.

5.5 BFRs in biota

Aquatic organisms are considered good indicators of environmental pollution, since they concentrate bioaccumulative pollutants in their bodies from water and sediments, besides the uptake from their diet (Wang et al. 2007). According to this, the biota matrix records the highest number of reports (41%) on levels of BFRs related to the marine and coastal environment together with the sediment matrix (43%) (Fig. 8). Also, the biota includes a wide variety of aquatic organisms such as macro-algae, mussels, fish, birds, and seals, which have been widely used to monitor the pollution of anthropogenic contaminants (Hong 2002). Results and comments published by Kim et al. (2018), Fernandes et al. (2018), Verreault et al. (2018), Zaccaroni et al. (2018), Aznar Alemany (2019) and Bjurlid et al. (2018) were selected as a sample of the most recent or relevant studies conducted on PBDEs in biota matrixes.

Kim et al. (2018) found PBDE levels in mosses and lichens from the South Shetland Islands Area, Antarctica, ranged from 3.2 to 71.5 pg g^{-1} dw and from 1.5 to 188 pg g^{-1} dw respectively. Despite the limited sample size and uncertainties arising from the structural and habitational differences of the mosses and lichens, this study depicted the spatial distribution of these contaminants in the South Shetland Islands, with higher levels near Antarctic research stations.

Fernandes et al. (2018) investigated the occurrence of PBDEs (sum of 17 congeners) in edible marine fish sampled mainly from the UK marine regions, along the coast of Norway and south to the Algarve (Portugal). In this study, contaminant occurrence varied with species and location, and all measured contaminants were detected in the fishes studied. The highest average values were observed for herring, sea bass, mackerel and sprat (2.08, 2.0, 1.45 and 1.27 µg kg^{-1} respectively), and Fernandes et al. (2018) observed that this data provides no evidence of a downward trend in PBDE concentrations in marine species, in comparison with previous reports (Fernandes et al. 2014).

Verreault et al. (2018) found that PBDE (sum of 38 congeners) concentrations in liver samples of glaucous gull (a top predator of the Canadian Arctic) were 3-fold higher in males (7.24 ± 3.07 ng g^{-1} ww) than in females (24.5 ± 5.46 ng g^{-1} ww), and that BDE-209 (principal constituent in the deca-BDE) was the dominant congener; the results suggest dietary exposure from the local marine food web and perhaps also from nearby community landfills. On the other hand, and with the exception of this BDE congener, concentrations of other halogenated FRs were found to be in the lower range in liver of Eastern Canadian Arctic glaucous gulls than in individuals from other circumpolar populations (Svalbard and Greenland).

Zaccaroni et al. (2018) found various PBDE congeners (47, 99, 100, 154, entering the composition of penta-BDE mixture) in blubber of adult female sperm whales (three individuals designated as SW1, SW2, and SW3) stranded in the southern Adriatic Sea coasts (Italy), with values (sum of 8 congeners) up to 183 ng g^{-1} lw tissue. BDE-47 is the most frequently detected congener in all the tissues except for liver samples. Moreover, this study showed a preferential accumulation in the blubber with respect to other tissues analyzed (Fig. 11).

Meanwhile, Aznar Alemany (2019) found a total PBDE concentration (sum of 5 congeners) from 0.27 to 2.21 ng g^{-1} lw in muscle and from undetected levels to 5.13 ng g^{-1} lw in adipose tissue of the southern elephant seal, and from undetected levels to 2.74 ng g^{-1} lw in adipose tissue of the Antarctic fur seal, and only BDE-28, BDE-47 and BDE-99 were above their limit of quantitation (the latter in just one sample).

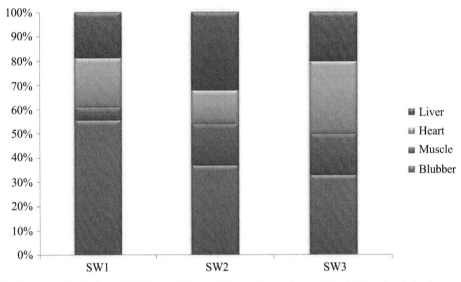

Fig. 11. Percentage distribution of PBDEs in relation to different tissues observed in adult female whales from southern Adriatic Sea coasts (Italy) (Zaccaroni et al. 2018).

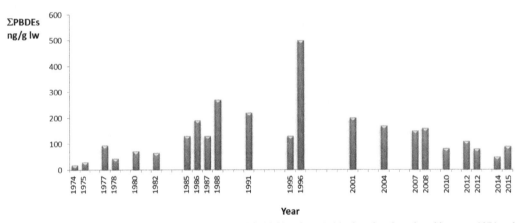

Fig. 12. Temporal trends of SPBDEs (sum of 9 congeners) in blubber from Baltic ringed seals gathered between 1974 and 2015 (Bjurlid et al. 2018).

On the other hand, the assessment of temporal trends of POPs in Arctic marine biota were analyzed by Rigét et al. (2019), and concentrations of BDE-47 congener showed a typical trend of increasing concentration up to approximately the mid-2000s followed by a decreasing concentration. The study suggested that some Arctic animal populations would have responded rapidly to reduction of the production and/or use of BFRs. Bjurlid et al. (2018) observed similar behavior for PBDEs (sum of 9 congeners) in ringed seals from the Baltic Sea, where the concentrations increased until the end of the 1990s and then decreased until the end of the study period between 1974 and 2015 (Fig. 12).

5.6 BFRs in sediment

Sediment has been identified as an ultimate sink and as a potential source of PBDEs (Lee and Kim 2015). Consistently, the higher percentage of studies of BFRs observed in our review (47%) was carried out on this matrix. In this case, Pozo et al. (2015), Ma et al. (2017), Tombesi et al. (2017), Zhu et al. (2018), Lee et al. (2018), and Dasgupta et al. (2018) depicted the state of the art in studies of PBDEs in sediments related to marine and coastal environments.

Ma et al. (2017) reported for the Bering Sea, Bering Strait, and Chukchi Sea that concentrations of BDE-47, BDE-99, and BDE-153 were generally in the range of tens of picograms per gram of dry weight, while concentrations of BDE-209 were in the range of hundreds of picograms per gram of dry weight. Also, Ma et al. (2017) observed that the concentrations of PBDEs (excluding BDE-209) seemed to decrease from 2008 to 2012 for this region. With respect to the Antarctic region, Wang et al. (2012) found levels of PBDEs ranged from 2.76 to 51.4 pg g^{-1} dw in soil and sediment, and the BDE-47 was the dominant congener. These results indicated that long-range atmospheric transport seemed to be the principal pathway of POPs to King George Island, while anthropogenic influence (e.g., from research station, tourism) could also influence the spatial distribution.

PBDEs (sum of 10 congeners) were analyzed by Tombesi et al. (2017) in four coastal surface sediments collected from the northern shore of Bahía Blanca estuary (southwest of Buenos Aires Province, Argentina). The results show PBDE concentration from 0.16 to 2.02 ng g^{-1} dw with dominance of BDE-209 and the highest concentration was detected close to an industrial effluent discharge point.

A special study for PBDEs in surficial sediments was carried out by Pozo et al. 2015 on Concepción Bay after the 2010 tsunami. The dominant congeners were PBDE-47, PBDE-99, PBDE-100 and PBDE-209, over a total of ten congeners analyzed, as recorded also in biota. Sum of PBDE-47, PBDE-99 and PBDE-100 ranged between 0.02 and 0.09 ng g^{-1} dw, and PBDE-209 displayed concentrations up to 21 ng g^{-1} dw. While the sum of PBDE-47, PBDE-99 and PBDE-100 was about 10 times lower than reported in a study (Barón et al. 2013b) previous to the 2010 tsunami (0.3 ng g^{-1} dw), PBDE-209 was 10 times higher (1.7 ng g^{-1} dw). Change in the composition of the PBDE technical mixture used in Chile after the tsunami, and massive urban debris removed by this special event, were argued by Pozo et al. (2015) as a possible explanation to this particular finding.

Zhu et al. (2018) observed for the south China region (Pearl River Delta region and Mirs Bay) that PBDE congeners were dominated by BDE-209, with the average value accounting for 86.8% of the total PBDEs in the sediment samples. The concentration of total PBDEs (sum of 12 congeners) ranged from 8.090 to 595.8 ng g^{-1} dw, and it was found that the highest concentration was associated with waste discharges of Guangzhou city.

Lee et al. (2018) reported that PBDE concentrations in surface marine sediments of the South Korean coast (near Tongyeong sewage treatment plant) were within the range of 2.18–307 ng g^{-1} dw. BDE-209 was the dominant congener, at 90% of the total PBDEs (sum of 19 congeners) investigated. Additionally, PBDE concentrations in sediments gradually decreased with increasing distance from inner bay including the sewage outfall, while PBDE concentrations in sediment cores exponentially decreased with increasing depth (Fig. 13) and the decrease of organic carbon content.

Dasgupta et al. (2018) also investigated PBDE concentration in core sample sediments, in this case from the southern Mariana Trench at water depths of 7,000–11,000 m. Values (sum of 10 congeners) between 245 and 591 pg g^{-1} dw were obtained, BDE-47 and BDE-153 being the more common, and showing how far anthropogenic contaminants can reach in the oceanic environment.

The information displayed in this chapter about the presence of BFRs, especially PBDEs, in coastal and deep ocean environments constitute evidence as to the distance these contaminants have reached. Thus, any type of pollutant will have long-term or brief implications, either ecologically damaging or toxic, depending on the scale of impact. As Dasgupta et al. (2018) stated, after finding anthropogenic pollutants, including PBDEs, in sediments of the deepest ocean on Earth, "possibly, there no longer exists 'pure land' that can completely be isolated from human activities in the Earth's ocean."

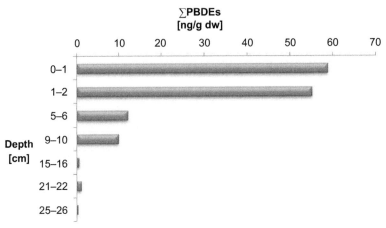

Fig. 13. Total concentration PBDEs (sum of 19 congeners) in sediment cores collected from Tongyeong Bay vertical profile (Lee et al. 2018).

References

Al-Omran, L. 2018. Physiochemical properties and environmental levels of legacy and novel brominated flame retardants. In Fire Control-Basic to Nanotechnology.

Andersson, Ö. and G. Blomkvist. 1981. Polybrominated aromatic pollutants found in fish in Sweden. Chemosphere 10(9): 1051–1060.

Andersson, Ö. and A. Wartanian. 1992. Levels of polychlorinated camphenes (toxaphene), chlordane compounds and polybrominated diphenyl ethers in seals from Swedish waters. Ambio, 550–552.

Anim, A.K., D.S. Drage, A. Goonetilleke, J.F. Mueller and G.A. Ayoko. 2017. Distribution of PBDEs, HBCDs and PCBs in the Brisbane River estuary sediment. Mar. Pollut. Bull. 120(1-2): 165–173.

Aschberger, K., I. Campia, L.Q. Pesudo, A. Radovnikovic and V. Reina. 2017. Chemical alternatives assessment of different flame retardants–a case study including multi-walled carbon nanotubes as synergist. Environ. Int. 101: 27–45.

Aznar-Alemany, Ò., X. Yang, M.B. Alonso, E.S. Costa, J.P.M. Torres, O. Malm et al. 2019. Preliminary study of long-range transport of halogenated flame retardants using Antarctic marine mammals. Sci. Total Environ. 650: 1889–1897.

Barón, E., I. Rudolph, G. Chiang, R. Barra, E. Eljarrat and D. Barceló. 2013a. Occurrence and behavior of natural and anthropogenic (emerging and historical) halogenated compounds in marine biota from the Coast of Concepcion (Chile). Sci. Total Environ. 461: 258–264.

Barón, E., P. Gago-Ferrero, M. Gorga, I. Rudolph, G. Mendoza, A.M. Zapata et al. 2013b. Occurrence of hydrophobic organic pollutants (BFRs and UV-filters) in sediments from South America. Chemosphere 92: 309–316.

Bayen, S., G.O. Thomas, H.K. Lee and J.P. Obbard. 2003. Occurrence of polychlorinated biphenyls and polybrominated diphenyl ethers in green mussels (Perna viridis) from Singapore, southeast Asia. Eviron. Toxicol. Chem. 22(10): 2432–2437.

Binelli, A., S.K. Sarkar, M. Chatterjee, C. Riva, M. Parolini, B. deb Bhattacharya et al. 2007. Concentration of polybrominated diphenyl ethers (PBDEs) in sediment cores of Sundarban mangrove wetland, northeastern part of Bay of Bengal (India). Mar. Pollut. Bull. 54(8): 1220–1229.

Bjurlid, F., A. Roos, I.E. Jogsten and J. Hagberg. 2018. Temporal trends of PBDD/Fs, PCDD/Fs, PBDEs and PCBs in ringed seals from the Baltic Sea (Pusa hispida botnica) between 1974 and 2015. Sci. Total Environ. 616: 1374–1383.

Boryniec, S. and W. Przygocki. 2001. Polymer combustion processes. 3. Flame retardants for polymeric materials. Progress in Rubber and Plastics Technology 17(2): 127.

Bottaro, C.S., J.W. Kiceniuk and A. Chatt. 1999. Spatial distribution of extractable organohalogens in northern pink shrimp in the North Atlantic. Biol. Trace Elem. Res. 71(1): 149–166.

Cappelletti, N., E. Speranza, L. Tatone, M. Astoviza, M.C. Migoya and J.C. Colombo. 2015. Bioaccumulation of dioxin-like PCBs and PBDEs by detritus-feeding fish in the Rio de la Plata estuary, Argentina. Environ. Sci. Pollut. Res. 22(9): 7093–7100.

Carleton, E. 1932. U.S. Patent No. 1,852,998. Washington, DC: U.S. Patent and Trademark Office.

Castro-Jiménez, J., B. Barhoumi, A. Paluselli, M. Tedetti, B. Jiménez, J. Muñoz-Arnanz et al. 2017. Occurrence, loading, and exposure of atmospheric particle-bound POPs at the african and european edges of the western mediterranean sea. Environ. Sci. Technol. 51(22): 13180–13189.

Chao, H.R., D.Y. Lin, K.Y. Chen, Y.Y. Gou, T.H. Chiou, W.J. Lee et al. 2014. Atmospheric concentrations of persistent organic pollutants over the Pacific Ocean near southern Taiwan and the northern Philippines. Sci. Total Environ. 491: 51–59.

Chen, S.J., X.J. Gao, B.X. Mai, Z.M. Chen, X.J. Luo, G.Y. Sheng et al. 2006. Polybrominated diphenyl ethers in surface sediments of the Yangtze River Delta: levels, distribution and potential hydrodynamic influence. Environ. Pollut. 144(3): 951–957.

Cheng, Z., T. Lin, W. Xu, Y. Xu, J. Li, C. Luo et al. 2015. A preliminary assessment of polychlorinated biphenyls and polybrominated diphenyl ethers in deep-sea sediments from the Indian Ocean. Mar. Pollut. Bull. 94(1-2): 323–328.

Choi, S.D., S.Y. Baek and Y.S. Chang. 2008. Atmospheric levels and distribution of dioxin-like polychlorinated biphenyls (PCBs) and polybrominated diphenyl ethers (PBDEs) in the vicinity of an iron and steel making plant. Atmos. Environ. 42(10): 2479–2488.

Dasgupta, S., X. Peng, S. Chen, J. Li, M. Du, Y.H. Zhou et al. 2018. Toxic anthropogenic pollutants reach the deepest ocean on Earth. Geochem. Persp. Let. 7: 22–26.

de Boer, J., P.G. Wester, H.J. Klamer, W.E. Lewis and J.P. Boon. 1998. Do flame retardants threaten ocean life? Nature 394(6688): 28.

Dias, P.S., C.V. Cipro, S. Taniguchi and R.C. Montone. 2013. Persistent organic pollutants in marine biota of Sao Pedro and Sao Paulo Archipelago, Brazil. Mar. Pollut. Bull. 74(1): 435–440.

Dickhut, R.M., A. Cincinelli, M. Cochran and H. Kylin. 2012. Aerosol-mediated transport and deposition of brominated diphenyl ethers to Antarctica. Environ. Sci. Technol. 46(6): 3135–3140.

Drage, D., J.F. Mueller, G. Birch, G. Eaglesham, L.K. Hearn and S. Harrad. 2015. Historical trends of PBDEs and HBCDs in sediment cores from Sydney estuary, Australia. Sci. Total Environ. 512: 177–184.

Dryfus, C. 1933. U.S. Patent 1,907,521. Derivatives of Cellulose Compositions of Low Inflammability, 3 pp.

Eljarrat, E., A. De La Cal, D. Larrazabal, B. Fabrellas, A.R. Fernandez-Alba, E. Borrull et al. 2005. Occurrence of polybrominated diphenylethers, polychlorinated dibenzo-p-dioxins, dibenzofurans and biphenyls in coastal sediments from Spain. Environ. Pollut. 136(3): 493–501.

Fernandes, A.R., D. Mortimer, M. Holmes, M. Rose, L. Zhihua, X. Huang et al. 2018. Occurrence and spatial distribution of chemical contaminants in edible fish species collected from UK and proximate marine waters. Environ. Int. 114: 219–230.

Fernandes, A., D. Mortimer, M. Rose, F. Smith and S. Panton. 2014. Brominated dioxins and PBDEs: occurrence trend in UK food. Organohalogen Compd. 76: 764–767.

Fromant, A., A. Carravieri, P. Bustamante, P. Labadie, H. Budzinski, L. Peluhet et al. 2016. Wide range of metallic and organic contaminants in various tissues of the Antarctic prion, a planktonophagous seabird from the Southern Ocean. Sci. Total Environ. 544: 754–764.

Gay-Lussac, J.L. 1821. Note on properties of salts for making fabrics incombustible. Ann. Chim. Phys. 18(2): 211.

Gevao, B., M.U. Beg, A.N. Al-Ghadban, A. Al-Omair, M. Helaleh and J. Zafar. 2006. Spatial distribution of polybrominated diphenyl ethers in coastal marine sediments receiving industrial and municipal effluents in Kuwait. Chemosphere 62(7): 1078–1086.

Guerra, P., M. Alaee, E. Eljarrat and D. Barceló. 2010. Introduction to brominated flame retardants: Commercially products, applications, and physicochemical properties. pp. 1–17. *In*: Brominated Flame Retardants. Springer, Berlin, Heidelberg. doi:10.1007/698_2010_93.

Haglund, P.S., D.R. Zook, H.R. Buser and J. Hu. 1997. Identification and quantification of polybrominated diphenyl ethers and methoxy-polybrominated diphenyl ethers in Baltic biota. Environ. Sci. Technol. 31(11): 3281–3287.

Hanson, E.R. 1936. U.S. Patent No. 2,028,715. Washington, DC: U.S. Patent and Trademark Office.

Hindersinn, R.R. 1990. Chapter 7: Historical aspects of polymer fire retardance. Fire and Polymers: Hazards Identification and Prevention, 87–96.

Hong, S.H. 2002. Environmental Occurrence and Bioaccumulation of Organochlorines in Korean Coastal Waters, Ph.D. Thesis, Ewha Womens University, Korea, 186 p

INCHEM. 1997. http://www.inchem.org/documents/ehc/ehc/ehc192.htm#PartNumber:1.

Innes, A. and J. Innes. 2012. Flame Retardants. Handbook of Environmental Degradation of Materials, 309–335. Doi: 10.1016/b978-1-4377-3455-3.00010-9.

Isobe, T., K. Ramu, N. Kajiwara, S. Takahashi, P.K. Lam, T.A. Jefferson et al. 2007. Isomer specific determination of hexabromocyclododecanes (HBCDs) in small cetaceans from the South China Sea–Levels and temporal variation. Mar. Pollut. Bull. 54(8): 1139–1145.

Jansson, B., L. Asplund and M. Olsson. 1987. Brominated flame retardants—ubiquitous environmental pollutants? Chemosphere 16(10-12): 2343–2349.

Jin, J., W. Liu, Y. Wang and X.Y. Tang. 2008. Levels and distribution of polybrominated diphenyl ethers in plant, shellfish and sediment samples from Laizhou Bay in China. Chemosphere 71(6): 1043–1050.

Johansson, I., K. Héas-Moisan, N. Guiot, C. Munschy and J. Tronczyński. 2006. Polybrominated diphenyl ethers (PBDEs) in mussels from selected French coastal sites: 1981–2003. Chemosphere 64(2): 296–305.

Ju, T., W. Ge, T. Jiang and C. Chai. 2016. Polybrominated diphenyl ethers in dissolved and suspended phases of seawater and in surface sediment from Jiaozhou Bay, North China. Sci. Total Environ. 557: 571–578.

Kajiwara, N., S. Kamikawa, K. Ramu, D. Ueno, T.K. Yamada, A. Subramanian et al. 2006. Geographical distribution of polybrominated diphenyl ethers (PBDEs) and organochlorines in small cetaceans from Asian waters. Chemosphere 64(2): 287–295.

Kannan, K., I. Watanabe and J.P. Giesy. 1998. Congener profile of polychlorinated/brominated dibenzo-p-dioxins and dibenzofurans in soil and sediments collected at a former chlor-alkali plant: Communicated by Toxicological and Environmental Chemistry (April 1997). Toxicol. Environ. Chem. 67(1-2): 135–146.

Kim, J.T., Y.J. Choi, M. Barghi, Y.J. Yoon, J.H. Kim, J.H. Kim et al. 2018. Occurrence and distribution of old and new halogenated flame retardants in mosses and lichens from the South Shetland Islands, Antarctica. Environ. Pollut. 235: 302–311.

Kuehl, D.W., R. Haebler and C. Potter. 1991. Chemical residues in dolphins from the US Atlantic coast including Atlantic bottlenose obtained during the 1987/88 mass mortality. Chemosphere 22(11): 1071–1084.

La Guardia, M.J., R.C. Hale and E. Harvey. 2006. Detailed polybrominated diphenyl ether (PBDE) congener composition of the widely used penta-, octa-, and deca-PBDE technical flame-retardant mixtures. Environ. Sci. Technol. 40(20): 6247–6254.

La Guardia, M.J., R.C. Hale and B. Newman. 2013. Brominated flame-retardants in Sub-Saharan Africa: burdens in inland and coastal sediments in the Thekwini metropolitan municipality, South Africa. Environ. Sci. Technol. 47(17): 9643–9650.

Lee, H.J. and G.B. Kim. 2015. An overview of polybrominated diphenyl ethers (PBDEs) in the marine environment. Ocean Sci. J. 50(2): 119–142.

Lee, H.J., H.J. Jeong, Y.L. Jang and G.B. Kim. 2018. Distribution, accumulation, and potential risk of polybrominated diphenyl ethers in the marine environment receiving effluents from a sewage treatment plant. Mar. Pollut. Bull. 129(1): 364–369.

Lindström, G., H. Wingfors, M. Dam and B.V. Bavel. 1999. Identification of 19 polybrominated diphenyl ethers (PBDEs) in long-finned pilot whale (Globicephala melas) from the Atlantic. Arch. Environ. Con. Tox. 36(3): 355–363.

Liu, Y., G.J. Zheng, H. Yu, M. Martin, B.J. Richardson, M.H. Lam et al. 2005. Polybrominated diphenyl ethers (PBDEs) in sediments and mussel tissues from Hong Kong marine waters. Mar. Pollut. Bull. 50(11): 1173–1184.

Luo, Q., Z.W. Cai and M.H. Wong. 2007. Polybrominated diphenyl ethers in fish and sediment from river polluted by electronic waste. Sci. Total Environ. 383(1-3): 115–127.

Ma, Y., Z. Xie, R. Lohmann, W. Mi and G. Gao. 2017. Organophosphate ester flame retardants and plasticizers in ocean sediments from the North Pacific to the Arctic Ocean. Environ. Sci. Technol. 51(7): 3809–3815.

Mai, B., S. Chen, S. Chen, X. Luo, L. Chen, L. Chen et al. 2005. Distribution of polybrominated diphenyl ethers in sediments of the Pearl River Delta and adjacent South China Sea. Environ. Sci. Technol. 39(10): 3521–3527.

Minh, N.H., T. Isobe, D. Ueno, K. Matsumoto, M. Mine, N. Kajiwara et al. 2007. Spatial distribution and vertical profile of polybrominated diphenyl ethers and hexabromocyclododecanes in sediment core from Tokyo Bay, Japan. Environ. Pollut. 148(2): 409–417.

Möller, A., Z. Xie, R. Sturm and R. Ebinghaus. 2011. Polybrominated diphenyl ethers (PBDEs) and alternative brominated flame retardants in air and seawater of the European Arctic. Environ. Pollut. 159(6): 1577–1583.

Moon, H.B., K. Kannan, S.J. Lee and M. Choi. 2007a. Atmospheric deposition of polybrominated diphenyl ethers (PBDEs) in coastal areas in Korea. Chemosphere 66(4): 585–593.

Moon, H.B., K. Kannan, M. Choi and H.G. Choi. 2007b. Polybrominated diphenyl ethers (PBDEs) in marine sediments from industrialized bays of Korea. Mar. Pollut. Bull. 54(9): 1402–1412.

Moon, H.B., K. Kannan, S.J. Lee and M. Choi. 2007c. Polybrominated diphenyl ethers (PBDEs) in sediment and bivalves from Korean coastal waters. Chemosphere 66(2): 243–251.

Moon, H.B., S.P. Yoon, R.H. Jung and M. Choi. 2008. Wastewater treatment plants (WWTPs) as a source of sediment contamination by toxic organic pollutants and fecal sterols in a semi-enclosed bay in Korea. Chemosphere 73(6): 880–889.

Nylund, K., L. Asplund, B. Jansson, P. Jonsson, K. Litzén and U. Sellström. 1992. Analysis of some polyhalogenated organic pollutants in sediment and sewage sludge. Chemosphere 24(12): 1721–1730.

Olsson, A., M. Vitinsh, M. Plikshs and Å. Bergman. 1999. Halogenated environmental contaminants in perch (Perca fluviatilis) from Latvian coastal areas. Sci. Total Environ. 239(1-3): 19–30.

Pan, J., Y.L. Yang, Q. Xu, D.Z. Chen and D.L. Xi. 2007. PCBs, PCNs and PBDEs in sediments and mussels from Qingdao coastal sea in the frame of current circulations and influence of sewage sludge. Chemosphere 66(10): 1971–1982.

Pegoraro, C.N., T. Harner, K. Su and M.S. Chiappero. 2016. Assessing levels of POPs in air over the South Atlantic Ocean off the coast of South America. Sci. Total Environ. 571: 172–177.

Pettersson, A., B. van Bavel, M. Engwall and B. Jimenez. 2004. Polybrominated diphenylethers and methoxylated tetrabromodiphenylethers in cetaceans from the Mediterranean Sea. Arch. Environ. Con. Tox. 47(4): 542–550.

Pettigrew, A. 1993. Halogenated flame retardants. Kirk-Othmer Encyclopedia of Chemical Technology, 4th ed. New York, NY: John Wiley and Sons 10: 954–976.

Pozo, K., P. Kukučka, L. Vaňková, P. Přibylová, J. Klánová, A. Rudolph et al. 2015. Polybrominated Diphenyl Ethers (PBDEs) in concepción bay, central Chile after the 2010 Tsunami. Mar. Pollut. Bull. 95(1): 480–483.

Ramu, K., N. Kajiwara, T. Isobe, S. Takahashi, E.Y. Kim, B.Y. Min et al. 2007a. Spatial distribution and accumulation of brominated flame retardants, polychlorinated biphenyls and organochlorine pesticides in blue mussels (Mytilus edulis) from coastal waters of Korea. Environ. Pollut. 148(2): 562–569.

Ramu, K., N. Kajiwara, A. Sudaryanto, T. Isobe, S. Takahashi, A. Subramanian et al. 2007b. Asian mussel watch program: contamination status of polybrominated diphenyl ethers and organochlorines in coastal waters of Asian countries. Environ. Sci. Technol. 41(13): 4580–4586.

Rauert, C., J.K. Schuster, A. Eng and T. Harner. 2018. Global atmospheric concentrations of brominated and chlorinated flame retardants and organophosphate esters. Environ. Sci. Technol. 52(5): 2777–2789.

Rigét, F., A. Bignert, B. Braune, M. Dam, R. Dietz, M. Evans et al. 2019. Temporal trends of persistent organic pollutants in Arctic marine and freshwater biota. Sci. Total Environ. 649: 99–110.

Rochman, C.M., R.L. Lewison, M. Eriksen, H. Allen, A.M. Cook and S.J. Teh. 2014. Polybrominated diphenyl ethers (PBDEs) in fish tissue may be an indicator of plastic contamination in marine habitats. Sci. Total Environ. 476: 622–633.

Salvadó, J.A., A. Sobek, D. Carrizo and O. Gustafsson. 2016. Observation-based assessment of PBDE loads in Arctic Ocean waters. Environ. Sci. Technol. 50(5): 2236–2245.

Sellström, U., B. Jansson, A. Kierkegaard, C. de Wit, T. Odsjö and M. Olsson. 1993. Polybrominated diphenyl ethers (PBDE) in biological samples from the Swedish environment. Chemosphere 26(9): 1703–1718.

Speight, J.G. 2017. Chapter 4—source and types of organic pollutants. Environmental Organic Chemistry for Engineers.

Strobel, A., P. Schmid, H. Segner, P. Burkhardt-Holm and M. Zennegg. 2016. Persistent organic pollutants in tissues of the white-blooded Antarctic fish Champsocephalus gunnari and Chaenocephalus aceratus. Chemosphere 161: 555–562.

Takigami, H., S. Sakai and A. Brouwer. 2005. Bio/chemical analysis of dioxin-like compounds in sediment samples from Osaka Bay, Japan. Environ. Technol. 26(4): 459–470.

Tanabe, S. 2008. Temporal trends of brominated flame retardants in coastal waters of Japan and South China: retrospective monitoring study using archived samples from es-Bank, Ehime University, Japan. Mar. Pollut. Bull. 57(6-12): 267–274.

Terauchi, H., S. Takahashi, P.K. Lam, B.Y. Min and S. Tanabe. 2009. Polybrominated, polychlorinated and monobromo-polychlorinated dibenzo-p-dioxins/dibenzofurans and dioxin-like polychlorinated biphenyls in marine surface sediments from Hong Kong and Korea. Environ. Pollut. 157(3): 724–730.

Tombesi, N., K. Pozo, M. Álvarez, P. Přibylová, P. Kukučka, O. Audy et al. 2017. Tracking polychlorinated biphenyls (PCBs) and polybrominated diphenyl ethers (PBDEs) in sediments and soils from the southwest of Buenos Aires Province, Argentina (South eastern part of the GRULAC region). Sci. Total Environ. 575: 1470–1476.

Troitzsch, J.H. 1990. International Plastics Flammability Handbook: Principles, Regulations, Testing and Approval, 2nd ed. München, Germany, Hanser Publishers.

UNEP (United Nations Environment Programme). Stockholm Convention on Persistent Organic Pollutants (POPs). http://chm.pops.int/default.aspx, accessed in December 2018.

Venier, M., A. Salamova and R.A. Hites. 2015. Halogenated flame retardants in the Great Lakes environment. Accounts Chem. Res. 48(7): 1853–1861.

Verreault, J., R.J. Letcher, M.L. Gentes and B.M. Braune. 2018. Unusually high Deca-BDE concentrations and new flame retardants in a Canadian Arctic top predator, the glaucous gull. Sci. Total Environ. 639: 977–987.

Wang, Y., G. Jiang, P.K.S. Lam and A. Li. 2007. Polybrominated diphenyl ether in the East Asian environment: a critical review. Environ. Int. 33: 963–973.

Wang, P., Q.H. Zhang, T. Wang, W.H. Chen, D.W. Ren, Y.M. Li and G.B. Jiang. 2012. PCBs and PBDEs in environmental samples from King George Island and Ardley Island, Antarctica. Rsc Advances 2(4): 1350–1355.

Wang, P., Y. Li, Q. Zhang, Q. Yang, L. Zhang, F. Liu et al. 2017. Three-year monitoring of atmospheric PCBs and PBDEs at the Chinese Great Wall Station, West Antarctica: levels, chiral signature, environmental behaviors and source implication. Atmos. Environ. 150: 407–416.

Wang, Z., X. Ma, Z. Lin, G. Na and Z. Yao. 2009. Congener specific distributions of polybrominated diphenyl ethers (PBDEs) in sediment and mussel (Mytilus edulis) of the Bo Sea, China. Chemosphere 74(7): 896–901.

Watanabe, I., T. Kashimoto and R. Tatsukawa. 1983. The flame retardant tetrabromobisphenol-A and its metabolite found in river and marine sediments in Japan. Chemosphere 12(11-12): 1533–1539.

Watanabe, I., T. Kashimoto and R. Tatsukawa. 1985. Brominated phenols and anisoles in river and marine sediments in Japan. Bull. Environ. Contam. Toxicol. 35(1): 272–278.

Watanabe, I., T. Kashimoto and R. Tatsukawa. 1986. Hexabromobenzene and its debrominated compounds in river and estuary sediments in Japan. Bull. Environ. Contam. Toxicol. 36: 778–784.

Watanabe, I., T. Kashimoto and R. Tatsukawa. 1987. Polybrominated biphenyl ethers in marine fish, shellfish and river and marine sediments in Japan. Chemosphere 16(10-12): 2389–2396.

Wurl, O., J.R. Potter, C. Durville and J.P. Obbard. 2006. Polybrominated diphenyl ethers (PBDEs) over the open Indian Ocean. Atmos. Environ. 40(29): 5558–5565.

Wyld, O. British Patent #551 (1735).

Xiang, C.H., X.J. Luo, S.J. Chen, M. Yu, B.X. Mai and E.Y. Zeng. 2007. Polybrominated diphenyl ethers in biota and sediments of the Pearl River Estuary, South China. Eviron. Toxicol. Chem. 26(4): 616–623.

Yu, M., X.J. Luo, J.P. Wu, S.J. Chen and B.X. Mai. 2009. Bioaccumulation and trophic transfer of polybrominated diphenyl ethers (PBDEs) in biota from the Pearl River Estuary, South China. Environ. Int. 35(7): 1090–1095.

Zaccaroni, A., R. Andreini, S. Franzellitti, D. Barceló and E. Eljarrat. 2018. Halogenated flame retardants in stranded sperm whales (Physeter macrocephalus) from the Mediterranean Sea. Sci. Total Environ. 635: 892–900.

Zhang, G., P. Chakraborty, J. Li, P. Sampathkumar, T. Balasubramanian, K. Kathiresan et al. 2008. Passive atmospheric sampling of organochlorine pesticides, polychlorinated biphenyls, and polybrominated diphenyl ethers in urban, rural, and wetland sites along the coastal length of India. Environ. Sci. Technol. 42(22): 8218–8223.

Zhu, B., J.C. Lam and P.K. Lam. 2018. Halogenated flame retardants (HFRs) in surface sediment from the Pearl River Delta region and Mirs Bay, South China. Mar. Pollut. Bull. 129(2): 899–904.

Index

Color Plate Section

Chapter 1

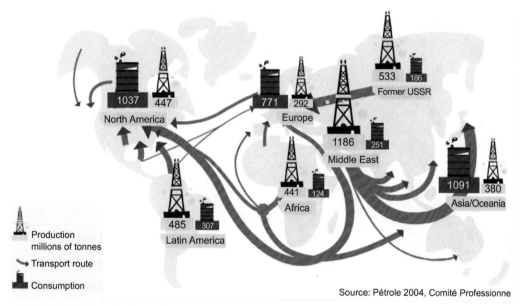

Fig. 2. Worldwide production and consumption in 2004 (millions of tonnes) (https://www.eia.gov/beta/international/).

Fig. 3. Worldwide sea-borne flow of oil in 2000 (modified from Newton 2002; other information sources include U.S. Geological Survey, U.S. Coast Guard, Minerals Management Service). Solid black dots indicate spills included in the average, annual (1990–1999) (NRC 2003).

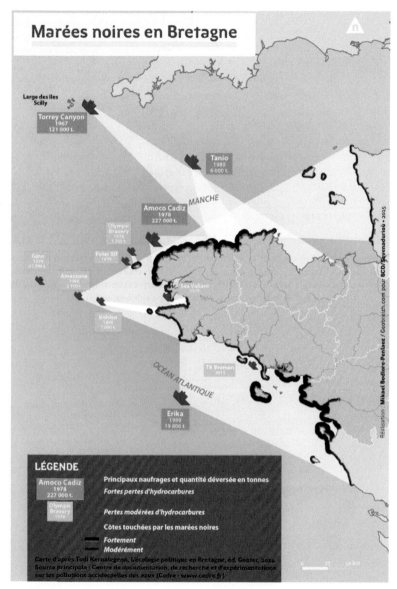

Fig. 8. Amoco Cadiz (http://bcd.bzh/becedia/fr/la-maree-noire-de-l-amoco-cadiz).

Fig. 9. Exxon Valdez. Field studies continue to examine the effects of the Exxon supertanker's disastrous grounding on Bligh Reef in Alaska's Prince William Sound in 1989. Photo courtesy Erik Hill, Anchorage Daily News (https://response.restoration.noaa.gov/about/media/10-photos-tell-story-exxon-valdez-oil-spill-and-its-impacts.html; https://aoghs.org/transportation/exxon-valdez-oil-spill/).

Chapter 2

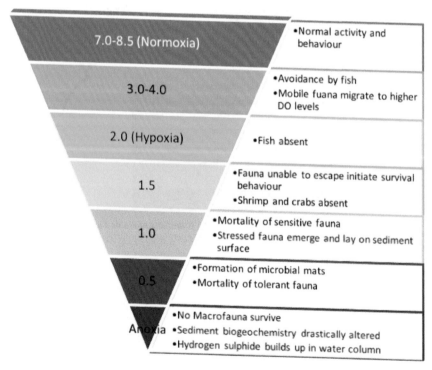

Fig. 1. General indication of harm caused to marine organisms at different dissolved oxygen concentrations (modified from Diaz et al. 2013).

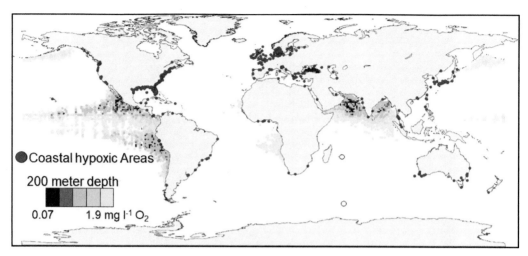

Fig. 2. Distribution of global coastal hypoxic areas (red circles) and oxygen minimum zones at a depth of 200 m (blue-shaded areas) (Breitburg et al. 2018).

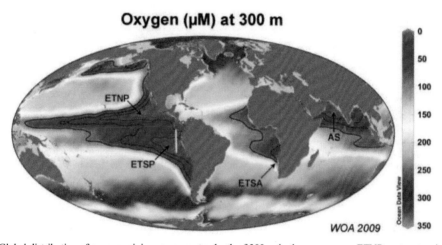

Fig. 3. Global distribution of oxygen minimum zones at a depth of 300 m in the open ocean. ETNP, eastern tropical North Pacific; ETSP, eastern tropical South Pacific; ETSA, eastern tropical South Atlantic; AS, Arabian Sea. Source: Max Planck Institute for Marine Microbiology, based on data from the World Ocean Atlas 2009.

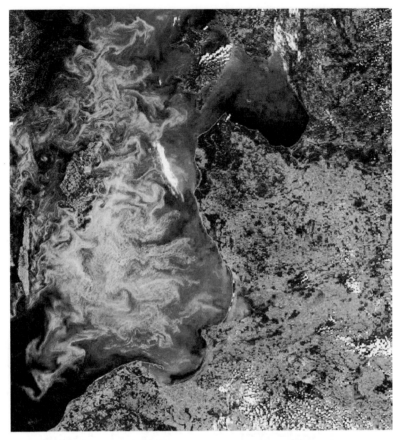

Fig. 4. A large cyanobacteria bloom in the Baltic Sea in summer 2005. The proliferation of such blooms leads to the consumption of dissolved oxygen in bottom waters following their collapse (Image: Envisat Meris 13 July 2005).

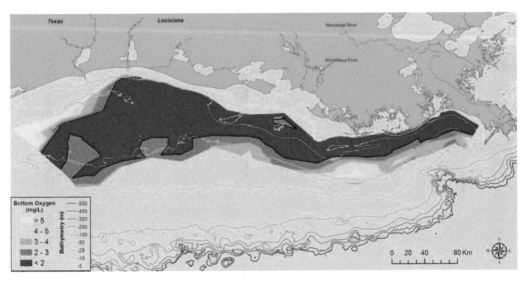

Fig. 5. Distribution of the Gulf of Mexico Dead Zone in summer 2017 (source: Gulfhypoxia.net).

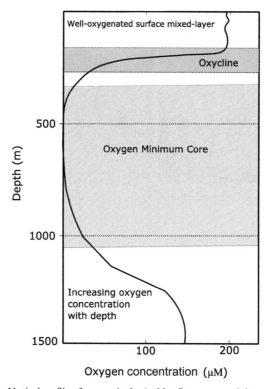

Fig. 6. Vertical profile of oxygen in the Arabian Sea oxygen minimum zone.

Chapter 3

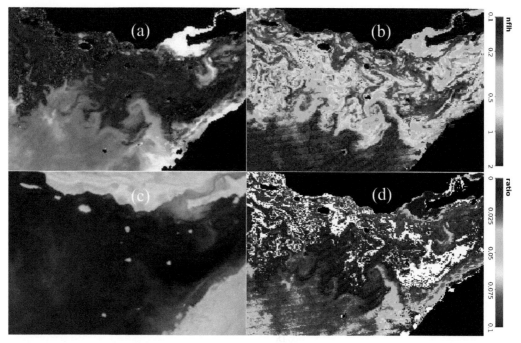

Fig. 2. Satellite-derived enhanced red-green-blue composite (ERGB) (a), normalized fluorescence line height (nflh) (b), true color image (c), the ratio of particulate backscattering (b_bp) to nflh for December 23, 2008, over the Arabian Gulf when an extensive HAB event occurred.

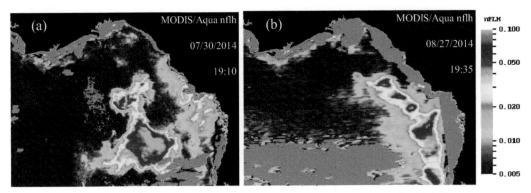

Fig. 3. MODIS/Aqua derived nflh imagery for July 30, 2014 (a), and August 27, 2014 (b).

Chapter 11

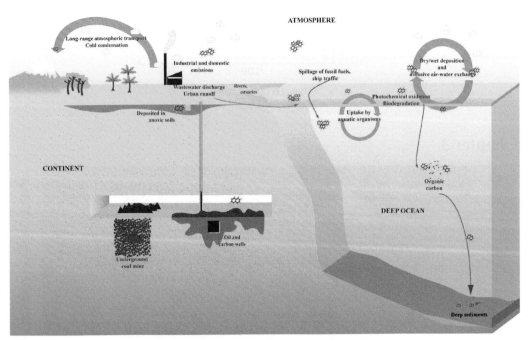

Fig. 1. Distribution and fate of PAHs in the marine environment.

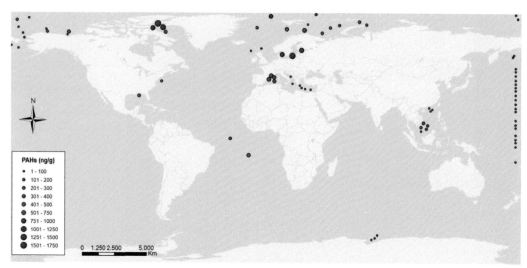

Fig. 3. Levels of PAHs in deep oceans (Domine et al. 1994, Takada et al. 1994, Tolosa et al 1996, Yunker et al. 1996, 2011, Guzzella and De Paolis 1994, Ohkouchi et al. 1999, Yang 2000, Pikkarainen 2004, Bouloubassi et al 2006, 2012, Cui et al. 2008, Boitsov et al. 2009, Shao et al. 2010, Zaborska et al. 2011, Parinos et al. 2013, Mandalakis et al. 2014, Foster et al. 2015, Romero et al. 2015, Xue et al. 2016, Zhao et al. 2016, Ma et al. 2017, Kaiser et al. 2018, Webster et al. 2018).

Printed and bound by CPI Group (UK) Ltd, Croydon, CR0 4YY

24/10/2024

01778290-0009